Numerical Techniques
Computing with C
and
MATLAB

Reena Garg PhD
Assistant Professor
Department of Mathematics
YMCA University of Science and Technology
Faridabad 121004
E-mail: reenagargymca@gmail.com

R S Goel MSc
Former Principal
Aggarwal College
Ballabhgarh, Faridabad

CBS Publishers & Distributors Pvt Ltd

New Delhi • Bengaluru • Chennai • Kochi • Kolkata • Mumbai
Hyderabad • Jharkhand • Nagpur • Patna • Pune • Uttarakhand

Numerical Techniques
Computing with C and MATLAB

ISBN: 978-93-87085-83-1

First Edition: 2018

Published by Satish Kumar Jain and produced by Varun Jain for

CBS Publishers & Distributors Pvt Ltd

4819/XI Prahlad Street, 24 Ansari Road, Daryaganj, New Delhi 110 002, India.
Ph: 23289259, 23266861, 23266867   Website: www.cbspd.com
Fax: 011-23243014   e-mail: delhi@cbspd.com; cbspubs@airtelmail.in.

Corporate Office: 204 FIE, Industrial Area, Patparganj, Delhi-110092
Ph: 4934 4934   Fax: 4934 4935   e-mail: publishing@cbspd.com; publicity@cbspd.com

Branches

- **Bengaluru:** Seema House 2975, 17th Cross, K.R. Road,
  Banasankari 2nd Stage, Bengaluru 560 070, Karnataka
  Ph: +91-80-26771678/79   Fax: +91-80-26771680   e-mail: bangalore@cbspd.com
- **Chennai:** 7, Subbaraya Street, Shenoy Nagar, Chennai 600 030, Tamil Nadu
  Ph: +91-44-26680620, 26681266   Fax: +91-44-42032115   e-mail: chennai@cbspd.com
- **Kochi:** Ashana House, No. 39/1904, AM Thomas Road, Valanjambalam,
  Ernakulam 682 018, Kochi, Kerala
  Ph: +91-484-4059061-65   Fax: +91-484-4059065   e-mail: kochi@cbspd.com
- **Kolkata:** 6/B, Ground Floor, Rameswar Shaw Road, Kolkata-700 014, West Bengal
  Ph: +91-33-22891126, 22891127, 22891128   e-mail: kolkata@cbspd.com
- **Mumbai:** 83-C, Dr E Moses Road, Worli, Mumbai-400018, Maharashtra
  Ph: +91-22-24902340/41   Fax: +91-22-24902342   e-mail: mumbai@cbspd.com

Representatives

- **Hyderabad** 0-9885175004
- **Jharkhand** 0-9811541605
- **Nagpur** 0-9021734563
- **Patna** 0-9334159340
- **Pune** 0-9623451994
- **Uttarakhand** 0-9716462459

Printed at: India Binding House, Noida, UP, India

to
our family members

Reena Garg
RS Goel

Foreword

I am glad to know that Dr Reena Garg, Assistant Professor, Mathematics at YMCA University of Science and Technology, Faridabad and Prof RS Goel, Principal (retired), Aggarwal College, Ballabhgarh have authored a book titled **'Numerical Techniques: Computing with C and MATLAB'**.

Mathematics has played a very significant role in the progress and expansion of Indian culture for centuries. Mathematical ideas that originated in the Indian subcontinent have had a thoughtful impact on the world. In ancient time, mathematics was primarily used in a supplementary or practical role. The most fundamental contribution of India in mathematics is the invention of decimal system of enumeration, including the invention of zero.

It has been proved that learning mathematics can be made easier and enjoyable if our curriculum includes mathematical activities and games. Maths puzzles and riddles encourage and attract an alert and open-minded attitude among youngsters and help them develop clarity in their thinking. Therefore, emphasis should be laid on development of clear concept in mathematics in a child, right from the primary classes.

I hope the book will serve the purpose for betterment of students and help to enhance their interest in the concerned area. I wish and hope that the book succeeds in achieving its objectives.

I applaud the efforts of learned authors for publishing the book and wish them success for their future endeavours.

Prof Kaptan Singh Solanki
Governor Haryana

Foreword

It gives me immense pleasure to know that Dr Reena Garg, Assistant Professor, Mathematics at YMCA University of Science and Technology, Faridabad and Prof RS Goel, Principal (retired), Aggarwal College, Ballabhgarh have brought out a book titled **'Numerical Techniques: Computing with C and MATLAB'** for the students of engineering and science at undergraduate and postgraduate levels.

India has the privilege of introducing decimal number system. The Indus Valley civilization exhibits rich knowledge of mathematics; Aryabhata, Brahmagupta, Bhaskaracharya and many more have contributed for the development of mathematical knowledge which forms the foundation of India's computation genius. I hope that this book will also add to existing knowledge domain of mathematics by providing easy and better understanding of the concerned subject.

I am sure the book will prove to be beneficial as a reference book dealing with the techniques of numerical analysis for the students pursuing studies of engineering and science.

I congratulate the learned authors for their efforts in bringing out this publication and wish them success.

Prof. Dinesh Kumar
Vice-Chancellor
YMCAUST, Faridabad

Preface

The classical analysis, which is three centuries old, has undergone radical changes through the contributions from innumerable researcher, mathematician and scientists in the field of engineering and technology with the development of superfast, efficient and digital computers. This has led mathematician to evolve numerical techniques which are effective tools for providing solutions to mathematical problems which are not solvable by analytical methods. Numerical techniques possibly started and developed with man's need of counting. The systematic development of numerical techniques started in seventeenth century by great mathematicians and scientists like Isaac Newton, Lagrange, Carl Friedrich Gauss, Bessel, Karl Weierstrass, C Runge, MW Kutta, etc.

Numerical analysis is a branch of mathematics that studies the numerical solutions to problems involving nonlinear equations, systems of linear equations, interpolation and approximation, empirical laws for curve fitting, differentiations, integrals, ordinary and partial differential equations, finite differences, etc. Numerical methods are normally being used to find the solution to a problem whose analytical solution is difficult to achieve, thus it is felt that a study in applied sciences and engineering is essential and found wide applications in all areas of science, technology and economics.

The book introduces techniques of numerical analysis for the students pursing studies in engineering and applied sciences at undergraduate and postgraduate levels in various universities. The basic objective of this presentation is to provide the fundamental mathematical concepts and problem-solving skills using numerical techniques with C and MATLAB.

For better understanding the book is divided into 11 chapters. Chapter 1 deals with errors in numerical calculations; Chapter 2 focuses on a detailed study of simultaneous linear algebraic equations and Eigenvalue problems; Chapters 3–7 introduce finite differences, interpolation, empirical laws and curve fitting, solution to numerical, algebraic and transcendental equations; Chapters 7–9 enumerate numerical differentiation and integration, numerical solution of ordinary and partial differential equations; Chapters 10 and 11 deal with computational techniques using C and MATLAB.

Each chapter is supplemented with elaborate illustrations, flowcharts along with program details, solved examples, and related problems with hints (wherever necessary) to provide an aid in comprehension of principles involved. At the end of the book, model question papers and some university papers have been included.

We hope this book will be of immense use to the teachers and students of technical institutes. Suggestions from students and teachers for improvement in future editions of this book are welcome.

Reena Garg
RS Goel

Acknowledgements

I express my sincere gratitude to friends, faculty members and wellwishers for their contribution, guidance and suggestions to write this book. I am especially indebted to my most respected Prof Dinesh Aggarwal, Vice Chancellor, YMCA University of Science and Technology, Faridabad, whose motivational guidance helped me a lot to do the best in my field. I am thankful to Prof Bani Singh (former Professor, IIT Roorkee); Prof Bhudev Sharma (former Professor, University of Atlanta, USA), Prof Babu Ram (former Professor, Maharshi Dayanand University, Rohtak); Dr Madhu Jain (Associate Professor, IIT Roorkee); Dr SC Malik (Professor, Maharshi Dayanand University, Rohtak); Dr Deepankar Sharma (Director, DJ College of Engineering, Modinagar); Prof Kuldeep Bansal (Professor, Guru Jambheshwar University of Science and Technology, Hisar). I am also thankful to my friend and colleague Mrs Hem Lata Aggarwal for helping me in computer programming.

Last but not the least, I want to thank my family members and two loving children, Vedant Garg and Vaishnavi Garg, without their constant support I would not be able to achieve this milestone of my life.

I am thankful to almighty God for giving me the opportunity to serve the society.

Reena Garg

We feel proud to express our sincere gratitude to friends, faculty members and wellwishers for their contribution, guidance and suggestions to write this book.

We are extremely thankful and grateful to Mr Satish Kumar Jain, CMD, CBS Publishers & Distributors, New Delhi, for his continued support. We are thankful to Mr YN Arjuna (Senior Vice President Publishing, Editorial and Publicity), and his team for bringing out the book in the present form.

Reena Garg
RS Goel

Contents

Errors in Numerical Calculations

1.1 INTRODUCTION

In practical applications, one would finally require results in a numerical form. For example, a set of tabulated data is given and inferences have to be drawn from it or a system of linear algebraic equations is given and one is required to solve them. The aim of numerical analysis is to provide efficient methods for obtaining numerical solutions to such problems.

While solving problems, one usually starts with some initial data and then computes, after some intermediate steps, gives the final results. The numerical data used are only approximate being true to two, three or more figures. Sometimes, the methods used are also approximate and therefore, the error in a computed result may be due to errors in the data, or the error in the method, or both. Thus, in this chapter attention has been focussed on some basic ideas concerning errors and their analysis.

1.2 NUMBERS AND THEIR ACCURACY

There are two kinds of numbers, namely *exact* and *approximate* numbers. Examples of exact numbers are: $1, 2, 3, ..., \dfrac{1}{2}, \dfrac{3}{2}, ...(\pi, e)$ etc. Approximate numbers are those that represent the number to a certain degree of accuracy. Thus, an approximate value of π is 3.1416 or the better approximation would be 3.14159265. Obviously, the exact value of π cannot be written.

The digits that are used to express a number are called *significant digits* or *significant figures*. Thus, the numbers 3.1416, 0.66667 and 4.0687 contains five significant digits each. The number 0.00023 has, however, only two significant digits 2 and 3, since the zeros serve only to fix the position of the decimal point.

Frequently, one comes across numbers with a large number of digits and it will be necessary to cut them to a useable number of figures. This process is called *rounding off*. In this book, numbers are rounded-off according to the following rule. To round-off a number to n significant digits, discard all digits to the right of the nth digit, and if this discarded number is

    a. less than half a unit in the nth place, leave the nth digit unaltered.
    b. greater than half a unit in the nth place, increase the nth digit by unity.
    c. exactly half a unit in the nth place, increase the nth digit by unity if it is odd, otherwise leave it unchanged.

Example: The numbers given below are rounded-off to four significant figures.

2.63583 to 2.636

20.0567 to 20.06

1.3 MATHEMATICAL PRELIMINARIES

Theorems:

If $f(x)$ is continuous in $a \leq x \leq b$, and if $f(a)$ and $f(b)$ are of opposite signs, then $f(x) = 0$ for at least one number x such that $a < x < b$.

Rolle's theorem: If

   i. $f(x)$ is continuous in $a \leq x \leq b$

   ii. $f(x)$ exists in $a < x < b$, and

  iii. $f(a) = f(b) = 0$

Then, there exists at least one value of x say c, such that

$$f'(c) = 0, a < c < b$$

Generalised Rolle's theorem: Let $f(x)$ be a function which is n times differentiable on (a, b). If $f(x)$ vanishes at the $(n + 1)$ distinct points $x_0, x_1, ..., x_n$ in (a, b), then there exists a number c in (a, b) such that $f^n(c) = 0$.

Intermediate value theorem: Let $f(x)$ be continuous in $[a, b]$ and let k be any number between $f(a)$ and $f(b)$. Then, there exists a number c in (a, b) such that $f(c) = k$ (Fig. 1.1)

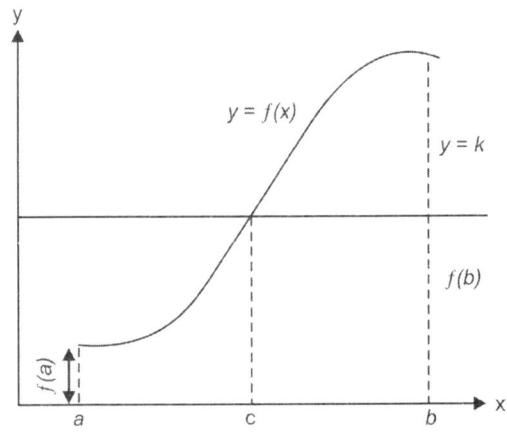

Fig. 1.1

Mean value theorem for derivatives: If

   i. $f(x)$ is continuous in $[a, b]$ and

  ii. $f(x)$ exists in (a, b)

Then there exists atleast one value of x, say c between a and b such that

$$f(c) = \frac{f(b) - f(a)}{b - a}, a < c < b$$

Putting $b = a + h$, this phenomenon takes the form

$$f(a + h) = f(a) + hf'(a + \theta h), 0 < \theta < 1$$

Taylor's series for a function of one variable: If $f(x)$ is continuous and possesses continuous derivatives of order n in an interval at $x = a$, is given by

$$f(x) = f[(a) + (x - a)]$$

$$= f(a) + (x - a) f'(a) + \frac{(x - a)^2}{2!} f''(a) + ... + \frac{(x - a)^{n-1}}{(n - 1)!} f^{n-1}(a) + R_n((x)$$

where $R_n(x)$, the remainder term can be expressed in the form

$$R_n(x) = \frac{(x - a)^n}{n!} f^n(\xi), \, a < \xi < x$$

Maclaurin's expansion can be obtained by putting $a = 0$ in Taylor's

$$f(x) = f(0) + x f'(0) + \frac{x^2}{2!} f''(0) + ... + \frac{x^n}{n!} f^n(0) + ...$$

Taylor's series for a function of two variables:

$$f(x_1 + \Delta x_1, x_2 + \Delta x_2) = f(x_1, x_2) + \frac{\partial f}{\partial x_1} \Delta x_1 + \frac{\partial f}{\partial x_2} \Delta x_2$$

$$+ \frac{1}{2} \left[\frac{\partial^2 f}{\partial x_1^2} (\Delta x_1)^2 + 2 \frac{\partial^2 f}{\partial x_1 \partial x_2} \Delta x_1 \Delta x_2 + \frac{\partial^2 f}{\partial x_2^2} (\Delta x_2)^2 \right] + ...$$

This can easily be generalized.

Taylor's series for a function of multiple variables:

$$f(x_1 + \Delta x_1, x_2 + \Delta x_2, ..., x_n + \Delta x_n) = f(x_1, x_2, ..., x_n) + \frac{\partial f}{\partial x_1} \Delta x_1 + \frac{\partial f}{\partial x_2} \Delta x_2 + ... + \frac{\partial f}{\partial x_n} \Delta x_n$$

$$+ \frac{1}{2} \frac{\partial^2 f}{\partial x_1^2} (\Delta x_1)^2 + ... + \frac{\partial^2 f}{\partial x_n^2} (\Delta x_n)^2 + 2 \frac{\partial^2 f}{\partial x_1 \partial x_2} \Delta x_1 \Delta x_2 + ...$$

$$2 \frac{\partial^2 f}{\partial x_{n-1} \partial x_n} \Delta x_{n-1} \Delta x_{n^2} + \frac{\partial^2 f}{\partial x_2} (\Delta x_2)^2 + ... + \frac{\partial^2 f}{\partial x_{n^2}} (\Delta x_n)^2 + ...$$

1.4 ERRORS AND THEIR COMPUTATION

In numerical analysis, one comes across three types of errors:

i. **Inherent errors:** Most numerical computations are inaccurate, either due to the given data been approximate, or like the limitations of the computing aids: mathematical tables, desk calculators or the digital computers. Due to this limitation, number have to be rounded-off, causing error which are called *rounding errors*. In computations, inherent errors can be minimized by obtaining accurate data by correcting obvious errors in the data and by using computing aids of higher precision. In hand computations, the round-off errors can be reduced by carrying the computations to more significant figures at each step of the computation, the useful rule is as follows.

At each step of the computation, retain at least one more significant figures that is given in the data, perform the last operation, and then round-off.

ii. **Truncation errors:** These are caused by using approximate results or on replacing an infinite process by a finite one. If one is using a decimal computer having a fixed word length of 4 digits, thus rounding-off of 13.658 gives 13.66, whereas, truncation gives 13.65.

For example, if $e^x = 1 + x + \dfrac{x^2}{2!} + \dfrac{x^3}{3!} + \dfrac{x^4}{4!} + \dots \infty = X$ (say) is replaced by

$$1 + x + \dfrac{x^2}{2!} + \dfrac{x^3}{3!} = X^n \text{ (say)}$$

then the truncation error is $X - X^n$.

Truncation error is a type of algorithm error.

iii. **Absolute, relative and percentage errors:** If X is the true value of a quantity and X^n is its approximate value, then $|X - X^n|$ is called the absolute error E_A.

The relative error is defined by $E_R = \left| \dfrac{X - X'}{X} \right|$

and percentage error is defined by $E_P = 100 E_R = 100 \left| \dfrac{X - X'}{X} \right|$

If $\bar{X}$ be such a number that

$$|X - X^n| \le \bar{X}$$

then $\bar{X}$ is an upper limit on the magnitude of absolute error and measures the absolute accuracy.

Notes:

1. The relative and percentage errors are independent units used while absolute error is expressed in terms of these units

2. If a number is correct to n decimal places then the error is $= \dfrac{1}{2} 10^{-n}$.

   For example, if the number is 3.1416, correct to 4 decimal places then the error

   $$= \dfrac{1}{2} 10^{-4} = 0.00005$$

3. If the first significant figure of a number is K and the number is correct to n significant figures, then the relative error $< \dfrac{1}{K \times 10^{n-1}}$.

Example 1.1: An approximate value of π is given by 3.1428571 and its true value is $X = 3.1415926$. Find absolute and relative errors.

Solution: $E_A = 3.1415926 - 3.1428571 = -0.0012645$

$$E_R = \dfrac{-0.0012645}{3.1415926} = -0.000402$$

Example 1.2: Find the relative error of the number 8.6, if both of its digits are correct, [where $E_A = 0.05$].

Solution: Relative error $E_R = \dfrac{0.05}{8.6} = 0.0058$

Example 1.3: Three approximate value of the number $1/3$ are given as $0.30, 0.33$ and 0.34. Which of these three is the best approximation?

Solution:
$$\left|\frac{1}{3} - 0.30\right| = \frac{1}{30}$$

$$\left|\frac{1}{3} - 0.33\right| = \frac{0.01}{3} = \frac{1}{300}$$

and
$$\left|\frac{1}{3} - 0.34\right| = \frac{0.02}{3} = \frac{1}{150}$$

This shows that 0.33 is the best approximation for $1/3$.

Example 1.4: Given that
$$a = 10.00 \pm 0.05$$
$$b = 0.0356 \pm 0.0002$$
$$c = 15300 \pm 100$$
$$d = 62000 \pm 500.$$

Find the maximum value of absolute error in $a + b + c + d$.

Solution: Absolute error in $a = |\pm 0.05| = 0.05$

Absolute error in $b = |\pm 0.0002| = 0.0002$

Absolute error in $c = |\pm 100| = 100$

Absolute error in $d = |\pm 500| = 500$

Thus, the maximum absolute error in
$$a + b + c + d = 0.05 + 0.0002 + 100 + 500$$
$$= 600.0502$$

Example 1.5: Find the sum $S = \sqrt{3} + \sqrt{5} + \sqrt{7}$ to 4 significant digits and find its absolute and relative errors.

Solution:
$$\sqrt{3} = 1.732$$
$$\sqrt{5} = 2.236$$
$$\sqrt{7} = 2.646$$
$$S = \sqrt{3} + \sqrt{5} + \sqrt{7}$$
$$= 1.732 + 2.236 + 2.646 = 6.614$$

Since $\sqrt{3} + \sqrt{5} + \sqrt{7}$ is correct to three decimal places so each has the error

$$= \frac{1}{2}(10^{-3}) = 0.0005.$$

Thus, the total absolute error
$$E_A = 0.0005 + 0.0005 + 0.0005 = 0.0015$$

This absolute error shows that the sum is correct to 2 decimal places. So S is correct to 3 significant digits only.

$\therefore$ Taking
$$S = 6.61$$

Relative error
$$E_R = \frac{E_A}{S} = \frac{0.0015}{6.61} = 0.0002$$

Example 1.6: Find the percentage error, if 625.483 is approximated to three significant figures.

Solution:

$$X = 625.483, \ X^n = 625$$

$$E_A = 625.483 - 625 = 0.483$$

$$E_R = \frac{0.483}{625.483} = 0.000772$$

$$E_P = 100 E_R = 100 \,(0.000772) = 0.0772$$

1.5 A GENERAL ERROR FORMULA

A general formula for the error committed in using a certain formula or a functional relation can be derived by taking a function $u = f(x_1, x_2, ..., x_n)$ of several variables $x_i \ i = 1, 2, ..., n$, and let the error in each x_i be Δx_i. Then, the error Δu in u is given by

$$u + \Delta u = f(x_1 + \Delta x_1, x_2 + \Delta x_2, ..., x_n + \Delta x_n) \tag{1.1}$$

Expanding the RHS of Eq. (1.1) by Taylor's series, we obtain

$$u + \Delta u = f(x_1, x_2, ..., x_n) + \sum_{i=1}^{n} \frac{\partial f}{\partial x_i} \Delta x_i + \text{terms involving } (\Delta x_i)^2 \text{ etc.}$$

Assuming that errors in x_i are small, so that $\dfrac{\Delta x_i}{x_i} < 1$, the squares and higher power of Δx_i can be neglected, the above relation yields

$$\Delta u = \sum_{i=1}^{n} \frac{\partial f}{\partial x_i} \Delta x_i = \frac{\partial f}{\partial x_1} \Delta x_1 + \frac{\partial f}{\partial x_2} \Delta x_2 + ... + \frac{\partial f}{\partial x_n} \Delta x_n \tag{1.2}$$

It is observed that this formula has the same form as that for the total differential of u. The formula for the relative error follows immediately,

$$E_R = \frac{\Delta u}{u} = \frac{\partial u}{\partial x_1} \frac{\Delta x_1}{u} + \frac{\partial u}{\partial x_2} \frac{\Delta x_2}{u} + ... + \frac{\partial u}{\partial x_n} \frac{\Delta x_n}{u} \tag{1.3}$$

Example 1.7: If $u = \dfrac{5xy^2}{z^3}$ and error in x, y, z be 0.001, compute the relative maximum error in u when $x = y = z = 1$.

Solution: Now

$$\frac{\partial u}{\partial x} = \frac{5y^2}{z^3}, \frac{\partial u}{\partial y} = \frac{10xy}{z^3}, \frac{\partial u}{\partial z} = \frac{-15xy^2}{z^4}$$

$$\Delta u = \frac{5y^2}{z^3} \Delta x + \frac{10xy}{z^3} \Delta y - \frac{15xy^2}{z^4} \Delta z$$

In general, the errors (Δx, Δy, and Δz) may be positive or negative and hence the absolute values of the terms are taken on RHS, giving

$$(\Delta u)_{max} = \left| \frac{5y^2}{z^3} \Delta x \right| + \left| \frac{10xy}{z^3} \Delta y \right| + \left| \frac{15xy^2}{z^4} \Delta z \right|$$

Now let $\Delta x = \Delta y = \Delta z = 0.001$ and $x = y = z = 1$. Then, the relative maximum error $(E_R)_{max}$ is given by $(E_R)_{max} = \dfrac{(\Delta u)_{max}}{u} = \dfrac{0.03}{5} = 0.006.$

Example 1.8: If $u = \dfrac{4x^2 y^3}{z^4}$ and error in x, y, z be 0.001, compute the relative maximum error in u when $x = y = z = 1$.

Solution:
$$\frac{\partial u}{\partial x} = \frac{8xy^3}{z^4}, \frac{\partial u}{\partial y} = \frac{12x^2 y^2}{z^4}, \frac{\partial u}{\partial z} = \frac{-16x^2 y^3}{z^5}$$

or
$$\Delta u = \frac{8xy^3}{z^4} \Delta x + \frac{12x^2 y^2}{z^4} \Delta y - \frac{16x^2 y^3}{z^5} \Delta z$$

Since, the errors Δx, Δy, Δz may be positive or negative, then by taking the absolute values of the terms on RHS, we have

$$(\Delta u)_{\max} = \left|\frac{8xy^3}{z^4} \Delta x\right| + \left|\frac{12x^2 y^2}{z^4} \Delta y\right| + \left|\frac{16x^2 y^3}{z^5} \Delta z\right|$$
$$= 8\,(0.001) + 12\,(0.001) + 16\,(0.001) = 0.036$$

Hence, the maximum relative error $(E_R)_{\max}$ is given by

$$(E_R)_{\max} = \frac{(\Delta u)_{\max}}{u} = \frac{0.036}{4} = 0.009$$

Example 1.9: Find the relative error in the function
$$u = ax_1^{m_1}, x_2^{m_2} \dots x_n^{m_n}$$

Solution:
$$\log u = \log a + m_1 \log x_1 + m_2 \log x_2 + \dots + m_n \log x_n$$
$$\frac{1}{u}\frac{\partial u}{\partial x_1} = \frac{m_1}{x_1}, \frac{1}{u}\frac{\partial u}{\partial x_2} = \frac{m_2}{x_2} \text{ etc.}$$

Hence
$$E_R \approx \frac{1}{u}\frac{\partial u}{\partial x_1}\Delta x_1 + \frac{1}{u}\frac{\partial u}{\partial x_2}\Delta x_2 + \dots + \frac{1}{u}\frac{\partial u}{\partial x_n}\Delta x_n$$
$$= \frac{m_1 \Delta x_1}{x_1} + \frac{m_2 \Delta x_2}{x_2} + \dots + m_n \frac{\Delta x_n}{x_n}$$

Since the errors Δx_1, Δx_2, ..., Δx_n may be positive or negative, by taking the absolute values of the terms on RHS, we have

$$(E_R)_{\max} \leq m_1 \left|\frac{\Delta x_1}{x_1}\right| + m_2 \left|\frac{\Delta x_2}{x_2}\right| + \dots + m_n \left|\frac{\Delta x_n}{x_n}\right|$$

Note: Taking $a = 1$, $m_1 = m_2 = \dots = m_n = 1$.

We have $\qquad u = x_1, x_2, \dots, x_n$

Then $\qquad E_R \approx \dfrac{\Delta x_1}{x_1} + \dfrac{\Delta x_2}{x_2} + \dots + \dfrac{\Delta x_n}{x_n}$

Thus, the relative error of a product of n number is approximately equal to the algebraic sum of their relative errors.

1.6 ERROR FORMULA TO THE FUNDAMENTAL OPERATIONS OF ARITHMETIC AND TO LOGARITHMS

i. Error formula for additions:

Let $\qquad X = x_1 + x_2 + \dots + x_n$

then $\qquad \delta X = \delta x_1 + \delta x_2 + \dots + \delta x_n$

$\therefore \qquad E_A = \delta x_1 + \delta x_2 + \dots + \delta x_n$

Thus, the absolute error of a sum of approximate numbers is equal to the algebraic sum of their absolute error.

Now relative error $\quad E_R = \dfrac{E_A}{X}$

$$= \frac{E_A}{X} = \frac{\delta x_1}{X} + \frac{\delta x_2}{X} + ... + \frac{\delta x_n}{X}$$

Maximum relative error $= \left|\dfrac{E_A}{X}\right| \le \left|\dfrac{\delta x_1}{X}\right| + \left|\dfrac{\delta x_2}{X}\right| + ... + \left|\dfrac{\delta x_n}{X}\right|$

ii. Error formula for subtraction:

Let $\qquad\qquad X = x_1 - x_2$

then $\qquad\qquad \delta X = \delta x_1 - \delta x_2$

$\therefore \qquad\qquad E_A = \delta x_1 - \delta x_2$

$$E_R = \frac{E_A}{X} = \frac{\delta X}{X} = \frac{\delta x_1}{X} - \frac{\delta x_2}{X}$$

Though the errors δx_1 and δx_2 may be either positive or negative, however, the sum of the absolute errors is taken in order to get the maximum error. Then, the result is that absolute error of the difference of two approximate numbers may equal to the sum of their absolute errors.

Maximum absolute error $= |\delta x| \le |\delta x_1| + |\delta x_2|$

Maximum relative error $= \left|\dfrac{\delta x}{X}\right| \le \left|\dfrac{\delta x_1}{X}\right| + \left|\dfrac{\delta x_2}{X}\right|$

iii. Error formula for multiplication:

Let $\qquad\qquad X = x_1 x_2 ... x_n$

Since, we know that when X is a function of $x_1, x_2, ..., x_n$, then

$$\delta X = \frac{\partial X}{\partial x_1}\delta x_1 + \frac{\partial X}{\partial x_2}\delta x_2 + ... + \frac{\partial X}{\partial x_n}\delta x_n$$

$$\frac{\delta x}{X} = \frac{1}{X}\frac{\partial X}{\partial x_1}\delta x_1 + \frac{1}{X}\frac{\partial X}{\partial x_2}\delta x_2 + ... + \frac{1}{X}\frac{\partial X}{\partial x_n}\delta x_n$$

Now $\qquad \dfrac{\partial X}{\partial x_1} = \dfrac{\partial}{\partial x_1}(x_1 x_2 ... x_n) = x_2 x_3 ... x_n$

So that $\qquad \dfrac{1}{X}\dfrac{\partial X}{\partial x_1} = \dfrac{1}{X}\dfrac{\partial}{\partial x_1} = \dfrac{x_2 x_3 ... x_n}{x_1 x_2 ... x_n} = \dfrac{1}{x_1}$

Similarly $\qquad \dfrac{1}{X}\dfrac{\partial X}{\partial x_2} = \dfrac{1}{x_2}, \dfrac{1}{X}\dfrac{\partial X}{\partial x_3} = \dfrac{1}{x_3}$

$$\frac{1}{X}\frac{\partial X}{\partial x_n} = \frac{1}{x_n}$$

$$\frac{\partial X}{X} = \frac{\delta x_1}{x_1} + \frac{\delta x_2}{x_2} + ... + \frac{\delta x_n}{x_n}$$

$$\text{Maximum relative error} = \left|\frac{\delta X}{X}\right| \leq \left|\frac{\delta x_1}{x_1}\right| + \left|\frac{\delta x_2}{x_2}\right| + ... + \left|\frac{\delta x_n}{x_n}\right|$$

$$\text{Maximum absolute error} = \left|\frac{\delta X}{X}\right| X$$

$$E_A = E_R \cdot X = \left|\frac{\delta X}{X}\right| \cdot (x_1, x_2, ..., x_n)$$

Thus, the relative error of a product of n approximate numbers is equal to the algebraic sum of the relative errors of the separate numbers. The absolute error, if required, can be found from the relation $E_A = E_R X$.

iv. Error formula for division:

Let
$$X = \frac{x_1}{x_2}$$

Since
$$\delta X = \frac{\partial X}{\partial x_1}\delta x_1 + \frac{\partial X}{\partial x_2}\delta x_2$$

$$\frac{\delta x}{X} = \frac{1}{X}\frac{\partial X}{\partial x_1}\delta x_1 + \frac{1}{X}\frac{\partial X}{\partial x_2}\delta x_2$$

Now
$$\frac{\partial X}{\partial x_1} = \frac{\partial}{\partial x_1}\left(\frac{x_1}{x_2}\right) = \frac{1}{x_2}$$

$$\frac{\partial X}{\partial x_2} = \frac{\partial}{\partial x_2}\left(\frac{x_1}{x_2}\right) = \frac{-x_1}{x_2^2}$$

$$\frac{\delta x}{X} = \frac{1}{x_2}\cdot\frac{1}{X}\delta x_1 - \frac{x_1}{x_2^2}\frac{x_2}{x_1}\delta x_2 = \frac{\delta x_1}{x_1} - \frac{\delta x_2}{x_2}$$

$$= \frac{1}{x_2},\frac{x_2}{x_1}\delta x_1 - \frac{x_1}{x_2^2}\frac{x_2}{x_1}\delta x_2 = \frac{\delta x_1}{x_1} - \frac{\delta x_2}{x_2}$$

$$\text{Relative error } E_R = \left|\frac{\delta x_1}{x_1}\right| + \left|\frac{\delta x_2}{x_2}\right|$$

$$\text{Absolute error } E_A = |\delta X| \leq \left|\frac{\delta X}{X}\right|\cdot X$$

v. Error formula for powers and roots:

Let
$$X = xm$$

$$E_R \leq m\left|\frac{\delta x}{x}\right|$$

Now for kth power of x, put $m = k$, and therefore

$$E_R \leq k\left|\frac{\delta x}{x}\right|$$

for pth root of x, put $m = \dfrac{1}{p}$ and therefore, we get

$$E_R = \frac{1}{p}\left|\frac{\delta x}{x}\right|$$

vi. Error formula for logarithms:

Let $\qquad\qquad\qquad X = \log_{10} x = 0.4429 \log_e x$

Then $\qquad\qquad\qquad \delta X = 0.43429\left(\dfrac{\delta x}{x}\right)$

or $\qquad\qquad\qquad \delta X \le \dfrac{1}{2}\left(\dfrac{\delta x}{x}\right)$

Thus, the absolute error in the common logarithm of a number is less than half the relative error of the given number.

Also $\qquad\qquad\qquad \delta x = 0.3026 \, x \, \delta x$

The error is the antilog may therefore be many times the error in the logarithm. For this reason, it is of the utmost importance that the logarithm of a result be free from error, as possible.

Example 1.10: $\sqrt{29} = 0.5385$ and $\sqrt{11} = 3.317$, correct to four significant figures, find the relative error in their sum and difference?

Solution: $\sqrt{29} = 0.5385$ and $\sqrt{11} = 3.317$ are correct to four significant figures, so that maximum error in each case

$$\Rightarrow \qquad\qquad \frac{1}{2} \times 10^{-3} = 0.0005$$

Let $\qquad\qquad\qquad x_1 = \sqrt{29} = 5.385, \; x_2 = \sqrt{11} = 3.317$

Then $\qquad\qquad\qquad \delta x_1 = \delta x_2 = 0.0005$

i. Relative error in their sum

$$\left|\frac{\delta x}{X}\right| \le \left|\frac{\delta x_1}{X}\right| + \left|\frac{\delta x_2}{X}\right| \le \left|\frac{0.0005}{8.702}\right| + \left|\frac{0.0005}{8.702}\right|$$

$$= \left|\frac{0.0005}{8.702}\right| = 1.149 \times 10^{-4}$$

Thus, the relative error in $(x_1 + x_2) < 1.149 \times 10^{-4}$

ii. Relative error in $(x_1 - x_2)$

$$\left|\frac{\delta x}{X}\right| \le \left|\frac{\delta x_1}{X}\right| + \left|\frac{\delta x_2}{X}\right|$$

Now $\qquad\qquad\qquad X = x_1 - x_2$

$$= 5.385 - 3.317 = 2.068$$

Relative error in their difference

$$\left|\frac{\delta x}{X}\right| \le \left|\frac{0.0005}{2.068}\right| + \left|\frac{0.0005}{2.068}\right| = 2\left|\frac{0.0005}{2.068}\right| = 4.835 \times 10^{-4}$$

Thus, the relative error in $(x_1 - x_2) < 4.835 \times 10^{-4}$.

Example 1.11: Find the relative error in calculation of $\dfrac{7.342}{0.241}$. Numbers are correct to three decimal places. Determine the smallest interval in which true result lies.

Solution: Let $x_1 = 7.342$, $x_2 = 0.241$, then

$$\delta x_1 = \frac{1}{2} \times 10^{-3} = 0.0005$$

$$\delta x_2 = \frac{1}{2} \times 10^{-3} = 0.0005$$

$$\text{Relative error} \leq \left|\frac{0.0005}{7.342}\right| + \left|\frac{0.0005}{0.241}\right|$$

$$\leq 0.0005\left(\frac{1}{7.342} + \frac{1}{0.241}\right)$$

$$= 0.0005\,(4.285580262) = 0.0021$$

$$\text{Absolute error} = 0.0021 \times \frac{x_1}{x_2} = 0.0021 \times \frac{7.3242}{0.241} = 0.0639$$

$$\frac{x_1}{x_2} = \frac{7.342}{0.241} = 30.4647$$

Thus, the true value of $\dfrac{x_1}{x_2}$ lies between $(30.4647 - 0.0639)$ and $(30.4647 + 0.0639)$, i.e. between 30.4008 and 30.5286.

1.7 ERROR IN A SERIES APPROXIMATION

The error committed in a series approximation can be evaluated by using the remainder after n terms. Taylor's series for $f(x)$ at $x = a$ is given by

$$f(x) = f(a) + (x - a)f'(a) + \frac{(x - a)^2}{2!} f''(a) + \ldots + \frac{(x - a)^{n-1}}{(n - 1)!} f^{(n-1)}(a) + R_n(x)$$

The last term is $R_n(x)$, is called the remainder. For a series which is convergent, $R_n(x)$ $\rightarrow 0$ as $n \rightarrow \infty$. Thus, if $f(x)$ is approximated to the first n terms of a series, then the maximum error committed in this approximation is given by the remainder term. Conversely, if the accuracy required is specified in advance, then it would be possible to find n, the number of terms, such that the finite series yields the required accuracy.

Example 1.12: Find the number of terms of the exponential series such that their sum gives the values of e^x correct to eight decimal places at $x = 1$.

Solution:
$$e^x = 1 + x + \frac{x^2}{2!} + \frac{x^3}{3!} + \ldots + \frac{x^{n-1}}{(n - 1)!} + R_n(x)$$

where
$$R_n(x) = \frac{x^{n-1}}{n!} e^{\theta}, 0 < \theta < x$$

Maximum absolute error at $(\theta - x) = \dfrac{x^n}{n!} e^x$

and the maximum relative error $= \dfrac{x^n}{n!}$

Hence $(E_R)_{max}$ at $x = 1$ is $\dfrac{1}{n!}$.

For a eight decimal accuracy at $x = 1$, $\dfrac{1}{n!} < \dfrac{1}{2} 10^{-8}$, i.e. $n! > 2 \times 10^8$, which gives $n = 12$.

Thus, 12th terms of the exponential series are needed to get the sum correct to 8 decimal places.

PROBLEMS 1.1

1. Find the relative error if 2/3 is approximated to 0.667.   [Ans. 0.0005]

2. Find the percentage error if 0.005998 is approximated to three significant figures.
[Ans. 0.03344]

3. Round-off the following numbers to four significant figures.
   0.0022218, 19.235101, 2.36425   [Ans. 0.002222, 19.24, 2.364]

4. Find the smaller root of the equation $x^2 - 400x + 1 = 0$ correct to 4 decimal places.
[Ans. 0.0025]

5. If $R = 10x^2y^2z^2$ and error in x, y, z are 0.03, 0.01, and 0.02 respectively at $x = 3$, $y = 1$, and $z = 2$. Calculate the absolute error and percentage relative error in evaluating R?   [Ans. 140.4, 13%]

6. If $R = 4xy^2/z^3$ and error in x, y, z be 0.001, show that the maximum relative error at $x = y = z = 1$ is 0.006.

7. Find the number of terms required in the series $\log(1 + x)$ to evaluate $\log 1.2$ correct to six decimal places.   [Ans. 10]

8. Prove that the relative error of a product of three non-zero numbers does not exceed the sum of the relative errors of the given numbers.

9. Prove that the function $f(x) = \tan^{-1}x$ can be expanded at

$$\tan^{-1} x = x - \frac{x^3}{3} + \frac{x^5}{5} + \dots + (-1)^{n-1} \frac{x^{2n-1}}{(2n-1)}.$$

find n such that the series determines $\tan^{-1}1$ correct to eight significant digits.
[Ans. $N = 108 + 1$]

10. If $r = 3h(h^6 - 2)$, find the percentage error in r at $h = 1$, if the percentage error in h is 5.
[Ans. 25%]

11. The error in the measurement of the area of a circle is allowed to exceed 0.1%. How accurately should be the diameter is measured?   [Ans. 0.05%]

2

Simultaneous Linear Algebraic Equations and Eigenvalue Problems

2.1 INTRODUCTION

Simultaneous linear algebraic equations are very common in various fields of engineering and science. Matrix inversion method or Cramer's rule is used to solve these equations in general. But these methods prove to be tedious when the system of equation contain a large number of unknows. To solve such equations, there are other numerical methods which are particularly suited for computer operations. These are of two types: Direct and iterative.

1. **Direct methods:** Gauss elimination method, Gauss–Jordan method, triangularisation method, and Crout's method.

2. **Iterative methods:** Gauss–Jacobi method, Gauss–Seidel iterative method and relaxation method.

2.2 GAUSS ELIMINATION METHOD (Direct Method)

Consider the system of equations

$$\left.\begin{array}{l} a_1 x + b_1 y + c_1 z = d_1 \\ a_2 x + b_2 y + c_2 z = d_2 \\ a_3 x + b_3 y + c_3 z = d_3 \end{array}\right\} \tag{2.1}$$

This system in matrix form is $AX = B$, where

$$A = \begin{pmatrix} a_1 & b_1 & c_1 \\ a_2 & b_2 & c_2 \\ a_3 & b_3 & c_3 \end{pmatrix}, \ B = \begin{pmatrix} d_1 \\ d_2 \\ d_3 \end{pmatrix} \text{ and } X = \begin{pmatrix} x \\ y \\ z \end{pmatrix}$$

Consider the augmented matrix $[A : B]$

$$[A : B] = \begin{pmatrix} a_1 & b_1 & c_1 & : & d_1 \\ a_2 & b_2 & c_2 & : & d_2 \\ a_3 & b_3 & c_3 & : & d_3 \end{pmatrix} \tag{2.2}$$

Now Eq. (2.2) can be reduced to an upper triangular matrix. Let $a_1 \neq 0$, then

$$
\begin{array}{c}
R_2 \rightarrow R_2 - \dfrac{a_2}{a_1} R_1 \\[2mm]
\sim \\[2mm]
R_3 \rightarrow R_3 - \dfrac{a_3}{a_1} R_1
\end{array}
\begin{pmatrix}
a_1 & b_1 & c_1 & : & d_1 \\
0 & b_2' & c_2' & : & d_2' \\
0 & b_3' & c_3' & : & d_3'
\end{pmatrix}
\tag{2.3}
$$

Here a_1 is called first pivot.

Now, take b_2' as the pivot $(b_2' \neq 0)$, then

$$
R_3 \rightarrow R_3 - \dfrac{b_3'}{b_2'} R_2 \sim
\begin{pmatrix}
a_1 & b_1 & c_1 & : & d_1 \\
0 & b_2' & c_2' & : & d_2' \\
0 & b_3' & c_3'' & : & d_3''
\end{pmatrix}
\tag{2.4}
$$

Now take $c_3'' \neq 0$ as the pivot from Eq. (2.4), the given system of linear equation is equivalent to

$$a_1 x + b_1 y + c_1 z = d_1$$

$$b_2' y + c_2' z = d_2'$$

$$c_s'' z = d_3''$$

or

$$z = \frac{d_3''}{c_3''}$$

By back substitution

$$y = \frac{(c_3'' d_2'' - c_2' d_3'')}{b_2' c_3''}$$

and

$$x = \frac{1}{a_1 b_2' c_3''}(d_1 b_2' c_3'' + b_1 c_2' d_3'' - b_2' c_1 d_3'' - b_1 c_3'' d_2')$$

Notes:

1. This method fails if anyone of the pivots a_1, b_2', c_3'' becomes zero. In such cases by interchanging the rows, we can get the nonzero pivots.
2. **Partial pivoting:** From the first column of Eq. (2.2), select the component with the largest absolute value. This is called *pivot*. Then at the second stage, i.e. from the second column of Eq. (2.3), select once again the component with largest absolute value as the pivot. Continue this process. This procedure is called partial pivoting.
3. **Complete pivoting:** If one is not interested in the elimination of x, y, z in a particular order, then choose at each stage numerically the largest coefficient of the entire coefficient matrix. This requires an interchange of equations and also an interchange of positions of the variables.

Example 2.1: Solve the system of equations $3x + y - z = 3$, $2x - 8y + z = -5$, $x - 2y + 9z = 8$, using Gauss elimination method

Solution: The given method is equivalent to $\begin{pmatrix} 3 & 1 & -1 \\ 2 & -8 & 1 \\ 1 & -2 & 9 \end{pmatrix} \begin{pmatrix} x \\ y \\ z \end{pmatrix} = \begin{pmatrix} 3 \\ -5 \\ 8 \end{pmatrix}$

The augmented matrix is

$$[A : B] = \begin{pmatrix} 3 & 1 & -1 & : & 3 \\ 2 & -8 & 1 & : & -5 \\ 1 & -2 & 9 & : & 8 \end{pmatrix}$$

$$\begin{aligned} R_2 &\to R_2 - \frac{2}{3}R_1 \\ R_3 &\to R_3 - \frac{1}{3}R_1 \end{aligned} \quad \sim \begin{pmatrix} 3 & 1 & -1 & : & 3 \\ 0 & \dfrac{-26}{3} & \dfrac{5}{3} & : & -7 \\ 0 & \dfrac{-7}{3} & \dfrac{28}{3} & : & 7 \end{pmatrix}$$

Choosing $\dfrac{-26}{3}$ as the pivot from the 2nd column, we have

$$R_3 \to R_3 - \frac{7}{26}R_2 \quad \sim \begin{pmatrix} 3 & 1 & -1 & : & 3 \\ 0 & \dfrac{-26}{3} & \dfrac{5}{3} & : & -7 \\ 0 & 0 & \dfrac{2079}{234} & : & \dfrac{231}{26} \end{pmatrix}$$

or

$$3x + y - z = 3$$
$$\frac{-26}{3}y + \frac{5}{3}z = -7$$
$$\frac{2079}{234}z = \frac{231}{26} \text{ or } z = \frac{231 \times 234}{26 \times 2079} = 1$$

Now by back substitution, $z = 1$
$$-26y = 26 \text{ or } y = 1$$
and
$$3x = 3x \text{ or } x = 1$$
$$x = 1, y = 1, z = 1$$

Example 2.2: Using Gauss elimination method, solve the system of equations.
$$3.15x - 1.96y + 3.85z = 12.95$$
$$2.13x + 5.12y - 2.89z = -8.61$$
$$5.92x + 3.05y + 2.15z = 6.88$$

Solution: The given system is equivalent to

$$\begin{pmatrix} 3.15 & -1.96 & 3.85 \\ 2.13 & 5.12 & -2.89 \\ 5.92 & 3.05 & 2.15 \end{pmatrix} \begin{pmatrix} x \\ y \\ z \end{pmatrix} = \begin{pmatrix} 12.95 \\ -8.61 \\ 6.88 \end{pmatrix}$$

$$AX = B$$

$$(A : B) = \begin{pmatrix} 3.15 & -1.96 & 3.85 & : & 12.95 \\ 2.13 & 5.12 & -2.89 & : & -8.61 \\ 5.92 & 3.05 & 2.15 & : & 6.88 \end{pmatrix}$$

Choosing 3.15 as pivot

$$R_2 \to R_2 - \frac{2.13}{3.15} R_1$$
$$R_3 \to R_3 - \frac{5.92}{3.15} R_1$$

$$\sim \begin{pmatrix} 3.15 & -1.96 & 3.85 & : & 12.95 \\ 0 & 6.4453 & -5.4933 & : & -17.3667 \\ 0 & 6.7335 & -5.0855 & : & -17.4578 \end{pmatrix}$$

Choosing 6.4453 as pivot

$$R_3 \to R_3 - \frac{6.7335}{6.4453} R_2$$

$$\sim \begin{pmatrix} 3.15 & -1.96 & 3.85 & : & 12.95 \\ 0 & 6.4453 & -5.4933 & : & -17.3667 \\ 0 & 0 & 0.6534 & : & 0.6854 \end{pmatrix}$$

$$3.15x - 1.96y + 3.85z = 12.95$$
$$6.4453y - 5.4933z = -17.3667$$
$$0.6534z = 0.6854$$

By back substitution

$$z = \frac{0.6854}{0.6534} = 1.04897459, \quad y = \frac{5.4933z - 17.3667}{6.4453} = -1.80043875$$

and

$$x = \frac{1.96y - 3.85 + 12.95}{3.15} = 1.70875806$$

Example 2.3: Solve the system of equations

$$x_1 + x_2 + x_3 + x_4 = 2$$
$$x_1 + x_2 + 3x_3 - 2x_4 = -6$$
$$2x_1 + 3x_2 - x_3 + 2x_4 = 7$$
$$x_1 + 2x_2 + x_3 - x_4 = -2$$

by Gauss elimination method.

Solution:

$$\begin{pmatrix} 1 & 1 & 1 & 1 \\ 1 & 1 & 3 & -2 \\ 2 & 3 & -1 & 2 \\ 1 & 2 & 1 & -1 \end{pmatrix} \begin{pmatrix} x_1 \\ x_2 \\ x_3 \\ x_4 \end{pmatrix} = \begin{pmatrix} 2 \\ -6 \\ 7 \\ -2 \end{pmatrix}$$

$$AX = B$$

$$[A : B] = \begin{pmatrix} 1 & 1 & 1 & 1 & : & 2 \\ 1 & 1 & 3 & -2 & : & -6 \\ 2 & 3 & -1 & 2 & : & 7 \\ 1 & 2 & 1 & -1 & : & -2 \end{pmatrix}$$

$$\begin{matrix} R_2 \to R_2 - R_1 \\ R_3 \to R_2 - 2R_1 \\ R_4 \to R_4 - R_1 \end{matrix} \sim \begin{pmatrix} 1 & 1 & 1 & 1 & : & 2 \\ 0 & 0 & 2 & -3 & : & -8 \\ 0 & 1 & -3 & 0 & : & 3 \\ 0 & 1 & 0 & -2 & : & -4 \end{pmatrix}$$

Since the element in second row, second column is zero, interchange second and third row to get pivot element 1, i.e.

$$R(2,3) \sim \begin{pmatrix} 1 & 1 & 1 & 1 & : & 2 \\ 0 & 1 & -3 & 0 & : & 3 \\ 0 & 0 & 2 & -3 & : & -8 \\ 0 & 1 & 0 & -2 & : & -4 \end{pmatrix}$$

$$R_4 \rightarrow R_4 - R_2 \sim \begin{pmatrix} 1 & 1 & 1 & 1 & : & 2 \\ 0 & 1 & -3 & 0 & : & 3 \\ 0 & 0 & 2 & -3 & : & -8 \\ 0 & 0 & 3 & -2 & : & -7 \end{pmatrix}$$

Now the pivot is 2, therefore

$$R_4 \rightarrow R_4 - \frac{3}{2} R_3 \sim \begin{pmatrix} 1 & 1 & 1 & 1 & : & 2 \\ 0 & 1 & -3 & 0 & : & 3 \\ 0 & 0 & 2 & -3 & : & -8 \\ 0 & 0 & 0 & \frac{5}{2} & : & 5 \end{pmatrix}$$

$$x_1 + x_2 + x_3 + x_4 = 2$$
$$x_2 - 3x_3 = 3$$
$$2x_3 - 3x_4 = -8$$
$$\frac{5}{2} x_4 = 5$$
$$x_4 = 2$$
$$x_3 = \frac{1}{2}(-8 + 3x_4) = \frac{1}{2}(-8 + 6) = -1$$

Now by back substitution

$$x_2 = 3 + 3x_3 = 3 - 3 = 0$$
$$x_1 = 2 - x_2 - x_3 - x_4 = 2 - 0 - (-1) - 2 = 1$$
$$x_1 = 1, x_2 = 0, x_3 = -1, x_4 = 2$$

PROBLEMS 2.1

Solve the system of equations by Gauss elimination method.

i. $x + 2y + z = 3$
    $2x + 3y + 3z = 10$
    $3x - y + 2x = 13$ [Ans. $x = 2, y = -1, z = 3$]

ii. $2x + 3y - z = 5$
    $4x + 4y - 3z = 3$
    $2x - 3y + 2z = 2$ [Ans. $x = 1, y = 2, z = 3$]

iii. $5x_1 + x_2 + x_3 + x_4 = 4$
    $x_1 + 7x_2 + x_3 + x_4 = 12$
    $x_1 + x_2 + 6x_3 + x_4 = -5$
    $x_1 + x_2 + x_3 + 4x_4 = 6$ [Ans. $x_1 = 1, x_2 = 2, x_3 = -1, x_4 = -2$]

2.3 GAUSS–JORDAN METHOD

This method is a modified form of Gauss elimination method. The coefficient matrix A of $AX = B$ is reduced to a *diagonal matrix* or *unit matrix* by making all the elements above and below the principal diagonal of A as zero. The time of back substitution is saved here, even though it involves additional computations.

Example 2.4: Solve the equations
$$10x + y + z = 12$$
$$2x + 10y + z = 13$$
$$x + y + 5z = 7$$
by Gauss–Jordan method.

Solution: The given system in matrix form is

$$\begin{pmatrix} 10 & 1 & 1 \\ 2 & 10 & 1 \\ 1 & 1 & 5 \end{pmatrix} \begin{pmatrix} x \\ y \\ z \end{pmatrix} = \begin{pmatrix} 12 \\ 13 \\ 7 \end{pmatrix}$$

$$AX = B$$

$$[A : B] = \begin{pmatrix} 10 & 1 & 1 & : & 12 \\ 2 & 10 & 1 & : & 13 \\ 1 & 1 & 5 & : & 7 \end{pmatrix}$$

$$R_1 \rightarrow R_1 - 9R_3 \sim \begin{pmatrix} 1 & -8 & -44 & : & -51 \\ 2 & 10 & 1 & : & 13 \\ 1 & 1 & 5 & : & 7 \end{pmatrix}$$

$$\begin{matrix} R_2 \rightarrow R_2 - 2R_1 \\ R_3 \rightarrow R_3 - R_1 \end{matrix} \sim \begin{pmatrix} 1 & -8 & -44 & : & -51 \\ 0 & 26 & 89 & : & 115 \\ 0 & 9 & 49 & : & 58 \end{pmatrix}$$

$$R_2 \rightarrow 3R_3 - R_2) \sim \begin{pmatrix} 1 & -8 & -44 & : & -51 \\ 0 & 1 & 58 & : & 59 \\ 0 & 9 & 49 & : & 58 \end{pmatrix}$$

$$\begin{matrix} R_1 \rightarrow R_1 + 8R_2 \\ R_3 \rightarrow R_3 - 9R_2 \end{matrix} \sim \begin{pmatrix} 1 & 0 & 420 & : & 421 \\ 0 & 1 & 58 & : & 59 \\ 0 & 0 & -473 & : & -473 \end{pmatrix}$$

$$R_3 \rightarrow -\frac{1}{473} R_3 \sim \begin{pmatrix} 1 & 0 & 420 & : & 421 \\ 0 & 1 & 58 & : & 59 \\ 0 & 0 & 1 & : & 1 \end{pmatrix}$$

$$\begin{matrix} R_1 \to R_1 - 420R_3 \\ R_2 \to R_2 - 58R_3 \end{matrix} \sim \begin{pmatrix} 1 & 0 & 0 & : & 1 \\ 0 & 1 & 0 & : & 1 \\ 0 & 0 & 1 & : & 1 \end{pmatrix}$$

The system $AX = B$ reduces to the form

$$\begin{pmatrix} 1 & 0 & 0 \\ 0 & 1 & 0 \\ 0 & 0 & 1 \end{pmatrix} \begin{pmatrix} x \\ y \\ z \end{pmatrix} = \begin{pmatrix} 1 \\ 1 \\ 1 \end{pmatrix}$$

i.e. $x = 1 = y = z$.

Example 2.5: Solve the equations

$$10x_1 + x_2 + x_3 = 12, \quad x_1 + 10x_2 - x_3 = 10$$

and $\qquad x_1 - 2x_2 + 10x_3 = 9$ by Gauss–Jordan method.

Solution: The given system in matrix form is

$$\begin{pmatrix} 10 & 1 & 1 \\ 1 & 10 & -1 \\ 1 & -2 & 10 \end{pmatrix} \begin{pmatrix} x_1 \\ x_2 \\ x_3 \end{pmatrix} = \begin{pmatrix} 12 \\ 10 \\ 9 \end{pmatrix}$$

$$AX = B$$

$$[A : B] = \begin{pmatrix} 10 & 1 & 1 & : & 12 \\ 1 & 10 & -1 & : & 10 \\ 1 & -2 & 10 & : & 9 \end{pmatrix}$$

$$R_1 \to R_1 - 9R_2 \sim \begin{pmatrix} 1 & -89 & 10 & : & -78 \\ 1 & 10 & -1 & : & 10 \\ 1 & -2 & 10 & : & 9 \end{pmatrix}$$

$$\begin{matrix} R_2 \to R_2 - R_1 \\ R_3 \to R_3 - R_1 \end{matrix} \sim \begin{pmatrix} 1 & -89 & 10 & : & -78 \\ 0 & 99 & -11 & : & 88 \\ 0 & 87 & 0 & : & 87 \end{pmatrix}$$

$$\begin{matrix} R_2 \to \dfrac{R_2}{11} \\ R_3 \to \dfrac{R_3}{87} \end{matrix} \sim \begin{pmatrix} 1 & -89 & 10 & : & -78 \\ 0 & 9 & -1 & : & 8 \\ 0 & 1 & 0 & : & 1 \end{pmatrix}$$

$$R_2 \to R_2 - 8R_3 \sim \begin{pmatrix} 1 & -89 & 10 & : & -78 \\ 0 & 1 & -1 & : & 0 \\ 0 & 1 & 0 & : & 1 \end{pmatrix}$$

$$\begin{matrix} R_1 \to R_1 + 89R_2 \\ R_3 \to R_3 - R_2 \end{matrix} \sim \begin{pmatrix} 1 & 0 & -79 & : & -78 \\ 0 & 1 & -1 & : & 0 \\ 0 & 0 & 1 & : & 1 \end{pmatrix}$$

$$\begin{matrix} R_1 \to R_1 + 79R_3 \\ R_2 \to R_2 - R_3 \end{matrix} \sim \begin{pmatrix} 1 & 0 & 0 & : & 1 \\ 0 & 1 & 0 & : & 1 \\ 0 & 0 & 1 & : & 1 \end{pmatrix}$$

The system $AX = B$ reduces to

$$\begin{pmatrix} 1 & 0 & 0 \\ 0 & 1 & 0 \\ 0 & 0 & 1 \end{pmatrix} \begin{pmatrix} x_1 \\ y_2 \\ x_3 \end{pmatrix} = \begin{pmatrix} 1 \\ 1 \\ 1 \end{pmatrix}$$

$\therefore x_1 = x_2 = x_3 = 1.$

Example 2.6: Solve the equations by Gauss–Jordan method

$$x + 2y + z - w = -2$$
$$2x + 3y - z + 2w = 7$$
$$x + y + 3z - 2w = -6$$
$$x + y + z + w = 2$$

Solution: The given system in matrix form is

$$\begin{pmatrix} 1 & 2 & 1 & -1 \\ 2 & 3 & -1 & 2 \\ 1 & 1 & 3 & -2 \\ 1 & 1 & 1 & 1 \end{pmatrix} \begin{pmatrix} x \\ y \\ z \\ w \end{pmatrix} = \begin{pmatrix} -2 \\ 7 \\ -6 \\ 2 \end{pmatrix}$$

$$AX = B$$

Now $\qquad [A : B] = \begin{pmatrix} 1 & 2 & +1 & -1 & : & -2 \\ 2 & 3 & -1 & 2 & : & 7 \\ 1 & 1 & 3 & -2 & : & -6 \\ 1 & 1 & 1 & 1 & : & 2 \end{pmatrix}$

$$\begin{matrix} R_2 \to -(R_2 - 2R_1) \\ R_3 \to R_3 - R_1 \\ R_4 \to R_4 - R_1 \end{matrix} \sim \begin{pmatrix} 1 & 2 & 1 & -1 & : & -2 \\ 0 & 1 & 3 & -4 & : & -11 \\ 0 & -1 & 2 & -1 & : & -4 \\ 0 & -1 & 0 & 2 & : & 4 \end{pmatrix}$$

$$\begin{matrix} R_1 \to R_1 - 2R_2 \\ R_3 \to \dfrac{1}{5}(R_3 + R_2) \\ R_4 \to R_4 + R_2 \end{matrix} \sim \begin{pmatrix} 1 & 0 & -5 & 7 & : & 20 \\ 0 & 1 & 3 & -4 & : & -11 \\ 0 & 0 & 1 & -1 & : & -3 \\ 0 & 0 & 3 & -2 & : & 7 \end{pmatrix}$$

$$\begin{matrix} R_1 \to R_1 + 5R_3 \\ R_2 \to R_2 - 3R_2 \\ R_4 \to R_4 + 3R_3 \end{matrix} \sim \begin{pmatrix} 1 & 0 & 0 & 2 & : & 5 \\ 0 & 1 & 0 & -1 & : & -2 \\ 0 & 0 & 1 & -1 & : & -3 \\ 0 & 0 & 0 & 1 & : & 6 \end{pmatrix}$$

∴ The system $AX = B$ reduces to form

$$\begin{matrix} R_1 \rightarrow R_1 - 2R_4 \\ R_2 \rightarrow R_2 + R_4 \\ R_3 \rightarrow R_3 + R_4 \end{matrix} \sim \begin{pmatrix} 1 & 0 & 0 & 0 \\ 0 & 1 & 0 & 0 \\ 0 & 0 & 1 & 0 \\ 0 & 0 & 0 & 1 \end{pmatrix} \begin{pmatrix} x \\ y \\ z \\ w \end{pmatrix} = \begin{pmatrix} -7 \\ 4 \\ 3 \\ 6 \end{pmatrix}$$

∴ $x = -7$, $y = 4$, $z = 3$ and $w = 6$.

Example 2.7: Solve the system of equations

$$5x - y - 2z = 142$$
$$x - 3y - z = -30$$
$$2x - y - 3z = -5$$

by Gauss–Jordan method.

Solution: Now $[A : B] \sim \begin{pmatrix} 5 & -1 & -2 & : & 142 \\ 1 & -3 & -1 & : & -30 \\ 2 & -1 & -3 & : & -5 \end{pmatrix}$

$$R_1 \leftrightarrow R_2 \sim \begin{pmatrix} 1 & -3 & -1 & : & -30 \\ 5 & -1 & -2 & : & 142 \\ 2 & -1 & -3 & : & -5 \end{pmatrix}$$

$$\begin{matrix} R_2 \rightarrow R_2 - 5R_1 \\ R_3 \rightarrow R_3 - 2R_1 \end{matrix} \sim \begin{pmatrix} 1 & -3 & -1 & : & -30 \\ 0 & 14 & 3 & : & 292 \\ 0 & 5 & -1 & : & 55 \end{pmatrix}$$

$$R_2 \rightarrow \frac{R_2}{14} \sim \begin{pmatrix} 1 & -3 & -1 & : & -30 \\ 0 & 1 & \dfrac{3}{14} & : & \dfrac{146}{7} \\ 0 & 5 & -1 & : & 55 \end{pmatrix}$$

$$\begin{matrix} R_2 \rightarrow R_1 + 3R_2 \\ R_3 \rightarrow R_3 - 5R_2 \end{matrix} \sim \begin{pmatrix} 1 & 0 & -\dfrac{5}{14} & : & \dfrac{228}{7} \\ 0 & 1 & \dfrac{3}{14} & : & \dfrac{146}{7} \\ 0 & 0 & -\dfrac{29}{14} & : & -\dfrac{345}{7} \end{pmatrix}$$

$$R_3 \rightarrow \left(-\dfrac{14}{29}\right) R_3 \sim \begin{pmatrix} 1 & 0 & -\dfrac{5}{14} & : & \dfrac{228}{7} \\ 0 & 1 & \dfrac{3}{14} & : & \dfrac{146}{7} \\ 0 & 0 & 1 & : & \dfrac{690}{29} \end{pmatrix}$$

$$R_1 \to R_1 + \frac{5}{14}R_3$$
$$R_2 \to R_2 - \frac{3}{14}R_3$$
$$\sim \begin{pmatrix} 1 & 0 & 0 & : & \dfrac{8337}{203} \\ 0 & 1 & 0 & : & \dfrac{3199}{203} \\ 0 & 0 & 1 & : & \dfrac{690}{29} \end{pmatrix}$$

$$\therefore \ x = \frac{8337}{203}, y = \frac{3199}{203}, z = \frac{690}{29}.$$

Example 2.8: Solve the system of equations

$$x + \frac{y}{2} + \frac{z}{3} = 1$$
$$\frac{x}{2} + \frac{y}{3} + \frac{z}{4} = 0$$
$$\frac{x}{3} + \frac{y}{4} + \frac{z}{5} = 0$$

by Gauss–Jordan method.

Solution: $6x + 3y + 2z = 6$
$$6x + 4y + 3z = 0$$
$$20x + 15y + 12z = 0$$

$$[A : B] = \begin{pmatrix} 6 & 3 & 2 & : & 6 \\ 6 & 4 & 3 & : & 0 \\ 20 & 15 & 12 & : & 0 \end{pmatrix}$$

$$R_1 \to \frac{R_1}{6} \sim \begin{pmatrix} 1 & \dfrac{1}{2} & \dfrac{1}{3} & : & 1 \\ 6 & 1 & 3 & : & 0 \\ 20 & 5 & 12 & : & 0 \end{pmatrix}$$

$$\begin{matrix} R_2 \to R_2 - 6R_1 \\ R_3 \to R_3 - 20R_1 \end{matrix} \sim \begin{pmatrix} 1 & \dfrac{1}{2} & \dfrac{1}{3} & : & 1 \\ 0 & 1 & 3 & : & -6 \\ 0 & 5 & \dfrac{16}{3} & : & -20 \end{pmatrix}$$

$$\begin{matrix} R_1 \to R_1 - \dfrac{1}{2}R_2 \\ R_3 \to R_3 - 5R_2 \end{matrix} \sim \begin{pmatrix} 1 & 0 & -\dfrac{1}{6} & : & 4 \\ 0 & 1 & 1 & : & -6 \\ 0 & 0 & \dfrac{1}{3} & : & 10 \end{pmatrix}$$

$$R_3 \to 3R_3 \sim \begin{pmatrix} 1 & 0 & -\dfrac{1}{6} & : & 4 \\ 0 & 1 & 1 & : & -6 \\ 0 & 0 & 1 & : & 30 \end{pmatrix}$$

$$R_1 \to R_1 + \frac{1}{6} R_3 \quad \sim \begin{pmatrix} 1 & 0 & 0 & : & 9 \\ 0 & 1 & 0 & : & -36 \\ 0 & 0 & 1 & : & 30 \end{pmatrix}$$
$$R_2 \to R_2 - R_3$$

$\therefore x = 9, y = -36, z = 30.$

PROBLEMS 2.2

Solve the system of equations by Gauss–Jordan method.

i. $x + y + z + w = 2$
$2x - y + 2z - w = -5$
$3x + 2y + 3z + 4w = 7$
$x - 2y - 3z + 2w = 5$
[Ans. $x = 0, y = 1, z = -1, w = 2$]

ii. $10x + y + z = 12$
$2x + 10y + z = 13$
$x + y + 5z = 7$
[Ans. $x = 1, y = 1, z = 1$]

iii. $3x + 4y + 5z = 18$
$2x - y + 8z = 13$
$5x - 2y + 7z = 20$
[Ans. $x = 3, y = 1$]

iv. $x + 2y + z - w = -2$
$2x + 3y - z + 2w = 7$
$x + y + 3z - 2w = 6$
$x + y + z + w = 2$
[Ans. $x = 1, y = 0, z = -1, w = 2$]

v. $10x + y + z = 18.141$
$x + 10y + z = 28.140$
$x + y + 10z = 38.139$
[Ans. $x = 1.234, y = 2.348, z = 3.455$]

2.4 GAUSS ELIMINATION METHOD FOR FINDING THE INVERSE OF A MATRIX

Let A be a nonsingular square matrix of order three. Then the inverse of A is a matrix X which satisfy the equation $AX = I$, where I is the unit matrix of order three. Now, we have to find the elements of the inverse matrix X.

Let
$$A = \begin{pmatrix} a_{11} & a_{12} & a_{13} \\ a_{21} & a_{22} & a_{23} \\ a_{31} & a_{32} & a_{33} \end{pmatrix}$$

and
$$X = \begin{pmatrix} x_{11} & x_{12} & x_{13} \\ x_{21} & x_{22} & x_{23} \\ x_{31} & x_{32} & x_{33} \end{pmatrix}$$

The equation becomes $AX = I$

$$\begin{pmatrix} a_{11} & a_{12} & a_{13} \\ a_{21} & a_{22} & a_{23} \\ a_{31} & a_{32} & a_{33} \end{pmatrix} \begin{pmatrix} x_{11} & x_{12} & x_{13} \\ x_{21} & x_{22} & x_{23} \\ x_{31} & x_{32} & x_{33} \end{pmatrix} = \begin{pmatrix} 1 & 0 & 0 \\ 0 & 1 & 0 \\ 0 & 0 & 1 \end{pmatrix}$$

This matrix is equivalent to three equations, which are equivalent to three system of equations

$$\begin{pmatrix} a_{11} & a_{12} & a_{13} \\ a_{21} & a_{22} & a_{23} \\ a_{31} & a_{32} & a_{33} \end{pmatrix} \begin{pmatrix} x_{11} \\ x_{21} \\ x_{31} \end{pmatrix} = \begin{pmatrix} 1 \\ 0 \\ 0 \end{pmatrix} \tag{2.5}$$

$$\begin{pmatrix} a_{11} & a_{12} & a_{13} \\ a_{21} & a_{22} & a_{23} \\ a_{31} & a_{32} & a_{33} \end{pmatrix} \begin{pmatrix} x_{12} \\ x_{22} \\ x_{32} \end{pmatrix} = \begin{pmatrix} 0 \\ 1 \\ 0 \end{pmatrix} \tag{2.6}$$

$$\begin{pmatrix} a_{11} & a_{12} & a_{13} \\ a_{21} & a_{22} & a_{23} \\ a_{31} & a_{32} & a_{33} \end{pmatrix} \begin{pmatrix} x_{13} \\ x_{23} \\ x_{33} \end{pmatrix} = \begin{pmatrix} 0 \\ 0 \\ 1 \end{pmatrix} \tag{2.7}$$

The systems (2.5), (2.6) and (2.7) of Eqs (2.5)–(2.7) can be solve by Gauss-elimination procedure. The solution set of each system Eqs (2.5), (2.6) and (2.7) will be the corresponding column of the inverse matrix X.

Note: Since the coefficient matrix is same in all the Eqs (2.5), (2.6) and (2.7), all can be simultaneously solved by forming a definite system

$$[A/I] = \begin{pmatrix} a_{11} & a_{12} & a_{13} & | & 1 & 0 & 0 \\ a_{21} & a_{22} & a_{23} & | & 0 & 1 & 0 \\ a_{31} & a_{32} & a_{33} & | & 0 & 0 & 1 \end{pmatrix}$$

Example 2.9: By Gauss elimination, find the inverse of $A = \begin{pmatrix} 0 & 1 & 1 \\ 1 & 2 & 0 \\ 3 & -1 & -4 \end{pmatrix}$.

Solution: The augmented system (A/I) is

$$(A/I) \sim \begin{pmatrix} 0 & 1 & 1 & | & 1 & 0 & 0 \\ 1 & 2 & 0 & | & 0 & 1 & 0 \\ 3 & -1 & -4 & | & 0 & 0 & 1 \end{pmatrix}$$

Since the element $a_{11} = 0$, we will interchange the first and second row, the reduced system is

$$(A/I) \sim \begin{pmatrix} 1 & 2 & 0 & | & 0 & 1 & 0 \\ 0 & 1 & 1 & | & 1 & 0 & 0 \\ 3 & -1 & -4 & | & 0 & 0 & 1 \end{pmatrix}$$

we get

$$R_3 \to R_3 + (-3)R_1 \sim \begin{pmatrix} 1 & 2 & 0 & | & 0 & 1 & 0 \\ 0 & 1 & 1 & | & 1 & 0 & 0 \\ 0 & -7 & -4 & | & 0 & -3 & 1 \end{pmatrix}$$

$$R_3 \to R_3 + 7R_2 \sim \begin{pmatrix} 1 & 2 & 0 & | & 0 & 1 & 0 \\ 0 & 1 & 1 & | & 1 & 0 & 0 \\ 0 & 0 & 3 & | & 7 & -3 & 1 \end{pmatrix}$$

$$x_{31} = \frac{7}{3}$$

Thus
$$\left.\begin{array}{r} x_{11} + 2x_{21} = 0 \\ x_{21} + x_{31} = 1 \\ 3x_{31} = 7 \end{array}\right\} \Rightarrow x_{21} = -\frac{4}{3}$$

$$x_{11} = \frac{8}{3}$$

$$\left.\begin{array}{r} x_{12} + 2x_{22} = 1 \\ x_{22} + x_{32} = 0 \\ 3x_{32} = -3 \end{array}\right\} \Rightarrow \begin{array}{l} x_{32} = -1 \\ x_{22} = 1 \\ x_{12} = -1 \end{array}$$

$$x_{33} = \frac{1}{3}$$

$$\left.\begin{array}{r} x_{13} + 2x_{23} = 0 \\ x_{23} + x_{33} = 0 \\ 3x_{33} = 1 \end{array}\right\} \Rightarrow x_{23} = -\frac{1}{3}$$

$$x_{13} = \frac{2}{3}$$

Hence
$$A^{-1} = \begin{pmatrix} \dfrac{8}{3} & -1 & \dfrac{2}{3} \\[2mm] -\dfrac{4}{3} & 1 & -\dfrac{1}{3} \\[2mm] \dfrac{7}{3} & -1 & \dfrac{1}{3} \end{pmatrix}$$

Example 2.10: Find by Gauss elimination method, the inverse of $A = \begin{pmatrix} 3 & -1 & 1 \\ -15 & 6 & -5 \\ 5 & -2 & 2 \end{pmatrix}$.

Solution:
$$[A/I] = \begin{pmatrix} 3 & -1 & 1 & | & 1 & 0 & 0 \\ -15 & 6 & -5 & | & 0 & 1 & 0 \\ 5 & -2 & 2 & | & 0 & 0 & 1 \end{pmatrix}$$

$$R_2 \to R_2 + 5R_1, R_3 \to R_3 + \left(-\frac{5}{3}\right)R_1 \sim \begin{pmatrix} 3 & -1 & 1 & | & 1 & 0 & 0 \\ 0 & 1 & 0 & | & 5 & 1 & 0 \\ 0 & \dfrac{-1}{3} & \dfrac{1}{3} & | & \dfrac{-5}{3} & 0 & 1 \end{pmatrix}$$

$$R_3 \to R_3 + \left(\frac{1}{3}\right)R_2 \sim \begin{pmatrix} 3 & -1 & 1 & | & 1 & 0 & 0 \\ 0 & 1 & 0 & | & 5 & 1 & 0 \\ 0 & 0 & \dfrac{1}{3} & | & 0 & \dfrac{1}{3} & 1 \end{pmatrix}$$

Now the system is equivalent to three systems.

$$\left(\begin{array}{ccc|c} 3 & -1 & 1 & 1 \\ 0 & 1 & 0 & 5 \\ 0 & 0 & \dfrac{1}{3} & 0 \end{array}\right)$$

$$\left(\begin{array}{ccc|c} 3 & -1 & 1 & 0 \\ 0 & 1 & 0 & 1 \\ 0 & 0 & \dfrac{1}{3} & \dfrac{1}{3} \end{array}\right)$$

and

$$\left(\begin{array}{ccc|c} 3 & -1 & 1 & 0 \\ 0 & 1 & 0 & 0 \\ 0 & 0 & \dfrac{1}{3} & 1 \end{array}\right)$$

$$\left.\begin{array}{r} 3x_{11} - x_{21} + x_{31} = 1 \\ x_{21} = 5 \\ \dfrac{1}{3}x_{31} = 0 \end{array}\right\} \Rightarrow \begin{array}{l} x_{31} = 0 \\ x_{21} = 5 \\ x_{11} = 2 \end{array}$$

$$\left.\begin{array}{r} 3x_{12} - x_{22} + x_{32} = 0 \\ x_{22} = 1 \\ \dfrac{1}{3}x_{32} = \dfrac{1}{3} \end{array}\right\} \Rightarrow \begin{array}{l} x_{32} = 1 \\ x_{22} = 1 \\ x_{12} = 0 \end{array}$$

$$\left.\begin{array}{r} 3x_{13} - x_{23} + x_{33} = 0 \\ x_{23} = 0 \\ \dfrac{1}{3}x_{33} = 1 \end{array}\right\} \Rightarrow \begin{array}{l} x_{33} = 3 \\ x_{23} = 0 \\ x_{13} = -1 \end{array}$$

$$A^{-1} = \begin{pmatrix} 2 & 0 & -1 \\ 5 & 1 & 0 \\ 0 & 1 & 3 \end{pmatrix}$$

PROBLEMS 2.3

1. Find the inverse of the following matrices by Gauss elimination method.

i. $\begin{pmatrix} 2 & 0 & -1 \\ 5 & 1 & 0 \\ 0 & 1 & 3 \end{pmatrix}$

Ans. $\begin{bmatrix} 3 & -1 & 1 \\ -15 & 6 & -5 \\ 5 & -2 & 2 \end{bmatrix}$

ii. $\begin{pmatrix} 8 & -1 & -3 \\ -5 & 1 & 2 \\ 10 & -1 & -4 \end{pmatrix}$

Ans. $\begin{bmatrix} 2 & 1 & -1 \\ 0 & 2 & 1 \\ 5 & 2 & 3 \end{bmatrix}$

iii. $\begin{pmatrix} -1 & 3 & 5 \\ -3 & 1 & 7 \\ 7 & -5 & -11 \end{pmatrix}$ $\qquad$ $\left[\text{Ans.} \begin{pmatrix} 3 & 1 & 2 \\ 2 & -3 & -1 \\ 1 & 2 & 1 \end{pmatrix} \right]$

2.5 GAUSS–JORDAN METHOD FOR FINDING THE INVERSE OF A MATRIX

Let A be square of order three and $|A| \neq 0$. Then the inverse of A is a matrix X which satisfies the equation $AX = I$, where I is the unit matrix of order three. Now, we have to find the elements of the inverse matrix X.

Let $\qquad A = \begin{pmatrix} a_{11} & a_{12} & a_{13} \\ a_{21} & a_{22} & a_{23} \\ a_{31} & a_{32} & a_{33} \end{pmatrix}$

and $\qquad X = \begin{pmatrix} x_{11} & x_{12} & x_{13} \\ x_{21} & x_{22} & x_{23} \\ x_{31} & x_{32} & x_{33} \end{pmatrix}$

Then, $\begin{pmatrix} a_{11} & a_{12} & a_{13} \\ a_{21} & a_{22} & a_{23} \\ a_{31} & a_{32} & a_{33} \end{pmatrix} \begin{pmatrix} x_{11} & x_{12} & x_{13} \\ x_{21} & x_{22} & x_{23} \\ x_{31} & x_{32} & x_{33} \end{pmatrix} = \begin{pmatrix} 1 & 0 & 0 \\ 0 & 1 & 0 \\ 0 & 0 & 1 \end{pmatrix}$

This equation is equivalent to three equations, which are equivalent to three systems of equations. Solve each system by Gauss–Jordan method. This solution set of each system will be corresponding column of the inverse matrix. Here also we can solve all the systems simultaneously by forming the augmented system.

$$[A/I] = \begin{pmatrix} a_{11} & a_{12} & a_{13} & | & 1 & 0 & 0 \\ a_{21} & a_{22} & a_{23} & | & 0 & 1 & 0 \\ a_{31} & a_{32} & a_{33} & | & 0 & 0 & 1 \end{pmatrix}$$

Example 2.11: Find the inverse of $A = \begin{pmatrix} 3 & -3 & 4 \\ 2 & -3 & 4 \\ 0 & -1 & 1 \end{pmatrix}$ by Gauss-Jordan method.

Solution: The augmented system (A/I) is

$$(A/B) = \begin{pmatrix} 3 & -3 & 4 & | & 1 & 0 & 0 \\ 2 & -3 & 4 & | & 0 & 1 & 0 \\ 0 & -1 & 1 & | & 0 & 0 & 1 \end{pmatrix}$$

$$R_1 \rightarrow \frac{1}{3} R_1 \quad \sim \begin{pmatrix} 1 & -1 & \dfrac{4}{3} & | & \dfrac{1}{3} & 0 & 0 \\ 2 & -3 & 4 & | & 0 & 1 & 0 \\ 0 & -1 & 1 & | & 0 & 0 & 1 \end{pmatrix}$$

$$R_2 \rightarrow R_2 - 2R_1 \sim \begin{pmatrix} 1 & -1 & \dfrac{4}{3} & \Big| & \dfrac{1}{3} & 0 & 0 \\ 0 & -1 & \dfrac{4}{3} & \Big| & -\dfrac{2}{3} & 1 & 0 \\ 0 & -1 & 1 & \Big| & 0 & 0 & 1 \end{pmatrix}$$

$$R_2 \rightarrow (-1)R_2 \sim \begin{pmatrix} 1 & -1 & \dfrac{4}{3} & \Big| & \dfrac{1}{3} & 0 & 0 \\ 0 & 1 & -\dfrac{4}{3} & \Big| & \dfrac{2}{3} & -1 & 0 \\ 0 & -1 & 1 & \Big| & 0 & 0 & 1 \end{pmatrix}$$

$$R_1 \rightarrow R_1 + R_2 \text{ and } R_3 \rightarrow R_3 + R_2 \sim \begin{pmatrix} 1 & 0 & 0 & \Big| & 1 & -1 & 0 \\ 0 & 1 & -\dfrac{4}{3} & \Big| & \dfrac{2}{3} & -1 & 0 \\ 0 & 0 & -\dfrac{1}{3} & \Big| & \dfrac{2}{3} & -1 & 1 \end{pmatrix}$$

$$R_3 \rightarrow (-3)R_3 \sim \begin{pmatrix} 1 & 0 & 0 & \Big| & 1 & -1 & 0 \\ 0 & 1 & -\dfrac{4}{3} & \Big| & \dfrac{2}{3} & -1 & 0 \\ 0 & 0 & 1 & \Big| & -2 & 3 & -3 \end{pmatrix}$$

$$R_2 \rightarrow R_2 + \dfrac{4}{3}R_3 \sim \begin{pmatrix} 1 & 0 & 0 & \Big| & 1 & -1 & 0 \\ 0 & 1 & 0 & \Big| & -2 & 3 & -4 \\ 0 & 0 & 1 & \Big| & -2 & 3 & -3 \end{pmatrix}$$

Inverse of $\qquad A = \begin{pmatrix} 1 & -1 & 0 \\ -2 & 3 & -4 \\ -2 & 3 & -3 \end{pmatrix}$

PROBLEMS 2.4

1. Find the inverse of the following matrices by Gauss–Jordan method.

i. $\begin{pmatrix} 2 & 0 & -1 \\ 5 & 1 & 0 \\ 0 & 1 & 8 \end{pmatrix}$
$\qquad$ Ans. $\begin{bmatrix} \begin{pmatrix} 3 & -1 & 1 \\ -15 & 6 & -5 \\ 5 & -2 & 2 \end{pmatrix} \end{bmatrix}$

ii. $\begin{pmatrix} 3 & -3 & 4 \\ 2 & -3 & 4 \\ 0 & -1 & 1 \end{pmatrix}$
$\qquad$ Ans. $\begin{bmatrix} \begin{pmatrix} 1 & -1 & 0 \\ -2 & 3 & -4 \\ -2 & 3 & -3 \end{pmatrix} \end{bmatrix}$

iii. $\begin{pmatrix} -1 & 3 & 5 \\ -3 & 1 & 7 \\ 7 & -5 & -11 \end{pmatrix}$
$\qquad$ Ans. $\begin{bmatrix} \begin{pmatrix} 3 & 1 & 2 \\ 2 & -3 & -1 \\ 1 & 2 & 1 \end{pmatrix} \end{bmatrix}$

iv. $\begin{pmatrix} 4 & 1 & 2 \\ 2 & 3 & -1 \\ 1 & -2 & 2 \end{pmatrix}$

Ans. $\begin{bmatrix} \begin{pmatrix} -\dfrac{4}{3} & 2 & \dfrac{7}{3} \\ \dfrac{5}{3} & -2 & -\dfrac{8}{3} \\ \dfrac{7}{3} & -3 & -\dfrac{10}{3} \end{pmatrix} \end{bmatrix}$

v. $\begin{pmatrix} 0 & 1 & 1 \\ 1 & 2 & 0 \\ 3 & -1 & -4 \end{pmatrix}$

Ans. $\begin{bmatrix} \begin{pmatrix} \dfrac{8}{3} & -1 & \dfrac{2}{3} \\ -\dfrac{4}{3} & 1 & -\dfrac{1}{3} \\ \dfrac{7}{3} & -1 & -\dfrac{1}{3} \end{pmatrix} \end{bmatrix}$

2.6 METHOD OF DECOMPOSITION/METHOD OF FACTORISATION/METHOD OF TRIANGULARISATION

Consider the following system of equations

$$a_{11}x_1 + a_{12}x_2 + a_{13}x_3 = b_1$$
$$a_{21}x_1 + a_{22}x_2 + a_{23}x_3 = b_2$$
$$a_{31}x_1 + a_{32}x_2 + a_{33}x_3 = b_3$$

This system is equivalent to $AX = B$ (2.8)

where $\quad A = \begin{pmatrix} a_{11} & a_{12} & a_{13} \\ a_{21} & a_{22} & a_{23} \\ a_{31} & a_{23} & a_{33} \end{pmatrix}, X = \begin{pmatrix} x_1 \\ x_2 \\ x_3 \end{pmatrix}, B = \begin{pmatrix} b_1 \\ b_2 \\ b_3 \end{pmatrix}$

Step 1: In this method the coefficient matrix (A) is decomposed or factored into lower (L) and upper (U) triangular matrices.

Now A will be factorized as the product of lower triangular matrix

$$L = \begin{pmatrix} 1 & 0 & 0 \\ l_{21} & 1 & 0 \\ l_{31} & l_{32} & 1 \end{pmatrix}$$ with 1's on the diagonal and an upper triangular matrix

$$U = \begin{pmatrix} u_{11} & u_{12} & u_{13} \\ 0 & u_{22} & u_{23} \\ 0 & 0 & u_{33} \end{pmatrix}$$

Step 2: Let $\quad\quad\quad LUX = B$ (2.9)

We write Eq. (2.9) as the following two system of equations

$$UX = Y \quad\quad\quad (2.10)$$

then Eq. (2.9) becomes, $LY = B$ (2.11)

i.e. $\quad\quad \begin{pmatrix} 1 & 0 & 0 \\ l_{21} & 1 & 0 \\ l_{31} & l_{32} & 1 \end{pmatrix} \begin{pmatrix} y_1 \\ y_2 \\ y_3 \end{pmatrix} = \begin{pmatrix} b_1 \\ b_2 \\ b_3 \end{pmatrix}$

$$y_1 = b_1$$
$$l_2 y_1 + y_2 = b_2$$
$$l_{31} y_1 + l_{32} y_2 + y_3 = b_3$$

and by forward substitution, we can find y_1, y_2, and y_3 in Eq. (2.10), and we get

$$\begin{pmatrix} u_{11} & u_{12} & u_{13} \\ 0 & u_{22} & u_{23} \\ 0 & 0 & u_{33} \end{pmatrix} \begin{pmatrix} x_1 \\ x_2 \\ x_3 \end{pmatrix} = \begin{pmatrix} y_1 \\ y_2 \\ y_3 \end{pmatrix}$$

$$\Rightarrow \quad u_{11} x_1 + u_{12} x_2 + u_{13} x_3 = y_1$$
$$u_{22} x_2 + u_{23} x_3 = y_2$$
$$u_{33} x_3 = y_3$$

By backward substitution, one can find x_1, x_2, x_3. Now L and U can be found from $LU = A$.

Step 3: Now we need the values of l_{21}, l_{31}, l_{32} and u_{11}, u_{12}, u_{13}, u_{22}, u_{23}, u_{33}.
Since $LU = A$, therefore

$$\begin{pmatrix} 1 & 0 & 0 \\ l_{21} & 1 & 0 \\ l_{31} & l_{22} & 1 \end{pmatrix} \begin{pmatrix} u_{11} & u_{12} & u_{13} \\ 0 & u_{22} & u_{23} \\ 0 & 0 & u_{33} \end{pmatrix} = \begin{pmatrix} a_{11} & a_{12} & a_{13} \\ a_{21} & a_{22} & a_{23} \\ a_{31} & a_{32} & a_{33} \end{pmatrix}$$

i.e. $\begin{pmatrix} u_{11} & u_{12} & u_{13} \\ l_{21} u_{11} & l_{21} u_{12} + u_{22} & l_{21} u_{13} + u_{23} \\ l_{31} u_{11} & l_{31} u_{12} + l_{22} u_{22} & u_{13} + l_{22} u_{23} + u_{33} \end{pmatrix} = \begin{pmatrix} a_{11} & a_{12} & a_{13} \\ a_{21} & a_{22} & a_{23} \\ a_{31} & a_{32} & a_{33} \end{pmatrix}$ (2.12)

Equating the corresponding elements in both sides of Eq. (2.12), we have

$$u_{11} = a_{11}; \; u_{12} = a_{12}; \; u_{13} = a_{13}$$

i.e. first row of U is the same as the first row of A

and $$l_{21} a_{21} \Rightarrow l_{21} = \frac{a_{21}}{u_{11}}$$

$$l_{21} u_{12} + u_{22} = a_{22} \Rightarrow u_{22} = a_{22} - \frac{a_{21}}{u_{11}} \cdot a_{12} = a_{22} - \frac{a_{21}}{a_{11}} \cdot a_{12}$$

$$l_{21} u_{13} + u_{23} = a_{23} \Rightarrow u_{23} = a_{23} - \frac{a_{21}}{a_{11}} \cdot a_{13}$$

Similarly the values of l_{12} and u_{33} are obtained.
∴ L and U are known.

Step 4: Substitute the value of L in Eq. (2.11) and find Y, and substitute the value of Y in Eq. (2.10), to get the value of X.

Note: L and U can be selected as

$$L = \begin{pmatrix} l_{11} & 0 & 0 \\ l_{21} & l_{22} & 0 \\ l_{31} & l_{32} & l_{33} \end{pmatrix} \text{ and } U = \begin{pmatrix} 1 & u_{12} & u_{13} \\ 0 & 1 & u_{23} \\ 0 & 0 & 1 \end{pmatrix}$$

Example 2.12: Use the method of factorization to solve the following equations.

$$3x_1 - 0.1x_2 - 0.2x_3 = 7.85$$
$$0.1x_1 + 7x_2 - 0.3x_3 = -19.3$$
$$0.3x_1 - 0.2x_2 + 10x_3 = 71.4$$

Solution: The given system is equivalent to $AX = B$

where
$$A = \begin{pmatrix} 3 & -0.1 & -0.2 \\ 0.1 & 7 & -0.3 \\ 0.3 & -0.2 & 10 \end{pmatrix}, X = \begin{pmatrix} x_1 \\ x_2 \\ x_3 \end{pmatrix}, B = \begin{pmatrix} 7.85 \\ -19.3 \\ 71.4 \end{pmatrix}$$

Now, factorize A into product of L and U

where
$$L = \begin{pmatrix} 1 & 0 & 0 \\ l_{21} & 1 & 0 \\ l_{31} & l_{32} & 1 \end{pmatrix}, \text{ and } U = \begin{pmatrix} u_{11} & u_{12} & u_{12} \\ 0 & u_{22} & u_{23} \\ 0 & 0 & u_{33} \end{pmatrix}$$

i.e. $$LU = A$$

i.e. $$\begin{pmatrix} 1 & 0 & 0 \\ l_{21} & 1 & 0 \\ l_{31} & l_{32} & 1 \end{pmatrix}\begin{pmatrix} u_{11} & u_{12} & u_{13} \\ 0 & u_{22} & u_{23} \\ 0 & 0 & u_{33} \end{pmatrix} = \begin{pmatrix} 3 & -0.1 & -0.2 \\ 0.1 & 7 & -0.3 \\ 0.3 & -0.2 & 10 \end{pmatrix}$$

i.e. $$\begin{pmatrix} u_{11} & u_{12} & u_{13} \\ l_{21}u_{11} & l_{21}u_{12} + u_{22} & l_{21}u_{13} + u_{23} \\ l_{31}u_{11} & l_{31}u_{12} + l_{32}u_{22} & l_{31}u_{13} + l_{32}u_{23} + u_{33} \end{pmatrix} = \begin{pmatrix} 3 & -0.1 & -0.2 \\ 0.1 & 7 & -0.3 \\ 0.3 & -0.2 & 10 \end{pmatrix}$$

Equating the corresponding elements, we get

$$u_{11} = 3, u_{12} = -0.1, u_{13} = -0.2$$

$$l_{21}u_{11} = 0.1 \Rightarrow l_{21} = \frac{0.1}{u_{11}} = \frac{0.1}{3} = 0.0333$$

$$l_{21}u_{12} + u_{22} = 7 \Rightarrow u_{22} = 7 - (-0.1 \times 0.0333) = 7.00333$$

$$l_{21}u_{13} + u_{23} = -0.3 \Rightarrow u_{23} = -0.3 - (0.0333)(-0.2) = -0.2933$$
$$= -0.3 + 0.00666$$

$$l_{31}u_{11} = 0.3 \Rightarrow l_{31} = \frac{0.3}{u_{11}} = \frac{0.3}{3} = 0.1000$$

$$l_{31}u_{12} + l_{32}u_{22} = 0.2 \Rightarrow l_{32} = \frac{-0.2 - (0.1)(-0.1)}{7.0033} = 0.0271$$

$$l_{31}u_{13} + l_{32}u_{23} + u_{33} = 10 \Rightarrow u_{33} = 10 - (l_{31}u_{13} + l_{32}u_{23})$$

$$= 10.0120 = 10 - (0.1) + (-0.2) + (-0.0271)(-0.2933)$$

$$A = LU = \begin{pmatrix} 1 & 0 & 0 \\ 0.0333 & 1 & 0 \\ 0.1000 & -0.0271 & 1 \end{pmatrix}\begin{pmatrix} 3 & -0.1 & -0.2 \\ 0 & 7.0033 & -0.2933 \\ 0 & 0 & 10.0120 \end{pmatrix}$$

Here $$AX = B \text{ becomes } LUX = B$$

i.e. $$LY = B \text{ where } UX = Y$$

Now $LY = B$ becomes

$$\begin{pmatrix} 1 & 0 & 0 \\ 0.0333 & 1 & 0 \\ 0.1000 & -0.0271 & 1 \end{pmatrix} \begin{pmatrix} y_1 \\ y_2 \\ y_3 \end{pmatrix} = \begin{pmatrix} 7.85 \\ -19.3 \\ 71.4 \end{pmatrix}$$

i.e.
$$y_1 = 7.85$$
$$0.0333y_1 + y_2 = -19.3$$
$$0.1000y_1 - 0.0271y_2 + y_3 = 71.4$$
$$y_1 = 7.85$$

i.e.
$$y_2 = -19.3 - (0.0333)(7.85) = -19.56$$
$$y_3 = 71.4 - (0.1000)(7.85) + (0.0271)(-19.56)$$
$$= 70.08 = 71.4 - 0.785 - 0.530076$$

and $UX = Y$ becomes

$$\begin{pmatrix} 3 & -0.1 & -0.2 \\ 0 & 7.0033 & -0.2933 \\ 0 & 0 & 10.0120 \end{pmatrix} \begin{pmatrix} x_1 \\ x_2 \\ x_3 \end{pmatrix} = \begin{pmatrix} 7.85 \\ -19.56 \\ 70.08 \end{pmatrix}$$

i.e.
$$3x_1 - 0.1x_2 - 0.2x_3 = 7.85$$
$$7.0033x_2 - 0.2933x_3 = -19.56$$
$$10.0120x_3 = 70.08$$

By back substitution

$$x_3 = \frac{70.08}{10.0120} = 6.99960048 \approx 7$$

$$x_2 = \frac{-19.56 + (0.2933)(7)}{7.0033} = -2.49980 \approx -2.5$$

$$x_1 = \frac{7.85 + (0.2)(7) + (0.1)(-2.5)}{3} = 3$$

$\therefore x_1 = 3; x_2 = -2.5; x_3 = 7.$

Example 2.13: Use the method of factorisation to solve the following system.
$$2x + y + 4z = 12$$
$$8x + 3y + 2z = 20$$
$$4x + 11y - z = 33$$

Solution: The given system is equivalent to $AX = B$

where
$$A = \begin{pmatrix} 2 & 1 & 4 \\ 8 & 3 & 2 \\ 4 & 11 & -1 \end{pmatrix}, X = \begin{pmatrix} x \\ y \\ z \end{pmatrix} \text{ and } B = \begin{pmatrix} 12 \\ 20 \\ 33 \end{pmatrix}$$

Now, factorize A into product of L and U

where
$$L = \begin{pmatrix} 1 & 0 & 0 \\ l_{21} & 1 & 0 \\ l_{31} & l_{32} & 1 \end{pmatrix}, \text{ and } u = \begin{pmatrix} u_{11} & u_{12} & u_{13} \\ 0 & u_{22} & u_{23} \\ 0 & 0 & u_{33} \end{pmatrix}$$

i.e.
$$LU = A$$

or
$$\begin{pmatrix} 1 & 0 & 0 \\ l_{21} & 1 & 0 \\ l_{31} & l_{32} & 1 \end{pmatrix} \begin{pmatrix} u_{11} & u_{12} & u_{13} \\ 0 & u_{22} & u_{23} \\ 0 & 0 & u_{33} \end{pmatrix} = \begin{pmatrix} 2 & 1 & 4 \\ 8 & 3 & 2 \\ 4 & 11 & -1 \end{pmatrix}$$

i.e.
$$\begin{pmatrix} u_{11} & u_{12} & u_{13} \\ l_{21}u_{11} & l_{21}u_{12} + u_{22} & l_{21}u_{13} + u_{23} \\ l_{31}u_{11} & l_{31}u_{12} + l_{32}u_{22} & l_{31}u_{13} + l_{32}u_{23} + u_{33} \end{pmatrix} = \begin{pmatrix} 2 & 1 & 4 \\ 8 & +3 & 2 \\ 4 & 11 & -1 \end{pmatrix}$$

Equating the corresponding elements, we get
$$u_{11} = 2, u_{12} = 1, u_{13} = 4$$
$$l_{21}u_{11} = 8, l_{21}u_{12} + u_{22} = +3, l_{21}u_{13} + u_{23} = 2$$
$$l_{31}u_{11} = 4, l_{31}u_{12} + l_{32}u_{22} = 11$$
$$l_{31}u_{13} + l_{32}u_{23} + u_{33} = -1$$
$$u_{33} = -135$$

Solving the above equations for l_{21} and u_{23}, we have
$$u_{11} = 2, u_{12} = 1, u_{13} = 4$$

we get
$$u_{22} = -1, u_{23} = -14, u_{23} = -14$$

and
$$l_{21} = 4, l_{31} = 2, l_{32} = -\frac{9}{1} = -9$$

$$A = \begin{pmatrix} 1 & 0 & 0 \\ 4 & 1 & 0 \\ 2 & -\dfrac{9}{1} & 1 \end{pmatrix} \begin{pmatrix} 2 & 1 & 4 \\ 0 & -1 & -14 \\ 0 & 0 & -135 \end{pmatrix}$$

Since
$$AX = B$$
$$LUX = B$$

Let
$$LY = B, \text{ where } UX = Y$$
$$LY = B \text{ becomes}$$

$$\begin{pmatrix} 1 & 0 & 0 \\ 4 & 1 & 0 \\ 2 & -\dfrac{9}{1} & 1 \end{pmatrix} \begin{pmatrix} y_1 \\ y_2 \\ y_3 \end{pmatrix} = \begin{pmatrix} 12 \\ 20 \\ 33 \end{pmatrix}$$

i.e.
$$y_1 = 12$$
$$4y_1 + y_2 = 20 \Rightarrow y_2 = -28$$

$$2y_1 - \frac{9}{7}y_2 + y_3 = 33 \Rightarrow 24 + 252 + y_3 = 33$$

or
$$y_3 = 33 - 276 = -243$$

Solving the above system by forward substitution, we get
$$y_1 = 12, y_2 = -28, y_3 = -243$$

Now $UX = Y$ becomes

$$\begin{pmatrix} 2 & 1 & 4 \\ 0 & -1 & -14 \\ 0 & 0 & -135 \end{pmatrix} \begin{pmatrix} x \\ y \\ z \end{pmatrix} = \begin{pmatrix} 12 \\ -28 \\ -27 \end{pmatrix}$$

$$2x + y + 4z = 12 \Rightarrow 2x = 12 + 2.8 - 4\left(\frac{9}{5}\right) = 3$$

i.e.

$$-1y - 14z = -28 \Rightarrow y = 28 - 14\left(\frac{9}{4}\right) = 2.8$$

$$-27z = -27 \Rightarrow -135z = -243 \text{ or } z = \frac{243}{135} = \frac{9}{5} = 1.8$$

Solving the above system by back substitution, we get $z = 1, y = 2, x = 3$.

Example 2.14: Using the method of triangularisation, solve the following system

$$x + 3y + z = 3$$
$$x + 4y + 2z = 3$$
$$x + 2y - 3z = 6$$

Solution: Given system can be written as $AX = B$

where

$$A = \begin{pmatrix} 1 & 3 & 1 \\ 1 & 4 & 2 \\ 1 & 2 & -3 \end{pmatrix}, X = \begin{pmatrix} x \\ y \\ z \end{pmatrix}, B = \begin{pmatrix} 3 \\ 3 \\ 6 \end{pmatrix}$$

In the method of triangularization, first we will factorize A as the product of L and U

where

$$L = \begin{pmatrix} 1 & 0 & 0 \\ l_{21} & 1 & 0 \\ l_{31} & l_{32} & 1 \end{pmatrix}$$

and

$$U = \begin{pmatrix} u_{11} & u_{12} & u_{13} \\ 0 & u_{22} & u_{23} \\ 0 & 0 & u_{33} \end{pmatrix}$$

or

$$LU = A$$

or

$$\begin{pmatrix} 1 & 0 & 0 \\ l_{21} & 1 & 0 \\ l_{31} & l_{32} & 1 \end{pmatrix} \begin{pmatrix} u_{11} & u_{12} & u_{13} \\ 0 & u_{22} & u_{23} \\ 0 & 0 & u_{33} \end{pmatrix} = \begin{pmatrix} 1 & 3 & 1 \\ 1 & 4 & 2 \\ 1 & 2 & -3 \end{pmatrix}$$

Equating the corresponding elements, we get

$$u_{11} = 1, u_{12} = 3, u_{13} = 1$$
$$l_{21}u_{11} = 1, l_{21}u_{12} + u_{22} = 4, l_{21}u_{13} + u_{23} = 2$$
$$l_{31}u_{11} = 1, l_{31}u_{12} + l_{32}u_{22} = 2, l_{31}u_{13} + l_{32}u_{23} + u_{33} = -3$$

Solving these equations, we get

$$u_{11} = 1, u_{12} = 3, u_{13} = 1$$
$$l_{21} = 1, u_{22} = 1, u_{23} = 1$$
$$l_{31} = 1, l_{32} = -1, u_{33} = -3$$

Since $LUX = B$, $LY = B$, where $UX = Y$

From $LY = B$, we have

$$\begin{pmatrix} 1 & 0 & 0 \\ 1 & 1 & 0 \\ 1 & -1 & 1 \end{pmatrix} \begin{pmatrix} y_1 \\ y_2 \\ y_3 \end{pmatrix} = \begin{pmatrix} 3 \\ 3 \\ 6 \end{pmatrix}$$

$\Rightarrow$

$$y_1 = 3$$
$$y_1 + y_2 = 3$$
$$y_1 - y_2 + y_3 = 6$$

$\Rightarrow y_1 = 3; y_2 = 0; y_3 = 3$

and $UX = Y$ gives

$$\begin{pmatrix} 1 & 3 & 1 \\ 1 & 1 & 1 \\ 1 & 0 & -3 \end{pmatrix} \begin{pmatrix} x \\ y \\ z \end{pmatrix} = \begin{pmatrix} 3 \\ 0 \\ 3 \end{pmatrix}$$

$$x + 3y + z = 3$$
$$y + z = 0$$
$$-3z = 3$$

By back substitution $z = -1, y = 1, x = 1$

$\therefore$ The required solution is $x = 1, y = 1, z = -1$.

PROBLEMS 2.5

Solve the following system of equations by method of factorisation, decomposition or triangularisation.

1. $2x + y + 3z = 13$
   $x + 5y + z = 14$
   $3x + y + 4z = 17$ [Ans. $x = 1; y = 2; z = 3$]

2. $x + y + 4z = 16$
   $2x - y + 3z = 16$
   $3x + y - z = -3$ [Ans. $x = 1; y = 2; z = 4$]

3. $3x + y + 2z = 3$
   $2x - 3y - z = -3$
   $x + 2y + z = 4$ [Ans. $x = 1; y = 2; z = -1$]

4. $5x - 2y + z = 4$
   $7x + y - 5z = 8$
   $3x + 7y + 4z = 4$ $\left[\text{Ans. } x = \dfrac{366}{327}; y = \dfrac{284}{327}; z = \dfrac{46}{327} \right]$

2.7 METHOD OF DECOMPOSITION FOR FINDING THE INVERSE OF A MATRIX

Suppose A is the given nonsingular (i.e. $|A| \neq 0$) square matrix of order three.

Step 1: Decompose the matrix A as the product of LO and U, where

$$L = \begin{pmatrix} 1 & 0 & 0 \\ l_{21} & 1 & 0 \\ k_{31} & l_{32} & 1 \end{pmatrix} \text{ and } U = \begin{pmatrix} u_{11} & u_{12} & u_{13} \\ 0 & u_{22} & u_{23} \\ 0 & 0 & u_{33} \end{pmatrix}$$

As explained earlier,

$$A = LU$$

Step 2: Let X be the inverse of A, then $AX = I$

Now $\qquad\qquad AX = I$ becomes $LU(X_1 X_2 X_3) = (I_1 I_2 I_3)$ (2.13)

where $\quad X_1 = \begin{pmatrix} x_{11} \\ x_{21} \\ x_{31} \end{pmatrix}, X_2 = \begin{pmatrix} x_{12} \\ x_{22} \\ x_{33} \end{pmatrix}, X_3 = \begin{pmatrix} x_{13} \\ x_{23} \\ x_{33} \end{pmatrix}$

and $\qquad I_1 = \begin{pmatrix} 1 \\ 0 \\ 0 \end{pmatrix}, I_2 = \begin{pmatrix} 0 \\ 1 \\ 0 \end{pmatrix}, I_3 = \begin{pmatrix} 0 \\ 0 \\ 1 \end{pmatrix}$ are columns of the matrices X and I respectively.

Also Eq. (2.13) is equivalent to the following three systems.

$$LUX_1 = I_1 \tag{2.14}$$
$$LUX_2 = I_2 \tag{2.15}$$
$$LUX_3 = I_3 \tag{2.16}$$

Step 3: In this method, inverse can be computed in a column-by-column fashion by generating solutions with unit vectors I_1, I_2, I_3 as the right-hand side constants.

Solve the systems (2.14), (2.15) and (2.16) for X_1, X_2, X_3 by the method explained earlier.

The resulting solutions X_1, X_2, X_3 corresponding to I_1, I_2, I_3 are the columns of inverse matrix of A respectively. Thus, it is ideal for evaluating the multiple unit vectors needed to compute the inverse.

Example 2.15: Find the inverse of $A = \begin{bmatrix} 3 & 1 & 2 \\ 2 & -3 & -1 \\ 1 & 2 & 1 \end{bmatrix}$ by the method of triangularisation.

Solution: Let $A = LU$, where $L \begin{pmatrix} 1 & 0 & 0 \\ l_{21} & 1 & 0 \\ l_{31} & l_{32} & 1 \end{pmatrix}$ and $U = \begin{pmatrix} u_{11} & u_{12} & u_{13} \\ 0 & u_{22} & u_{23} \\ 0 & 0 & u_{33} \end{pmatrix}$

Then $\qquad \begin{pmatrix} 1 & 0 & 0 \\ l_{21} & 1 & 0 \\ l_{31} & l_{32} & 1 \end{pmatrix} \begin{pmatrix} u_{11} & u_{12} & u_{13} \\ 0 & u_{22} & u_{23} \\ 0 & 0 & u_{33} \end{pmatrix} = \begin{pmatrix} 3 & 1 & 2 \\ 2 & -3 & -1 \\ 1 & 2 & 1 \end{pmatrix}$

$$\Rightarrow \begin{pmatrix} u_{11} & u_{12} & u_{13} \\ l_{21}u_{11} & l_{21}u_{12} + u_{22} & l_{21}u_{13} + u_{23} \\ l_{31}u_{11} & l_{31}u_{12} + l_{32}u_{22} & l_{31}u_{13} + l_{32}u_{23} + u_{33} \end{pmatrix} = \begin{pmatrix} 3 & 1 & 2 \\ 2 & -3 & -1 \\ 1 & 2 & 1 \end{pmatrix}$$

Equating the corresponding elements, we get

$$u_{11} = 3, u_{12} = 1, u_{13} = 2$$

$$l_{21}u_{11} = 2, l_{21}u_{12} + u_{22} = -3, l_{21}u_{13} + u_{23} = -1$$

$$\Rightarrow \qquad l_{21} = \frac{2}{3} \Rightarrow u_{22} = -3 - \frac{2}{3} \times 1 = -\frac{11}{3} \Rightarrow u_{23} = -1 - \frac{2}{3} \times 2 = -\frac{7}{3}$$

$$l_{31}u_{11} = 1, l_{31}u_{12} + l_{32}u_{22} = 2$$

$$\Rightarrow \qquad l_{31} = \frac{1}{3}, l_{32} = \frac{(2 - l_{31}u_{12})}{u_{22}} = \left[2 - \frac{1}{3} \times 1\right] \Big/ \left[-\frac{11}{3}\right] = -\frac{5}{11}$$

$$l_{31}u_{13} + l_{32}u_{23} + u_{33} = 1$$

$$\Rightarrow \qquad u_{33} = 1 - (l_{31}u_{13} + l_{32}u_{23})$$

$$= 1 - \left[\left(\frac{1}{3} \times 2\right) + \left(\frac{5}{11} \times -\frac{7}{3}\right)\right] = 1 - \left[\frac{2}{3} + \frac{35}{33}\right] = -\frac{24}{33} = -\frac{8}{11} \quad \text{(i)}$$

Thus $\qquad\qquad LU = A$ becomes

$$\begin{pmatrix} 1 & 0 & 0 \\ \dfrac{2}{3} & 1 & 0 \\ \dfrac{1}{3} & -\dfrac{5}{11} & 1 \end{pmatrix} \begin{pmatrix} 3 & 1 & 2 \\ 0 & -\dfrac{11}{3} & -\dfrac{7}{3} \\ 0 & 0 & -\dfrac{24}{33} \end{pmatrix} = \begin{pmatrix} 3 & 1 & 2 \\ 2 & -3 & -1 \\ 1 & 2 & 1 \end{pmatrix}$$

Let X be the inverse of A, then

$$AX = I$$

$$\Rightarrow \qquad LU(X_1X_2X_3) = (I_1I_2I_3) \qquad\qquad\qquad \text{(ii)}$$

where $\qquad X_1 = \begin{pmatrix} x_{11} \\ x_{21} \\ x_{31} \end{pmatrix}, X_2 = \begin{pmatrix} x_{12} \\ x_{22} \\ x_{32} \end{pmatrix}, X_3 = \begin{pmatrix} x_{13} \\ x_{23} \\ x_{33} \end{pmatrix} \qquad\qquad \text{(iii)}$

and $\qquad I_1 = \begin{pmatrix} 1 \\ 0 \\ 0 \end{pmatrix}, I_2 = \begin{pmatrix} 0 \\ 1 \\ 0 \end{pmatrix}, I_3 = \begin{pmatrix} 0 \\ 0 \\ 1 \end{pmatrix}$ are columns of X and I respectively.

Equation (ii) is equivalent to the following three systems

$$LUX_1 = I_1$$

$$LUX_2 = I_2$$

$$LUX_3 = I_3$$

Solving the above system Eqs (i)–(iii) for X_1, X_2 and X_3, as explained earlier, we get,

$$X_1 = \begin{pmatrix} -1 \\ -3 \\ 7 \end{pmatrix}, X_2 = \begin{pmatrix} 3 \\ 1 \\ -5 \end{pmatrix}, X_3 = \begin{pmatrix} 5 \\ 7 \\ -11 \end{pmatrix}$$

Inverse of A is
$$X = \begin{pmatrix} -1 & 3 & 5 \\ -3 & 1 & 7 \\ 7 & -5 & -11 \end{pmatrix}$$

PROBLEMS 2.6

Using decomposition method, find the inverse of the following.

i. $\begin{pmatrix} 3 & -0.1 & -0.2 \\ 0.1 & 7 & -0.3 \\ 0.3 & -0.2 & 10 \end{pmatrix}$

Ans. $\begin{bmatrix} 0.332 & 0.004 & 0.006 \\ -0.005 & 0.143 & 0.004 \\ -0.010 & 0.002 & 0.099 \end{bmatrix}$

ii. $\begin{pmatrix} 3 & -3 & 4 \\ 2 & -3 & 4 \\ 0 & 1 & 1 \end{pmatrix}$

Ans. $\begin{bmatrix} 1 & -1 & 0 \\ -2 & 3 & -4 \\ -2 & 3 & -3 \end{bmatrix}$

2.8 CROUT'S METHOD

Here, we decompose the coefficient matrix A as product of L and U and proceed. In the factorization method, the L matrix has 'l' on the diagonal. But an alternative approach involves the U matrix with l's on the diagonal. This is called Crout's method or Crout's decomposition.

Consider the following system of equations

$$\alpha_{11} x_1 + \alpha_{12} x_2 + \alpha_{13} x_3 = b_1$$
$$\alpha_{21} x_1 + \alpha_{22} x_2 + \alpha_{23} x_3 = b_2$$
$$\alpha_{31} x_1 + \alpha_{32} x_2 + \alpha_{33} x_3 = b_1 \tag{2.17}$$

The system is equivalent to $AX = B$

where
$$A = \begin{pmatrix} \alpha_{11} & \alpha_{12} & \alpha_{13} \\ \alpha_{21} & \alpha_{22} & \alpha_{23} \\ \alpha_{31} & \alpha_{32} & \alpha_{33} \end{pmatrix}, X = \begin{pmatrix} x_1 \\ x_2 \\ x_3 \end{pmatrix}, \text{and } B = \begin{pmatrix} b_1 \\ b_2 \\ b_3 \end{pmatrix}$$

Step 1: Here the coefficient matrix A is decomposed or factored into lower (L) and upper (U) triangular matrices.

Now, A will be factorized as the product of lower triangular matrix.

$$L = \begin{pmatrix} l_{11} & 0 & 0 \\ l_{21} & l_{22} & 0 \\ l_{31} & l_{32} & l_{33} \end{pmatrix}$$

and
$$U = \begin{pmatrix} 1 & u_{12} & u_{13} \\ 0 & 1 & u_{23} \\ 0 & 0 & 1 \end{pmatrix} \text{ with 1's on the diagonal, so that}$$

$$LUX = B \tag{2.18}$$

Step 2: Let $\qquad\qquad UX = Y \tag{2.19}$

then Eq. (2.18) becomes $LY = B \tag{2.20}$

Now $LU = A$ reduces to

$$\begin{pmatrix} l_{11} & 0 & 0 \\ l_{21} & l_{22} & 0 \\ l_{31} & l_{32} & l_{33} \end{pmatrix} \begin{pmatrix} 1 & u_{12} & u_{13} \\ 0 & 1 & u_{23} \\ 0 & 0 & 1 \end{pmatrix} = \begin{pmatrix} \alpha_{11} & \alpha_{12} & \alpha_{13} \\ \alpha_{21} & \alpha_{22} & \alpha_{23} \\ \alpha_{31} & \alpha_{32} & \alpha_{33} \end{pmatrix}$$

i.e. $\begin{pmatrix} l_{11} & l_{11}u_{12} & l_{11}u_{13} \\ l_{21} & l_{21}u_{12} + l_{22} & l_{21}u_{13} + l_{22}u_{23} \\ l_{31} & l_{31}u_{12} + l_{32} & l_{31}u_{13} + l_{32}u_{23} + l_{33} \end{pmatrix} = \begin{pmatrix} \alpha_{11} & \alpha_{12} & \alpha_{13} \\ \alpha_{21} & \alpha_{22} & \alpha_{23} \\ \alpha_{31} & \alpha_{32} & \alpha_{33} \end{pmatrix}$

equating the corresponding terms, we get

$$l_{11} = \alpha_{11}; \, l_{21} = \alpha_{21}; \, l_{31} = \alpha_{31}$$

i.e. $$l_{i1} = \alpha_{i1} \text{ for } i = 1, 2, ..., n$$

$$l_{11}u_{12} = \alpha_{12} \Rightarrow u_{12} = \frac{\alpha_{12}}{\alpha_{11}} \qquad [\because l_{11} = \alpha_{11}]$$

$$l_{11}u_{13} = \alpha_{13} \Rightarrow u_{13} = \frac{\alpha_{13}}{\alpha_{11}}$$

i.e. $$u_{1j} = \frac{\alpha_{1j}}{\alpha_{11}} \text{ for } j = 2, 3, ..., n$$

$$l_{21}u_{12} + l_{22} = \alpha_{22} \Rightarrow l_{22} = \alpha_{22} - l_{21}u_{12} = \alpha_{22} - \frac{\alpha_{21}\alpha_{12}}{\alpha_{11}}$$

$$l_{31}u_{12} + l_{32} = \alpha_{32} \Rightarrow l_{32} = \alpha_{32} - l_{31}u_{12} = \alpha_{32} - \frac{\alpha_{31}\alpha_{12}}{\alpha_{11}}$$

i.e. $$l_{ij} = \alpha_{ij} - \sum_{k=1}^{j-1} l_{ik}u_{kj}, \text{ for } i = j, j+1, ..., n$$

$$l_{12}u_{13} + l_{22}u_{23} = \alpha_{23} \Rightarrow u_{23} = \frac{\alpha_{23} - l_{21}u_{13}}{l_{22}}$$

i.e. $$u_{jk} = \frac{\alpha_{jk} - \sum_{i=1}^{j-1} l_{ji}u_{ik}}{l_{ji}}, \text{ for } k = j+1, j+2, ..., n$$

and $$l_{31}u_{13} + l_{32}u_{23} + l_{33} = \alpha_{33}$$
$$l_{33} = \alpha_{33} - (l_{31}u_{13} + l_{32}u_{23})$$

$$l_{mn} = \alpha_m - \sum_{k=1}^{n-1} l_{nk}u_{1n}$$

i.e. $\therefore$ L and U are known.

Step 3: Then Eq. (2.18) becomes

$$\begin{pmatrix} l_{11} & 0 & 0 \\ l_{21} & l_{22} & 0 \\ l_{31} & l_{32} & l_{33} \end{pmatrix} \begin{pmatrix} y_1 \\ y_2 \\ y_3 \end{pmatrix} = \begin{pmatrix} b_1 \\ b_2 \\ b_3 \end{pmatrix}$$

$$l_{11}y_1 = b_1$$

i.e. $$l_{21}y_1 + l_{22}y_2 = b_2$$
$$l_{31}y_1 + l_{32}y_2 + l_{33}y_3 = b_3$$

By forward substitution

$$y_1 = \frac{b_1}{l_{11}}$$

$$y_2 = \frac{b_2 - l_{21}y_1}{l_{22}}$$

$$y_3 = \frac{b_3 - (l_{31}y_1 + l_{32}y_2)}{l_{33}}$$

$\therefore y_1, y_2, y_3$ are known.

Substitute Y in Eq. (2.19) and use back substitution, we can get X.

Example 2.16: Use the Crout's method to solve the given system

$$3x_1 - 0.1x_2 - 0.2x_3 = 7.85$$
$$0.1x_1 + 7x_2 - 0.3x_3 = -19.3$$
$$0.3x_1 - 0.2x_2 - 10x_3 = 71.4$$

Solution: The given system is equivalent to $AX = B$

where $\quad A = \begin{pmatrix} 3 & -0.1 & -0.2 \\ 0.1 & 7 & -0.3 \\ 0.3 & -0.2 & 10 \end{pmatrix}, X = \begin{pmatrix} x_1 \\ x_2 \\ x_3 \end{pmatrix}, B = \begin{pmatrix} 7.85 \\ -19.3 \\ 71.4 \end{pmatrix}$ (i)

Now, we factorize A into product of L and U where

$$L = \begin{pmatrix} l_{11} & 0 & 0 \\ l_{21} & l_{22} & 0 \\ l_{31} & l_{32} & l_{33} \end{pmatrix}$$

and $\quad U = \begin{pmatrix} 1 & u_{12} & u_{13} \\ 0 & 1 & u_{23} \\ 0 & 0 & 1 \end{pmatrix}$

i.e. $\quad\quad LU = A$

i.e. $\begin{pmatrix} l_{11} & 0 & 0 \\ l_{21} & l_{22} & 0 \\ l_{31} & l_{32} & l_{33} \end{pmatrix} \begin{pmatrix} 1 & u_{12} & u_{13} \\ 0 & 1 & u_{23} \\ 0 & 0 & 1 \end{pmatrix} = \begin{pmatrix} 3 & -0.1 & -0.2 \\ 0.1 & 7 & -0.3 \\ 0.3 & -0.2 & 10 \end{pmatrix}$

i.e. $\begin{pmatrix} l_{11} & l_{11}u_{12} & l_{11}u_{13} \\ l_{21} & l_{21}u_{12} + l_{22} & l_{21}u_{13} + l_{22}u_{23} \\ l_{31} & l_{31}u_{12} + l_{32} & l_{31}u_{13} + l_{32}u_{23} + l_{33} \end{pmatrix} = \begin{pmatrix} 3 & -0.1 & -0.2 \\ 0.1 & 7 & -0.3 \\ 0.3 & -0.2 & 10 \end{pmatrix}$

Equating the corresponding elements, we get

$$l_{11} = 3; \ l_{11}u_{12} = -0.1; \ l_{11}u_{13} = -0.2$$
$$l_{21} = 0.1; \ l_{21}u_{12} + l_{22} = 7; \ l_{21}u_{13} + l_{22}u_{23} = -0.3$$
$$l_{31} = 0.3; \ l_{31}u_{12} + l_{32} = -0.2; \ l_{31}u_{13} + l_{32}u_{23} + l_{33} = 10$$

Solving the above equations, we get

$$l_{11} = 8; \quad u_{12} = \frac{-0.1}{3} = -0.033; \quad u_{13} = \frac{-0.2}{3} = -0.066$$

$$l_{21} = 0.1; \quad l_{22} = 7 - (0.1)(-0.033) = 7.003$$

$$l_{31} = 0.3; \quad l_{32} = -0.2 - (0.3)(-0.033) = -0.190$$

$$l_{21}u_{13} + l_{22}u_{23} = -0.3 \Rightarrow u_{23} = \frac{-0.3 - (0.1)(-0.066)}{7.003} = -0.042$$

$$l_{31}u_{13} + l_{32}u_{23} + l_{33} = 10 \Rightarrow l_{33} = 10 - (0.3)(-0.066) - (0.190)(-0.042) = 10.012$$

∴ Equation (i) becomes

$$\begin{bmatrix} 3 & 0 & 0 \\ 0.1 & 7.003 & 0 \\ 0.3 & -0.190 & 10.012 \end{bmatrix} \begin{bmatrix} 1 & -0.033 & -0.066 \\ 0 & 1 & -0.042 \\ 0 & 0 & 1 \end{bmatrix} \begin{pmatrix} x_1 \\ x_2 \\ x_3 \end{pmatrix} = \begin{pmatrix} 7.85 \\ -19.3 \\ 71.4 \end{pmatrix} \qquad (ii)$$

i.e. $LUX = B$

Let $LY = B$ $\qquad\qquad$ (iii)

where $UX = Y$ $\qquad\qquad$ (iv)

Now, Eq. (iii) becomes

$$\begin{pmatrix} 3 & 0 & 0 \\ 0.1 & 7.003 & 0 \\ 0.3 & -0.190 & 10.012 \end{pmatrix} \begin{pmatrix} y_1 \\ y_2 \\ y_3 \end{pmatrix} = \begin{pmatrix} 7.85 \\ -19.3 \\ 71.4 \end{pmatrix} \qquad \text{[By Eq. (iii)]}$$

$$3y_1 = 7.85$$

i.e. $\qquad 0.1y_1 + 7.003y_2 = -19.3$

$$0.3y_1 - 0.190y_2 + 10.012y_3 = 71.4$$

By forward substitution, we get

$$y_1 = \frac{7.85}{3} = 2.62$$

$$y_2 = \frac{-19.3 - (0.1)((2.62)}{7.003} = -2.79$$

$$y_3 = \frac{71.4 - (0.3)(2.62) + (0.190)(-2.79)}{10.012} = 7$$

and Eq. (iv) becomes

$$\begin{pmatrix} 1 & -0.033 & -0.066 \\ 0 & 1 & -0.042 \\ 0 & 0 & 1 \end{pmatrix} \begin{pmatrix} x_1 \\ x_2 \\ x_3 \end{pmatrix} = \begin{pmatrix} 2.62 \\ -2.79 \\ 7 \end{pmatrix} \qquad \text{[By Eq. (iv)]}$$

$$x_1 - 0.033x_2 - 0.066x_3 = 2.62$$

i.e. $\qquad\qquad x_2 - 0.042x_3 = -2.79$

$$x_3 = 7$$

By back substitution, we get

$$x_3 = 7$$
$$x_2 = -2.79 + (0.042 \times 7) = -2.5$$
$$x_1 = 2.62 + (0.033)(-2.5) + (0.066)(7) = 2.9995$$

$\therefore x_1 = 3; x_2 = -2.5; x_3 = 7.$

PROBLEMS 2.7

Use Crout's method to solve the following systems.

i. $2x + y + 4z = 12$
$8x - 3y + 2z = 20$
$4x + 11y - z = 33$ [Ans. $x = 3; y = 2; z = 1$)

ii. $2x + y + 3z = 13$
$x + 5y + z = 14$
$3x + y + 4z = 17$ [Ans. $x = 1; y = 2; z = 3$)

iii. $3x + y + 2z = 3$
$2x - 3y - z = -3$
$x + 2y + z = 4$ [Ans. $x = 1; y = 2; z = -1$)

iv. $x + y + 5z = 16$
$2x + 3y + z = 4$
$4x + y - z = 4$ [Ans. $x = 2; y = -1; z = 3$)

v. $x + y + z = 6$
$x + 2y + 3z = 14$
$x - 2y + 3z = 6$ [Ans. $x = 1; y = 2; z = 3$)

vi. $x + y + z = 1$
$3x + y - 3z = 5$
$x - 2y - 5z = 10$ [Ans. $x = 2; y = 6; z = -7$)

2.9 CROUT'S METHOD FOR FINDING THE INVERSE OF A GIVEN MATRIX

This method is same as the method of decomposition for finding the inverse of a given matrix. But here, the matrix U has 1 on the diagonal. This method has been explained in the following problem.

Suppose A is the given nonsingular square matrix. We have seen already, that A can be decomposed into $A = LU$, where L is the lower triangular matrix and U is the upper triangular matrix. Then

$$A^{-1} = (LU)^{-1} = U^{-1}L^{-1}$$

Since L is lower triangular matrix, L^{-1} is also lower triangular matrix and since U is upper triangular matrix, U^{-1} is also upper triangular matrix.

Since L is known and $LL^{-1} = 1$, we can find L^{-1}. Similarly, since U is known and $UU^{-1} = 1$, we can find U^{-1}.

Having known L^{-1} and U^{-1}, we get $A^{-1} = U^{-1}L^{-1}$.

Example 2.17: By Crout's method, find the inverse of $A = \begin{pmatrix} 1 & -2 & 3 \\ 0 & -1 & 4 \\ -2 & 2 & 0 \end{pmatrix}$.

Solution: Method I: In this method first factorize A and LU

where
$$L = \begin{pmatrix} l_{11} & 0 & 0 \\ l_{21} & l_{22} & 0 \\ l_{31} & l_{32} & l_{33} \end{pmatrix} \text{ and } U = \begin{pmatrix} 1 & u_{12} & u_{13} \\ 0 & 1 & u_{23} \\ 0 & 0 & 1 \end{pmatrix}$$

i.e.
$$LU = A$$

i.e.
$$\begin{pmatrix} l_{11} & 0 & 0 \\ l_{21} & l_{22} & 0 \\ l_{31} & l_{32} & l_{33} \end{pmatrix}\begin{pmatrix} 1 & u_{12} & u_{13} \\ 0 & 1 & u_{23} \\ 0 & 0 & 1 \end{pmatrix} = \begin{pmatrix} 1 & -2 & 3 \\ 0 & -1 & 4 \\ -2 & 2 & 0 \end{pmatrix}$$

i.e.
$$\begin{pmatrix} l_{11} & l_{11}u_{12} & l_{11}u_{13} \\ l_{21} & l_{21}u_{12} + l_{22} & l_{21}u_{13} + l_{22}u_{23} \\ l_{31} & l_{31}u_{12} + l_{32} & l_{31}u_{13} + l_{32}u_{23} + l_{33} \end{pmatrix} = \begin{pmatrix} 1 & -2 & 3 \\ 0 & -1 & 4 \\ -2 & 2 & 0 \end{pmatrix}$$

Equating the corresponding elements, we get

$$l_{11} = 1;\ u_{12} = \frac{-2}{1} = -2;\ u_{13} = \frac{3}{1} = 3$$

$$l_{21} = 0;\ l_{22} = -1 - 0 = -1;\ u_{23} = \frac{4-0}{-1} = -4$$

$$l_{31} = -2;\ l_{32} = 2 - 4 = -2;\ l_{33} = 0 - (-2)(3) - (-2)(-4) = -2$$

$$L = \begin{pmatrix} 1 & 0 & 0 \\ 0 & -1 & 0 \\ -2 & -2 & -2 \end{pmatrix}, U = \begin{pmatrix} 1 & -2 & 3 \\ 0 & 1 & -4 \\ 0 & 0 & 1 \end{pmatrix}$$

Let
$$A^{-1} = \begin{pmatrix} x_{11} & x_{12} & x_{13} \\ x_{21} & x_{22} & x_{23} \\ x_{31} & x_{32} & x_{33} \end{pmatrix} \text{ such that } AA^{-1} = I$$

$$\begin{pmatrix} 1 & -2 & 3 \\ 0 & -1 & 4 \\ -2 & 2 & 0 \end{pmatrix}\begin{pmatrix} x_{11} & x_{12} & x_{13} \\ x_{21} & x_{22} & x_{23} \\ x_{31} & x_{32} & x_{33} \end{pmatrix} = \begin{pmatrix} 1 & 0 & 0 \\ 0 & 1 & 0 \\ 0 & 0 & 1 \end{pmatrix}$$

i.e. the above equation is equivalent to the following system

$$\begin{pmatrix} 1 & -2 & 3 \\ 0 & -1 & 4 \\ -2 & 2 & 0 \end{pmatrix}\begin{pmatrix} x_{11} \\ x_{21} \\ x_{31} \end{pmatrix} = \begin{pmatrix} 1 \\ 0 \\ 0 \end{pmatrix}$$

$$\begin{pmatrix} 1 & -2 & 3 \\ 0 & -1 & 4 \\ -2 & 2 & 0 \end{pmatrix} \begin{pmatrix} x_{12} \\ x_{22} \\ x_{32} \end{pmatrix} = \begin{pmatrix} 0 \\ 1 \\ 0 \end{pmatrix}$$

$$\begin{pmatrix} 1 & -2 & 3 \\ 0 & -1 & 4 \\ -2 & 2 & 0 \end{pmatrix} \begin{pmatrix} x_{13} \\ x_{23} \\ x_{33} \end{pmatrix} = \begin{pmatrix} 0 \\ 0 \\ 1 \end{pmatrix}$$

Each equation represents a system. Solve the above equations by Crout's method explained earlier, we get

$$\left. \begin{array}{l} x_{11} = -4; \ x_{12} = 3; \ x_{13} = -\dfrac{5}{2} \\[2mm] x_{21} = -4; \ x_{22} = 3; \ x_{23} = -2 \\[2mm] x_{31} = -1; \ x_{32} = 1; \ x_{33} = -\dfrac{1}{2} \end{array} \right] \qquad \text{(verify it)}$$

or

$$A^{-1} = \begin{pmatrix} -4 & 3 & -\dfrac{5}{2} \\[2mm] -4 & 3 & -2 \\[2mm] -1 & 1 & -\dfrac{1}{2} \end{pmatrix}$$

$A = LU$ becomes

Method II: $\quad \begin{pmatrix} 1 & -2 & 3 \\ 0 & -1 & 4 \\ -2 & 2 & 0 \end{pmatrix} = \begin{pmatrix} 1 & 0 & 0 \\ 0 & -1 & 1 \\ -2 & -2 & -2 \end{pmatrix} \begin{pmatrix} 1 & -2 & 3 \\ 0 & 1 & -4 \\ 0 & 0 & 1 \end{pmatrix}$

Since $\qquad\qquad\qquad A = LU$

$$A^{-1} = (LU)^{-1} = U^{-1}L^{-1}$$

To find U^{-1}: Since U is upper triangular, U^{-1} is also upper triangular.

Let $\qquad\qquad\qquad U^{-1} = \begin{pmatrix} 1 & a_{12} & a_{13} \\ 0 & 1 & a_{23} \\ 0 & 0 & 1 \end{pmatrix}$ such that $UU^{-1} = I$

Then $\quad \begin{pmatrix} 1 & -2 & 3 \\ 0 & 1 & -4 \\ 0 & 0 & 1 \end{pmatrix} \begin{pmatrix} 1 & a_{12} & a_{13} \\ 0 & 1 & a_{23} \\ 0 & 0 & 1 \end{pmatrix} = \begin{pmatrix} 1 & 0 & 0 \\ 0 & 1 & 0 \\ 0 & 0 & 1 \end{pmatrix}$

$\Rightarrow \quad \begin{pmatrix} 1 & a_{12}-2 & a_{13}-2a_{23}+3 \\ 0 & 1 & a_{23}-4 \\ 0 & 0 & 1 \end{pmatrix} = \begin{pmatrix} 1 & 0 & 0 \\ 0 & 1 & 0 \\ 0 & 0 & 1 \end{pmatrix}$

Equating the corresponding elements, we get

$$a_{12} - 2 = 0; \ a_{13} - 2a_{23} + 3 = 0; \ a_{23} - 4 = 0$$

$$a_{12} = 2; \ a_{23} = 4; \ a_{13} = 5$$

$$U^{-1} = \begin{pmatrix} 1 & 2 & 5 \\ 0 & 1 & 4 \\ 0 & 0 & 1 \end{pmatrix}$$

To find L^{-1}: Let $L = \begin{pmatrix} b_{11} & 0 & 0 \\ b_{21} & b_{22} & 0 \\ b_{31} & b_{32} & b_{33} \end{pmatrix}$ such that $LL^{-1} = 1$

Then

$$\begin{pmatrix} 1 & 0 & 0 \\ 0 & -1 & 0 \\ -2 & -2 & -2 \end{pmatrix} \begin{pmatrix} b_{11} & 0 & 0 \\ b_{21} & b_{22} & 0 \\ b_{31} & b_{32} & b_{33} \end{pmatrix} = \begin{pmatrix} 1 & 0 & 0 \\ 0 & 1 & 0 \\ 0 & 0 & 1 \end{pmatrix}$$

i.e.

$$\begin{pmatrix} b_{11} & 0 & 0 \\ -b_{21} & -b_{22} & 0 \\ -2b_{11} - 2b_{21} - 2b_{31} & -2b_{22} - 2b_{32} & -2b_{21} \end{pmatrix} = \begin{pmatrix} 1 & 0 & 0 \\ 0 & 1 & 0 \\ 0 & 0 & 1 \end{pmatrix}$$

Equating the corresponding elements and solving the equations, we get

$$b_{11} = 1$$

$$b_{21} = 0; \ b_{22} = -1$$

$$b_{31} = -1; \ b_{32} = 1; \ b_{33} = -\frac{1}{2}$$

$$L^{-1} = \begin{pmatrix} 1 & 0 & 0 \\ 0 & -1 & 0 \\ -1 & 1 & -\frac{1}{2} \end{pmatrix}$$

$$A^{-1} = U^{-1}L^{-1} = \begin{pmatrix} 1 & 2 & 5 \\ 0 & 1 & 4 \\ 0 & 0 & 1 \end{pmatrix} \begin{pmatrix} 1 & 0 & 0 \\ 0 & -1 & 0 \\ -1 & 1 & -\frac{1}{2} \end{pmatrix} = \begin{pmatrix} -4 & 3 & -\frac{5}{2} \\ -4 & 3 & -2 \\ -1 & 1 & -\frac{1}{2} \end{pmatrix}$$

PROBLEMS 2.8

Find the inverse of the following matrices by Crout's method:

i. $\begin{pmatrix} 1 & 1 & 1 \\ 1 & 2 & -3 \\ 2 & -1 & 3 \end{pmatrix}$

$\left[\text{Ans.} \dfrac{1}{11} \begin{pmatrix} -3 & 4 & 5 \\ 9 & -1 & -4 \\ 5 & -3 & -1 \end{pmatrix} \right]$

ii. $\begin{pmatrix} 2 & 1 & 1 \\ 3 & 2 & 3 \\ 1 & 4 & 9 \end{pmatrix}$

Ans. $\begin{bmatrix} -3 & \dfrac{5}{2} & -\dfrac{1}{2} \\ 12 & -\dfrac{17}{2} & \dfrac{3}{2} \\ -5 & \dfrac{7}{2} & -\dfrac{1}{2} \end{bmatrix}$

iii. $\begin{pmatrix} 2 & 1 & -1 \\ 0 & 2 & 1 \\ 5 & 2 & -3 \end{pmatrix}$

Ans. $\begin{bmatrix} 8 & -1 & -3 \\ -5 & 1 & 2 \\ 10 & -1 & -4 \end{bmatrix}$

2.10 ITERATIVE METHODS FOR SOLUTION (Indirect Method)

For large systems, iterative methods are better than the direct methods.

An iterative method is one in which we start with an approximation to the actual solution and obtain better and better approximation. In this method, number of iterations depends on the degree of accuracy required. Iteration is a self-correcting process and any error made at any stage of computation gets automatically corrected in subsequent steps.

The method of iteration is not applicable to all system of equations.

The solution of the system of linear equations will exist by iterative procedure if "the absolute value of the largest coefficient is greater than sum of the absolute value of all remaining coefficient in each equation (condition of convergence).

2.11 GAUSS–JACOBI METHOD OF ITERATION

Consider the following system of equations:

$$\left.\begin{array}{l} a_1 x + b_1 y + c_1 z = d_1 \\ a_2 x + b_2 y + c_2 z = d_2 \\ a_3 x + b_3 y + c_3 z = d_3 \end{array}\right\} \text{(solution to linear equation exists)}$$

Condition of convergence must satisfy, that is

$$|a_1| > |b_1| + |c_1|;\ |b_2| > |a_2| + |c_2|;\ |c_3| > |a_3| + |b_3|$$

Then,
$$x = \frac{1}{a_1}(d_1 - b_1 y - c_1 z)$$

$$y = \frac{1}{b_2}(d_2 - a_2 x - c_2 z)$$

$$z = \frac{1}{c_3}(d_3 - a_3 x - b_3 y)$$

Let $x_0,\ y_0,\ z_0$ be initial approximation, then the 1st approximation is given by

$$x_1 = \frac{1}{a_1}(d_1 - b_1 y_0 - c_1 z_0)$$

$$y_1 = \frac{1}{b_2}(d_2 - a_2 x_0 - b_2 z_0)$$

$$z_1 = \frac{1}{c_3}(d_3 - a_3 x_0 - b_3 y_0)$$

Second approximation

$$x_2 = \frac{1}{a_1}(d_1 - b_1y_1 - c_1z_1)$$

$$y_2 = \frac{1}{b_2}(d_2 - a_2x_1 - c_2z_1)$$

$$z_2 = \frac{1}{c_2}(d_3 - a_3x_1 - b_3y_1)$$

Following the same procedure, we have

$$x_r + 1 = \frac{1}{a_1}(d_1 - b_1y_r - c_1z_r)$$

$$y_r + 1 = \frac{1}{b_2}(d_2 - a_2x_r - c_2z_r)$$

$$z_r + 1 = \frac{1}{c_3}(d_3 - a_3x_r - b_3y_r)$$

Note: If one is unable to make initial approximation appropriately, then take
$$x_0 = y_0 = z_0 = 0$$

2.12 GAUSS–SEIDAL METHOD OF ITERATION

This is only a refinement of Gauss–Jacobi method.
$$a_1x + b_1y + c_1z = d_1, a_2x + b_2y + c_2z = d_2, a_3x + b_3y + c_3z = d_3$$

is written as
$$x = \frac{1}{a_1}(d_1 - b_1y - c_1z) \tag{2.21}$$

$$y = \frac{1}{b_2}(d_2 - a_2x - c_2z) \tag{2.22}$$

$$z = \frac{1}{c_3}(d_3 - a_3x - b_3y) \tag{2.23}$$

We start with initial approximation $y_0 = z_0 = 0$.
Substituting y_0 and z_0 in Eq. (2.21), we get

$$x_1 = \frac{1}{a_1}(d_1 - b_1y_0 - c_1z_0)$$

Substituting $x = x_1, z = z_0$ in Eq. (2.22)

$$y_1 = \frac{1}{b_2}(d_2 - a_2x_1 - c_2z_0)$$

Put $x = x_1, y = y_1$ in Eq. (2.23)

$$z_1 = \frac{1}{c_3}(d_3 - a_3x_1 - b_3y_1)$$

This process is continued till the value of x, y, z are obtained to the desired degree of accuracy. In general, if x_k, y_k, z_k are kth iterates then

$$x_{k+1} = \frac{1}{a_1}(d_1 - b_1 y_k - c_1 z_k)$$

$$y_{k+1} = \frac{1}{b_2}(d_2 - a_2 x_{k+1} - c_2 z_k)$$

$$z_{k+1} = \frac{1}{c_3}(d_3 - a_3 x_{k+1} - b_3 y_{k+1})$$

Since the current values of unknown at each stage of iteration are used in proceeding to next stage of iteration, this method is more rapid in convergence than Gauss–Jacobi method.

Notes: 1. The rate of convergence of Gauss-Seidal method is roughly twice than that of Gauss-Jacobi, but condition of convergent is same as Jacobi method.

2. Gauss-Seidel method converges only for special system of equations.

In general the round of error will be small in iteration method. Moreover, these are self-correcting methods, i.e. any error made in computation will be corrected.

Example 2.18: Solve the following system by Gauss–Jacobi and Gauss–Seidel method
$$10x - 5y - 2z = 3; \ 4x - 10y + 3z = -3; \ x + 6y + 10z = -3$$

Solution: The coefficient matrix $\begin{pmatrix} 10 & -5 & -2 \\ 4 & -10 & 3 \\ 1 & 6 & 10 \end{pmatrix}$ is diagonally dominant, since

$|10| > |-5| + |-2|, \ |-10| > |4| + |3|$ and $|10| > |1| + |6|$.

∴ By Gauss–Jacobi method

$$x = \frac{1}{10}(3 + 5y + 2z) \tag{i}$$

$$y = \frac{1}{10}(3 + 4x + 3z) \tag{ii}$$

$$z = \frac{1}{10}(-3 - x - 6y) \tag{iii}$$

First iteration: Let the initial values be $(0, 0, 0)$
Using these initial values in Eqs (i)–(iii), we get

$$x^{(1)} = \frac{1}{10}[3 + 5(0) + 2(0)] = 0.3$$

$$y^{(1)} = \frac{1}{10}[3 + 4(0) + 3(0)] = 0.3$$

$$z^{(1)} = \frac{1}{10}[-3 - (0) - 6(0)] = -0.3$$

Second iteration: Using these values in Eqs (i)–(iii), we get

$$x^{(2)} = \frac{1}{10}[3 + 5(0.3) + 2(-0.3)] = 0.39$$

$$y^{(2)} = \frac{1}{10}[3 + 4(0.3) + 3(-0.3)] = 0.33$$

$$z^{(2)} = \frac{1}{10}[-3 - (0.3) - 6(0.3)] = -0.51$$

Third iteration: Using the values of $x^{(2)}$, $y^{(2)}$, $z^{(2)}$ in Eqs (i)–(iii), we get

$$x^{(3)} = \frac{1}{10}[3 + 5(0.33) + 2(-0.51)] = 0.363$$

$$y^{(3)} = \frac{1}{10}[3 + 4(0.39) + 3(-0.51)] = 0.303$$

$$z^{(3)} = \frac{1}{10}[-3 - (0.39) - 6(0.33)] = -0.537$$

Fourth iteration:

$$x^{(4)} = \frac{1}{10}[3 + 5(0.303) + 2(-0.537)] = 0.3441$$

$$y^{(4)} = \frac{1}{10}[3 + 4(0.363) + 3(-0.537)] = 0.2841$$

$$z^{(4)} = \frac{1}{10}[-3 - (0.363) - 6(0.303)] = -0.5181$$

Fifth iteration:

$$x^{(5)} = \frac{1}{10}[3 + 5(0.2841) + 2(-0.5181)] = 0.33843$$

$$y^{(5)} = \frac{1}{10}[3 + 4(0.3441) + 3(-0.5181)] = 0.2822$$

$$z^{(5)} = \frac{1}{10}[-3 - (0.3441) - 6(0.2841)] = -0.50487$$

Sixth iteration:

$$x^{(6)} = \frac{1}{10}[3 + 5(0.2822) + 2(-0.50487)] = 0.340126$$

$$y^{(6)} = \frac{1}{10}[3 + 4(0.33843) + 3(-0.50487)] = 0.283911$$

$$z^{(6)} = \frac{1}{10}[-3 - (0.33843) - 6(0.2822)] = -0.503163$$

Seventh iteration:

$$x^{(7)} = \frac{1}{10}[3 + 5(0.283911) + 2(-0.503163)] = 0.3413229$$

$$y^{(7)} = \frac{1}{10}[3 + 4(0.340126) + 3(-0.503163)] = 0.2851015$$

$$z^{(7)} = \frac{1}{10}[-3 - (0.340126) - 6(0.283911)] = -0.5043592$$

Eighth iteration:

$$x^{(8)} = \frac{1}{10}[3 + 5(0.2851015) + 2(-0.5043592)] = 0.34167891$$

$$y^{(8)} = \frac{1}{10}[3 + 4(0.3413229) + 3(-0.5043592)] = 0.2852214$$

$$z^{(8)} = \frac{1}{10}[-3 - (0.3413229) - 6(0.2851015)] = -0.50519319$$

Ninth iteration:

$$x^{(9)} = \frac{1}{10}[3 + 5(0.2852214) + 2(-0.50519319)] = 0.341572062$$

$$y^{(9)} = \frac{1}{10}[3 + 4(0.34167891) + 3(-0.50519319)] = 0.285113607$$

$$z^{(9)} = \frac{1}{10}[-3 - (0.34167891) - 6(-0.2852214)] = -0.505300731$$

Hence correct to three decimal places, the values are
$$x = 0.342, y = 0.285, z = -0.505$$

Gauss–Seidel method: Initial values: $y = 0, z = 0$.
First iteration

$$x^{(1)} = \frac{1}{10}[3 + 5(0) + 2(0)] = 0.3$$

$$y^{(1)} = \frac{1}{10}[3 + 4(0.3) + 3(0)] = 0.42$$

$$z^{(1)} = \frac{1}{10}[-3 - (0.3) - 6(0.42)] = -0.582$$

Second iteration:

$$x^{(2)} = \frac{1}{10}[3 + 5(0.42) + 2(-0.582)] = 0.3936$$

$$y^{(2)} = \frac{1}{10}[3 + 4(0.3936) + 3(-0.582)] = 0.28284$$

$$z^{(2)} = \frac{1}{10}[-3 - (0.3936) - 6(0.282884)] = -0.509064$$

Third iteration:

$$x^{(3)} = \frac{1}{10}[3 + 5(0.28284) + 2(-0.509064)] = 0.3396072$$

$$y^{(3)} = \frac{1}{10}[3 + 4(0.3396072) + 3(-0.509064)] = 0.28312368$$

$$z^{(3)} = \frac{1}{10}[-3 - (0.3396072) - 6(0.28312368)] = -0.503834928$$

Fourth iteration:

$$x^{(4)} = \frac{1}{10}[3 + 5(0.28312368) + 2(-0.503834928)] = 0.34079485$$

$$y^{(4)} = \frac{1}{10}[3 + 4(0.34079485) + 3(-0.503834928)] = 0.285167464$$

$$z^{(4)} = \frac{1}{10}[-3 - (0.34079485) - 6(0.28516746] = -0.50517996$$

Fifth iteration:

$$x^{(5)} = \frac{1}{10}[3 + 5(0.28516746) + 2(-0.50517996)] = 0.34155477$$

$$y^{(5)} = \frac{1}{10}[3 + 4(0.34155477) + 3(-0.50517996)] = 0.28506792$$

$$z^{(5)} = \frac{1}{10}[-3 - (0.34155477) - 6(0.28506792)] = -0.505196229$$

Sixth iteration:

$$x^{(6)} = \frac{1}{10}[3 + 5(0.28506792) + 2(-0.505196229)] = 0.341494714$$

$$y^{(6)} = \frac{1}{10}[3 + 4(0.341494714) + 3(-0.505196229)] = 0.285039017$$

$$z^{(6)} = \frac{1}{10}[-3 - (0.341494714) - 6(0.285039017)] = -0.5051728$$

Seventh iteration:

$$x^{(7)} = \frac{1}{10}[3 + 5(0.285039017) + 2(-0.5051728)] = 0.3414849$$

$$y^{(7)} = \frac{1}{10}[3 + 4(0.3414849) + 3(-0.5051728)] = 0.28504212$$

$$z^{(7)} = \frac{1}{10}[-3 - (0.3414849) - 6(0.28504212)] = -0.5051737$$

The values at each iteration by both methods are tabulated below.

Iteration	Gauss–Jacobi method			Gauss–Seidel method		
	x	y	z	x	y	z
1	0.3	0.3	-0.3	0.3	0.42	-0.582
2	0.39	0.33	-0.51	0.3936	0.28284	-0.509064
3	0.363	0.303	-0.537	0.3396072	0.28312364	-0.503834928
4	0.3441	0.2841	-0.5181	0.34079485	0.28516746	-0.50517996
5	0.33843	0.2822	-0.50487	0.3415547	0.28506792	-0.505196229
6	0.340126	0.283911	-0.503163	0.3414947	0.2850390	-0.5051728
7	0.3413229	0.2851015	-0.5043592	0.3414849	0.28504212	-0.5051737
8	0.34167891	0.2852214	-0.50519319			
9	0.341572062	0.285113607	-0.505300731			

The values correct to three decimal places are

$$x = 0.342, y = 0.285, z = -0.505$$

Note: After getting the values of the unknowns, substitute these values in the given equations, and check the correctness of the result.

Example 2.19: Solve the following system of equations by using Gauss–Jacobi and Gauss–Seidel methods (correct to three decimal places):

$$8x - 3y + 2z = 20$$
$$4x + 11y - z = 33$$
$$6x + 3y + 12z = 35$$

Solution: Since the diagonal elements are dominant in the coefficient matrix, we write x, y, z as follows.

$$x = \frac{1}{8}[20 + 3y - 2z] \tag{i}$$

$$y = \frac{1}{11}[33 - 4x + z] \tag{ii}$$

$$z = \frac{1}{12}[35 - 6x - 3y] \tag{iii}$$

Gauss–Jacobi method

First iteration: Let the initial values be $x = 0$, $y = 0$, $z = 0$.

Using the values $x = 0$, $y = 0$, $z = 0$ in Eqs (i)–(iii), we get

$$x^{(1)} = \frac{1}{8}[20 + 3(0) - 2(0)] = 2.5$$

$$y^{(1)} = \frac{1}{11}[33 - 4(0) + 0] = 3.0$$

$$z^{(1)} = \frac{1}{12}[35 - 6(0) - 3(0)] = 2.916666$$

Second iteration: Using these values $x^{(1)}$, $y^{(1)}$, $z^{(1)}$ again in Eqs (i)–(iii), we get

$$x^{(2)} = \frac{1}{8}[20 + 3(3.0) - 2(2.916666)] = 2.895833$$

$$y^{(2)} = \frac{1}{11}[33 - 4(2.5) + (2.916666)] = 2.356060$$

$$z^{(2)} = \frac{1}{12}[35 - 6(2.5) - 3(3.0)] = 0.916666$$

Third iteration:

$$x^{(3)} = \frac{1}{8}[20 + 3(2.356060) - 2(0.916666)] = 3.154356$$

$$y^{(3)} = \frac{1}{11}[33 - 4(2.895833) + (0.916666)] = 2.030303$$

$$z^{(3)} = \frac{1}{12}[35 - 6(2.895833) - 3(2.356060)] = 0.879735$$

Fourth iteration:

$$x^{(4)} = \frac{1}{8}[20 + 3(2.030303) - 2(0.879735)] = 3.041430$$

$$y^{(4)} = \frac{1}{11}[33 - 4(3.154356) + (0.879735)] = 1.932937$$

$$z^{(4)} = \frac{1}{12}[35 - 6(3.154356) - 3(2.030303)] = 0.831919$$

Fifth iteration:

$$x^{(5)} = \frac{1}{8}[20 + 3(1.932937) - 2(0.831913)] = 3.016873$$

$$y^{(5)} = \frac{1}{11}[33 - 4(3.041430) + (0.831913)] = 1.969654$$

$$z^{(5)} = \frac{1}{12}[35 - 6(3.041430) - 3(1.932937)] = 0.912717$$

Sixth iteration:

$$x^{(6)} = \frac{1}{8}[20 + 3(1.969654) - 2(0.912717)] = 3.010441$$

$$y^{(6)} = \frac{1}{11}[35 - 4(3.016873) + (0.912717)] = 1.985930$$

$$z^{(6)} = \frac{1}{12}[35 - 6(3.016873) - 3(1.969654)] = 0.915817$$

Seventh iteration:

$$x^{(7)} = \frac{1}{8}[20 + 3(1.985930) - 2(0.915817)] = 3.015770$$

$$y^{(7)} = \frac{1}{11}[33 - 4(3.01044) + (0.915817)] = 1.988550$$

$$z^{(7)} = \frac{1}{12}[35 - 6(3.010441) - 3(1.985930)] = 0.914964$$

Eighth iteration:

$$x^{(8)} = \frac{1}{8}[20 + 3(1.988550) - 2(0.914964)] = 3.016946$$

$$y^{(8)} = \frac{1}{11}[33 - 4(3.015770) + (0.914964)] = 1.986535$$

$$z^{(8)} = \frac{1}{12}[35 - 6(3.015770) - 3(1.988550)] = 0.911644$$

Ninth iteration:

$$x^{(9)} = \frac{1}{8}[20 + 3(1.986535) - 2(0.911644)] = 3.017039$$

$$y^{(9)} = \frac{1}{11}[33 - 4(3.016946) + (0.911644)] = 1.985805$$

$$z^{(9)} = \frac{1}{12}[35 - 6(3.016946) - 3(1.986535)] = 0.911560$$

Tenth iteration:

$$x^{(10)} = \frac{1}{8}[20 + 3(1.985805) - 2(0.911560)] = 3.016786$$

$$y^{(10)} = \frac{1}{11}[33 - 4(3.017039) + (0.911560)] = 1.985764$$

$$z^{(10)} = \frac{1}{12}[35 - 6(3.017039) - 3(1.985805)] = 0.911696$$

In eighth, ninth and tenth iterations, the values of x, y, z are the same, correct to three decimal places. Hence, we stop at this level.

Gauss–Seidel method: We take the initial values as $y = 0$, $z = 0$ and use Eq. (i)

First iteration:

$$x^{(1)} = \frac{1}{8}[20 + 3(0) - 2(0)] = 2.5$$

$$y^{(1)} = \frac{1}{11}[33 - 4(2.5) + 0] = 2.090909$$

$$z^{(1)} = \frac{1}{12}[35 - 6(2.5) - 3(2.090909] = 1.143939$$

Second iteration:

$$x^{(2)} = \frac{1}{8}[20 + 3(2.090909) - 2(1.143939)] = 2.998106$$

$$y^{(2)} = \frac{1}{11}[33 - 4(2.998106) + (1.143939)] = 2.013774$$

$$z^{(2)} = \frac{1}{12}[35 - 6(2.998106) - 3(2.013774)] = 0.914170$$

Third iteration:

$$x^{(3)} = \frac{1}{8}[20 + 3(2.013774) - 2(0.914170)] = 3.026623$$

$$y^{(3)} = \frac{1}{11}[33 - 4(3.026623) + (0.914170)] = 1.872516$$

$$z^{(3)} = \frac{1}{12}[35 - 6(3.026623) - 3(1.9825161)] = 0.907726$$

Fourth iteration:

$$x^{(4)} = \frac{1}{8}[20 + 3(1.982516) - 2(0.907726)] = 3.016512$$

$$y^{(4)} = \frac{1}{11}[33 - 4(3.016512) + (0.907726)] = 1.985607$$

$$z^{(4)} = \frac{1}{12}[35 - 6(3.016512) - 3(1.985607)] = 0.912009$$

Fifth iteration:

$$x^{(5)} = \frac{1}{8}[20 + 3(1.985607) - 2(0.912009)] = 3.016600$$

$$y^{(5)} = \frac{1}{11}[33 - 4(3.016600) + (0.912009)] = 1.985964$$

$$z^{(5)} = \frac{1}{12}[35 - 6(3.016600) - 3(1.985964)] = 0.911876$$

Sixth iteration:

$$x^{(6)} = \frac{1}{8}[20 + 3(1.985964) - 2(0.911876)] = 3.016767$$

$$y^{(6)} = \frac{1}{11}[33 - 4(3.016767) + (0.911876)] = 1.985892$$

$$z^{(6)} = \frac{1}{12}[35 - 6(3.016767) - 3(1.985892)] = 0.911810$$

(The values of x, y, z by Jacobi method correct to three decimal places are obtained even in the 6th iteration by Gauss–Seidel method.)

Seventh iteration:

$$x^{(7)} = \frac{1}{8}[20 + 3(1.985892) - 2(0.911810)] = 3.016757$$

$$y^{(7)} = \frac{1}{11}[33 - 4(3.016757) + (0.911810)] = 1.985889$$

$$z^{(7)} = \frac{1}{12}[35 - 6(3.016757) - 3(1.985889)] = 0.911816$$

Since the seventh and eighth iterations give the same values for x, y, z correct to four decimal places, we stop here.

$\therefore$ $x = 3.0168$, $y = 1.9859$, $z = 0.9118$

The values of x, y, z by both methods at each iteration are tabulated below:

Iteration	Gauss–Jacobi method			Gauss–Seidel method		
	x	y	z	x	y	z
1	2.5	3.0	2.916666	2.5	2.090909	1.143939
2	2.895833	2.236060	0.916666	2.998106	2.013774	0.914170
3	3.154356	2.030303	0.879735	3.026623	1.982516	0.907726
4	3.041430	1.932937	0.831913	3.016512	1.985607	0.912009
5	3.016873	1.969654	0.912717	3.016600	1.985964	0.911876
6	3.010441	1.985930	0.915817	3.016767	1.985892	0.911810
7	3.015770	1.988550	0.914964	3.016757	1.985889	0.911816
8	3.016946	1.986535	0.911644			
9	3.017039	1.985805	0.911560			
10	3.016786	1.985764	0.911696			

This shows that the convergence is rapid in Gauss-Seidel method when compared to Gauss-Jacobi method. We see that 10 iterations are necessary in Jacobi method to get the same accuracy as got by seventh iteration in Gauss-Seidel method.

Example 2.20: Solve the following system of equations by Gauss-Jacobi and Gauss-Seidel methods correct to three decimal places.

$$x + y + 54z = 110$$
$$27x + 6y - z = 85$$
$$6x + 15y + 2z = 72$$

Solution: As the coefficient matrix is not diagonally dominant as it is, we rewrite the equation as noted below, so that the coefficient matrix becomes diagonally dominant.

$$27x + 6y - z = 85$$
$$6x + 15y + 2z = 72$$
$$x + y + 54z = 110$$

Solving for x, y, z, we get

$$x = \frac{1}{27}[85 - 6y + z] \tag{i}$$

$$y = \frac{1}{15}[72 - 6x - 2z] \tag{ii}$$

$$z = \frac{1}{54}[110 - x - y] \tag{iii}$$

Starting with the initial value $x = 0, y = 0, z = 0$ and using Eqs (i)–(iii) and repeating the process, we get the values of x, y, z as tabulated by both methods (Gauss-Jacobi and Gauss-Seidel).

Iteration	Gauss–Jacobi method			Gauss–Seidel method		
	x	y	z	x	y	z
1	3.14815	4.8	2.0370	3.14815	3.54074	1.91317
2	2.15693	3.26913	1.88985	2.43218	3.57204	1.92585
3	2.49167	3.68525	1.93655	2.42569	3.57294	1.92595
4	2.40093	3.54513	1.92265	2.42549	3.57301	1.92595
5	2.43155	3.58327	1.92692	2.42548	3.57301	1.92595
6	2.42323	3.57046	1.92565	2.42548	3.57301	1.92595
7	2.42603	3.57395	1.92604			
8	2.42527	3.57278	1.92593			

Hence $x = 2.425$, $y = 3.573$, and $z = 1.926$ (correct to three decimal places).

Example 2.21: Solve, by Gauss–Seidel method, the following system:

$$28x + 4y - z = 32$$
$$x + 3y + 10z = 24$$
$$2x + 17y + 4z = 35$$

Solution: Since the diagonal elements in the coefficient matrix are not dominant, we rearrange the equations, as follows, such that the elements in the coefficient matrix are dominant.

$$28x + 4y - z = 32$$
$$2x + 17y + 4z = 35$$
$$x + 3y + 10z = 24$$

Hence,
$$x = \frac{1}{28}[32 - 4y + z]$$

$$y = \frac{1}{17}[35 - 2x - 4z]$$

$$z = \frac{1}{10}[24 - x - 3y]$$

Setting $y = 0$, $z = 0$, we get

First iteration:

$$x^{(1)} = \frac{1}{28}[32 - 4(0) + 0] = 1.1429$$

$$y^{(1)} = \frac{1}{17}[35 - 2(1.1429) - 4(0)] = 1.9244$$

$$z^{(1)} = \frac{1}{10}[24 - 1.1429 - 3(1.9244)] = 1.8084$$

Second iteration:

$$x^{(2)} = \frac{1}{28}[32 - 4(1.9244) + 1.8084)] = 0.9325$$

$$y^{(2)} = \frac{1}{17}[35 - 2(0.9325) - 4(1.8084)] = 1.5236$$

$$z^{(2)} = \frac{1}{10}[24 - 0.9325 - 3(1.5236)] = 1.8497$$

Third iteration:

$$x^{(3)} = \frac{1}{28}[32 - 4(1.5236) + 1.8497] = 0.9913$$

$$y^{(3)} = \frac{1}{17}[35 - 2(0.9913) - 4(1.8497)] = 1.5070$$

$$z^{(3)} = \frac{1}{10}[24 - 0.9913 - 3(1.5070)] = 1.8488$$

Fourth iteration:

$$x^{(4)} = \frac{1}{28}[32 - 4(1.5070) + 1.8488] = 0.9936$$

$$y^{(4)} = \frac{1}{17}[35 - 2(0.9936) - 4(1.8488)] = 1.5069$$

$$z^{(4)} = \frac{1}{10}[24 - 0.9936 - 3(1.5069)] = 1.8486$$

Fifth iteration:

$$x^{(5)} = \frac{1}{28}[32 - 4(1.5069) + 1.8486] = 0.9936$$

$$y^{(5)} = \frac{1}{17}[35 - 2(0.9936) - 4(1.8486)] = 1.5069$$

$$z^{(5)} = \frac{1}{10}[24 - 0.9936 - 3(1.5069)] = 1.8486$$

Since the values of x, y, z in the fourth and fifth iterations are same, we stop the process here.

Hence $x = 0.9936$, $y = 1.5069$, $z = 1.8486$.

PROBLEMS 2.9

Solve the following system of equations by (a) Gauss–Jacobi method, and (b) Gauss–Seidel method.

1. $30x - 2y + 3z = 75$
$2x + 2y + 18z = 30$
$x + 17y - 2z = 48$ [Ans. 2.5796, 2.7976, 1.0693]

2. $3.122x + 0.5756y - 0.1565z - 0.0067t = 1.571$
$0.5756x + 2.938y + 0.1103z - 0.0015t = -0.9275$
$-0.1565x + 0.1103y + 4.127z + 0.2051t = -0.0652$
$-0.0067x - 0.0015y + 0.2051z + 4.133t = -0.0178$
[Ans. 0.5835, −0.4307, 0.0181, −0.0044]

3. $7.6x - 2.4y + 1.3z = 20.396$
$3.7x - 7.9y - 2.5z = 35.866$
$1.9x - 4.3y + 8.2z = 32.514$ [Ans. 3.23, 4.85, 5.76]

4. $83x + 11y - 4z = 95$
$7x + 52y + 13z = 104$
$3x + 8y + 29z = 71$ [Ans. 1.06, 1.37, 1.96]

2.13 RELAXATION METHOD

We will consider a system of three equations in three unknowns as given below for the sake of simplicity. The method is applicable even for more number of equations.

Consider the system of equations

$$\left. \begin{array}{l} a_1x + b_1y + c_1z = d_1 \\ a_2x + b_2y + c_2z = d_2 \\ a_3x + b_3y + c_3z = d_3 \end{array} \right\} \tag{2.24}$$

Define the residuals r_1, r_2, r_3 by the relation

$$\left. \begin{array}{l} r_1 = a_1x + b_1y + c_1z - d_1 \\ r_2 = a_2x + b_2y + c_2z - d_2 \\ r_3 = a_3x + b_3y + c_3z - d_3 \end{array} \right\} \tag{2.25}$$

If we can find the values of x, y, z so that $r_1 = 0 = r_2 = r_3$, then those values of x, y, z are the exact values of the system. If it is not possible to make $r_1 = 0 = r_2 = r_3$, then we make simultaneously the values of r_1, r_2, r_3 as close to zero as possible. In other words, we "liquidate" the residuals r_1, r_2, r_3 by taking better approximate values of x, y, z. If a slight change is made in the values of x, y, z, what will be the corresponding changes in the residuals r_1, r_2, r_3? We give below an 'operation table' from which we can easily know the corresponding changes in r_1, r_2, r_3 for a change of 1 unit in x, while there is no change in y and z, for a change of 1 unit in y while there is no change in x and z; for a change of 1 unit in z while there is no change in y and x.

Operation table						
Operation	Changes in (or increment in)					
	x	y	z	r_1	r_2	r_3
R_1	1	0	0	a_1	a_2	a_3
R_2	0	1	0	b_1	b_2	b_3
R_3	0	0	1	c_1	c_2	c_3

What is the meaning of the above table?

The operator R_1 increases the value of x by 1, y by zero, z by zero (no change in y and z) and this operation increase the residuals r_1 by a_1, r by a_2 and r_3 by a_3 (the increase in r_1, r_2, r_3 are nothing but the coefficient of x in the equations given). Similarly, R_3 increases the value of z by 1 (while x, y are kept constant) and the effect of this operation increases the values of r_1, r_2, r_3 by c_1, c_2, c_3 respectively.

One can easily see that the operation table consists of the unit matrix I and the transpose of the matrix A namely A', where A is coefficient matrix of the system of equations.

Convergence of the relaxation method: If the method should converge, the diagonal elements of the coefficient matrix A should be dominant, i.e. A is diagonally dominant. Referring to the system of equation given above, the system can be solved by this method successfully only if

$$|a_1| \geq |b_1| + |c_1|$$
$$|b_2| \geq |a_2| + |c_2|$$
$$|c_3| \geq |a_3| + |b_3|$$

where at least once the strict inequality holds.

Example 2.22: By relaxtion method, solve:

$$12x + y + z = 31$$
$$2x + 8y - z = 24$$
$$3x + 4y + 10z = 58$$

Solution: The coefficient matrix is diagonally dominant. Hence, we will use relaxation method with confidence. The residuals are

$$r_1 = 12x + y + z - 31$$
$$r_2 = 2x + 8y - z - 24$$
$$r_3 = 3x + 4y + 10z - 58$$

			Operation table			
	x	y	z	r_1	r_2	r_3
R_1	1	0	0	12	2	3
R_2	0	1	0	1	8	4
R_3	0	0	1	1	-1	10

We will start with the initial values $x = 0$, $y = 0$, $z = 0$. Relaxation procedure is given below.

		x	y	z	r_1	r_2	r_3	
Initial		0	0	0	-31	-24	-58	(1)
$6R_3$	$\rightarrow$	0	0	6	-25	-30	2	(2)
$4R_2$	$\rightarrow$	0	4	0	-21	2	18	(3)
$2R_1$	$\rightarrow$	2	0	0	3	6	24	(4)
$-2R_3$	$\rightarrow$	0	0	-2	1	8	4	(5)
$-1R_2$	$\rightarrow$	0	-1	0	0	0	0	(6)
		2	3	4	0	0	0	(7)

Since all residuals are made to zero, the solution is $x = 2$, $y = 3$, $z = 4$.

Explanation: In line (1), giving $x = 0$, $y = 0$, $z = 0$, we get $r_1 = -31$, $r_2 = -24$, $r_3 = -58$ is numerically the greatest residual. Hence, we liquidate $r_3 = -58$ by suitable multiple of R_3. In R_3, a change of 1 in z will cause an addition 10 in r_3. To nullify -58, we add 6 in z so that a change of 60 is effected in r_3. Therefore, using $6R_3$, changes in r_1, r_2, r_3 are $6, -6, 60$. Hence, new $r_1 = -31 + 6 = -25$, $r_2 = -24 + (-6) = 30$, $r_3 = -58 + 60 = 2$, which is line number (2). In line (2), the numerically greatest residual is $r_2 = -30$, we liquidate this by a proper multiple of R_2. By R_2, an increase of 1 in y makes changes 1 in r_1, 8 in r_2 and 4 in r_3. To liquidate -30, we do operation $4R_2$, so that $4R_2$ causes changes 4, 32, 16 in r_1, r_2, r_3 respectively. Hence, by $4R_2$.

Now, $r_1 = -25 + (4) = -21$, $r_2 = -30 + 32 = 2$, $r_3 = 2 + 16 = 18$.

Therefore, in line (3), $r_1 = -21$, $r_2 = 2$, $r_3 = 18$. Now $r_1 = -21$ is numerically the greatest residual. Hence, we now liquidate the value of $r_1 = -21$. An increase of 1 in x (i.e. R_1 operation) increases 12 in r_1. Hence, we do operation $2R_1$ to liquidate $r_2 = -21$. This causes an increase in the value of r_3 by -20. After the operation of $-2R_3$, we get $r_1 = 1$, $r_2 = 8$, $r_3 = 4$. Therefore, now we liquidate $r_2 = 8$ by $-1R_2$ and we get the resultant $r_1 = 0$, $r_2 = 0$, $r_3 = 0$. Adding all the values of x, y, z, we get $x = 2$, $y = 3$, $z = 4$ as the exact solution.

Example 2.23: Use relaxation method to solve the system.

$$8x + y + z + w = 14$$
$$2x + 10y + 3z + w = -8$$
$$x - 2y - 20z + 3w = 111$$
$$3x + 2y + 2z + 19w = 53$$

Solution: Evidently the coefficient matrix $A = \begin{pmatrix} 8 & 1 & 1 & 1 \\ 2 & 10 & 3 & 1 \\ 1 & -2 & -20 & 3 \\ 3 & 2 & 2 & 18 \end{pmatrix}$ is diagonally dominant,

we have the residuals r_1, r_2, r_3, r_4.

$$r_1 = 8x + y + z + w - 14$$
$$r_2 = 2x + 10y + 3z + w + 8$$
$$r_3 = x - 2y - 20z + 3w - 111$$
$$r_4 = 3x + 2y + 2z + 19w - 53$$

				Operation table				
	x	y	z	w	r_1	r_2	r_3	r_4
R_1	1	0	0	0	8	2	1	3
R_2	0	1	0	0	1	10	-2	2
R_3	0	0	1	0	1	3	-20	2
R_4	0	0	0	1	1	1	1	19

We will start with $x = 0$, $y = 0$, $z = 0$, $w = 0$ as initial values.
Relaxation procedure is given below.

	x	y	z	w	r_1	r_2	r_3	r_4
	0	0	0	0	-14	8	-111	-53
$-5R_3 \rightarrow$	0	0	-5	0	-19	-7	-11	-63
$3R_4 \rightarrow$	0	0	0	3	-16	-4	-2	-6
$2R_1 \rightarrow$	2	0	0	0	0	0	0	0
Adding	2	0	-5	3	0	0	0	0

Since all residuals are zero, the exact solution is $x = 2$, $y = 0$, $z = -5$, $w = +3$.

Note: In all the above procedure, we always liquidate the numerically largest residual to zero or nearer to zero.

PROBLEMS 2.10

Solve the following system of equations by relaxation method.

i. $2x - 3y + 10z = 3$
$-x + 4y + 2z = 20$
$5x + 4y + z = -12$ [Ans. $-4, 3, 2$]

ii. $50x + 2y - 3z = 196$
$3x + 65y + 2z = 81$
$-x + y + 33z = 63$ [Ans. $4, 1, 2$]

iii. $10x - 2y + z = 12$
$x + 9y - z = 10$
$2x - y + 11z = 20$ [Ans. $1.3, 1.2, 1.7$]

iv. $10x + 2y - w = 11$
$-x + 20y + 2z = 49.5$
$-x + 10y - w = 27.5$
$-y + 2z + 20w = 92.4$ [Ans. $x = 1.1$, $y = 2.2$, $z = 3.3$, $w = 4.4$]

2.14 POWER METHOD

Let A be the given square matrix of order n and let $X_1, X_2, ..., X_n$ be the eigen vectors corresponding to the distinct eigen values $\lambda_1, \lambda_2, ..., \lambda_n$. So we have the relation

$$AX_i = \lambda_i X_i \qquad i = 1, 2, ..., n \tag{2.26}$$

Equation (2.26) forms the basis for an iterative solution technique that eventually yields the highest eigen value and its associated eigen vector.

Note:

1. To find the numerically smallest eigen value of A, obtain the numerically greatest eigen value of A^{-1} and then take its reciprocal.
2. The eigen value of $A - K1$ are $\lambda_i - K$ where λ_i are eigen values of A.
3. Trace of a matrix is equal to the sum of the diagonal elements of the matrix.
4. To find the numerically smallest eigen value of A, obtain the dominant eigen value of A and then find $B = A - \lambda_1 I$ and find the dominant eigen value of B. Then, the smallest eigen value of A is equal to the dominant eigen value of $B + \lambda_1$.

Example 2.24: Find the numerically largest eigen value of $A = \begin{pmatrix} 25 & 1 & 2 \\ 1 & 3 & 0 \\ 2 & 0 & -4 \end{pmatrix}$ and the corresponding eigen vector.

Solution: Let $X_1 = \begin{pmatrix} 1 \\ 0 \\ 0 \end{pmatrix}$ be an arbitrary initial eigen vector.

$$AX_1 = \begin{pmatrix} 25 & 1 & 2 \\ 1 & 3 & 0 \\ 2 & 0 & -4 \end{pmatrix}\begin{pmatrix} 1 \\ 0 \\ 0 \end{pmatrix} = \begin{pmatrix} 25 \\ 1 \\ 2 \end{pmatrix} = 25\begin{pmatrix} 1 \\ 0.04 \\ 0.08 \end{pmatrix} = 25X_2$$

$$AX_2 = \begin{pmatrix} 25 & 1 & 2 \\ 1 & 3 & 0 \\ 2 & 0 & -4 \end{pmatrix}\begin{pmatrix} 1 \\ 0.04 \\ 0.08 \end{pmatrix} = \begin{pmatrix} 25.2 \\ 1.12 \\ 1.68 \end{pmatrix} = 25.2\begin{pmatrix} 1 \\ 0.0444 \\ 0.0667 \end{pmatrix} = 25.2X_3$$

$$AX_3 = \begin{pmatrix} 25 & 1 & 2 \\ 1 & 3 & 0 \\ 2 & 0 & -4 \end{pmatrix}\begin{pmatrix} 1 \\ 0.0444 \\ 0.0667 \end{pmatrix} = \begin{pmatrix} 25.1778 \\ 1.1332 \\ 1.7332 \end{pmatrix} = 25.1778\begin{pmatrix} 1 \\ 0.0450 \\ 0.0688 \end{pmatrix} = 25.1778X_4$$

$$AX_4 = \begin{pmatrix} 25.1826 \\ 1.135 \\ 1.7248 \end{pmatrix} = 25.1826\begin{pmatrix} 1 \\ 0.0451 \\ 0.0685 \end{pmatrix} = 25.1826X_5$$

$$AX_5 = \begin{pmatrix} 25.1821 \\ 1.1353 \\ 1.7260 \end{pmatrix} = 25.1821\begin{pmatrix} 1 \\ 0.0451 \\ 0.0685 \end{pmatrix} = 25.1821X_6$$

$\lambda_1 = 25.1821$ and the corresponding eigen vector is $\begin{pmatrix} 1 \\ 0.0451 \\ 0.0685 \end{pmatrix}$.

Example 2.25: Find the dominant eigen value and the corresponding eigen vector of

$A = \begin{pmatrix} 1 & 6 & 1 \\ 1 & 2 & 0 \\ 0 & 0 & 3 \end{pmatrix}$. Find also the least latent root and hence the third eigen value also.

Solution: Let $X_1 = \begin{pmatrix} 1 \\ 0 \\ 0 \end{pmatrix}$ be an approximate eigen value.

$$AX_1 = \begin{pmatrix} 1 & 6 & 1 \\ 1 & 2 & 0 \\ 0 & 0 & 3 \end{pmatrix} \begin{pmatrix} 1 \\ 0 \\ 0 \end{pmatrix} = \begin{pmatrix} 1 \\ 1 \\ 0 \end{pmatrix} = 1 \begin{pmatrix} 1 \\ 1 \\ 0 \end{pmatrix} = 1X_2$$

$$AX_2 = \begin{pmatrix} 1 & 6 & 1 \\ 1 & 2 & 0 \\ 0 & 0 & 3 \end{pmatrix} \begin{pmatrix} 1 \\ 1 \\ 0 \end{pmatrix} = \begin{pmatrix} 7 \\ 3 \\ 0 \end{pmatrix} = 7 \begin{pmatrix} 1 \\ 0.4286 \\ 0 \end{pmatrix} = 7X_3$$

$$AX_3 = \begin{pmatrix} 1 & 6 & 1 \\ 1 & 2 & 0 \\ 0 & 0 & 3 \end{pmatrix} \begin{pmatrix} 1 \\ 0.4286 \\ 0 \end{pmatrix} = \begin{pmatrix} 3.5714 \\ 1.8572 \\ 0 \end{pmatrix} = 3.5714 \begin{pmatrix} 1 \\ 0.52 \\ 0 \end{pmatrix} = 3.5714X_4$$

$$AX_4 = \begin{pmatrix} 1 & 6 & 1 \\ 1 & 2 & 0 \\ 0 & 0 & 3 \end{pmatrix} \begin{pmatrix} 1 \\ 0.52 \\ 0 \end{pmatrix} = \begin{pmatrix} 4.12 \\ 2.04 \\ 0 \end{pmatrix} = 4.12 \begin{pmatrix} 1 \\ 0.4951 \\ 0 \end{pmatrix} = 4.12X_5$$

$$AX_5 = \begin{pmatrix} 1 & 6 & 1 \\ 1 & 2 & 0 \\ 0 & 0 & 3 \end{pmatrix} \begin{pmatrix} 1 \\ 0.4951 \\ 0 \end{pmatrix} = \begin{pmatrix} 3.9706 \\ 1.9902 \\ 0 \end{pmatrix} = 3.9706 \begin{pmatrix} 1 \\ 0.5012 \\ 0 \end{pmatrix} = 3.9706X_6$$

$$AX_6 = \begin{pmatrix} 1 & 6 & 1 \\ 1 & 2 & 0 \\ 0 & 0 & 3 \end{pmatrix} \begin{pmatrix} 1 \\ 0.5012 \\ 0 \end{pmatrix} = \begin{pmatrix} 4.0072 \\ 2.0024 \\ 0 \end{pmatrix} = 4.0072 \begin{pmatrix} 1 \\ 0.4997 \\ 0 \end{pmatrix} = 4.0072X_7$$

$$AX_7 = \begin{pmatrix} 1 & 6 & 1 \\ 1 & 2 & 0 \\ 0 & 0 & 3 \end{pmatrix} \begin{pmatrix} 1 \\ 0.4997 \\ 0 \end{pmatrix} = \begin{pmatrix} 3.9982 \\ 1.9994 \\ 0 \end{pmatrix} = 3.9982 \begin{pmatrix} 1 \\ 0.5000 \\ 0 \end{pmatrix} = 3.9982X_8$$

$$AX_8 = \begin{pmatrix} 1 & 6 & 1 \\ 1 & 2 & 0 \\ 0 & 0 & 3 \end{pmatrix} \begin{pmatrix} 1 \\ 0.5 \\ 0 \end{pmatrix} = \begin{pmatrix} 4 \\ 2 \\ 0 \end{pmatrix} = 4 \begin{pmatrix} 1 \\ 0.5 \\ 0 \end{pmatrix} = 4X_9$$

Dominant eigen value = 4, corresponding eigen vector is $(1, 0.5, 0)$.

To find the least eigen value, let $B = A - 4I$ since $\lambda_1 = 4$

$$B = \begin{pmatrix} 1 & 6 & 1 \\ 1 & 2 & 0 \\ 0 & 0 & 2 \end{pmatrix} - \begin{pmatrix} 4 & 0 & 0 \\ 0 & 4 & 0 \\ 0 & 0 & 4 \end{pmatrix} = \begin{pmatrix} -3 & 6 & 1 \\ 1 & -2 & 2 \\ 0 & 0 & -2 \end{pmatrix}$$

We will find the dominant eigen value of B.

Let $Y_1 = \begin{pmatrix} 1 \\ 0 \\ 0 \end{pmatrix}$ be initial vector

$$BY_1 = \begin{pmatrix} -3 & 6 & 1 \\ 1 & -2 & 2 \\ 0 & 0 & -2 \end{pmatrix}\begin{pmatrix} 1 \\ 0 \\ 0 \end{pmatrix} = \begin{pmatrix} -3 \\ 1 \\ 0 \end{pmatrix} = -3\begin{pmatrix} 1 \\ -0.3333 \\ 0 \end{pmatrix} = -3Y_2$$

$$BY_2 = \begin{pmatrix} -3 & 6 & 1 \\ 1 & -2 & 2 \\ 0 & 0 & -2 \end{pmatrix}\begin{pmatrix} 1 \\ -0.3333 \\ 0 \end{pmatrix} = \begin{pmatrix} -5 \\ 1.6666 \\ 0 \end{pmatrix} = -5\begin{pmatrix} 1 \\ -0.3333 \\ 0 \end{pmatrix} = -5Y_3$$

$$BY_3 = \begin{pmatrix} -3 & 6 & 1 \\ 1 & -2 & 2 \\ 0 & 0 & -2 \end{pmatrix}\begin{pmatrix} 1 \\ -0.3333 \\ 0 \end{pmatrix} = \begin{pmatrix} -5 \\ 1.6660 \\ 0 \end{pmatrix} = -5\begin{pmatrix} 1 \\ -0.3333 \\ 0 \end{pmatrix} = -5y_4$$

Dominant eigen value of $B = -5$.

Adding 4, smallest eigen value of $A = -5 + 4 = -1$.

Sum of eigen values = Trace of $A = 1 + 2 + 3 = 6$

$$4 + (-1) + \lambda_3 = 6$$
$$\lambda_3 = 3$$

All the three eigen values are $4, 3, -1$.

PROBLEMS 2.11

Using power method, find the largest eigen value and corresponding eigen vector.

i. $\begin{pmatrix} -0.5 & 1.5 & 0.17 \\ 0.25 & -0.25 & -0.08 \\ 0 & 0 & 0.33 \end{pmatrix}$ [Ans. $-1, (1, -0.33, 0)$]

ii. $\begin{pmatrix} 3 & 2 & 4 \\ -1 & 4 & 10 \\ 1 & 3 & -1 \end{pmatrix}$ $\left[\text{Ans. } 7.4, \begin{pmatrix} 0.8 \\ 1 \\ 0.4 \end{pmatrix}\right]$

iii. $\begin{pmatrix} 1 & 3 & -1 \\ 3 & 2 & 4 \\ -1 & 4 & 10 \end{pmatrix}$ $\left[\text{Ans. } 11.66, \begin{pmatrix} 0.025 \\ 0.422 \\ 1 \end{pmatrix}\right]$

iv. $\begin{pmatrix} 4 & 1 \\ 1 & 3 \end{pmatrix}$

$$\left[\text{Ans. } 4.6179, \begin{pmatrix} 1 \\ 0.618 \end{pmatrix} \right]$$

v. $\begin{pmatrix} 5 & 0 & 1 \\ 0 & -2 & 0 \\ 1 & 0 & 5 \end{pmatrix}$

$$\left[\text{Ans. } 5.997, \begin{pmatrix} 1 \\ 0 \\ 0.9980 \end{pmatrix} \right]$$

3

3.1 INTRODUCTION

The calculus of finite differences deals with the change that takes place in the value of a function (dependent variable) due to finite change in the independent variable.

Let $y = f(x)$ be a function of x and let $y = y_0, y_1, y_2, ..., y_n$ be the values of y corresponding to the equispaced values of $x = x_0, x_0 + h, x_0 + 2h, ..., x_0 + nh$. The values of the independent variable x are called the *argument* and the corresponding functional values y are known as the *entry*.

To determine the values of $f(x)$, etc. for some intermediate argument, the following three types of differences are found useful.

a. Forward differences
b. Backward differences
c. Central differences.

3.2 FORWARD DIFFERENCES

The differences $y_1 - y_0, y_2 - y_1, y_3 - y_2, ..., y_n - y_{n-1}$ when denoted by $\Delta y_0, \Delta y_1, \Delta y_2, ..., \Delta y_{n-1}$ respectively are called first forward differences and Δ is called the difference operator. The first forward differences are given by

$$\Delta y_0 = y_1 - y_0$$
$$\Delta y_1 = y_2 - y_1$$
$$\Delta y_2 = y_3 - y_2$$
$$... = ...$$
$$... = ...$$
$$\Delta y_{n-1} = y_n - y_{n-1}$$

or
$$\Delta y_r = y_{r+1} - y_r$$

Similarly, second forward differences are

$$\Delta^2 y_r = \Delta(\Delta y_r) = \Delta[y_{r+1} - y_r] = \Delta y_{r+1} - \Delta y_r$$
$$= (y_{r+2} - y_{r+1}) - (y_{r+1} - y_r)$$
$$= y_{r+2} - 2y_{r+1} + y_r$$
$$= {}^2C_0 y_{r+2} - {}^2C_1 y_{r+1} + {}^2C_r y_r$$

Similarly, third order differences are

$$\Delta^3 y_r = \Delta^2 y_{r+1} - \Delta^2 y_r = (\Delta y_{r+2} - \Delta y_{r+1})(\Delta y_{r+1} - \Delta y_r)(\Delta y_{r+2} - 2\Delta y_{r+1} + \Delta y_r)$$
$$= (y_{r+3} - y_{r+2}) - 2(y_{r+2} - y_{r+1}) - (y_{r+1} - y_r)$$
$$= y_{r+3} - 3y_{r+2} + 3y_{r+1} + y_r$$
$$= {}^3C_0 y_{r+3} - {}^3C_1 y_{r+2} + {}^3C_2 y_{r+1} - {}^3C_3 y_r$$

Hence, $\Delta^n y_r = {}^nC_0 y_{r+n} - {}^nC_1 y_{r+n-1} + {}^nC_2 y_{r+n-2} - {}^nC_3 y_{r+n-3}, ..., (-1)^n {}^nC_n y_r$

The forward differences are arranged in Table 3.1.

Table 3.1

Value of x argument	Value of $y = f(x)$ entry	First difference	Second difference	Third difference	Fourth difference	Fifth difference
x_0	y_0					
$x_0 + h$	y_1	$y_1 - y_0 = \Delta y_0$				
$x_0 + 2h$	y_2	$y_2 - y_1 = \Delta y_1$	$\Delta y_1 - \Delta y_0 = \Delta^2 y_0$	$\Delta^3 y_0$		
$x_0 + 3h$	y_3	$y_3 - y_2 = \Delta y_2$	$\Delta y_2 - \Delta y_1 = \Delta^2 y_1$	$\Delta^3 y_1$	$\Delta^4 y_0$	$\Delta^5 y_0$
$x_0 + 4h$	y_4	$y_4 - y_3 = \Delta y_3$	$\Delta y_3 - \Delta y_2 = \Delta^2 y_2$	$\Delta^3 y_2$	$\Delta^4 y_1$	
$x_0 + 5h$	y_5	$y_5 - y_4 = \Delta y_4$	$\Delta y_4 - \Delta y_3 = \Delta^2 y_3$			

y_0 is called the *leading term* and the differences $\Delta y_0, \Delta y_1, \Delta y_2, ...$ etc. are called *leading differences*.

Notes:

1. We know that

$$\Delta y_0 = y_1 - y_0$$
$$\Delta^2 y_0 = \Delta(\Delta y_0) = \Delta y_1 - \Delta y_0 = (y_2 - y_1) - (y_1 - y_0) = y_2 - 2y_1 + y_0$$
$$= {}^2C_0 y_2 - {}^2C_1 y_1 + {}^2C_2 y_0$$
$$\Delta^3 y_0 = \Delta^2(\Delta y_0) = \Delta^2(y_1 - y_0) = \Delta^2 y_1 - \Delta^2 y_0 = (y_3 - 2y_2 + y_1) - (y_2 - 2y_1 + y_0)$$
$$= y_3 - 3y_2 + 3y_1 - y_0$$
$$= {}^3C_0 y_3 - {}^3C_1 y_2 + {}^3C_2 y_1 - {}^3C_3 y_0$$

Similarly

$$\Delta^4 y_0 = {}^4C_0 y_4 - {}^4C_1 y_3 + {}^4C_2 y_2 - {}^4C_3 y_1 + {}^4C_4 y_0$$

Hence

$$\Delta^n y_0 = {}^nC_0 y_n - {}^nC_1 y_{n-1} + {}^nC_2 y_{n-2}, ..., (-1)^n {}^nC_n y_0$$

2. We know that

$$y_1 - y_0 = \Delta y_0$$
$$y_1 = (\Delta + 1) y_0$$
$$y_2 = (\Delta + 1) y_1 = (\Delta + 1)^2 y_0$$
$$y_3 = (\Delta + 1)^3 y_0$$
$$y_4 = (\Delta + 1)^4 y_0$$
$$... = ...$$

$$\dots = \dots$$

$$y_n = (\Delta + 1)^n \, y_0$$
$$= (1 + \Delta)^n \, y_0 = {}^nC_0 y_0 + {}^nC_1 \Delta y_0 + {}^nC_2 \Delta^2 y_0 + \dots + {}^nC_n \Delta^n y_0$$
$$y_n = {}^nC_0 y_0 + {}^nC_1 \Delta y_0 + {}^nC_2 \Delta^2 y_0 + \dots + {}^nC_n \Delta^n y_0$$

3.3 BACKWARD DIFFERENCES

The differences $y_1 - y_0, y_2 - y_1, y_3 - y_2, \dots, y_n - y_{n-1}$, which denoted by $\nabla y_1, \nabla y_2, \nabla y_3, \dots, \nabla y_n$ respectively are called the first *backward differences*, where ∇ is the *backward operator*.

The first differences are given by

$$\nabla y_1 = y_1 - y_0$$
$$\nabla y_2 = y_2 - y_1$$
$$\nabla y_3 = y_3 - y_2$$
$$\nabla y_r = y_r - y_{r-1}$$

Second differences are

$$\nabla^2 y_r = \nabla(\nabla y_r) = \nabla(y_r - y_{r-1}) = \nabla y_r - \nabla y_{r-1}$$
$$= y_r - y_{r-1} - y_{r-1} + y_{r-2} = y_r - 2y_{r-1} + y_{r-2}$$
$$= {}^2C_0 y_r - {}^2C_1 y_{r-1} + {}^2C_2 y_{r-2}$$

Similarly

$$\nabla^3 y_r = {}^3C_0 y_r - {}^3C_1 y_{r-1} + {}^3C_2 y_{r-2} - {}^3C_3 y_{r-3}$$
$$\nabla^n y_r = {}^nC_0 y_r - {}^nC_1 y_{r-1} + {}^nC_2 y_{r-2} + \dots + (-1)\,{}^nC_n y_{r-n}$$

The backward differences are arranged in Table 3.2.

			Table 3.2			
Value of x argument	*Value of y = f(x) entry*	*First difference*	*Second difference*	*Third difference*	*Fourth difference*	*Fifth difference*
x_0	y_0					
$x_0 + h$	y_1	∇y_1				
		∇y_2	$\nabla^2 y_2$			
$x_0 + 2h$	y_2		$\nabla^2 y_3$	$\nabla^3 y_3$		
		∇y_3		$\nabla^3 y_4$	$\nabla^4 y_4$	
$x_0 + 3h$	y_3		$\nabla^2 y_4$			$\nabla^5 y_5$
		∇y_4		$\nabla^3 y_5$	$\nabla^4 y_5$	
$x_0 + 4h$	y_4		$\nabla^2 y_5$			
		∇y_5				
$x_0 + 5h$	y_5					

Note: We know that

$$y_n - y_{n-1} = \nabla y_n$$
$$y_{n-1} = (1 - \nabla) \, y_n$$
$$y_{n-2} = (1 - \nabla)^2 \, y_n$$
$$y_{n-3} = (1 - \nabla)^3 \, y_n$$
$$\dots = \dots$$
$$\dots = \dots$$
$$y_{n-k} = (1 - \nabla)^k \, y_n$$
$$= {}^kC_0 y_n - {}^kC_1 \nabla y_n + {}^kC_2 \nabla^2 y_n, \dots, (-1)^k \, {}^kC_k \nabla^k y_n$$

3.4 CENTRAL DIFFERENCES

The differences $y_1 - y_0,\ y_2 - y_1,\ y_3 - y_2,\ ...,\ y_n - y_{n-1}$, which is denoted by $\delta y_{\frac{1}{2}},\ \delta y_{\frac{3}{2}},\ \delta y_{\frac{5}{2}}$, ..., $\delta y_{\frac{2n-1}{2}}$ respectively are called central differences, where δ is called the *central difference operator*. The central differences are given by

$$y_1 - y_0 = \delta y_{\frac{1}{2}}$$

$$y_2 - y_1 = \delta y_{\frac{3}{2}}$$

$$y_3 - y_2 = \delta y_{\frac{5}{2}}$$

$$... = ...$$

$$... = ...$$

$$y_n - y_{n-1} = \delta y_{\frac{2n-1}{2}}$$

These are first central differences, other central differences are

$$\delta^2 y_1 = \delta y_{\frac{3}{2}} - \delta y_{\frac{1}{2}}$$

$$\delta^2 y_2 = \delta y_{\frac{5}{2}} - \delta y_{\frac{3}{2}}$$

$$\delta^3 y_1 = \delta^2 y_{\frac{3}{2}} - \delta^2 y_{\frac{1}{2}}$$

$$\delta^3 y_2 = \delta^2 y_{\frac{5}{2}} - \delta^2 y_{\frac{3}{2}}$$

The central difference is given in Table 3.3.

Table 3.3

Value of x	Value of $y = f(x)$	First difference	Second difference	Third difference	Fourth difference	Fifth difference
x_0	y_0					
		$\delta y_{\frac{1}{2}}$				
$x_0 + h$	y_1		$\delta^2 y_1$			
		$\delta y_{\frac{3}{2}}$		$\delta^3 y_{\frac{3}{2}}$		
$x_0 + 2h$	y_2		$\delta^2 y_2$		$\delta^4 y_2$	
		$\delta y_{\frac{5}{2}}$		$\delta^3 y_{\frac{5}{2}}$		$\delta^5 y_{\frac{5}{2}}$
$x_0 + 3h$	y_3		$\delta^2 y_3$		$\delta^4 y_3$	
		$\delta y_{\frac{7}{2}}$		$\delta^3 y_{\frac{7}{2}}$		
$x_0 + 4h$	y_4		$\delta^2 y_4$			
		$\delta y_{\frac{9}{2}}$				
$x_0 + 5h$	y_5					

3.5 DIFFERENCES OF POLYNOMIAL

Theorem: The nth differences of a polynomial of nth degree are constant and all higher order differences are zero.

Proof: Let the polynomial of nth degree be

$$f(x) = a_0 x^n + a_1 x^{n-1} + a_2 x^{n-2} + ... + a_{n-1} x + a_n$$

where $a_0, a_1, a_2, ..., a_n$ are all constants and $a_0 \neq 0$

$$f(x + h) = a_0(x + h)^n + a_1(x + h)^{n-1} + a_2(x + h)^{n-2} + ... + a_{n-1}(x + h) + a_n$$

$$\Delta f(x) = f(x + h) - f(x)$$

$$= a_0[(x + h)^n - x^n] + a_1[(x + h)^{n-1} - x^{n-1}]$$

$$+ a_2[(x + h)^{n-2} - x^{n-2}] - ... a_{n-1}[(x + h) - x]$$

$$= a_0[x^n + {}^nC_1 x^{n-1} h + {}^nC_2 x^{n-2} h^2 + ... - x^n]$$

$$+ a_1[x^{n-1} + {}^{n-1}C_1 x^{n-2} h + {}^{n-1}C_2 x^{n-3} h^2 + ... - x^{n-1}]$$

$$+ a_2[x^{n-2} + {}^{n-2}C_1 x^{n-3} h + {}^{n-2}C_2 x^{n-4} h^2 + ... - x^{n-2}] + ... + a_{n-1}h$$

$$= a_0 nh\, x^{n-1} + (a_0 {}^nC_2 h^2 + a_1 {}^{n-1}C_1 h)x^{n-2}$$

$$+ (a_0 {}^nC_3 h^3 + a_1 {}^{n-1}C_2 h^2 + a_2 {}^{n-2}C_1 h)x^{n-3} + ... + a_{n-1}h$$

$$= a_0 nh\, x^{n-1} + b_1 x^{n-2} + b_2 x^{n-3} + ... + b_{n-2} + b_{n-1}$$

where $b_1, b_2, b_3, ..., b_{n-1}$ are new constants.

Thus first difference of a polynomial of nth degree is a polynomial of $(n - 1)$th degree.

Similarly, the second difference

$$\Delta^2 f(x) = \Delta[\Delta f(x)] = \Delta[f(x + h) - f(x)] = \Delta f(x + h) - \Delta f(x)$$

$$= a_0 nh[(x + h)^{n-1} - x^{n-1}] + b_1[(x + h)^{n-2} - x^{n-2}]$$

$$+ b_2[(x + h)^{n-3} - x^{n-3}] + ... + b_{n-2}[x + h - x]$$

$$= a_0 nh[h(n - 1)x^{n-2}] + b_1' x^{n-3} + b_2' x^{n-4} ... + b_{n-2}'h$$

when $b_1', b_2', ..., b_{n-2}'$ are constants.

$$\Delta^2 f(x) = a_0 n(n - 1)h^2 x^{n-2} + b_1' x^{n-3} + b_2' x^{n-4} + ... + b_{n-2}'h$$

Thus the second differences of a polynomial of nth degree is a polynomial of $(n - 2)$th degree.

Proceeding in the same say, nth difference of a polynomial of nth degree will be a polynomial of $(n - n)$, i.e. zero degree

$$\Delta^n f(x) = a_0 n\, (n - 1)\, (n - 2) ... [n - (n - 1)]\, h^n x^{n - n}$$

$$= a_0 n\, (n - 1)\, (n - 2) ...\, 1 h^n x^0$$

$$\Delta^n f(x) = a_0 n!\, h^n$$

Also

$$\Delta^{n+1} f(x) = \Delta[\Delta^n f(x)] = \Delta[a_0 n!\, h^n]$$

$$= a_0 n\, !\, [h^n] - a_0 n!\, [h^n] = 0$$

$$\Delta^{n+1} f(x) = 0$$

$$\Delta^{n+2} f(x) = 0$$

which completes the proof of the theorem.

Example 3.1: Evaluate $\Delta^{10}(1-x)(1-2x^2)(1-3x^3)(1-4x^4)$, $h=2$.

Solution:

$$\Delta^{10}(1-x)(1-2x^2)(1-3x^3)(1-4x^4) = \Delta^{10}[(x-1)(2x^2-1)(3x^3-1)(4x^4-1)]$$
$$= \Delta^{10}[24x^{10} + ()x^9 + ()x^8 + ... + 1] = 24 \times 10!h^{10}$$
$$= 24 \times 10!2^{10} \quad [\because h=2]$$

3.6 FACTORIAL NOTATION

A product of the form $x(x-1)...(x-r+1)$ denoted by $[x]^r$ is called a factorial of order r.

In particular $[x] = x$, $[x]^2 = x(x-1)$, $[x]^3 = x(x-1)(x-2)$ is the factorial notation

In general $[x]^n = x(x-1)(x-2)...(x-n+1)$ here 1 is the interval

If h is the interval $[x]^n = x(x-h)(x-2h)...(x-\{n-1\}h)$

Example 3.2: Show that $\Delta^n[x]^n = n!\,h^n$

Solution:

$$\Delta[x]^n = [x+h]^n - [x]^n$$
$$= (x+h)(x+h-h)(x+h-2h)...[x+h-(n-1)h]$$
$$\quad - x(x-h)(x-2h)...\{x-(n-2)h\}\{x-(n-1)h\}$$
$$= \{(x+h)x(x-h)...(x-nh+2h)\}$$
$$\quad - \{x(x-h)(x-2h)...(x-nh+2h)(x-nh+h)\}$$
$$= x(x-h)(x-2h)...(x-nh+2h)\{x+h-x+nh-h\}$$
$$= x(x-h)(x-2h)...(x-nh+2h)(nh)$$
$$= x(x-h)(x-2h)...(x-(n-2)h)(nh)$$
$$= nh[x]^{n-1}$$

$$\Delta^2[x]^n = \Delta[\Delta(x)^n] = \Delta[nh(x)^{n-1}]$$
$$= nh\,\Delta(x)^{n-1}$$
$$= n(n-1)h^2[x]^{n-2}$$

Similarly $\Delta^n[x]^n = n(n-1)...[n-(n-1)][x]^{n-n}h^n$
$$= n(n-1)(n-2)...1 \cdot h^n = n!h^n$$

Note: The result of differencing $[x]^n$ is analogous to that of the differentiation of x^n when $h=1$.

Theorem: Every polynomial of degree n can be expressed as polynomial of the same degree and vice-versa.

Proof: Let $f(x) = a_0 + a_1x + a_2x^2 + ... + a_nx^n$ be a polynomial of nth degree to be expressed in factorial polynomial.

Let $$f(x) = A_0 + A_1[x] + A_2[x]^2 + A_3[x]^3 + ... + A_n[x]^n$$

where $[x], [x]^2, [x]^3$ ...are factorials and $A_0, A_1, A_2, ..., A_n$ are constants to be determined.

$$\Delta f(x) = A_1 + 2A_2[x] + 3A_3[x]^2 + \dots + nA_n[x]^{n-1}$$
$$\Delta^2 f(x) = 2A_2 + 6A_3[x] + \dots + n(n-1) A_n[x]^{n-2}$$
$$\Delta^3 f(x) = 6A_3 + \dots + n(n-1)(n-2) A_n[x]^{n-3}$$
$$\dots = \dots$$
$$\Delta^n f(x) = n(n-1)(n-2)\dots(n-(n-1))[x]^{n-n} A_n$$
$$= n(n-1)(n-2)\dots 2\cdot 1 A_n$$
$$= n! A_n$$

Putting $x = 0$ in the above relations, we have

$$f(0) = A_0 \qquad\qquad A_0 = f(0)$$
$$\Delta f(0) = A_1 \qquad\qquad A_1 = \Delta f(0)$$
$$\Delta^2 f(0) = 2A_2 \qquad\qquad A_2 = \frac{1}{2!}\Delta^2 f(0)$$
$$\Delta^3 f(0) = 6A_3 \qquad\qquad A_3 = \frac{1}{3!}\Delta^3 f(0)$$
$$\dots = \dots \qquad\qquad \dots = \dots$$
$$\Delta^n f(0) = n! A_n \qquad\qquad A_n = \frac{1}{n!}\Delta^n f(0)$$

$$f(x) = f(0) + \Delta f(0)[x] + \frac{1}{2!}\Delta^2 f(0)[x]^2 + \frac{1}{3!}\Delta^3 f(0)[x]^3 + \dots + \frac{1}{n!}\Delta^n f(0)[x]^n$$

Illustrative Examples

Example 3.3: Evaluate: a. $\Delta \log f(x)$, b. $\Delta \tan^{-1}x$, c. $\Delta^2 \sin(px + q)$, d. $\Delta^n e^{ax + b}$.

Solution: a. $\quad \Delta \log f(x) = \log f(x + h) - \log f(x) \quad [\Delta f(x) = f(x + h) - f(x)]$

$$= \log\left[\frac{f(x + h)}{f(x)}\right] = \log\left[\frac{f(x + \Delta f(x))}{f(x)}\right]$$

$$= \log\left[1 + \frac{\Delta f(x)}{f(x)}\right]$$

b. $\qquad\qquad \Delta \tan^{-1}x = \tan^{-1}(x + h) - \tan^{-1}x$

$$= \tan^{-1}\left[\frac{x + h - x}{1 + x(x + h)}\right] = \tan^{-1}\frac{h}{1 + x(x + h)}$$

c. $\Delta^2 \sin(px + q)$

Now $\qquad \Delta \sin(px + q) = \sin[p(x + h) + 1] - \sin[px + q]$

$$= 2\cos\left(px + q + \frac{ph}{2}\right)\sin\frac{ph}{2}$$

$$= 2\sin\frac{ph}{2}\sin\left[\frac{\pi}{2} + px + q + \frac{ph}{2}\right]$$

$$\Delta^2\sin(px+q) = 2\sin\frac{ph}{2}\Delta\left[\sin\left\{px+1+\frac{1}{2}(\pi+ph)\right\}\right]$$

$$= \left(2\sin\frac{ph}{2}\right)^2\left[\sin(px+1)+2\times\frac{1}{2}(\pi+ph)\right]$$

$$= \left(2\sin\frac{ph}{2}\right)^2\left[\sin(px+1+\pi+ph)\right]$$

d. $\qquad \Delta e^{ax+b} = e^{a(x+h)+b} - e^{ax+b} = e^{ax+b}[e^{ah}-1]$

$$\Delta^2 e^{ax+b} = \Delta\left[e^{ax+b}\left(e^{ah}-1\right)\right] = \left(e^{ah}-1\right)\Delta\left[e^{ax+b}\right]$$

$$= (e^{ah}-1)^2\, e^{ax+b}$$

$$\Delta^n e^{ax+b} = (e^{ah}-1)^n\, e^{ax+b}$$

Example 3.4: Evaluate

a. $\Delta^n\sin(ax+h)$

b. $\Delta\left(\dfrac{2^x}{x!}\right)$ (taking interval of difference as unity)

c. $\Delta^2\left(\dfrac{1}{x^2+5x+6}\right)$ (taking interval of difference as unity)

Solution: a. $\quad \Delta^n\sin(ax+b) = \sin[a(x+h)+b] - \sin(ax+b)$

$$= \sin(ax+b+ah) - \sin(ax+b)$$

$$= 2\cos\left(ax+b+\frac{ah}{2}\right)\sin\left(\frac{ah}{2}\right)$$

$$= 2\sin\frac{ah}{2}\sin\left(\frac{\pi}{2}+(ax+b)+\frac{ah}{2}\right)$$

$$= 2\sin\frac{ah}{2}\sin\left[ax+b+\frac{1}{2}(\pi+ah)\right]$$

$$\Delta^2\sin(ax+b) = 2\sin\frac{ah}{2}\Delta\sin\left[ax+b+\frac{1}{2}(\pi+ah)\right]$$

$$= \left(2\sin\frac{ah}{2}\right)^2\sin\left[ax+b+2\times\frac{1}{2}(\pi+ah)\right]$$

Similarly, $\Delta^3\sin(ax+b) = \left(2\sin\frac{ah}{2}\right)^3\sin\left[ax+b+3\times\frac{1}{2}(\pi+ah)\right]$

$$... = ...$$

$$\Delta^n\sin(ax+b) = \left(2\sin\frac{ah}{2}\right)^n\sin\left[ax+b+\frac{n}{2}(\pi+ah)\right]$$

b.
$$\Delta\left(\frac{2^x}{x!}\right) = \frac{2^{x+1}}{x+1!} - \frac{2^n}{x!}$$

$$= \frac{2^x}{x!}\left[\frac{2}{x+1} - 1\right] = \frac{2^x}{x!}\left[\frac{2-x-1}{x+1}\right] = \frac{2^x}{(x+1)!}(1-x)$$

c.
$$\Delta\left(\frac{1}{x^2+5x+6}\right) = \Delta\left[\frac{1}{(x+3)(x+2)}\right]$$

$$= \Delta\left[\frac{1}{x+2} - \frac{1}{x+3}\right] = \left[\left(\frac{1}{x+3} - \frac{1}{x+2}\right) - \left(\frac{1}{x+4} - \frac{1}{x+3}\right)\right]$$

$$= \left[\frac{-1}{(x+3)(x+2)} - \frac{-1}{(x+3)(x+4)}\right] = \frac{1}{x+3}\left[\frac{1}{x+4} - \frac{1}{x+2}\right]$$

$$= \frac{1}{(x+3)(x+4)} - \frac{1}{(x+2)(x+3)}$$

$$\Delta^2\left(\frac{1}{x^2+5x+6}\right) = \left[\frac{1}{(x+4)(x+5)} - \frac{1}{(x+3)(x+4)}\right] - \left[\frac{1}{(x+3)(x+4)} - \frac{1}{(x+2)(x+3)}\right]$$

$$= \frac{1}{x+4}\left[\frac{1}{x+5} - \frac{1}{x+3}\right] - \frac{1}{x+3}\left[\frac{1}{x+4} - \frac{1}{x+2}\right]$$

$$= \frac{-2}{(x+3(x+4)(x+5)} - \frac{-2}{(x+2(x+3)(x+4)}$$

$$= \frac{2}{(x+3)(x+4)}\left[\frac{1}{x+2} - \frac{1}{x+5}\right]$$

$$= \frac{2}{(x+3)(x+4)} \times \frac{3}{(x+2)(x+6)} = \frac{6}{(x+2)(x+3)(x+4)(x+5)}$$

Example 3.5: Prove that $\Delta^4 y = 136080 (3x + 13)(3x + 16)(3x + 19)(3x + 22)$ given $y = (3x + 1)(3x + 4) \ldots (3x + 22)$

Solution:
$$y = 3^8\left(x+\frac{1}{3}\right)\left(x+\frac{4}{3}\right)\left(x+\frac{7}{3}\right)\ldots\left(x+\frac{22}{3}\right) = 3^8\left[x+\frac{22}{3}\right]^8$$

$$\Delta y = 3^8 \cdot 8\left[x+\frac{22}{3}\right]^7$$

$$\Delta^2 y = 3^8 \cdot 8.7\left[x+\frac{22}{3}\right]^6$$

$$\Delta^3 y = 3^8 \cdot 8 \cdot 7.6\left[x+\frac{22}{3}\right]^5$$

$$\Delta^4 y = 3^8 \cdot 8.7 \cdot 6.5 \left[x + \frac{22}{3} \right]^4$$

$$= 11022480 \left(x + \frac{22}{3} \right) \left(x + \frac{22}{3} - 1 \right) \left(x + \frac{22}{3} - 2 \right) \left(x + \frac{22}{3} - 3 \right)$$

$$= 136080 \, (3x + 22) \, (3x + 19) \, (3x + 16) \, (3x + 13)$$

Example 3.6: Prove that

i. $y_3 = y_2 + \Delta y_1 + \Delta^2 y_0 + \Delta^3 y_0$

ii. $\nabla^2 y_8 = y_8 - 2y_7 + y_6$

iii. $\delta^2 y_5 = y_6 - 2y_5 + y_4$

Solution: Method I:

i. $\qquad y_3 - y_2 = \Delta y_2$

$$y_3 = y_2 + \Delta y_2 = y_2 + \Delta[y_1 + \Delta y_1] \qquad \begin{cases} \Delta y_1 = y_2 - y_1 \\ y_2 = y_1 + \Delta y_1 \\ \Delta y_0 = y_1 - y_0 \\ y_1 = y_0 + \Delta y_0 \end{cases}$$

$$= y_2 + \Delta y_1 + \Delta^2 y_1$$

$$= y_2 + \Delta y_1 + \Delta^2 [y_0 + \Delta y_0]$$

$$= y_2 + \Delta y_1 + \Delta^2 y_0 + \Delta^3 y_0$$

Method II: $\Delta^3 y_0 = \Delta^2 y_1 - \Delta^2 y_0$

$$\Delta^3 y_0 + \Delta^2 y_0 = \Delta^2 y_1$$

$$= (\Delta y_2 - \Delta y_1)$$

$$\Delta^3 y_0 + \Delta^2 y_0 + \Delta y_1 = \Delta y_2 = y_3 - y_2$$

$$y_3 = y_2 + \Delta y_1 + \Delta^2 y_0 + \Delta^3 y_0$$

ii. $\qquad \nabla^2 y_8 = \nabla y_8 - \nabla y_7$

$$= (y_8 - y_7) - (y_7 - y_6) = y_8 - 2y_7 + y_6$$

iii. $\qquad$ RHS $= y_6 - 2y_5 + y_4$

$$= (y_6 - y_5) - (y_5 - y_4)$$

$$= \delta y_{\frac{11}{2}} - \delta y_{\frac{9}{2}} \text{ [using the result of central differences, } see \text{ Section 3.4]}$$

$$= \delta^2 y_5 = \text{LHS (proved)}$$

Example 3.7: Construct a forward difference table for the following

x	0	1	2	3	4
$f(x)$	1	1.5	2.2	3.1	4.6

Hence evaluate (i) $\Delta^3 y_1$, (ii) y_x and (iii) y_5.

Solution: Difference table is

x	$f(x)$	Δy	$\Delta^2 y$	$\Delta^3 y$	$\Delta^4 y$
$x_0 = 0$	$y_0 = 1$				
$x_0 = 1$	$y_1 = 1.5$	$0.5 = \Delta y_0$			
$x_2 = 2$	$y_2 = 2.2$	$0.7 = \Delta y_1$	$0.2 = \Delta^2 y_0$		
			$0.2 = \Delta^2 y_1$	$0 = \Delta^3 y_0$	
$x_3 = 3$	$y_3 = 3.1$	$0.9 = \Delta y_2$		$0.4 = \Delta^3 y_1$	$0.4 = \Delta^4 y_0$
			$0.6 = \Delta^2 y_2$		
$x_4 = 4$	$y_4 = 4.6$	$1.5 = \Delta y_3$			

Now (i)

$$\Delta^3 y_0 = {}^3C_0 y_3 - {}^3C_1 y_2 + {}^3C_2 y_1 - {}^3C_3 y_0$$

$$\Delta^3 y_1 = {}^3C_0 y_4 - {}^3C_1 y_3 + {}^3C_2 y_2 - {}^3C_3 y_1$$

$$= 4.6 - 3(3.1) + 3(2.2) - (1.5) = 0.4$$

(ii) It is known that

$$y_n = {}^nC_0 y_0 + {}^nC_1 \Delta y_0 + {}^nC_2 \Delta^2 y_0 + {}^nC_3 \Delta^3 y_0 + \dots + {}^nC_n \Delta^n y_0$$

$$y_x = {}^xC_0 y_0 + {}^xC_1 \Delta y_0 + {}^xC_2 \Delta^2 y_0 + {}^xC_3 \Delta^3 y_0 + \dots + {}^xC_n \Delta^n y_0$$

$$= y_0 + x\Delta y_0 + \frac{x(x-1)}{2!}\Delta^2 y_0 + \frac{x(x-1)(x-2)}{3!}\Delta^2 y_0$$

$$+ \frac{x(x-1)(x-2)(x-3)}{4!}\Delta^4 y_0$$

$$= 1 + x(0.5) + \frac{x(x-1)}{2!}(0.2) + \frac{x(x-1)(x-2)}{3!}(0) + \frac{x(x-1)(x-2)(x-3)}{4!}(0.4)$$

$$y_x = 1 + \frac{1}{2}x + \frac{1}{10}(x^2 - x) + \frac{1}{60}(x^4 - 6x^3 + 11x^2 - 6x)$$

$$= \frac{1}{60}[x^4 - 6x^3 + 17x^2 + 18x + 60]$$

(iii) $\quad y_5 = \dfrac{1}{60}[(5)^4 - 6(5)^3 + 17(5)^2 + 18(5) + 60] = 7.5$

Example 3.8: The values of $f(x)$ at $x = 0, 1, 2, 3, 4$ are 9, 10, 37, 54, 49. Find the polynomial $f(x)$. Also prove that $\Delta^4 f(x) = 24$.

Solution: The difference table is

x	$f(x)$	$\Delta f(x)$	$\Delta^2 f(x)$	$\Delta^3 f(x)$	$\Delta^4 f(x)$
0	9				
1	10	$1 = \Delta f(0)$	$26 = \Delta^2 f(0)$	$-36 = \Delta^3 f(0)$	
2	37	27	-10	-12	$24 = \Delta^4 f(0)$
3	54	17	-22		
4	49	-5			

The polynomial in factorial notation is

$$f(x) = f(0) + \frac{\Delta f(0)}{1!}[x] + \frac{\Delta^2 f(0)}{2!}[x]^2 + \frac{\Delta^3 f(0)}{3!}[x]^3 + \frac{\Delta^4 f(0)}{4!}[x]^4$$

$$= 9 + \frac{1}{1!}[x] + \frac{26}{2!}[x]^2 - \frac{36}{3!}[x]^3 + \frac{24}{4!}[x]^4$$

$$= 9 + [x] + 13[x]^2 - 6[x]^3 + [x]^4$$

$$\Delta f(x) = 1 + 26[x] - 18[x]^2 + 4[x]^3$$

$$\Delta^2 f(x) = 26 - 36[x] + 12[x]^2$$

$$\Delta^3 f(x) = -36 + 24[x]$$

$$\Delta^4 f(x) = 24$$

$f(x)$ can also be expressed as a polynomial

$$f(x) = 9 + x + 13x(x-1) - 6x(x-1)(x-2) + x(x-1)(x-2)(x-3)$$
$$= 9 + x + 13x^2 - 13x - 6(x^3 - 3x^2 + 2x) + x^4 - 6x^3 + 11x^2 - 6x$$
$$= 9 + x + 13x^2 - 13x - 6x^3 + 18x^2 - 12x + x^4 - 6x^3 + 11x^2 - 6x$$
$$= 9 - 30x + 42x^2 - 12x^3 + x^4$$

Example 3.9: Write down the polynomial of lowest degree which satisfy the following set of numbers 0, 7, 26, 63, 124, 215, 342, 511.

Solution: Difference table is

x	$f(x)$	$\Delta f(x)$	$\Delta^2 f(x)$	$\Delta^3 f(x)$	$\Delta^4 f(x)$
0	0				
		$7 = \Delta f(0)$			
1	7		$12 = \Delta^2 f(0)$		
		19		$6 = \Delta^3 f(0)$	
2	26		18		$0 = \Delta^4 f(0)$
		37		6	
3	63		24		0
		61		6	
4	124		30		0
		91		6	
5	215		36		0
		127		6	
6	342		42		
		169			
7	511				

$$f(x) = f(0) + \frac{\Delta f(0)}{1!}[x] + \frac{\Delta^2 f(0)}{2!}[x]^2 + \frac{\Delta^3 f(0)}{3!}[x]^3$$

$$= 0 + 7\{x\} + \frac{12}{2!}[x]^2 + \frac{6}{3!}[x]^3$$

$$= 7x + 6(x)(x-1) + (x)(x-1)(x-2) = x^3 + 3x^2 + 3x$$

Example 3.10: Find the first term of the series whose second and subsequent term are 8, 3, 0, −1, 0.

Solution: $f(2) = 8, f(3) = 3, f(4) = 0, f(5) = -1, f(6) = 0.$

Let the first term be y

Difference table is

x	$f(x)$	$\Delta f(x)$	$\Delta^2 f(x)$	$\Delta^3 f(x)$
1	y			
		$8 - y$		
2	8		$-13 + y$	
		-5		$15 - y$
3	3		2	
		-3		0
4	0		2	
		-1		0
5	-1		2	
		1		
6	0			

Since only three significant entries are given, the function can be represented by a second degree polynomial and hence

$$\Delta^3 f(1) = 0$$
$$15 - y = 0, \qquad y = 15$$

First term is 15.

Example 3.11: Express $f(x) = x^4 - 12x^3 + 42x^2 - 30x + 9$ in factorial notation where interval of difference is unity.

Solution: Let
$$y = f(x) = [x]^4 + A[x]^3 + B[x]^2 + C[x] + D$$

By synthetic division.

Dividing the given polynomial by $x, x-1, x-2, x-3, x-4$

0	1	-12	42	-30	9
	0	0	0	0	
1	1	-12	42	-30	9 = D
		1	-11	31	
2	1	-11	31	1 = C	
		2	-18		
3	1	-9	13 = B		
		3			
4	1	-6 = A			
	0				
	1				

$$f(x) = [x]^4 - 6[x]^3 + 13[x]^2 + [x] + 9$$

Example 3.12: Form a table of difference for function $f(x) = x^3 + 5x - 7$ for $x = -1, 0, 1, 2, 3, 4, 5$ and 6, hence find $f(6)$.

Solution: The difference table is

x	$f(x)$	$\Delta f(x)$	$\Delta^2 f(x)$	$\Delta^3 f(x)$
-1	-13			
		6		
0	-7		0	
		6		6
1	-1		6	
		12		6
2	11		12	
		24		6
3	35		18	
		42		6
4	77		24	
		66		$f(6) - 233$
5	143		$f(6) - 209$	
		$f(6) - 143$		
6	$f(6)$			

Since the equation is of third degree, $\Delta^3 f(x) = $ constant
$$f(6) - 233 = 6$$
$$f(6) = 239$$

Example 3.13: Find $f(6)$ given $f(0) = -3$
$$f(1) = 6$$
$$f(2) = 8$$
$$f(3) = 12 \quad \text{difference being constant.}$$

Solution: [**Hint:** $f(a + nh) = f(a) + {}^nC_1 \Delta f(a) + {}^nC_2 \Delta^2 f(a) + {}^nC_3 \Delta^3 f(a) + ...]$
Here $a = 0, h = 1, n = 6$.

The difference table is

x	$f(x)$	$\Delta f(x)$	$\Delta^2 f(x)$	$\Delta^3 f(x)$
0	$-3 = f(0)$	$9 = -\Delta f(0)$		
1	6		$-7 = \Delta^2 f(0)$	
		2		$9 = \Delta^3 f(0)$
2	8		2	
		4		
3	12			

Now
$$f(x) = f(0) + n\Delta f(0) + \frac{n(n-1)}{2!}\Delta^2 f(0) + \frac{n(n-1)(n-2)}{3!}\Delta^3 f(0)$$
$$f(6) = -3 + 6 \times 9 + 15(-7) + 20 \times 9 = 126$$

Example 3.14: Estimate the missing terms in the following table.

Solution:

x	$f(x)$	$\Delta f(x)$	$\Delta^2 f(x)$	$\Delta^3 f(x)$	$\Delta^4 f(x)$
0	$y_0 = 1$				
		2			
1	$y_1 = 3$		4		
		6		$y_3 - 19$	
2	$y_2 = 9$		$y_3 - 15$		$-4y_3 + 124$
		$y_3 - 9$		$105 - 3y_3$	
3	y_3		$90 - 2y_3$		
		$81 - y_3$			
4	81				

Since four entries are given, the function will be a polynomial of degree 3

∴
$$\Delta^4 y = 0$$
$$-4y_3 + 124 = 0 \qquad y_3 = 31$$

Example 3.15: Obtain the function whose first difference is $2x^3 + 3x^2 - 5x + 4$.

Solution: Now the first difference is a polynomial of degree 3.

Let us take $\qquad \Delta f(x) = 2[x]^3 + A[x]^2 + B[x] + C$

where A, B, C are calculated as below.

0	2	3	-5	4
	0	0	0	
1	2	3	-5	$4 = C$
	2	2	5	
2	2	5	$0 = B$	
	4			
3	2	$9 = A$		
	2			

$$\Delta f(x) = 2[x]^3 + 9[x]^2 + 4$$

Integrating the above function

$$= \frac{1}{2}x(x-1)(x-2)(x-3) + 3x(x-1)(x-2) + 4x + C$$

Example 3.16: Assuming that the following values of y belong to a polynomial of degree '4', compute the next three values.

$x = 0$	1	2	3	4	5	6	7
$y = 1$	-1	1	-1	1	–	–	–

Solution: Difference table is

x	y	Δy	$\Delta^2 y$	$\Delta^3 y$	$\Delta^4 y$
0	1				
		$-2 = y_0$			
1	-1		$4 = \Delta^2 y_0$		
		$2 = \Delta y_1$		$-8 = \Delta^3 y_0$	
2	1		$-4 = \Delta^2 y_1$		$16 = \Delta^4 y_0$
		$-2 = \Delta y_2$		$8 = \Delta^3 y_1$	
3	-1		$4 = \Delta^2 y_2$		$y_5 - 15 = \Delta^4 y_1 = 16$
		$2 = \Delta y_3$		$y_5 - 7 = \Delta^3 y_2$	
4	1		$y_5 - 3 = \Delta^2 y_3$		$y_6 - 4y_5 + 11 = \Delta^4 y_2 = 16$
		$y_5 - 1 = \Delta y_4$		$y_6 - 3y_5 + 4 = \Delta^3 y_3$	
5	y_5		$y_6 - 2y_5 + 1 = \Delta^2 y_4$		$y_7 - 4y_6 + 6y_5 - 5 = \Delta^4 y_3 = 16$
		$y_6 - y_5 = \Delta y_5$		$y_7 - 3y_6 + 3y_5 - 1 = \Delta^3 y_4$	
6	y_6		$y_7 - 2y_6 + y_5 = \Delta^2 y_5$		
		$y_7 - y_6 = \Delta y_6$			
7	y_7				

Since the polynomial is of '4' degree: $\Delta^4 y$ = constant.

So all the terms in the column below $\Delta^4 y$ will be the same, i.e. 16

$$\Delta^4 y_1 = 16 = \Delta^3 y_2 - \Delta^3 y_1$$
$$\Delta^3 y_2 = 16 + \Delta^3 y_1 = 16 + 8 = 24$$
$$\Delta^2 y_3 = \Delta^3 y_2 + 4 = 24 + 4 = 28$$
$$\Delta y_4 = \Delta^2 y_3 + 2 = 28 + 2 = 30$$
$$y_5 = \Delta y_4 + 1 = 30 + 1 = 31$$
$$... = ...$$
$$\Delta^3 y_3 = \Delta^3 y_2 + 16 = 24 + 16 = 40$$
$$\Delta^2 y_4 = \Delta^2 y_3 + 40 = 28 + 40 = 68$$
$$\Delta y_5 = \Delta^2 y_4 + \Delta y_4 = 68 + 30 = 98$$
$$y_6 = \Delta y_5 + y_5 = 98 + 31 = 129$$
$$... = ...$$
$$\Delta^3 y_4 = \Delta^3 y_3 + 16 = 40 + 16 = 56$$
$$\Delta^2 y_5 = \Delta^2 y_4 + \Delta^3 y_4 = 68 + 56 = 124$$
$$\Delta y_6 = \Delta^2 y_5 + \Delta y_5 = 124 + 98 = 222$$
$$y_7 = \Delta y_6 + y_6 = 222 + 129 = 351$$

Hence $\qquad y_5 = 31, y_6 = 129, y_7 = 351$

Example 3.17: Explain the following table to two more terms on either side by constructing the following table:

$x = -0.2$	0.0	0.2	0.4	0.6	0.8	1.0
$y = 2.6$	3.0	3.4	4.28	7.08	14.2	29.0

Solution: Difference table is

x	$y = f(x)$	Δy	$\Delta^2 y$	$\Delta^3 y$	$\Delta^4 y$
-0.6	y_{-3}				
		$y_{-2} - y_{-3}$			
-0.4	y_{-2}		$2.6 - 2y_{-2} + y_{-3}$		
		$2.6 - y_{-2}$		$-4.8 + 3y_{-2} - y_{-3}$	
-0.2	2.6		$-22 \times y_{-2}$		$7 - 4y_{-2} + y_{-3}$
		0.4		$2.2 - y_{-2}$	
0.0	3.0		0		$-1.72 + y_{-2}$
		0.4		0.48	
0.2	3.4		0.48		0.96
		0.88		1.44	
0.4	4.28		1.92		0.96
		2.80		2.40	
0.6	7.08		4.32		0.96
		7.12		3.36	
0.8	14.2		7.68		$y_6 - 54.84$
		14.8		$y_6 - 51.48$	
1.0	29.0		$y_6 - 43.8$		$y_7 - 4y_6 + 124.28$
		$y_6 - 29$		$y_7 - 3y_6 + 72.8$	
1.2	y_6		$y_7 - 2y_6 + 29$		
		$y_7 - y_6$			
1.4	y_7				

$$7 - 4y_{-2} + y_{-3} = 0.96$$
$$-1.72 + y_{-2} = 0.96$$
$$y_{-2} = 2.68$$
$$y_{-3} = 0.96 - 7 + 4y - 2$$
$$= 0.96 - 7 + 10.72 = 4.68$$
$$y_6 - 54.84 = 0.96$$
$$y_6 = 54.84 + 0.96 = 55.80$$
$$y_7 - 4y_6 + 124.28 = 0.96$$
$$y_7 = 0.96 - 124.28 + 4\,(55.80)$$
$$= 0.96 - 124.28 + 233.20 = 99.88$$

PROBLEMS 3.1

1. Find (a) $\Delta^{10}(1 - ax)(1 - bx^2)(1 - cx^3)(1 - dx^4)$ if $h = 1$.　　　　　[Ans. $abcd\ 10!$]
   (b) $\Delta^3(1 - x)(1 - 2x)(1 - 3x)$ if $h = 1$.　　　　　[Ans. -36]

2. Evaluate (a) $\Delta^2 \dfrac{5x + 12}{x^2 + 5x + 6}$.　　　$\left[\text{Ans. } \dfrac{-2}{(x+2)(x+3)} - \dfrac{3}{(x+3)(x+4)} \right]$

   (b) $\Delta^2 \cos 2x$　　　　　[Ans. $-4 \sin^2 h \cos^2(x + h)$]
   (c) $\Delta^n e^x$　　　　　[Ans. $(e - 1)^n e^x$]

3. Find the missing values in the following table

$x = 45$	50	55	60	65
$y = 3.0$	–	2.0	–	-2.4

   [Ans. $y_1 = 2.925,\ y_3 = 0.225$]

4. To prove that $f(4) = f(3) + \Delta f(2) + \Delta^2 f(1) + \Delta^3 f(1)$.

5. Find the function whose first difference is e^x.　　　$\left[\text{Ans. } f(x) = \dfrac{e^x}{e^k - 1} \right]$

6. Evaluate (a) $\Delta[f(x)\,g(x)]$. [Ans. $f(x+h)\,\Delta g(x) + g(x)\,\Delta f(x)$]

(b) $\Delta\left[\dfrac{f(x)}{g(x)}\right]$. $\left[\text{Ans. } \dfrac{g(x)\,\Delta f(x) - f(x)\,\Delta g(x)}{g(x+h)\,g(x)}\right]$

7. Assuming that the following values of y belong to a polynomial degree 4. Compute the next two values.

$x = 2$	4	6	8	10	12	14
$y = 2$	3	5	8	9	–	–

[Ans. 2, –22]

8. Given log 100 = 2, log 101 = 2.0043, log 103 = 2.0128, log 104 = 2.0170. Find log 102.

[Ans. 2.0086]

9. If $u_0 = 3$, $u_1 = 12$, $u_2 = 81$, $u_3 = 2000$, $u_4 = 100$. Calculate $\Delta^4 u_0$. [Ans. 7459]

10. Show that $\Delta^3 y_i = y_{i+3} - 3y_{i+2} + 3y_{i+1} - y_i$.

3.7 OTHER DIFFERENCE OPERATORS

We have already defined the operators Δ, ∇ and δ. Now we shall introduce the operators E and μ.

1. Shift operator E is such that $Ef(x) = f(x+h)$, $E^2 f(x) = f(x+2h)$ etc.
   Inverse shift operator E^{-1} is such that $E^{-1}f(x) = f(x-h)$, $E^{-2}f(x) = f(x-2h)$ etc.

2. Averaging operator μ is defined as $\mu y_x = \dfrac{1}{2}\left[y_{x+\frac{1}{2}h} + y_{x-\frac{1}{2}h}\right]$.

3.7.1 Relation between the Operators

Example 3.18:   i. $\Delta = E - 1$ ii. $\nabla = 1 - E^{-1}$ iii. $\delta = E^{1/2} - E^{-1/2}$
   iv. $\mu = 1/2\,[E^{1/2} + E^{-1/2}]$ v. $\Delta = E\nabla = \nabla E = \delta E^{1/2}$ vi. $E = e^{hD}$

Solution: i. $\Delta y_x = y_{x+h} - y_x = Ey_x - y_x = (E-1)y_x$

$\Delta = E - 1$

ii. $\nabla y_x = y_x - y_{x-h} - y_x - E^{-1}y_x = (1 - E^{-1})y_x$

$\nabla = 1 - E^{-1}$

iii. $\delta y_x = y_{x+h/2} - y_{x-h/2} = E^{1/2}y_x - E^{-1/2}y_x = (E^{1/2} - E^{-1/2})y_x$

$\delta = E^{1/2} - E^{-1/2}$

iv. $\mu y_x = \dfrac{1}{2}(y_{x+h/2} + y_{x-h/2}) = \dfrac{1}{2}\left[E^{\frac{1}{2}}y_x + E^{-\frac{1}{2}}y_x\right] = \dfrac{1}{2}\left(E^{\frac{1}{2}} + E^{-\frac{1}{2}}\right)y_x$

$\Rightarrow$ $\mu = \dfrac{1}{2}\left(E^{\frac{1}{2}} + E^{-\frac{1}{2}}\right)$

v. $E\nabla y_x = E(y_x - y_{x-h}) = Ey_x - Ey_{x-h} = y_{x+h} - y_x = \Delta y_x$

$E\nabla = \Delta$

$\delta E^{1/2}y_x = \delta y_{x+h/2} = y_{\left(x+\frac{h}{2}\right)+\frac{h}{2}} - y_{\left(x+\frac{h}{2}\right)-\frac{h}{2}}$

$$= y_{x+h} - y_x = \Delta y_x$$

$$\delta E^{1/2} = \Delta$$

hence

$$\nabla = E\nabla = \nabla E = \delta E^{1/2}$$

vi.

$$Ef(x) = f(x+h)$$

$$= f(x) + hf'(x) + \frac{h^2}{2!}f''(x) + ... \text{ [by Taylor's theorem]}$$

$$= \left(1 + hD + \frac{h^2}{2!}D^2 + ...\right) f(x) = e^{hD} f(x)$$

$$E = e^{hD}$$

or

$$E = 1 + \Delta = e^{hD}$$

Example 3.19: Prove that

i. $\nabla \Delta = \Delta - \nabla = \Delta \nabla = \delta^2$

ii. $\Delta = \mu\delta + \dfrac{1}{2}\delta^2$

iii. $\Delta = \dfrac{1}{2}\delta^2 + \delta\sqrt{1 + \dfrac{\delta^2}{4}}$

iv. $hD = \log(1 + \Delta) = -\log(1 - \nabla) = \sinh^{-1}(\mu\delta)$

v. $1 + \mu^2\delta^2 = \left(1 + \dfrac{1}{2}\delta^2\right)^2$

vi. $\nabla^3 y_2 = \nabla^3 y_5$

Solution: i.

$$\nabla\Delta = (1 - E^{-1})(E - 1)$$

$$= E + E^{-1} - 2 = (E^{1/2} - E^{-1/2})^2 = \delta^2$$

Similarly

$$\nabla\Delta = \delta^2$$

$$\nabla - \Delta = (E - 1) - (1 - E^{-1}) = E + E^{-1} - 2 = \delta^2$$

ii.

$$\Delta = \mu\delta + \frac{1}{2}\delta^2 = \frac{1}{2}\left[E^{\frac{1}{2}} + E^{-\frac{1}{2}}\right]\left[E^{\frac{1}{2}} - E^{-\frac{1}{2}}\right] + \frac{1}{2}\left[E^{\frac{1}{2}} - E^{-\frac{1}{2}}\right]^2$$

$$= \frac{1}{2}[E - E^{-1}] + \frac{1}{2}[E + E^{-1} - 2]$$

$$= \frac{1}{2}[E + E - E^{-1} + E^{-1} - 2] = \frac{1}{2}[2E - 2] = E - 1 = \Delta$$

iii.

$$\Delta = \frac{1}{2}\delta^2 + \delta\sqrt{1 + \frac{\delta^2}{4}}$$

or

$$= \frac{1}{2}\left[\left(E^{\frac{1}{2}} - E^{-\frac{1}{2}}\right)\right]^2 + \left(E^{\frac{1}{2}} - E^{-\frac{1}{2}}\right)\sqrt{1 + \frac{\left(E^{\frac{1}{2}} - E^{-\frac{1}{2}}\right)^2}{4}}$$

$$= \frac{1}{2}[E + E^{-1} - 2] + \left(E^{\frac{1}{2}} - E^{-\frac{1}{2}}\right)\sqrt{1 + \left(\frac{E + E^{-1} - 2}{4}\right)}$$

$$= \frac{1}{2}\left[E + E^{-1} - 2\right] + \left(E^{\frac{1}{2}} - E^{-\frac{1}{2}}\right)\sqrt{\left(\frac{E + E^{-1} - 2}{4}\right)}$$

$$= \frac{1}{2}\left[E + E^{-1} - 2\right] + \left[\frac{E^{\frac{1}{2}} - E^{-\frac{1}{2}}}{2}\right]\sqrt{\left(E^{\frac{1}{2}} + E^{-\frac{1}{2}}\right)}$$

$$= \frac{1}{2}\left[E + E^{-1} - 2 + E - E^{-1}\right] = \frac{1}{2}\left[2E - 2\right] = E - 1 = \Delta$$

iv. $\qquad hD = \log(1 + \Delta) = -\log(1 - \nabla) = \sin h^{-1}(\mu\delta)$

We know that $\qquad e^{hD} = E = 1 + \Delta$

$$hD = \log(1 + \Delta)$$

$$e^{-hD} = 1 - \nabla$$

$$-hD = \log(1 - \nabla)$$

or $\qquad hD = -[\log(1 - \nabla)]$

$$\mu\delta = \frac{1}{2}\left(E^{\frac{1}{2}} + E^{-\frac{1}{2}}\right)\left(E^{\frac{1}{2}} - E^{-\frac{1}{2}}\right)$$

$$= \frac{1}{2}\left(E - E^{-1}\right) = \frac{1}{2}\left(e^{hD} - e^{-hD}\right)$$

$$= \sin h(hD)$$

$$hD = \sin h^{-1}(\mu\delta)$$

v. Prove that $\qquad 1 + \mu\delta^2 = \left(1 + \frac{1}{2}\delta^2\right)^2$

$$1 + \mu^2\delta^2 = 1 + \left(\frac{E^{\frac{1}{2}} + E^{-\frac{1}{2}}}{2}\right)^2 \left[E^{\frac{1}{2}} - E^{-\frac{1}{2}}\right]^2$$

$$= 1 + \frac{(E - E^{-1})^2}{4} = 1 + \frac{E^2 - E^{-2} - 2}{4}$$

$$= \frac{E^2 + E^{-2} + 2}{4} = \left[\frac{E + E^{-1}}{2}\right]^2$$

$$\left[1 + \frac{1}{2}\delta^2\right]^2 = \left[1 + \frac{1}{2}\cdot\frac{1}{2}(E - E^{-1})^2\right]^2 = \left[1 + \frac{1}{4}(E - E^{-1})^2\right]^2$$

$$= \left[\frac{4 + E^2 + E^{-2} - 2}{4}\right] = \left(\frac{E + E^{-1}}{2}\right)^2$$

$$1 + \mu^2\delta^2 = \left(1 + \frac{1}{2}\delta^2\right)^2 = \left[\frac{E + E^{-1}}{2}\right]^2$$

vi. Prove that
$$\Delta^3 y_2 = \nabla^3 y_5$$
$$\Delta^3 = (E-1)^3 = E^3 - 1 - 3E^2 + 3E$$
$$\Delta^3 y_2 = E^3 y_2 - y_2 - 3E^2 y_2 + 3E y_2$$
$$= y_5 - 3y_4 + 3y_3 - y_2$$
$$\nabla^3 y_5 = (1 - E^{-1})^3 y_5 = (1 - E^{-3} - 3E^{-1} + 3E^{-2}) y_5$$
$$= y_5 - y_2 - 3y_4 + 3y_3$$
$$\Delta^3 y_2 = \nabla^3 y_5$$

Example 3.20: Prove that

i. $\Delta \log f(x) = \log\left[1 + \dfrac{\Delta f(x)}{f(x)}\right]$

ii. $\Delta(\tan^{-1} x) = \tan^{-1}\left[\dfrac{h}{1 + x(x + h)}\right]$

iii. $(\nabla + \Delta)^2 (x^2 + x) = 8h^2$

iv. $\left(\dfrac{\Delta^2}{E}\right) x^3 = 6x^2 h$

v. $\left(\dfrac{\Delta^2}{E}\right) e^x \cdot \dfrac{E(e^x)}{\Delta^2(e^x)} = e^x$

Solution: i. $\Delta \log f(x) = \log f(x + h) - \log f(x) = \log \dfrac{f(x + h)}{f(x)}$

$$= \log\left[\dfrac{f(x) + f(x + h) - f(x)}{f(x)}\right] = \log\left[1 + \dfrac{\Delta f(x)}{f(x)}\right]$$

ii. $\Delta[\tan^{-1} x] = \tan^{-1}(x + h) - \tan^{-1} x$

$$= \tan^{-1} \dfrac{(x + h) - x}{1 + (x + h)x} = \tan^{-1}\left[\dfrac{h}{1 + x(x + h)}\right]$$

iii. $(\nabla + \Delta)^2 (x^2 + x) = [(E - 1) + (1 - E^{-1})]^2 (x^2 + x)$

$$= (E - E^{-1})^2 (x^2 + x)$$
$$= (E^2 + E^{-2} - 2)(x^2 + x) = E^2 x^2 + E^{-2} x^2 + E^2 x + E^{-2} x - 2x - 2x^2$$
$$= (x + 2h)^2 + (x - 2h)^2 - 2x^2 + (x + 2h) + (x - 2h) - 2x$$
$$= 2[x^2 + 4h^2] - 2x^2 = 8h^2$$

iv. $\left(\dfrac{\Delta^2}{E}\right) x^3 = (\Delta^2 E^{-1}) x^3 = \Delta^2 (x - h)^3$

$$= \Delta[x^3 - (x - h)^3] = \Delta x^3 - \Delta(x - h)^3$$
$$= [(x + h)^3 - x^3] - [x^3 - (x - h)^3]$$

$$= [(x^3 + h^3 + 3x^2h + 3xh^2) - x^3] - [x^3 - (x^3 - h^3 - 3x^2h + 3xh^2)]$$
$$= (h^3 + 3x^2h + 3xh^2) - (h^3 + 3x^2h - 3xh^2)$$
$$= h^3 + 3x^2h + 3xh^2 - h^3 - 3x^2h + 3xh^2 = 6xh^2$$

v. $\left(\dfrac{\Delta^2}{E}\right) e^x \cdot \dfrac{E(e^x)}{\Delta^2(e^x)} = (\Delta^2 E^{-1}) e^x \cdot \dfrac{E(e^x)}{\Delta^2(e^x)}$

$$= \Delta^2 e^{x-h} \cdot \dfrac{e^{x+h}}{(e^{2h} - 2e^h + 1)e^x} = \Delta^2 e^{x-h} \dfrac{e^h}{(e^h - 1)^2}$$

$$= e^x [\Delta^2 e^{-h}] \cdot \dfrac{e^h}{(e^h - 1)^2} = e^x [\Delta(e^0 - e^{-h})] \dfrac{e^h}{(e^h - 1)^2}$$

$$= e^x [e^h - e^0) - (e^0 - e^{-h})] \dfrac{e^h}{(e^h - 1)^2}$$

$$= e^x \left[e^h - 2 + \dfrac{1}{e^h} \right] \dfrac{e^h}{(e^h - 1)^2} = e^x \dfrac{(e^h - 1)^2}{e^h} \dfrac{e^h}{(e^h - 1)^2} = e^x$$

Example 3.21: Evaluate: i. $\Delta x (e^{x+b})$ ii. $\Delta^n \cos(ax + b)$ iii. $\dfrac{\Delta^2}{E} \sin(x + h) + \dfrac{\Delta^2 \delta(x + h)}{E\delta(x + b)}$.

Solution: i. $\Delta(e^{ax+b}) = e^{a(x+h)+b} - e^{ax+b} = e^{ax+b} [e^{ah} - 1]$

$\Delta^2 (e^{ax+b}) = \Delta[e^{ax+b}(e^{ah} - 1)] = (e^{ah} - 1)(e^{ah} - 1)e^{ax+b}$

$\qquad\qquad = (e^{ah} - 1)^2 e^{ax+b}$

Similarly $\Delta^3 (e^{ax+b}) = (e^{ah} - 1)^3 e^{ax+b}$

$\qquad\qquad ... = ...$

$\qquad \Delta^n (e^{ax+b}) = (e^{ah} - 1)^n e^{ax+b}$

ii. $\Delta^n \cos(ax + b) = \cos[a(x + h) + b] - \cos(ax + b)$

$\qquad\qquad = \cos[(ax + b) + ah] - \cos[ax + b]$

$\qquad\qquad = -2\sin\left(ax + b + \dfrac{ah}{2} \right) \sin\dfrac{ah}{2}$

$\qquad\qquad = 2\sin\dfrac{ah}{2}\left[\cos\left(\dfrac{\pi}{2} + ax + b + \dfrac{ah}{2} \right) \right]$

$\qquad\qquad = 2\sin\dfrac{ah}{2}\left[\cos\left(ax + b + \dfrac{\pi + ah}{2} \right) \right]$

$\Delta^2 \cos(ax + b) = \Delta[\Delta \cos(ax + b)]$

$\qquad\qquad = \Delta\left[2\sin\dfrac{ah}{2}\cos\left(ax + b + \dfrac{\pi + ah}{2} \right) \right]$

$\qquad\qquad = 2\sin\dfrac{ah}{2}\left[\Delta\left\{ \cos\left(ax + b + \dfrac{\pi + ah}{2} \right) \right\} \right]$

$$= \left(2\sin\frac{ah}{2}\right)\left(2\sin\frac{ah}{2}\right)\left[\cos\left\{(ax+b)+2\left(\frac{\pi+ah}{2}\right)\right\}\right]$$

$$= \left(2\sin\frac{ah}{2}\right)^2 \cos\left[(ax+b)+2\left(\frac{\pi+ah}{2}\right)\right]$$

Similarly $\Delta^3 \cos(ax+b) = \left(2\sin\frac{ah}{2}\right)^3 \cos\left[ax+b+3\left(\frac{\pi+ah}{2}\right)\right]$

$$... = ...$$
$$... = ...$$

$$\Delta^n \cos(ax+b) = \left(2\sin\frac{ah}{2}\right)^n \cos\left[ax+b+n\left(\frac{\pi+ah}{2}\right)\right]$$

iii. $\dfrac{\Delta^2}{E}\sin(x+h)+\dfrac{\Delta^2\sin(x+h)}{E\sin(x+h)}$

$$= \Delta^2 E^{-1}\sin(x+h)+\frac{\Delta^2\sin(x+h)}{\sin(x+2h)} = \Delta^2\sin x+\frac{\Delta^2\sin(x+h)}{\sin(x+2h)}$$

$$= \Delta[\sin(x+h)-\sin x]+\frac{\Delta[\sin(x+2h)-\sin(x+h)]}{\sin(x+2h)}$$

$$= \sin(x+2h)-2\sin(x+h)+\sin x+\frac{\sin(x+3h)-2\sin(x+2h)+\sin(x+h)}{\sin(x+2h)}$$

$$= \sin(x+2h)+\sin x-2\sin(x+h)+\frac{\sin(x+3h)+\sin(x+h)-2\sin(x+2h)}{\sin(x+2h)}$$

$$= 2\sin(x+h)\cos h-2\sin(x+h)+\frac{2\sin(x+2h)\cos h-2\sin(x+2h)}{\sin(x+2h)}$$

$$= 2(\cos h-1)\sin(x+h)+2\cos h-2$$

$$= 2(\cos h-1)[\sin(x+h)+1]$$

Example 3.22: Prove that:

i. $\nabla^2 = h^2\Delta^2 - h^3\Delta^3 + \dfrac{7}{12}h^4\Delta^4 - ...$

ii. $\displaystyle\sum_{k=0}^{n-1}\Delta^n f_k = \Delta f_n - \Delta f_0$

iii. $\nabla^r f_k \Delta^r f_{k-2}$

iv. $\Delta(fk)^2 = (f_k + f_{k+1})\Delta f_k$

Solution: a

$$\nabla^2 = (1-E^{-1})^2 = \left(1-\frac{1}{E}\right)^2 = \left(\frac{E-1}{E}\right)^2 = \left(\frac{e^{hD}-1}{e^{hD}}\right)^2$$

$$= \left(1-\frac{1}{e^{hD}}\right)^2 = (1-e^{-hD})^2$$

$$= \left[1 - \left(1 - hD + \frac{h^2D^2}{2!} - \frac{h^3D^3}{3!} + \frac{h^4D^4}{4!} - \ldots \right) \right]^2$$

$$= \left(hD - \frac{h^2D^2}{2!} + \frac{h^3D^3}{3!} - \frac{h^4D^4}{4!} - \ldots \right)^2$$

$$= h^2D^2 - \frac{2h^3D^3}{2!} + \frac{h^4D^4}{(2!)^2} + \frac{2h^4D^4}{3!} - \ldots$$

$$= h^2D^2 - h^3D^3 + \frac{3h^4D^4 + 4h^4D^4}{12} - \ldots$$

$$= h^2D^2 - h^3D^3 + \frac{7}{12}h^4D^4 \ldots$$

ii.
$$\sum_{k=0}^{x=1} \Delta^2 f_k = \Delta^2 f_0 + \Delta^2 f_1 + \Delta^2 f_2 + \ldots + \Delta^2 f_{n-1}$$

$$= (\Delta f_1 - \Delta f_0) + (\Delta f_2 - \Delta f_1) + (\Delta f_3 - \Delta f_2) + \ldots + (\Delta f_n - \Delta f_{n-1})$$

$$= \Delta f_n - \Delta f_0$$

iii.
$$\nabla^r f_k = (1 - E^{-1})^r f_k = \frac{(E-1)^r}{E^r} f_k = (E-1)^r E^{-r} f_k = (E-1)^r f_{k-r}$$

$$= (E-1)^r f_{k-r} = \Delta^r f_{k-r}$$

iv.
$$\Delta(f_k^2) = f_{k+h}^2 - f_k^2 = (f_{k+h} - f_k)(f_{k+h} + f_k)$$
$$= \Delta f_k (f_{k+h} + f_k) \quad [\text{put } h = 1]$$
$$= \Delta f_k (f_{k+1} + f_k)$$

Example 3.23: Prove that
$$u_0 + u_1 + u_2 + \ldots + u_n = {}^{n+1}C_1 u_0 + {}^{n+1}C_2 \Delta u_0 + {}^{n+1}C_3 \Delta^2 u_0 + \ldots + {}^{n+1}C_{n+1} \Delta^n u_0$$

Solution: $u_0 + u_1 + u_2 + \ldots + u_n = u_0 + E u_0 + E^2 u_0 + \ldots + E^n u_0$

$$= (1 + E + E^2 + \ldots + E^n) u_0$$

$$= \frac{(1 - E^{n+1})}{1 - E} u_0 = \frac{E^{n+1} - 1}{E - 1} u_0 = \left[\frac{(1 + \Delta)^{n+1}}{\Delta} - \frac{1}{\Delta} \right] u_0$$

$$= \frac{({}^{n+1}C_0 + {}^{n+1}C_1 \Delta + {}^{n+1}C_2 \Delta^2 + \ldots + {}^{n+1}C_{n+1} \Delta^{n+1} - 1) u_0}{\Delta}$$

$$= ({}^{n+1}C_1 + {}^{n+1}C_2 \Delta + \ldots + {}^{n+1}C_{n+1} \Delta^n) u_0$$

$$= {}^{n+1}C_1 u_0 + {}^{n+1}C_2 \Delta u_0 + {}^{n+1}C_3 \Delta^2 u_0 + \ldots + {}^{n+1}C_{n+1} \Delta^n u_0$$

Example 3.24: Prove that $\Delta^n u_x = u_{x+n} - {}^n C_1 u_{x+n-1} + {}^n C_2 u_{x+n-2} - \ldots (-1)^n u_x$.

Solution:
$$\Delta u_x = u_{x+1} - u_x = {}^1 C_0 u_{x+1} - {}^1 C_1 u_x$$
$$\Delta^2 u_x = u_{x+2} - 2u_{x+1} + u_x = {}^2 C_0 u_{x+2} - {}^2 C_1 u_{x+1} + {}^2 C_2 u_x$$

$$\Delta^3 u_x = u_{x+3} - 3u_{x+2} + 3u_{x+1} - u_x$$
$$= {}^3C_0 u_{x+3} - {}^3C_1 u_{x+2} + {}^3C_2 u_{x+1} - {}^3C_3 u_x$$
$$\ldots = \ldots$$
$$\ldots = \ldots$$
$$\Delta^n u_n = {}^n C_0 u_{x+n} - {}^n C_1 u_{x+n-1} + {}^n C_2 u_{x+n-2} - \ldots + {}^n C_n (-1)^n u_x$$
$$= u_{x+n} - {}^n C_1 u_{x+n-1} + {}^n C_2 u_{x+n-2} - \ldots + (-1)^n u_x$$

Example 3.25: Prove that $y_x = y_n - {}^{n-x}C_1 \Delta y_{n-1} + {}^{n-x}C_2 \Delta^2 y_{n-2} + \ldots + (-1)^{n-x} \Delta^{n-x} y_{n-(x-r)}.$

Solution:
$$\text{RHS} = y_n - {}^{n-x}C_1 \Delta y_{n-1} + {}^{n-x}C_2 \Delta^2 y_{n-2} + \ldots + (-1)^{n-x} \Delta^{n-x} y_{n-(x-r)}$$
$$= y_n - {}^{n-x}C_1 \Delta E^{-1} y_n + {}^{n-x}C_2 \Delta^2 E^{-2} y_n + \ldots + (-1)^{n-x} \Delta^{n-x} E^{n-x} y_n$$
$$= (1 - \Delta E^{-1})^{n-x} y_n$$
$$= \left(1 - \frac{\Delta}{E}\right)^{n-x} y_n = \left(\frac{E - \Delta}{E}\right)^{n-x} y_n = \left(\frac{1}{E}\right)^{n-x} y_n$$
$$= y_n E^{-(n-x)} = y_{n-(n-x)} = y_x \text{ LHS proved}$$

Example 3.26: Prove that $y_k = \displaystyle\sum_{i=0}^{k} {}^k C_i \Delta^i y_0.$

Solution:
$$y_k = E^k y_0 = (1 + \Delta)^k y_0 = y_0 + {}^k C_1 \Delta y_0 + {}^k C_2 \Delta^2 y_0 + \ldots + {}^k C_n \Delta^k y_0$$
$$= (1 + {}^k C_1 \Delta + {}^k C_2 \Delta^2 + \ldots + {}^k C_4 \Delta^k) y_0$$
$$= \sum_{i=0}^{k} {}^k C_i \Delta^i y_0$$

Example 3.27: $\Delta^n u_x = u_{x+n} - {}^n C_1 u_{x+n-1} + {}^n C_2 u_{x+n-2} + \ldots + (-1)^n u_x.$

Solution:
$$\Delta u_x = u_{x+1} - u_x$$
$$\Delta^2 u_x = u_{x+2} - 2u_{x+1} + u_x = {}^2C_0 u_{x+2} - {}^2C_1 u_{x+1} + {}^2C_2 u_x$$
$$\Delta^3 u_x = u_{x+3} - 3u_{x+2} + 3u_{x+1} - u_x$$
$$= {}^3C_0 u_{x+3} - {}^3C_1 u_{x+2} + {}^3C_2 u_{x+1} - {}^3C_3 u_3$$
$$\ldots = \ldots$$
$$\Delta^n u_x = {}^n C_0 u_{x+n} - {}^n C_1 u_{x+n-1} + {}^n C_2 u_{x+n-2} - \ldots + {}^n C_n (-1)^n u_{x+n-n}$$
$$= {}^n C_0 u_{x+n} - {}^n C_1 u_{x+n-1} + {}^n C_2 u_{x+n-2} - \ldots + (-1)^n \, {}^n C_n u_x$$

Example 3.28: Sum the series and prove that

$$u_0 + u_1 x + u_2 x^2 + \ldots = \frac{u_0}{1-x} + \frac{x \Delta u_0}{(1-x)^2} + \frac{x^2 \Delta^2 u_0}{(1-x)^3} + \ldots \infty.$$

Hence find the sum of the series $1.2 + 2.3x + 3.4x^2 + \ldots$

Solution:
$$u_0 + u_1 x + u_2 x^2 + \ldots = (1 + xE + x^2 E^2 + x^3 E^3 + \ldots) u_0$$
$$= \frac{u_0}{1 - xE} = \frac{1}{1 - x(1 + \Delta)} u_0 = \frac{1}{(1 - x) - x\Delta} u_0$$

$$= \frac{1}{1-x}\left(1 - \frac{x}{1-x}\Delta\right)^{-1} u_0$$

$$= \frac{1}{1-x}\left[1 + \frac{\Delta x}{1-x} + \frac{\Delta^2 x^2}{(1-x)^2} + \dots\right] u_0$$

Now let us construct the difference table for the coefficient of the given series.

u	Δu	$\Delta^2 u$	$\Delta^3 u$
$u_0 = 2$			
	4		
$u_1 = 6$		2	
	6		0
$u_2 = 12$		2	
	8		0
$u_3 = 20$		2	
	10		
$u_4 = 30$			

This show that $u_0 = 2$, $\Delta u_0 = 4$, $\Delta^2 u_0 = 2$, $\Delta^3 u_0 = \Delta^4 u_0$, etc. all $= 0$

Thus, $1.2 + 2.3x + 3.4x^2 + \dots \infty = \dfrac{u_0}{1-x} + \dfrac{x}{(1-x)^2}\Delta u_0 + \dfrac{x^2}{(1-x)^3}\Delta^2 u_0 + \dots \infty$

$$= \frac{2}{1-x} + \frac{4x}{(1-x)^2} + \frac{2x^2}{(1-x)^3} = \frac{2}{(1-x)^3}$$

3.8 APPLICATION TO SUMMATION OF SERIES

An important approach of finite differences for the summation of series.

Example 3.29: Sum the following series $1^3 + 2^3 + 3^3 + \dots + n^3$.

Solution: Denoting $1^3, 2^3, 3^3, \dots$ by $u_0, u_1, u_2 \dots$ respectively.

The sum of the series is

$$S = u_0 + u_1 + u_2 + \dots + u_{n-1}$$

$$= (1 + E + E^2 + \dots + E^{n-1})u_0$$

$$= \frac{E^n - 1}{E - 1}u_0 = \frac{(1+\Delta)^n - 1}{\Delta}u_0 = \frac{1 + n\Delta + \dfrac{n(n-1)}{2!}\Delta^2 + \dots + \Delta^{n-1}}{\Delta}u_0$$

$$= \left(n + \frac{n(n-1)}{2!}\Delta + \frac{n(n-1)(n-2)}{3!}\Delta^2 + \dots + \Delta^{n-1}\right)u_0$$

$$= nu_0 + \frac{n(n-1)}{2!}\Delta u_0 + \frac{n(n-1)(n-2)}{3!}\Delta^2 u_0 + \dots$$

Now $\qquad \Delta u_0 = u_1 - u_0 = 2^3 - 1^3 = 7$

$$\Delta^2 u_0 = u_2 - 2u_1 + u_0 = 3^3 - 2.2^3 + 1^3 = 12$$

$$\Delta^3 u_0 = u_3 - 3u_2 + 3u_1 - u_0 = 4^3 - 3.3^3 + 3.2^3 - 1^3 = 6$$

Now $\Delta^4 u_0, \Delta^5 u_0 \ldots$ are all zero as $u_r = r^3$ is a polynomial of third degree.

Hence
$$S = n + \frac{n(n-1)}{2} \cdot 7 + \frac{n(n-1)(n-2)}{6} \cdot 12 + \frac{n(n-1)(n-2)(n-3)}{24} \cdot 6$$

$$= \frac{n^2}{4}(n^2 + 2n + 1) = \left[\frac{n(n+1)}{2}\right]^2$$

Example 3.30: Prove that $u_0 + \frac{u_1 x}{1!} + \frac{u_2}{2!}x^2 + \ldots = e^x \left\{ u_0 + x\Delta u_0 + \frac{x^2}{2!}\Delta^2 u_0 \ldots \right\}$ and hence

sum the series $1 + \frac{4x}{1!} + \frac{10x^2}{2!} + \frac{20x^3}{3!} + \frac{35x^4}{4!} + \frac{56x^5}{5!} + \ldots$ up to infinity.

Solution: Now $u_0 + \frac{u_1 x}{1!} + \frac{u_2 x^2}{2!} + \ldots = u_0 + \frac{x}{1!}Eu_0 + \frac{x^2}{2!}E^2 u_0 + \ldots$

$$= \left(1 + \frac{xE}{1!} + \frac{x^2 E^2}{2!} + \ldots\right) u_0$$

$$= e^{xE}u_0 = e^{x(1+\Delta)}u_0 = e^x \cdot e^{x\Delta}u_0$$

$$= e^x \left[1 + x\Delta + \frac{x^2\Delta^2}{2!} + \frac{x^3\Delta^3}{3!} + \ldots\right] u_0$$

$$= e^x \left[u_0 + x\Delta u_0 + \frac{x^2}{2!}\Delta^2 u_0 + \ldots\right] \qquad (i)$$

To find sum of $1 + \frac{4x}{1!} + \frac{10x^2}{2!} + \frac{20x^3}{3!} + \frac{35x^4}{4!} + \frac{56x^5}{5!} + \ldots$, we proceed as below.

u	Δu	$\Delta^2 u$	$\Delta^3 u$	$\Delta^4 u$
$u_0 = 1$				
	3			
$u_1 = 4$		3		
	6		1	
$u_2 = 10$		4		0
	10		1	
$u_3 = 20$		5		0
	15		1	
$u_4 = 35$		6		
	21			
$u_5 = 56$				

We find that $u_0 = 1, \Delta u_0 = 3, \Delta^2 u_0 = 3, \Delta^3 u_0 = 1, \Delta^4 u_0 = \ldots = 0.$

By using Eq. (i)

$$1 + \frac{4x}{1!} + \frac{10x^2}{2!} + \frac{20x^3}{3!} + \frac{35x^4}{4!} + \frac{56x^5}{5!} = e^x \left[1 + 3x + \frac{3}{2}x^2 + \frac{1}{6}x^3\right] = \frac{e^x}{6}[x^3 + 9x^2 + 18x + 6]$$

PROBLEMS 3.2

1. Evaluate a. $\dfrac{\Delta}{E}\sin x$ b. $\delta = 2\sin h\left(\dfrac{h\Delta}{2}\right)$. [Ans. $2\,(\cos h - 1)\sin x$]

2. Prove that a. $\delta = \Delta(1+\Delta)^{-\frac{1}{2}} = \nabla(1-\nabla)^{-\frac{1}{2}}$ b. $\delta = \Delta E^{-\frac{1}{2}} = \nabla E^{\frac{1}{2}}$.

3. Prove that (a) $\delta = E = \displaystyle\sum_{i=0}^{\infty}\nabla^i$ (b) $\left(E^{\frac{1}{2}} + E^{-\frac{1}{2}}\right)(1+\Delta)^{\frac{1}{2}} = 2 + \Delta$.

4. Using the method of finite difference, sum the series

   a. $1.2.3 + 2.3.4 + 3.4.5 + \ldots$ to n term. $\left[\text{Ans. } \dfrac{n(n+1)(n+2)(n+3)}{4}\right]$

   b. $\displaystyle\sum_{x=1}^{n} x(x+2)(x+4)$. $\left[\text{Ans. } \dfrac{n(n+1)(n+4)(n+5)}{4}\right]$

5. a. Sum the series $1.2\Delta x^n - 2.3\Delta^2 x^n + 3.4\Delta^3 x^n - 4.5\Delta^4 x^n + \ldots$

   $\left[\text{Ans. } 2\,(x-2)^n - (x-3)^n\right]$

   b. Show that $\Delta x^n - \dfrac{1}{2}\Delta^2 x^n + \dfrac{1\cdot 3}{2\cdot 4}\Delta^3 x^n - \dfrac{1\cdot 3\cdot 5}{2\cdot 4\cdot 6}\Delta^4 x^n + \ldots = \left(x + \dfrac{1}{2}\right)^n - \left(x - \dfrac{1}{2}\right)^n$

4

Interpolation

4.1 INTRODUCTION

Interpolation is the process of finding intermediate values of a function which is not exactly known from a given set of tabular values of the function. Let $y_0, y_1, y_2, ..., y_n$ be the values of y corresponding to $x_0, x_1, x_2, ..., x_n$. Then, the process of finding the values of y corresponding to any value of x lying between x_0 and x_n is called the *interpolation*, while the process of computing the value of y corresponding to any value of x outside the set of given values is called the *extrapolation*.

If $f(x)$ is completely known, then y is easily found corresponding to any value of x but if $f(x)$ is not known, it is difficult to find the value of $f(x)$ with the help of tabulated set of values. In such cases, we replace $f(x)$ by a simpler function $\phi(x)$ which assumes the same value of $f(x)$ as assumed by $f(x)$ at the tabulated points. Any other values may be computed from $\phi(x)$ which is known as the interpolating function but if $\phi(x)$ is a polynomial then $\phi(x)$ is called the *interpolating polynomial*.

4.2 INTERPOLATION FORMULA FOR EQUAL INTERVALS

4.2.1 Gregory–Newton's Forward Interpolation Formula or Newton's Forward Interpolation Formula for Equal Intervals

Let the function $y = f(x)$ takes the values $y_0, y_1, y_2, ..., y_n$ corresponding to the values $x_0, x_1, x_2, ..., x_n$. Let the values of x be equispaced. Suppose, it is required to evaluate y_p for $x = x_0 + ph$ (where p is any real number), that is near the beginning of the tabulated values of x. Now $y_p = f(x_0 + ph)$

$$y_p = f(x_0 + ph) = E^p f(x_0)$$

$$= E^p y_0 = (1 + \Delta)^p y_0 = \left(1 + p\Delta + \frac{p(p-1)}{2!}\Delta^2 + \frac{p(p-1)(p-2)}{3!}\Delta^3 + ...\right) y_0$$

$$= \left(y_0 + p\Delta y_0 + \frac{p(p-1)}{2!}\Delta^2 y_0 + \frac{p(p-1)(p-2)}{3!}\Delta^3 y_0 + ...\right)$$

which is called the Newton's formula for interpolation for equal intervals,

where $$p = \frac{x - x_0}{h}$$

Note:

1. Since Newton's formula utilizes y_0 and the forward differences of y_0, it is mainly used for interpolating the values of y near the beginning of a set of tabulated values.
2. This formula is applicable only if the incremental difference is a constant.

4.2.2 Gregory-Newton's Backward Interpolation Formula for Equal Intervals

Let the function $y = f(x)$ takes the values $y_0, y_1, y_2, ..., y_n$ corresponding to the values of $x = x_0, x_1, x_2, ..., x_n$. Suppose it is required to evaluate $y_p = f(x_n + ph)$ for $x = x_n + ph$ where p is any real number and the value x is nearer the end of a set of tabulated values. Now

$$y_p = f(x_n + ph) = E^p f(x_n) = (E^{-1})^{-p} y_n$$

$$= (1 - \nabla)^{-p} y_n = \left(1 + p\nabla + \frac{p(p+1)}{2!}\nabla^2 + \frac{p(p+1)(p+2)}{3!}\nabla^3 + ...\right) y_n$$

where

$$p = \frac{x - x_n}{h}$$

$$= y_n + p\nabla y_n + \frac{p(p+1)}{2!}\nabla^2 y_n + \frac{p(p+1)(p+2)}{3!}\nabla^3 y_n + ...$$

This is called the *Newton's backward formula*.

Note: This formula is used to interpolate the values of y nearer to the end of a set of tabulated values and also for extrapolating values of y a little ahead (to the right) of y_n.

Solved Problems

Example 4.1: Using Newton's forward interpolating formula, find y at $x = 8$ or $f(8)$ for the following.

x	0	5	10	15	20	25
y	7	11	14	18	24	32

Solution: Here $h = 5$, $y_0 = 7$, $x = 8$,

$$x = x_0 + ph$$
$$8 = 0 + 5p$$
$$p = 8/5 = 1.6$$
$$f(8) = f(x_0 + ph)$$

$$= y_0 + p\Delta y_0 + \frac{p(p-1)}{2!}\Delta^2 y_0 + \frac{p(p-1)(p-2)}{3!}\Delta^3 y_0 + ...$$

x	y	Δy	$\Delta^2 y$	$\Delta^3 y$	$\Delta^4 y$	$\Delta^5 y$
$0 = x_0$	$7 = y_0$					
		$4 = \Delta y_0$				
5	11		$-1 = \Delta^2 y_0$			
		3		$2 = \Delta^3 y_0$		
10	14		1		$-1 = \Delta^4 y_0$	
		4		0		0
15	18		2		-1	
		6		0		
20	24		2			
		8				
25	32					

$$f(8) = 7 + (1.6)(4) + \frac{(1.6)(0.6)}{2!}(-1) + \frac{(1.6)(0.6)(-0.4)}{3!}(2) + \frac{(1.6)(0.6)(-0.4)(-1.4)}{4!}(-1)$$

$$= 7 + 6.4 - 0.48 - 0.128 - 0.0224$$

$$= 12.7696$$

Example 4.2: Find the number of men getting wages between Rs. 10 and Rs. 15 from the data.

Wages (in Rs)	0–10	10–20	20–30	30–40
Frequency	9	30	25	42

Solution: x = wages in rupees, y = number of men getting wages < x.

x	$y = f(n)$	Δy	$\Delta^2 y$	$\Delta^3 y$
$10 = x_0$	$9 = y_0$			
		30		
20	39		5	
		35		2
39	74		7	
		42		
40	116			

Rs. 10 to Rs. 15 is nearer to the beginning, apply forward interpolation formula.

Here $h = 10$, $x_0 = 10$, $x = 15$, then $x_0 + ph = x$, $10 + 10p = 15 \Rightarrow p = 0.5$.

From Newton forward interpolating formula, we get

$$y = f(x) = f(x_0 + ph) = y_0 + p\Delta y_0 + \frac{p(p-1)}{2!}\Delta^2 y_0 + \frac{p(p-1)(p-2)}{3!}\Delta^3 y_0$$

$$= 9 + 0.5(30) - \frac{0.5(0.5)}{2}(5) + \frac{0.5(0.5)(0.5)}{6}(2)$$

$$= 9 + 15 - 0.625 + 0.125 = 23.5 = 24 \text{ approximately.}$$

Then men getting wages between Rs. 10 and Rs. 15 = 24 – 9 = 15.

Example 4.3: Given sin 45° = 0.7071, sin 50° = 0.776, sin 55° = 0.8192, sin 60° = 0.8660, find sin 52°.

Solution: Put

$$x = 45°, 50°, 55°, 60°$$
$$y = f(x) = \sin x$$

x	$y = f(x) = \sin x$	Δy	$\Delta^2 y$	$\Delta^3 y$
$45° = x_0$	$0.7071 = y_0$			
		$0.0589 = \Delta y_0$		
50°	0.766		$-0.0057 = \Delta^2 y_0$	
		0.0532		$-0.007 = \Delta^3 y_0$
55°	0.8192		-0.0064	
		0.0468		
60°	0.8660			

Here

$$x_0 + ph = 52$$
$$45 + 5p = 52$$
$$p = 1.4$$

$$\sin 52° = f(x_0 + ph) = y_0 + p\Delta y_0 + \frac{p(p-1)}{2!}\Delta^2 y_0 + \frac{p(p-1)(p-2)}{3!}\Delta^3 y_0$$

$$= 0.7071 + 1.4(0.589) + \frac{(1.4)(0.4)}{2!}(-0.0057) + \frac{(1.4)(0.4)(-0.6)}{3!}(-0.07)$$

$$= 0.7880 \text{ (approximately)}$$

Example 4.4: The following data gives the melting point of an alloy of lead and zinc where t is the temperature in centigrade and p is the percentage of lead in the alloy. Find the melting point of alloy containing 84% lead from the table.

p (%age)	60	70	80	90
t (°C)	226	250	276	304

Solution: Let

$$x = p \text{ (percentage of lead)}$$
$$y = t = \text{melting point of alloy in degree centigrade}$$

x	$y = t$	Δy	$\Delta^2 y$	$\Delta^3 y$
60	226			
		24		
70	250		2	
		26		0
80	276		2	
		28		
90	304			

Since $x = 84$ is nearer to the end of the table, we use Newton's backward interpolation formula.

Taking $x_n = 90$, $p = \dfrac{x - x_n}{h} = \dfrac{84 - 90}{10} = -0.6$

Using the line of backward difference $y_n = 304$, $\nabla y_n = 28$, $\nabla^2 y_n = 2$.

Newton's backward formula gives

$$y_{84} = y_{90} + p\nabla y_{90} = \frac{p(p+1)}{2!}\nabla^2 y_{90}$$

Now $y_{90} = 304$, $\nabla y_{90} = 28$, $\nabla^2 y_{90} = 2$.

$$y_{84} = 304 - (0.6)(28) + \frac{(-0.6 \times 0.4)}{2} \times 2$$

$$= 304 - 16.8 - 0.24 = 286.96$$

Example 4.5: The terms of a series of which 23.6 is the sixth term, find the first and the tenth term of the series.

x	3	4	5	6	7	8	9
y	4.8	8.4	14.5	23.6	36.2	52.8	73.9

Solution:

x	$y = f(x)$	Δy	$\Delta^2 y$	$\Delta^3 y$
$x_0 = 3$	$y_0 = 4.8$			
		3.6		
4	8.4		2.5	
		6.1		0.5
5	14.5		3.0	
		9.1		0.5
6	23.6		3.5	
		12.6		0.5
7	36.2		4.0	
		16.6		0.5
8	52.8		4.5	
		21.1		
9	73.9			

To find the first term which is nearer to 3.

$$x_0 = 3, x = 1, h = 1, p = \frac{x - x_0}{h} = \frac{1 - 3}{1} = -2$$

$$f(1) = f(x_0 + ph)$$

$$= f(x_0) + p\Delta f(x_0) + \frac{p(p-1)}{2!}\Delta^2 f(x_0) + \frac{p(p-1)(p-2)}{3!}\Delta^3 f(x_0) + \ldots$$

$$= y_0 + p\Delta y_0 + \frac{p(p-1)}{2!}\Delta^2 y_0 + \frac{p(p-1)(p-2)}{3!}\Delta^3 y_0 + \ldots$$

$$= 4.8 + (-2)(3.6) + \frac{(-2)(-3)}{2!}(2.5) + \frac{(-2)(-3)(-4)}{3!} \times 0.5$$

$$= 4.8 - 7.2 + 7.5 - 2 = 3.1$$

To find the 10th term which is nearer to the end of the tabular values

$$x_n = 9, h = 1, x = 10$$
$$x = x_n + ph, \ 10 = 9 + ph, \ p = 1$$
$$f(10) = f(x_n + ph)$$

$$= f(x_n) + p\nabla f(x_n) + \frac{p(p+1)}{2!}\nabla^2 f(x_n) + \frac{p(p+1)(p+2)}{3!}\nabla^3 f(x_n)$$

$$= y_n + p\nabla y_n + \frac{p(p+1)}{2!}\nabla^2 y_n + \frac{p(p+1)(p+2)}{3!}\nabla^3 y_n$$

$$= 73.9 + 21.1 + \frac{1.2}{2!}(4.5) + \frac{1 \cdot 2 \cdot 3}{3!} \times 0.5 = 100$$

Example 4.6: Apply Newton's backward formula to the data given below to obtain a fourth degree polynomial in x.

x	1	2	3	4	5
y	1	−1	1	−1	1

Solution: As we have to obtain a polynomial of fourth degree, we have to restrict ourselves to $\Delta^4 y$.

	x	y	Δy	$\Delta^2 y$	$\Delta^3 y$	$\Delta^4 y$
x_0	1	1				
			-2			
x_1	2	-1		4		
			2		-8	
x_2	3	1		-4		16
			-2		$8 = \nabla^3 y_4$	
x_3	4	-1		$4 = \nabla^2 y_4$		
			$2 = \nabla y_4$			
x_4	5	$1 = y_4$				

Here $x_n = x_4 = 5, h = 1$

$$x = x_n + ph = 5 + p, \, p = x - 5$$
$$f(x) = f(x_n + ph) = f(x_4 + ph)$$
$$= y_4 + p\nabla y_4 + \frac{p(p+1)}{2!}\nabla^2 y_4 + \frac{p(p+1)(p+2)}{3!}\nabla^3 y_4 + \frac{p(p+1)(p+2)(p+3)}{4!}\nabla^4 y_4$$
$$= 1 + (x-5)\times 2 \frac{(x-5)(x-4)}{2!}\times 4 + \frac{(x-5)(x-4)(x-3)}{3!}\times 8$$
$$+ \frac{(x-5)(x-4)(x-3)(x-2)}{4!}\times 16$$
$$= x^4 - 14x^3 + 71x^2 - 114x + 120$$

Example 4.7: The following table gives the population of a town during last six censuses. Estimate the increase in population during the period from 1976 to 1978.

x	1941	1951	1961	1971	1981	1991
y (crores)	12	15	20	27	39	52

Solution: x = year, y = population (in crores)

	x	y	Δy	$\Delta^2 y$	$\Delta^3 y$	$\Delta^4 y$	$\Delta^5 y$
x_0	1941	12					
			3				
x_1	1951	15		2			
			5		0		
x_2	1961	20		2		$+3$	
			7		3		$-10 = \nabla^4 y_n$
x_3	1971	27		5		$-7 = \nabla^4 y_n$	
			12		$-4 = \nabla^3 y_n$		
x_4	1981	39		$1 = \nabla^2 y_n$			
			$13 = \nabla y_n$				
x_5	1991	$52 = y_n$					

For 1976: Since x = 1976 is near the end of the tabulated values, we shall use Newton's backward interpolation formula.

Here $\qquad x_n = 1991, x = 1976, 1976 = 1991 + p \times 10$

$$p = -15/10 = -1.5$$

$$y(1976) = 52 + p \times 13 + \frac{p(p+1)}{2!} + \frac{p(p+1)(p+2)}{3!}(-4)$$
$$+ \frac{p(p+1)(p+2)(p+3)}{4!}(-7) + \frac{p(p+1)(p+2)(p+3)(p+4)}{5!}(-10)$$

$$= 52 + (-1.5) \times 13 + \frac{(-1.5)(-0.5)}{2!} \times 1 + \frac{(-1.5)(-0.5)(0.5)}{3!}(-4)$$

$$+ \frac{(-1.5)(-0.5)(0.5)(1.5)}{4!} \times (-7) + \frac{(-1.5)(-0.5)(0.5)(1.5)(2.5)}{5!}(-10)$$

$$= 52 - 19.5 + 0.375 + 0.25 - 0.1640 - 0.1171 = 32.843$$

Similarly for $p = -13/10 = -1.3$

$$y(1978) = 52 + (-1.3) \times 13 + \frac{(-1.3)(-0.3)}{2!} + \frac{(-1.3)(-0.3)(0.7)}{3!}(-4)$$

$$+ \frac{(-1.3)(-0.3)(0.7)(1.3)}{4!}(-7) + \frac{(-1.3)(-0.3)(0.7)(1.3)(2.3)}{5!}(-10)$$

$$= 52 - 16.9 + 0.195 - 0.182 - 0.1035 - 0.0156 = 34.99$$

Population for the period 1976 to 1978
$$y(1978) - y(1976) = 34.99 - 32.843 = 2.6$$

Example 4.8: From the following table, estimate the number of persons having income between 2000 to 2500.

Income	Below 500	500–1000	1000–2000	2000–3000	3000–4000
Person	6000	4250	3600	1500	650

Solution: Here x = income, y_x = number of persons having income $< x$.

	x	y_x	Δy_x	$\Delta^2 y_x$
x_0	2000	$y_0 = 13850$		
			$1500 = \Delta y_0$	
x_1	3000	$y_1 = 15350$		$-850 = \Delta^2 y_0$
			650	
x_2	4000	$y_2 = 16000$		
	$x = 2500$	$x_0 = 2000$	$h = 1000$	

$$2500 = 2000 + 1000\, p, \ p = 0.5, \ x_0 = 2000, \ x = 2500, \ h = 1000$$

By Newton's forward formula

$$y(2500) = y_0 + p\Delta y_0 + \frac{p(p-1)}{2!}\Delta^2 y_0 + \frac{p(p-1)(p-2)}{3!}\Delta^3 y_0$$

$$= 13850 + (90.5)(1500) + \frac{(0.5)(-1.5)}{2!}(-850)$$

$$= 13850 + 750 + 106.25$$

$$= 14706 \text{ (approximately)}$$

Required value = persons having income < 2500 – income < 2000

$$= 14706 - 13850 = 856$$

Example 4.9: Using polynomial of third degree, complete the record given below of the export of a certain commodity during five years.

Year	1989	1990	1991	1992	1993
Export (in tons)	443	384	–	397	467

Solution: Here x = year, y = export (in tons)

x	y	Δy	$\Delta^2 y$	$\Delta^3 y$	$\Delta^4 y$
x_0	1989	$443 = y_0$				
			-059			
x_1	1990	$384 = y_1$		$y_2 - 325$		
			$y_2 - 384$		$-3y_2 + 1106$	
x_2	1991	$- = y_2$		$-2y_2 + 781$		$6y_2 - 2214$
			$397 - y_2$		$3y_2 - 1108$	
x_3	1992	$397 = y_2$		$-327 + y_2$		
			70			
x_4	1993	$467 = y_4$				

From the table $6y_2 - 2214 = 0$, $y_2 = 2214/6 = 369$ tons.

Example 4.10: Given $u_1 = 40$, $u_3 = 45$, $u_5 = 54$. Find u_2 and u_4.

Solution:

x	u_x	Δu_x	$\Delta^2 u_x$	$\Delta^3 u_x$
1	40			
		$u_2 - 40$		
2	u_2		$85 - 2u_2$	
		$45 - u_2$		$u_4 - 175 + 3u_2$
3	45		$u_4 - 90 + u_2$	
		$u_4 - 45$		$189 - 3u_4 - u_2$
4	u_4		$99 - 2u_4$	
		$54 - u_4$		
5	54			

By the question
$$u_4 - 175 + 3u_2 = 0 \Rightarrow 3u_4 + 9u_2 = 525 \tag{i}$$
$$189 - 3u_4 - u_2 = 0 \Rightarrow 3u_4 + u_2 = 189 \tag{ii}$$

From Eqs (i) and (ii)
$$8u_2 = 336$$
$$u_2 = 42$$
$$3u_4 = 189 - 42 = 147 \Rightarrow u_4 = 49$$

Example 4.11: If $u_{-1} = 10$, $u_1 = 8$, $u_2 = 10$, $u_4 = 50$, find u_0 and u_3.

Solution:

x	u_x	Δu_x	$\Delta^2 u_x$	$\Delta^3 u_x$	$\Delta^4 u_x$
−1	10				
		$x_0 - 10$			
0	x_0		$18 - 2x_0$		
		$8 - x_0$		$3x_0 - 24$	
1	8		$x_0 - 6$		$x_3 - 4x_0 + 18$
		2		$x_3 - x_0 - 6$	
2	10		$x_3 - 12$		$78 - 4x_3 + x_0$
		$x_3 - 10$		$72 - 3x_3$	
3	x_3		$60 - 2x_3$		
		$50 - x_3$			
4	50				

From the table $x_3 - 4x_0 + 18 = 0$
$$78 - 4x_3 + x_0 = 0$$

or $\qquad 4x - 16x_2 + 72 = 0$ (i)

$\qquad\qquad -4x_3 + x_0 + 78 = 0$ (ii)

from Eqs (i) and (ii)

$$-15x_0 + 150 = 0, x_0 = 10$$
$$78 - 4x_3 + 10 = 0$$
$$4x_3 = 88$$
$$x_3 = 22$$

Example 4.12: Given $\sum\limits_{1}^{10} f(x) = 500426$, $\sum\limits_{4}^{10} f(x) = 329240$, $\sum\limits_{7}^{10} f(x) = 175212$, $f(10) = 40365$. Find $f(1)$.

Solution: Now, $f(1) = \sum\limits_{1}^{10} f(x) - \sum\limits_{2}^{10} f(x)$, so we have to find $\sum\limits_{2}^{10} f(x)$.

Also we have $y_x = \sum\limits_{x}^{10} f(x)$.

x	y_x	Δy_x	$\Delta^2 y_x$	$\Delta^3 y_x$
$x_0 = 1$	500426			
		$-171186 = \Delta y_0$		
4	329240		$17158 = \Delta^2 y_0$	
		-154028		$2023 = \Delta^3 y_0$
7	175212		19181	
		-134847		
10	40365			

$\qquad x = 2, x_0 = 1, h = 3$, we have $x = x_0 + ph$

$\qquad 2 = 1 + 3p, p = 1/3$

$$y_x = y_0 + p\Delta y_0 + \frac{p(p-1)}{2!}\Delta^2 y_0 + \frac{p(p-1)(p-2)}{3!}\Delta^3 y_0$$

$$y_2 = 500426 + \frac{1}{3}(-171186) + \frac{\frac{1}{3}\left(-\frac{2}{3}\right)}{2!}(17158) + \frac{\frac{1}{3}\left(-\frac{2}{3}\right)\left(-\frac{5}{3}\right)}{3!}(2023)$$

$$= 500426 - 57062 - 1906.444 + 124.8765$$

$$= 441582.432 = 441582.432$$

$$\sum\limits_{2}^{10} f(x) = 441582.432$$

$$f(1) = 500426 - 441582.432 = 58843.468$$

Example 4.13: The area A of a circle of diameter d is given for the following values

d	80	85	90	95	100
A	5026	5674	6362	7088	7854

Calculate area of a circle of diameter 105.

Solution: 105 is nearer to the end of table, so we have $x_n = x - 5p$ or $x_n = 100$, $x = 105$,

$n = 5$, $\dfrac{x - x_n}{5} = \dfrac{105 - 100}{5} = 1$.

$x = d$	$A = y$	Δy	$\Delta^2 y$	$\Delta^3 y$	$\Delta^4 y$
80	5026				
		648			
85	5674		40		
		688		-2	
90	6362		38		$4 = \nabla^4 y_{100}$
		726		$2 = \nabla^3 y_{100}$	
95	7088		$40 = \nabla^2 y_{100}$		
		$766 = \nabla y_{100}$			
100	$7854 = y_{100}$				

$$y_{105} = y_{100} + p\nabla y_{100} + \frac{p(p+1)}{2!}\nabla^2 y_{100} + \frac{p(p+1)(p+2)}{3!}\nabla^3 y_{100}$$

$$+ \frac{p(p+1)(p+2)(p+3)}{4!}\nabla^4 y_{100}$$

$$= 7854 + 766 + 40 + 2 + 4 = 8666$$

PROBLEMS 4.1

1. From the following table, find the number of students who obtain less than 45 marks.

Mark	No. of students
30–40	31
40–50	42
50–60	51
60–70	35
70–80	31

[Ans. 47.86]

2. Find the value of sin 54° from the given table.

	0°	45°	50°	55°	60°
$\sin\theta$		0.7071	0.7660	0.8192	0.8660

[Ans. 0.7880032]

3. Given that

x	1	2	3	4	5	6	7	8
$f(x)$	1	8	27	64	125	216	343	512

Find $f(7.5)$ using Newton's backward difference formula.   [Ans. 421.875]

4. The probability integral has following values.

x	1.000	1.05	1.10	1.15	1.20	1.25
ρ	0.682689	0.706282	0.728668	0.749856	0.769861	0.788700

Calculate ρ for $x = 1.235$ [Ans. 0.783172]

5. Calculate the value of tan 48°15′ from the following table:

$x°$	45	46	47	48	49	50
$\tan x°$	1.00000	1.03053	1.07237	1.11061	1.15037	1.19175

[Ans. 1.120402867]

4.3 CENTRAL DIFFERENCE INTERPOLATION FORMULA

These are most suitable when one has to interpolate near the center of difference table.

Let $y = f(x)$ be any function.

Let $x = x_0 - 2h$, $x_0 - h$, x_0, $x_0 + h$, $x_0 + 2h$ and corresponding values of y be y_{-2}, y_{-1}, y_0, y_1, y_2.

Then one can represent the difference table in two notations using operator $\delta = \Delta E^{1/2}$ known as central difference table.

x	y	First difference	Second difference	Third difference	Fourth difference
$x_0 - 2h$	y_{-2}				
		$\Delta y_{-2} = (\delta y_{-3/2})$			
$x_0 - h$	y_{-1}		$\Delta^2 y_{-2} = (\delta^2 y_{-1})$		
		$\Delta y_{-1} = (\delta y_{-1/2})$		$\Delta^3 y_{-2} = (\delta^3 y_{-1/2})$	
x_0	y_0		$\Delta^2 y_{-1} = (\delta^2 y_0)$		$\Delta^4 y_{-2} = (\delta^4 y_0)$
		$\Delta y_0 = (\delta y_{1/2})$		$\Delta^3 y_{-1} = (\delta^3 y_{1/2})$	
$x_0 + h$	y_1		$\Delta^2 y_0 = (\delta^2 y_1)$		
		$\Delta y_1 = (\delta y_{3/2})$			
$x_0 + 2h$	y_2				

4.3.1 Gauss Forward Interpolation Formula

Newton's forward interpolation formula

$$y_p = y_0 + p\Delta y_0 + \frac{p(p-1)}{2!}\Delta^2 y_0 + \frac{p(p-1)(p-2)}{3!}\Delta^3 y_0 + \dots \quad (4.1)$$

where
$$p = \frac{x - x_0}{h}$$

Also, we have
$$\Delta^2 y_0 - \Delta^2 y_{-1} = \Delta^3 y_{-1}$$

or
$$\Delta^2 y_0 = \Delta^2 y_{-1} + \Delta^3 y_{-1} \quad (4.2)$$

Similarly
$$\Delta^3 y_0 = \Delta^3 y_{-1} + \Delta^4 y_{-1} \quad (4.3)$$

$$\Delta^4 y_0 = \Delta^4 y_{-1} + \Delta^5 y_{-1} \quad (4.4)$$

$$\Delta^3 y_{-1} - \Delta^3 y_{-2} = \Delta^4 y_{-2}$$

$$\Delta^3 y_{-1} = \Delta^4 y_{-2} + \Delta^3 y_{-2}$$

$$\Delta^4 y_{-1} = \Delta^5 y_{-2} + \Delta^4 y_{-2} \text{ etc.} \quad (4.5)$$

Substituting the values of $\Delta^2 y_0$, $\Delta^3 y_0$, $\Delta^4 y_0$ in Eq. (4.1), we have

$$y_p = y_0 + p\Delta y_0 + \frac{p(p-1)}{2!}(\Delta^2 y_{-1} + \Delta^3 y_{-1}) + \frac{p(p-1)(p-2)}{3!}(\Delta^3 y_{-1} + \Delta^4 y_{-1})$$

$$+ \frac{p(p-1)(p-2)(p-3)}{4!}(\Delta^4 y_{-1} + \Delta^5 y_{-1}) + \dots$$

$$= \left[y_0 + p\Delta y_0 + \frac{p(p-1)}{2!}\Delta^2 y_{-1} \right] + \left[\frac{p(p-1)}{2!} + \frac{p(p-1)(p-2)}{3!} \right]\Delta^3 y_{-1}$$

$$+ \left[\frac{p(p-1)(p-2)}{3!} + \frac{p(p-1)(p-2)(p-3)}{4!} \right]\Delta^4 y_{-1} + \dots$$

$$= y_0 + p\Delta y_0 + \frac{p(p-1)}{2!}\Delta^2 y_{-1} + \frac{(p+1)p(p-1)}{3!}\Delta^3 y_{-1}$$

$$+ \frac{(p+1)p(p-1)(p-2)}{4!}\Delta^4 y_{-1} + \dots$$

$$= y_0 + p\Delta y_0 + \frac{p(p-1)}{2!}\Delta^2 y_{-1} + \frac{(p+1)p(p-1)(p-2)}{3!}\Delta^3 y_{-1}$$

$$+ \frac{(p+1)p(p-1)(p-2)}{4!}(\Delta^4 y_{-2} + \Delta^5 y_{-2})$$

$$+ \frac{(p+1)p(p-1)(p-2)(p-3)}{5!}(\Delta^5 y_{-2} + \Delta^6 y_{-2}) + \dots \text{ using Eq. (4.5)}$$

$$= y_0 + p\Delta y_0 + \frac{p(p-1)}{2!}\Delta^2 y_{-1} + \frac{(p+1)p(p-1)}{3!}\Delta^3 y_{-1}$$

$$+ \frac{(p+1)p(p-1)(p-2)}{4!}\Delta^4 y_{-2} + \frac{(p+2)p(p+1)(p-1)(p-2)}{5!}\Delta^5 y_{-2} + \dots$$

is called Gauss forward interpolation formula.

Notes: 1. In central difference notation, this formula can be written as

$$y_p = y_0 + p\delta y_{1/2} + \frac{p(p-1)}{2!}\delta^2 y_0 + \frac{(p+1)p(p-1)}{3!}\delta^3 y_{1/2} + \frac{(p+1)p(p-1)(p-2)}{4!}\delta^4 y_0 + \dots$$

2. It employs odd differences just below the central line and even differences on the central line as shown in Fig. 4.1.

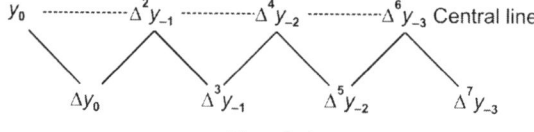

Fig. 4.1

3. This formula is used to interpolate the value of y for p $(0 < p < 1)$ measured forwardly from the origin.

4.3.2 Gauss Backward Interpolation Formula

$$y_p = y_0 + p\Delta y_0 + \frac{p(p+1)}{2!}\Delta^2 y_0 + \frac{(p+1)p(p-1)}{3!}\Delta^3 y_0 + \frac{(p+2)(p+1)p(p-1)}{4!}\Delta^4 y_0 + \dots$$

$$\tag{4.6}$$

As
$$\Delta y_0 - \Delta y_{-1} = \Delta^2 y_{-1}$$

$$\Delta y_0 = \Delta^2 y_{-1} + \Delta^2 y_{-1} \tag{4.7}$$

Similarly
$$\Delta^2 y_0 = \Delta^2 y_{-1} + \Delta^3 y_{-1} \tag{4.8}$$

$$\Delta^3 y_0 = \Delta^3 y_{-1} + \Delta^4 y_{-1} \text{ etc.} \tag{4.9}$$

Also
$$\Delta^3 y_{-1} - \Delta^3 y_{-2} = \Delta^4 y_{-2}$$

i.e.
$$\Delta^3 y_{-1} = \Delta^3 y_{-2} + \Delta^4 y_{-2} \tag{4.10}$$

Similarly
$$\Delta^4 y_{-1} = \Delta^4 y_{-2} + \Delta^5 y_{-2} \text{ etc.}$$

Substituting $\Delta y_0, \Delta^2 y_0, \Delta^3 y_0$ ... from Eqs (4.7)–(4.10) in Eq. (4.6), we get

$$y_p = y_0 + p(\Delta y_{-1} + \Delta^2 y_{-1}) + \frac{p(p-1)}{1\cdot 2}(\Delta^2 y_{-1} + \Delta^3 y_{-1})$$

$$+ \frac{p(p-1)(p-2)}{1\cdot 2\cdot 3}(\Delta^3 y_{-1} + \Delta^4 y_{-1}) + \frac{p(p-1)(p-2)(p-3)}{1\cdot 2\cdot 3\cdot 4}(\Delta^4 y_{-1} + \Delta^5 y_{-1}) + ...$$

$$= y_0 + p\Delta y_{-1} + \frac{(p-1)p}{1\cdot 2}\Delta^2 y_{-1} + \frac{(p+1)p(p-1)}{1\cdot 2\cdot 3}\Delta^3 y_{-1}$$

$$+ \frac{(p+1)p(p-1)(p-2)}{1\cdot 2\cdot 3\cdot 4}\Delta^4 y_{-1} + \frac{p(p-1)p(p-1)(p-2)}{1\cdot 2\cdot 3\cdot 4}\Delta^5 y_{-1} + ...$$

$$= y_0 + p\Delta y_{-1} + \frac{(p+1)p}{1\cdot 2}\Delta^2 y_{-1} + \frac{(p+1)p(p-1)}{1\cdot 2\cdot 3}(\Delta^3 y_{-2} + \Delta^4 y_{-2})$$

$$+ \frac{(p+1)p(p-1)(p-2)}{1\cdot 2\cdot 3\cdot 4}(\Delta^4 y_{-2} + \Delta^5 y_{-2}) + ...$$

$$= y_0 + p\Delta y_{-1} + \frac{(p+1)p}{2!}\Delta^2 y_{-1} + \frac{(p+1)p(p-1)}{3!}\Delta^3 y_{-2}$$

$$+ \frac{(p+2)(p+1)p(p-1)}{4!}\Delta^4 y_{-2} + ... \tag{4.11}$$

which is called *Gauss's backward interpolation formula*.

Notes:

1. In central differences notation, this formula will be

$$y_p = y_0 + p\delta y_{-\frac{1}{2}} + \frac{(p+1)p}{2!}\delta^2 y_0 + \frac{(p+1)p(p-1)}{3!}\delta^3 y_{-\frac{1}{2}}$$

$$+ \frac{(p+2)(p+1)p(p-1)}{4!}\delta^2 y_0 + ...$$

2. This formula contains odd differences above the central line and even differences on the central line as shown in Fig. 4.2.

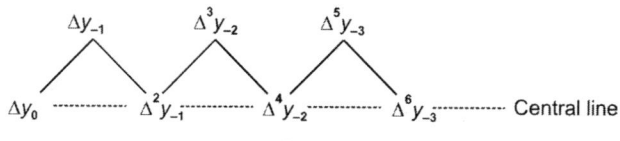

Fig. 4.2

3. It is used to interpolate the value of y for a negative value of p lying between -1 and 0.

4. Gauss' forward and backward formulae are not of much practical use, however, these serve as intermediate steps for obtaining the required formula.

4.3.3 Bessel's Formula

Gauss' forward interpolation formula is

$$y_p = y_0 + p\Delta y_0 + \frac{p(p-1)}{2!}\Delta^2 y_{-1} + \frac{(p+1)p(p-1)}{3!}\Delta^3 y_{-1}$$

$$+ \frac{(p+1)p(p-1)(p-2)}{4!}\Delta^4 y_{-2} + ... \tag{4.12}$$

We have $\quad \Delta^2 y_0 - \Delta^2 y_{-1} = \Delta^3 y_{-1}$ $\hspace{4cm}$ (4.12a)

$$\Delta^2 y_{-1} = \Delta^2 y_0 - \Delta^3 y_{-1} \tag{4.13}$$

Similarly $\quad \Delta^4 y_{-2} = \Delta^4 y_{-1} - \Delta^5 y_{-2}$ etc. $\hspace{2.5cm}$ (4.14)

Now Eq. (4.12) can be written as

$$y_p = y_0 + p\Delta y_0 + \frac{p(p-1)}{2!}\left(\frac{1}{2}\Delta^2 y_{-1} + \frac{1}{2}\Delta^2 y_{-1}\right) + \frac{p(p^2-1)}{3!}\Delta^3 y_{-1}$$

$$+ \frac{p(p^2-1)(p-2)}{4!}\frac{1}{2}\left[\Delta^4 y_{-2} + \frac{1}{2}\Delta^4 y_{-2}\right] + ...$$

$$= y_0 + p\Delta y_0 + \frac{1}{2}\left[\frac{p(p-1)}{2!}\right]\Delta^2 y_{-1} + \frac{1}{2}\left[\frac{p(p-1)}{3!}\right](\Delta^2 y_0 - \Delta^3 y_{-1})$$

$$+ \frac{p(p^2-1)}{3!}\Delta^3 y_{-1} + \frac{1}{2}\left[\frac{p(p^2-1)(p-2)}{4!}\Delta^4 y_{-2}\right] + ...$$

$$+ \left[\frac{1}{2}\frac{p(p^2-1)(p-2)}{4!}\right](\Delta^4 y_{-1} - \Delta^5 y_{-2})\text{ using Eqs (4.7 and 4.8) etc.}$$

$$= y_0 + p\Delta y_0 + \frac{p(p-1)}{2!}\frac{\Delta^2 y_{-1} + \Delta^2 y_0}{2} + \frac{p(p-1)}{2!}\times\left(\frac{p+1}{3}-\frac{1}{2}\right)\Delta^3 y_{-1}$$

$$+ \frac{p(p^2-1)(p-2)}{4!}\cdot\frac{\Delta^4 y_{-2} + \Delta^4 y_{-1}}{2} + ...$$

Hence $\quad y_p = y_0 + p\Delta y_0 + \dfrac{p(p-1)}{2!}\cdot\dfrac{\Delta^2 y_{-1} + \Delta^2 y_0}{2} + \dfrac{\left(p-\dfrac{1}{2}\right)p(p-1)}{3!}\Delta^3 y_{-1}$

$$+ \frac{p(p+1)(p-1)(p-2)}{4!}\cdot\frac{\Delta^4 y_{-2} + \Delta^4 y_{-1}}{2} + ... \tag{4.15}$$

which is known as *Bessel's formula*.

Notes: 1. In the central difference Eq. (4.15) becomes

$$y_p = y_0 + p\delta y_{\frac{1}{2}} + \frac{p(p-1)}{2!}\mu\delta^2 y_{\frac{1}{2}} + \frac{\left(p-\dfrac{1}{2}\right)p(p-1)}{3!}\mu\delta^3 y_{\frac{1}{2}}$$

$$+ \frac{(p+1)p(p-2)}{4!}\mu\delta^4 y_{\frac{1}{2}}$$

For $\quad \dfrac{1}{2}(\Delta^2 y_{-1} + \Delta^2 y_0) = \mu\delta^2 y_{\frac{1}{2}}, \dfrac{1}{2}(\Delta^4 y_{-2} + \Delta^4 y_{-1}) = \mu\delta^4 y_{\frac{1}{2}}$ etc.

2. It involves odd differences below the central line and mean of even differences below this line as shown.

$$\dots y_0 \dots \begin{Bmatrix} \Delta^2 y_{-1} \\ \Delta y_0 \end{Bmatrix} \begin{matrix} \dots \\ \Delta^2 y_0 \end{matrix} \begin{matrix} \dots \\ \Delta^3 y_{-1} \end{matrix} \begin{Bmatrix} \Delta^4 y_{-2} \\ \Delta^4 y_{-1} \end{Bmatrix} \begin{matrix} \dots \\ \Delta^5 y_{-2} \end{matrix} \begin{Bmatrix} \Delta^6 y_{-3} \\ \Delta^6 y_{-2} \end{Bmatrix}$$

3. If interpolation is required for 'p' lying between $1/3$ and $3/4$ use Bessel's formula.

4.3.4 Stirling's Formula

Taking mean of Gauss' forward and backward interpolation formulae, we have

$$y_p = y_0 + \frac{p(\Delta y_0 + \Delta y_1)}{2} + \frac{p^2}{2!} \Delta^2 y_{-1}$$

$$+ \frac{p(p^2 - 1)}{3!} \frac{(\Delta^3 y_{-1} + \Delta^3 y_{-2})}{2} + \frac{p^2 (p^2 - 1)}{4!} \delta^4 y_{-2} + \dots \quad (4.16)$$

which is called *Stirling's formula*.

Notes: 1. In central difference notation Eq. (4.16) becomes

$$y_p = y_0 + p\mu\delta y_0 + \frac{p^2}{2!} \delta^2 y_0 + \frac{p(p^2 - 1)}{3!} \mu\delta^3 y_0 + \frac{p^2(p^2 - 1)}{4!} \Delta^4 y_0$$

As $\quad \frac{1}{2}(\Delta y_0 + \Delta y_{-1}) = \frac{1}{2}\left(\delta y_{\frac{1}{2}} + \delta y_{-\frac{1}{2}}\right) = \mu\delta y_0$

$$\frac{1}{2}(\Delta^3 y_{-1} + \Delta^3 y_{-2}) = \frac{1}{2}\left(\delta^3 y_{\frac{1}{2}} + \delta^3 y_{-\frac{1}{2}}\right) = \mu\delta^3 y_0 \text{ etc.}$$

2. For $-\frac{1}{4} < p < \frac{1}{4}$, we use Stirling formula

$$\dots y_0 \dots \begin{Bmatrix} \Delta y_{-1} \\ \Delta y_0 \end{Bmatrix} \dots \Delta^2 y_{-1} \dots \begin{Bmatrix} \Delta^3 y_{-2} \\ \Delta^3 y_{-1} \end{Bmatrix} \dots \Delta^4 y_{-2} \dots \begin{Bmatrix} \Delta^5 y_{-3} \\ \Delta^5 y_{-2} \end{Bmatrix}$$

4.4 CHOICE OF AN INTERPOLATION FORMULA

The right choice of an interpolation formula depends upon position of interpolated values in the given data.

The following *rules* will be found useful:

1. To find an interpolated value near the beginning of table, use Newton's forward formula.

2. To find an interpolated value near the end of table, use Newton's backward formula.

3. To find an interpolated value near the centre of the table, use Stirling's or Bessel's formulae.

If interpolation is required for p lying between $-\frac{1}{4}$ and $\frac{1}{4}$, prefer Stirling's formula. If interpolation is desired for p lying between $\frac{1}{4}$ and $\frac{3}{4}$, use Bessel's formula.

Example 4.14: Interpolate by means of Gauss' backward formula, the population of a town for the year 1974, from the given data.

Year	1939	1949	1959	1969	1979	1989
Population	12	15	20	27	39	52

Solution: Let us take $x_0 = 1969, p = \dfrac{1974 - 1969}{10} = 0.5.$

x		p	y_p	Δy_p	$\Delta^2 y_p$	$\Delta^3 y_p$	$\Delta^4 y_p$	$\Delta^5 y_p$
x_{-3}	1939	-3	12					
				3				
x_{-2}	1949	-2	15		2			
				5		0		
x_{-1}	1959	-1	20		2		3	
				7		3		-10
x_0	1969	0	27		5		-7	
				12		-4		
x_1	1979	1	39		1			
				13				
x_2	1989	2	52					

Gauss' backward formula is

$$y_p = y_0 + p\Delta y_{-1} + \frac{(p+1)p}{2!}\Delta^2 y_{-1} + \frac{(p+1)p(p-1)}{3!}\Delta^3 y_{-2}$$

$$+ \frac{(p+2)(p+1)p(p-1)}{4!}\Delta^4 y_{-2} + \frac{(p+2)(p+1)p(p-1)(p-2)}{5!}\Delta^5 y_{-3}$$

$$y_5 = 27 + 0.5 \times 7 + \frac{(0.5+1)(0.5)}{2!}(5) + \frac{(0.5+1)(0.5)(0.5-1)}{3!}(3)$$

$$+ \frac{(0.5+2)(0.5+1)(0.5)(0.5-1)}{4!}(-7) + \frac{(0.5+2)(0.5+1)(0.5)(0.5-1)(0.5-2)}{5!}(-10)$$

$$= 32.345$$

Example 4.15: Using Gauss' forward formula, evaluate $f(3.75)$ from the given table.

x	2.5	3.0	3.5	4.0	4.5	5.0
y	24.145	22.043	20.225	18.644	17.262	16.047

Solution: Here $h = 0.5$, taking $x_0 = 3.5$ as the origin $x = 3.75$

$$p = \frac{x - x_0}{h} = \frac{3.75 - 3.5}{0.5} = 0.5,\ \text{which lies between 0 and 1.}$$

Therefore, the central difference table is:

x	$y_p = f(x)$	Δy_p	$\Delta^2 y_p$	$\Delta^3 y_p$	$\Delta^4 y_p$	$\Delta^5 y_p$
$x_{-2} = 2.5$	24.145					
		-2.102				
$x_1 = 3.0$	22.043		0.284			
		-1.818		-0.047		
$x_0 = 3.5$	20.225		0.237		0.009	
		-1.581		-0.038		-0.003
$x_1 = 4.0$	18.644		0.199		0.006	
		-1.382		-0.032		
$x_2 = 4.5$	17.262		0.167			
		-1.215				
$x_3 = 5.0$	16.047					

By Gauss' forward interpolation formula

$$y(3.75) = 20.225 + 0.5(-1.584) + \frac{(0.5)(0.5-1)}{2!}(0.237) + \frac{(1.5)(0.5)(0.5-1)}{3!}(-0.038)$$

$$+ \frac{(1.5)(0.5)(0.5-1)(0.5-2)}{4!}(0.009) + \frac{(2.5)(1.5)(0.5)(0.5-1)(0.5-2)}{5!}(-0.003)$$

$$= 20.225 - 0.7905 - 0.0296 + 0.0024 + 0.0002 - 0.00004 + 0.0000$$

$$= 19.40746$$

Example 4.16: Use Gauss' backward interpolation formula to find $y(32)$ from the following table:

x	25	30	35	40
y	0.2707	0.3027	0.3386	0.3794

Solution: Here $h = 0.5$, taking $x_0 = 35$ as the origin, $x = 32$.

$$p = \frac{32-35}{5} = -\frac{3}{5} = -0.6, \text{ which lies between } -1 \text{ and } 0.$$

Let prepare the following central difference table.

x	y	Δy	$\Delta^2 y$	$\Delta^3 y$
$x_{-2} = 25$	0.2707			
		0.0320		
$x_{-1} = 30$	0.3027		0.0039	
		0.0359		0.0010
$x_0 = 35$	0.3386		0.0049	
		0.0408		
$x_1 = 40$	0.3794			

Gauss' backward interpolation formula

$$y(x) = y_0 + p\Delta y_{-1} + \frac{p(p+1)}{2!}\Delta^2 y_{-1} + \frac{(p+1)p(p-1)}{3!}\Delta^3 y_{-2}$$

$$y(32) = 0.3386 + (-0.6)(0.0359) + \frac{(-0.6+1)(-0.6)}{2!}(0.0049)$$

$$+ \frac{(-0.6+1)(-0.6)(-0.6-1)}{3!}(0.0010)$$

$$= 0.3386 - 0.0215 - 0.00069 + 0.00006 = 0.31657$$

Example 4.17: Using Stirling's formula, find y_{35}, given $y_{20} = 512$, $y_{30} = 439$, $y_{40} = 346$, $y_{50} = 246$, where y represents the number of persons at age x years in a life table.

Solution: Here $h = 10$, taking $x_0 = 30$ as the origin, $x = 35$.

$$\therefore \qquad p = \frac{35-30}{10} = 0.5$$

The central difference table is

	x	y	Δy	$\Delta^2 y$	$\Delta^3 y$
x_{-1}	20	512			
			-73		0
x_0	30	439		-20	
			-93		10
x_1	40	346		-10	
			-103		
x_2	50	243			

By Stirling formula, we have

$$y_{35} = y_0 + p\left(\frac{\Delta y_0 + \Delta y_{-1}}{2}\right) + \frac{p^2}{2!}\Delta^2 y_{-1} + \frac{p(p^2-1)}{3!}\left(\frac{\Delta^3 y_{-1} + \Delta^3 y_{-2}}{2}\right)$$

$$= 439 + 0.5\left(\frac{-73-93}{2}\right) + \frac{(0.5)^2}{2!}(-20) + \frac{(0.5)[(0.5)^2-1]}{6}\left(\frac{0+10}{2}\right) = 394.68$$

Example 4.18: Given the table

x	310	320	330	340	350	360
$\log x$	2.49136	2.50515	2.51851	2.53148	2.54407	2.55630

find the value of log 337.5 by Gauss forward interpolation formula.

Solution: Here $h = 10$, $x_0 = 330$, $x = 337.5$, then

$$p = \frac{337.5 - 330}{10} = 0.75$$

Therefore the central difference table:

x	p	y_p	Δy_p	$\Delta^2 y_p$	$\Delta^3 y_p$	$\Delta^4 y_p$	$\Delta^5 y_p$
310	-2	2.4914					
			0.0138				
320	-1	2.5052		-0.0005			
			0.0133		0.0002		
330	0	2.5185		-0.0003		-0.0003	
			0.0130		-0.0001		0.0004
340	1	2.5315		-0.0004		0.0001	
			0.0126		0.0000		
350	2	2.5441		-0.0004			
			0.0122				
360	3	2.5563					

Now from the central difference table, Gauss' interpolation formula is

$$y_p = y_0 + p\Delta y_0 + \frac{p(p-1)}{2!}\Delta^2 y_{-1} + \frac{(p+1)p(p-1)}{3!}\Delta^3 y_{-1}$$

$$+ \frac{(p+1)p(p-1))(p-2)}{4!}\Delta^4 y_{-2} + \frac{(p+2)(p+1)p(p-1)(p-2)}{5!}\Delta^5 y_{-2} + \dots$$

$$y_{0.75} = (2.518 + 0.75)(0.0130) + \frac{(0.75)(0.75-1)}{2!}(-0.0003)$$

$$+ \frac{(0.75+1)(0.75)(0.75-1)}{3!}(-0.0001) + \frac{(0.75+1)(0.75)(0.75-1)(0.75-2)}{4!}(-0.0003)$$

$$+ \frac{(0.75+2)(0.75+1)(0.75)(0.75-1)(0.75-2)}{5!}(0.0004)$$

$$= 2.5185 + 9.75 \times 10^{-3} + 2.8125 \times 10^{-3} + 5.46875 \times 10^{-6}$$

$$- 5.1269531 \times 10^{-6} + 3.7597656 \times 10^{-4}$$

$$= 2.5282822, \quad \therefore \log 337.5 = 2.5283$$

Example 4.19: Interpolate by means of Gauss' backward interpolation formula, the sale of a concern for year 1976 from the given data.

x	1940	1950	1960	1970	1980	1990
y = sale (in lakhs)	17	20	27	32	36	38

Solution: Taking 1970 as origin and $n = 10$, $p = \dfrac{x - x_0}{h} = \dfrac{1976 - 1970}{10} = 0.6$.

Therefore the central difference table is as below.

x	p	y_p	Δy_p	$\Delta^2 y_p$	$\Delta^3 y_p$	$\Delta^4 y_p$	$\Delta^5 y_p$
1940	-3	17					
			3				
1950	-2	20		4			
			7		-6		
1960	-1	27		-2		7	
			5		1		-9
1970	0	32		-1		-2	
			4		-1		
1980	1	36		-2			
			2				
1990	2	38					

Gauss' backward interpolation formula

$$y_p = y_0 + p\Delta y_{-1} + \frac{(p+1)p}{2!}\Delta^2 y_{-1} + \frac{(p+1)p(p-1)}{3!}\Delta^3 y_{-2}$$
$$+ \frac{(p+2)(p+1)p(p-1)}{4!}\Delta^4 y_{-2} + \frac{(p+2)(p+1)p(p-1)(p-2)}{5!}\Delta^5 y_{-3}$$

Here $x = 1976$, $x_0 = 1970$, $p = 0.6$.

$$y_{0.6} = 32 + (0.6)(5) + \frac{(0.6+1)(0.6)}{2!}(-1) + \frac{(0.6+1)(0.6)(0.6-1)}{3!}(1)$$

$$+ \frac{(0.6+2)(0.6+1)(0.6)(0.6-1)}{4!}(-2) + \frac{(0.6+2)(0.6+1)(0.6)(0.6-1)(0.6-2)}{5!}(-9)$$

$$= 32 + 3 - 4.8 - 0.064 + 0.0832 - 0.104832$$
$$= 30.114368$$

Example 4.20: Apply Bessel formula to obtain y_{25}, given that $y_{20} = 2854$, $y_{24} = 3162$, $y_{28} = 3544$, $y_{32} = 3992$.

Solution: Here $x_0 = 24$, $p = \dfrac{x - 24}{4}$, where $h = 4$.

The central difference table is as below.

x	p	y_p	Δy_p	$\Delta^2 y_p$	$\Delta^3 y_p$
20	-1	2854			
			308		
24	0	3162		74	
			382		-8
28	1	3544		66	
			448		
32	2	3992			

Here $x = 25$, $p = \dfrac{25 - 24}{4} = 0.25$, $y_0 = 3162$.

Putting the values from the table in Bessel formula, we get

$$y_{25} = 3162 + (0.25)(3.82) + \frac{(0.25 - 1)(0.25)}{2!}\left(\frac{74 + 66}{2}\right) + \frac{(0.25 - 0.5)(0.25)(0.25 - 1)}{3!}(-8)$$

$$= 3162 + 95.5 - 0.65625 - 0.0625 = 3256.7813$$

$$y_{25} = 3256.7813$$

Example 4.21: Employ Stirling formula to compute $y_{12.2}$ from the following table. ($y_x = 1 + \log_{20} \sin x$)

$x°$	10	11	12	13	14
y_x	0.23967	0.28060	0.31780	0.35201	0.38360

Solution:

x	p	y_x	Δy_x	$\Delta^2 y_x$	$\Delta^3 y_x$	$\Delta^4 y_x$
x_{-2} 10	-2	0.23967				
			0.04093			
x_{-1} 11	-1	0.28060		-0.00365		
			0.03728		0.00058	
x_0 12	0	0.31780		0.00307		0.00013
			0.03421		0.00045	
x_1 13	1	0.35201		-0.00262		
			0.03159			
x_2 14	3	0.38360				

Taking the origin at $x_0 = 12$, $h = 1$, $p = x - 12$, $x = 12.2$, $x = x_0 + ph \Rightarrow 12.2 = 12 + p \Rightarrow$
$p = 0.2$ which lies between $-\dfrac{1}{4}$ to $\dfrac{1}{4}$. By Stirling's formula

$$y_{12.2} = y_p = y_0 + p\left(\frac{\Delta y_{-1} + \Delta y_0}{2}\right) + \frac{p^2}{2!}\Delta^2 y_{-1}$$

$$+ \frac{p(p^2 - 1)}{3!}\left(\frac{\Delta^3 y_{-2} + \Delta^3 y_{-1}}{2}\right) + \frac{p^2(p^2 - 1)}{4!}\Delta^4 y_{-2}$$

$$= 0.31780 + 0.2\left[\frac{0.03728 + 0.3421}{2!}\right] + \frac{(0.2)^2}{2!}(-0.00307)$$

$$+ \frac{(0.2[(0.2)^2 - 1]}{3!}\left[\frac{0.00058 + 0.00045}{2}\right] + \frac{(0.2)^2[(0.2)^2 - 1)]}{4!}(-0.00013)$$

$$= 0.32497$$

PROBLEMS 4.2

1. Apply Gauss' backward formula to find $\sin 45°$ from the table.

θ	20	30	40	50	60	70	80
$\sin \theta$	0.34202	0.502	0.64279	0.76604	0.86603	0.93969	0.98481

[Ans. 0.70711]

2. Using Stirling's formula, find y_{53}, given $y_{20} = 512$, $y_{30} = 439$, $y_{40} = 346$, $y_{50} = 243$, where y_x represents the number of persons at six years age in a life table. [Ans. 395]

3. Find $\sqrt{12516}$ using Gauss backward interpolation formula given that

$\sqrt{12500} = 111.8033$, $\sqrt{12510} = 111.8981$, $\sqrt{12520} = 111.8928$ and $\sqrt{12530} = 111.9374$.

[Ans. 111.8749]

4. The pressure p of wind corresponding to velocity v is given by the following data. Estimate p when $v = 25$

v	10	20	30	40
p	1.1	2	4.4	7.9

[Ans. 3.0375]

5. Given

θ	0°	5°	10°	15°	20°	25°	30°
$\tan\theta$	0	0.875	0.1763	0.2679	0.3640	0.4663	0.5774

Using Stirling's formula, show that $\tan 16° = 0.2867$.

6. Apply Bessel's formula to get the value of $e^{0.45}$ given

x	0.40	0.42	0.44	0.46	0.48
p	1.4918	1.5220	1.5523	1.5841	1.6161

[Ans. 3.0375]

7. Using (i) Stirling's formula, (ii) Bessel's formula, find $f(1.22)$ from the table.

x	0.5	1.0	1.5	2.0	2.5	3.0
$f(x)$	0.19146	0.34134	0.43319	0.47725	0.49379	0.49865

[Ans. (i) 0.38891, (ii) 0.38873]

8. Using Bessel's formula, find (46.24) from the given data.

x	41	45	49	53
$x^{1/3}$	3.4482	3.5569	3.6593	3.7563

[Ans. 3.5893]

4.5 INTERPOLATION WITH UNEQUAL INTERVALS

Here, we shall discuss (i) Lagrange's interpolation formula (ii) Newton's general interpolation formula with divided differences for unequal intervals.

4.5.1 Lagrange's Interpolation Formula with Unequal Intervals

If $y = f(x)$ takes the value $y_0, y_1, y_2, ..., y_n$ corresponding to $x = x_0, x_1, x_2, ..., x_n$, then

$$f(x) = \frac{(x - x_1)(x - x_2)...(x - x_n)}{(x_0 - x_1)(x_0 - x_2)...(x_0 - x_n)} y_0 + \frac{(x - x_0)(x - x_2)...(x - x_n)}{(x_1 - x_0)(x_1 - x_2)...(x_1 - x_n)} y_1 + ...$$

$$+ \frac{(x - x_0)(x - x_1)...(x - x_{n-1})}{(x_n - x_0)(x_n - x_1)...(x_n - x_{n-1})} y_n \qquad (4.17)$$

This is Lagrange's interpolation formula for unequal intervals.

Proof: Let the function $y = f(x)$ takes the value $y_0 = f(x_0)$, $y_1 = f(x_1)$, ..., $y_n = f(x_n)$ corresponding to the value $x_0, x_1, x_2, ..., x_n$. Since there are $(n + 1)$ values of y corresponding to $(n + 1)$ values of x, we can represent the function $f(x)$ by a polynomial in x of degree n. Let the polynomial be of the form

$$y = f(x) = a_0(x - x_1)(x - x_2)(x - x_3)...(x - x_n)$$

$$+ a_1(x - x_0)(x - x_2)(x - x_3)...(x - x_n)$$

$$+ a_2(x - x_0)(x - x_1)(x - x_3)...(x - x_n)$$

$$+ ....$$

$$+ a_{n-1}(x - x_0)(x - x_1)(x - x_2)...(x - x_n)$$

$$+ a_n(x - x_0)(x - x_1)(x - x_2)...(x - x_{n-1}) \qquad (4.18)$$

Putting $x = x_0$, $y = y_0$ in Eq. (4.18), we get

$$y_0 = a_0(x_0 - x_1)(x_0 - x_2)...(x_0 - x_n)$$

or

$$a_0 = \frac{y_0}{(x_0 - x_1)(x_0 - x_2)...(x_0 - x_n)}$$

similarly putting $x = x_1$, $y = y_1$ in Eq. (4.18), we get

$$a_1 = \frac{y_0}{(x_0 - x_1)(x_0 - x_2)...(x_0 - x_n)}$$

Proceeding in the same way, we find $a_2, a_3, ..., a_n$.

Substituting the values of $a_0, a_1, ..., a_n$ in Eq. (4.18), we get Eq. (4.17).

Notes:
1. This formula can be used even when the intervals are equally spaced, but there is a drawback, it is quite cumbersome to apply.
2. This formula can also be used to split the given function into partial fractions. On dividing both sides of Eq. (4.18) by $(x - x_0)(x - x_1)...(x - x_n)$, we get

$$\frac{f(x)}{(x - x_0)(x - x_1)...(x - x_n)} = \frac{y_0}{(x_0 - x_1)(x_0 - x_2)...(x_0 - x_n)} \cdot \frac{1}{x - x_0}$$

$$= \frac{y_1}{(x_1 - x_0)(x_1 - x_2)...(x_1 - x_n)} \cdot \frac{1}{x - x_1}$$

$$= \frac{y_2}{(x_2 - x_0)(x_2 - x_1)...(x_2 - x_n)} \cdot \frac{1}{x - x_2}$$

$$= \frac{y_n}{(x_n - x_0)(x_n - x_1)...(x_n - x_{n-1})} \cdot \frac{1}{x - x_n}$$

4.5.2 Divided Differences

The Lagrange's interpolation formula has a drawback as it is lengthy, cumbersome and time consuming to overcome this, we use a most convenient formula known as *divided difference formula* which is derived as follows. Before deriving this formula, we first define divided differences.

If (x_0, y_0), (x_1, y_1), (x_2, y_2) ... be given points, then the first divided difference for the arguments x_0, x_1 is defined by the relation

$$[x_0, x_1] = \frac{y_1 - y_0}{x_1 - x_0}$$

or

$$f(x_0, x_1) = \frac{f(x_1) - f(x_0)}{x_1 - x_0}$$

Similarly

$$[x_1, x_2] = \frac{y_2 - y_1}{x_2 - x_1}$$

and

$$[x_2, x_3] = \frac{y_3 - y_2}{x_3 - x_2}$$

The second divided difference for x_0, x_1, x_2 is defined as

$$[x_0, x_1, x_2] = \frac{[x_1, x_2] - [x_0, x_1]}{x_2 - x_0}$$

The third divided difference x_0, x_1, x_2, x_3 is defined as

$$[x_0, x_1, x_2, x_3] = \frac{[x_1, x_2, x_3] - [x_0, x_1, x_2]}{x_3 - x_0} \text{ and so on}$$

Notes: 1. We have

$$[x_0, x_1] = \frac{y_1 - y_0}{x_1 - x_0} = \frac{y_0 - y_1}{x_0 - x_1}$$

$$= \frac{y_0}{x_0 - x_1} + \frac{y_1}{x_1 - x_0} = [x_1, x_0]$$

$$[x_0, x_1, x_2] = \frac{[x_1, x_2] - [x_0, x_1]}{x_2 - x_0}$$

$$= \frac{y_0}{(x_0 - x_1)(x_0 - x_2)} + \frac{y_1}{(x_1 - x_0)(x_1 - x_2)} + \frac{y_2}{(x_2 - x_0)(x_2 - x_1)}$$

$$= [x_1, x_2, x_0] = [x_2, x_0, x_1]$$

2. The nth divided difference of a polynomial of the nth degree is constant.

Let the arguments be equally spaced so that $(x_1 - x_0) = (x_2 - x_1) = ... (x_n - x_{n-1}) = h$, then

$$[x_0, x_1] = \frac{y_1 - y_0}{x_1 - x_0} = \frac{\Delta y_0}{h}$$

$$[x_0, x_1, x_2] = \frac{[x_1, x_2] - [x_0, x_1]}{x_2 - x_0} = \frac{1}{2h}\left[\frac{\Delta y_1}{h} - \frac{\Delta y_0}{n}\right] = \frac{1}{2!h^2}\Delta^2 y_0$$

In general

$$[x_0, x_1, x_2, ..., x_n] = \frac{1}{n!h^n}\Delta^n y_0$$

If the tabulated function is a nth degree polynomial, then $\Delta^n y_0$ will be constant. Hence nth divided difference will be constant.

4.5.3 Newton's Divided Difference Formula

Let $y_0, y_1, ..., y_n$ be the value of $y = f(x)$ corresponding to the arguments $x_0, x_1, ..., x_n$, then from the definition of divided differences, we have

$$[x, x_0] = \frac{y - y_0}{x - x_0}$$

so that

$$y = y_0 + (x - x_0)[x, x_0] \qquad (4.19)$$

Again

$$[x, x_0, x_1] = \frac{[x, x_0] - [x_0, x_1]}{x - x_1}$$

which gives

$$[x, x_0] = [x_0, x_1] + (x - x_1)[x, x_0, x_1]$$

Substituting this value of $[x, x_0]$ in Eq. (4.19), we get

$$y = y_0 + (x - x_0)[x_0, x_1] + (x - x_0)(x - x_1)[x, x_0, x_1] \qquad (4.20)$$

Also

$$[x, x_0, x_1] = [x_0, x_1, x_2] + (x - x_2)[x, x_0, x_1, x_2]$$

Substituting this value of $[x, x_0, x_1]$ in Eq. (4.18), we obtain

$$y = y_0 + (x - x_0)[x_0, x_1] + (x - x_0)(x - x_1)[x_0, x_1, x_2]$$
$$+ (x - x_0)(x - x_1)(x - x_2)[x_0, x_1, x_2, x_3] + ...$$
$$+ (x - x_0)(x - x_1)...(x - x_n)[x, x_0, x_1, ..., x_{n+1}]$$

which is called Newton's general interpolation formula with divided differences.

Example 4.22: Use Lagrange's interpolation formula to find the value of $f(x)$ corresponding to $x = 27$ from the following data.

x	14	17	31	35
$f(x)$	68.7	64.0	44.0	39.1

Solution: By Lagrange's interpolation formula

$$y = f(x) = \frac{(x - x_1)(x - x_2)(x - x_3)}{(x_0 - x_1)(x_0 - x_2)(x_0 - x_3)} y_0 + \frac{(x - x_0)(x - x_2)(x - x_3)}{(x_1 - x_0)(x_1 - x_2)(x_1 - x_3)} y_1$$
$$+ \frac{(x - x_0)(x - x_1)(x - x_3)}{(x_2 - x_0)(x_2 - x_1)(x_2 - x_3)} y_2 + \frac{(x - x_0)(x - x_1)(x - x_2)}{(x_3 - x_0)(x_3 - x_1)(x_3 - x_2)} y_3$$

Here $x_0 = 14$, $x_1 = 17$, $x_2 = 31$, $x_3 = 35$, $y_0 = 68.7$, $y_1 = 64.0$, $y_2 = 44.0$, $y_3 = 39.1$

$$y = f(x) \frac{(x - 17)(x - 31)(x - 35)}{(14 - 17)(14 - 31)(14 - 35)}(68.7) + \frac{(x - 14)(x - 31)(x - 35)}{(17 - 14)(17 - 31)(17 - 35)}(64.0)$$
$$+ \frac{(x - 14)(x - 17)(x - 35)}{(31 - 14)(31 - 17)(31 - 35)}(44.0) + \frac{(x - 14)(x - 17)(x - 31)}{(35 - 14)(35 - 17)(35 - 31)}(39.1)$$

$$y(x = 27) = -\frac{68.7}{1071}(27 - 17)(27 - 31)(27 - 35) + \frac{64}{756}(27 - 14)(27 - 31)(27 - 35)$$

$$- \frac{44}{952}(27 - 14)(27 - 17)(27 - 35) + \frac{39.1}{1512}(27 - 14)(27 - 17)(27 - 31)$$

$$= -20.52 + 35.22 + 48.07 - 13.45 = 49.3$$

Example 4.23: Find the equation of the cubic curve that passes through the points $(-1, -8)$, $(0, 3)$, $(2, 1)$ and $(3, 2)$ using Lagrange's interpolation formula.

Solution: Putting $x_0 = -1$, $x_1 = 0$, $x_2 = 2$, $x_3 = 3$, $y_0 = -8$, $y_1 = 3$, $y_2 = 1$, $y_3 = 2$.

By Lagrange's formula, we have

$$y = \frac{(x-0)(x-2)(x-3)}{(-1-0)(-1-2)(-1-3)}(-8) + \frac{(x+1)(x-2)(x-3)}{(0+1)(0-2)(0-3)}(3)$$

$$+ \frac{(x+1)(x-0)(x-3)}{(2+1)(2-0)(2-3)}(1) + \frac{(x+1)(x-0)(x-2)}{(3+1)(3-0)(3-2)}(2)$$

$$= \frac{2}{3}(x^3 - 5x^2 + 6x) + \frac{1}{2}(x^3 - 4x^2 + x + 6) - \frac{1}{6}(x^3 - 2x^2 - 3x) + \frac{1}{6}(x^3 - x^2 - 2x)$$

$$y = \frac{7}{6}x^3 - \frac{31}{6}x^2 + \frac{14}{3}x + 3$$

The equation of the cubic curve is $6y = 7x^3 - 31x^2 + 28x + 18$.

Example 4.24: Using Lagrange's formula, express the function $\dfrac{x^2 + 6x - 1}{(x^2 - 1)(x - 4)(x - 6)}$ as a sum of partial fraction.

Solution: Let

$$f(x) = x^2 + 6x - 1$$

$$x^2 + 6x - 1 = f(x) = y = \frac{(x - x_1)(x - x_2)(x - x_3)}{(x_0 - x_1)(x_0 - x_2)(x_0 - x_3)} y_0 + \frac{(x - x_0)(x - x_2)(x - x_3)}{(x_1 - x_0)(x_1 - x_2)(x_1 - x_3)} y_1$$

$$+ \frac{(x - x_0)(x - x_1)(x - x_3)}{(x_2 - x_0)(x_2 - x_1)(x_2 - x_3)} y_2 + \frac{(x - x_0)(x - x_1)(x - x_2)}{(x_3 - x_0)(x_3 - x_1)(x_3 - x_2)} y_3 \qquad \text{(i)}$$

Let $\qquad x_0 = -1$, $x_1 = 1$, $x_2 = 4$, $x_3 = 6$

$$y_0 = -6, y_1 = 6, y_2 = 39, y_3 = 71$$

Dividing Eq. (i) by $(x - x_0)(x - x_1)(x - x_2)(x - x_3)$ and putting the values of x_0, x_1, x_2, x_3 and y_0, y_1, y_2, y_3, we get

$$\frac{x^2 + 6x - 1}{(x + 1)(x - 1)(x - 4)(x - 6)} = \frac{6}{(-1 - 1)(-1 - 4)(-1 - 6)} \cdot \frac{1}{x + 1} + \frac{6}{(1 + 1)(1 - 4)(1 - 6)} \cdot \frac{1}{x - 1}$$

$$+ \frac{39}{(4 + 1)(4 - 1)(4 - 6)} \cdot \frac{1}{x - 4} + \frac{71}{(6 + 1)(6 - 1)(6 - 4)} \cdot \frac{1}{x - 6}$$

$$= \frac{6}{-2 \times -5 \times -7} \cdot \frac{1}{x + 1} + \frac{6}{2 \times -3 \times -5} \cdot \frac{1}{x - 1} +$$

$$\frac{39}{5 \times 3 \times -2} \cdot \frac{1}{x - 4} + \frac{71}{7 \times 5 \times 2} \cdot \frac{1}{x - 6}$$

$$= \frac{3}{35} \cdot \frac{1}{x + 1} + \frac{1}{5} \cdot \frac{1}{x - 1} - \frac{13}{10} \cdot \frac{1}{x - 4} + \frac{71}{70} \cdot \frac{1}{x - 6}$$

Example 4.25: Given $\log_{10} 654 = 2.8156$, $\log_{10} 658 = 2.8182$, $\log_{10} 659 = 2.8189$, $\log_{10} 661 = 2.8202$. Find by using Lagrange's formula, the value of $\log_{10} 656$.

Solution: Putting $\quad x_0 = 654$, $x_1 = 658$, $x_2 = 659$, $x_3 = 661$

$$y_0 = \log_{10} 654 = 2.8156,\ y_1 = \log_{10} 658 = 2.8182$$
$$y_2 = \log_{10} 659 = 2.8189,\ y_3 = \log_{10} 661 = 2.8202$$

$$y = \log_{10} x = \frac{(x-x_1)(x-x_2)(x-x_3)}{(x_0-x_1)(x_0-x_2)(x_0-x_3)} y_0 + \frac{(x-x_0)(x-x_1)(x-x_3)}{(x_1-x_0)(x_1-x_2)(x_1-x_3)} y_1$$

$$+ \frac{(x-x_0)(x-x_1)(x-x_3)}{(x_2-x_0)(x_2-x_1)(x_2-x_3)} y_2 + \frac{(x-x_0)(x-x_1)(x-x_2)}{(x_3-x_0)(x_3-x_1)(x_1-x_2)} y_3$$

$$= \log_{10} 656 = \frac{(656-658)(656-659)(656-661)}{(654-658)(654-659)(654-661)} \times (2.8156)$$

$$+ \frac{(656-654)(656-659)(656-661)}{(658-654)(658-659)(658-661)} \times (2.8182)$$

$$+ \frac{(656-654)(656-658)(656-661)}{(659-654)(659-651)(659-661)} \times (2.8189)$$

$$+ \frac{(656-654)(656-658)(656-659)}{(661-654)(661-658)(661-659)} \times (2.8202)$$

$$\log_{10} 656 = 0.6033 + 7.0455 - 5.6378 + 0.8058 = 2.8168$$

Example 4.26: Solve problem 4.25 by using Newton's divided difference formula.

Solution: Divided difference table:

x	$y = f(x)$ $= \log_{10} x$	$\Delta f(x)$ (1st divided difference)	$\Delta^2 f(x)$ (2nd divided difference)	$\Delta^3 f(x)$ 3rd divided difference
654	2.8156	$\frac{2.8182 - 2.8156}{658 - 654} = 6.5 \times 10^{-4}$		
			$\frac{(7-6.5)\times 10^{-4}}{659-654} = 1 \times 10^{-5}$	
658	2.8182	$\frac{2.8189 - 2.8182}{659 - 658} = 7 \times 10^{-4}$		$\frac{(1.6-1)\times 10^{-5}}{661-654} = 1.3 \times 10^{-5}$
			$\frac{(6.5-7)\times 10^{-4}}{661-658} = -1.6 \times 10^{-5}$	
659	2.8189	$\frac{2.8202 - 2.8189}{661 - 659} = 6.5 \times 10^{-4}$		
661	2.8202			

Newton's divided difference formula is

$$f(x) = y_0 + (x-x_0)\ [x_0, x_1] + (x-x_0)\ (x-x_1)\ [x_0, x_1, x_2] + (x-x_0)$$
$$(x-x_1)\ (x-x_2)\ [x_0, x_1, x_2, x_3]$$

$$= 2.8156 + (656-654)\ (0.00065) + (656-654)\ (656-658)\ (0.00001)$$
$$+ (656-654)\ (656-658)\ (656-659)\ (-0.000013)$$

$$= 2.8156 + 0.0013 - 0.00004 - 0.000156$$

$$= 2.816704$$

$$\log_{10} 656 = 2.8167$$

Note: $[x_0, x_1]$ = 1st divided difference

$[x_0, x_1, x_2]$ = 2nd divided difference

$[x_0, x_1, x_2, x_3]$ = 3rd divided difference

Example 4.27: From the following table, find $f(x)$ as a polynomial in x and find $f(9)$ using Newton's divided difference formula.

x	4	5	7	10	11	13
$f(x)$	48	100	294	900	1210	2028

Solution: Divided difference table:

x	$f(x)$	$\Delta f(x)$ (1st divided diff.)	$\Delta^2 f(x)$ (2nd divided diff.)	$\Delta^3 f(x)$ (3rd divided diff.)	$\Delta^4 f(x)$ (4th divided diff.)
4	48	$\dfrac{100-48}{5-4}=52$			
			$\dfrac{97-52}{7-4}=15$		
5	100	$\dfrac{294-100}{7-5}=97$		$\dfrac{21-15}{10-4}=1$	
			$\dfrac{202-97}{10-5}=21$		0
7	294	$\dfrac{900-294}{10-7}=202$		$\dfrac{27-21}{11-5}=1$	
			$\dfrac{310-202}{11-7}=27$		0
10	900	$\dfrac{1210-900}{11-10}=310$		$\dfrac{33-27}{13-7}=1$	
			$\dfrac{409-310}{13-10}=33$		
11	1210	$\dfrac{2028-1210}{13-11}=409$			
13	2028				

By Newton's divided difference formula

$$f(x) = y = 48 + (x-4)\times 52 + (x-4)(x-5)\times 15 + (x-4)(x-5)(x-7)\times 1$$
$$= 48 + 52x - 208 + 15x^2 - 135x + 3x + x^3 - 16x^2 + 83x - 140$$
$$= x^3 - x^2$$
$$f(9) = 93 - 92 = 729 - 81 = 648$$

PROBLEMS 4.3

1. Use Lagrange's interpolation formula to find the value of y when $x = 10$, from the given table.

x	5	6	9	11
y	12	13	14	16

[Ans. 14.666667]

2. Given that $\log_{10} 300 = 2.4771$, $\log_{10} 304 = 2.4829$, $\log_{10} 305 = 2.4843$ and $\log_{10} 307 = 2.4871$, find by using Lagrange's formula, the value of $\log_{10} 310$. [Ans. 2.4786]

3. Find the equation of the cubic curve which passes through the point $(4, -43)$, $(7, 83)$, $(9, 327)$, $(12, 1053)$. [Ans. $y = x^3 - x^2 + 4x - 6)/6$]

4. The following table gives the viscosity of an oil as a function of temperature. Use Lagrange's formula to find the viscosity of oil at a temperature of $140°$.

Temp	110	130	160	190
Viscosity	10.8	8.1	5.5	4.8

[Ans. 7.03]

5. Given that $f(-1) = -2, f(0) = -1, f(2) = 1, f(3) = 4$ fit a polynomial of third degree.

[Ans. $f(x) = (x^3 - x^2 + 4x - 6)/6$]

6. Given $y_1 = 22, y_2 = 30, y_4 = 82, y_7 = 106, y_8 = 206$, find y_6.   [Ans. 83.515]

7. Using a polynomial of third degree, complete the record given below of the export of a given commodity during five years:

Years	1917	1918	1919	1920	1921	
Export (in tons)	443	384	–	397	467	[Ans. 369]

8. Find a polynomial satisfied by the following table:

x	−4	−1	0	2	5
$f(x)$	1245	33	5	9	1335

[Ans. $f(x) = 3x^4 - 5x^3 + 6x^2 - 14x + 5$]

9. If $f(x) = x - 1$, show that $f(x_0, x_1, ..., x_n) = \dfrac{(-1)^n}{x_0 \cdot x_1 \cdot ... \cdot x_n}$.

10. Using Newton's divided difference formula, evaluate $f(8)$ and $f(15)$, from the given table.

x	4	5	7	10	11	13	
$f(x)$	48	100	294	900	1210	2028	[Ans. 448, 3150]

11. Given the values

x	5	7	11	13	17
$f(x)$	150	392	1452	2366	5202

evaluate $f(9)$, using Lagrange's and Newton's divided difference formula.

[Ans. 810]

12. Find the third divided difference with arguments 2, 4, 9, 10 of the function $f(x) = x^3 - 2x$.   [Ans. 1]

4.6 INVERSE INTERPOLATION

So far, we have been finding the value of y for a certain value of x, from the given set of values of x and y. However, we may want to find a value of x corresponding to the value of y (which is not in the table). The process of finding such a value of x is called the *inverse interpolation*. When the values of x are unequally spaced, Lagrange's method is used and when the values of x are equally spaced, the iterative method should be used.

4.6.1 Lagrange's Interpolation

In this method, we choose y as independent variable and x as dependent variable in Lagrange's formula already explained, we obtain

$$x = f(y) = \frac{(y - y_1)(y - y_2) \cdots (y - y_n)}{(y_0 - y_1)(y_0 - y_2) \cdots (y_0 - y_n)} x_0 + \frac{(y - y_0)(y - y_2) \cdots (y - y_n)}{(y_1 - y_0)(y_1 - y_2) \cdots (y_1 - y_n)} x_1$$

$$+ \cdots + \frac{(y - y_0)(y - y_1) \cdots (y - y_{n-1})}{(y_n - y_0)(y_n - y_1) \cdots (y_n - y_{n-1})} x_n \tag{4.21}$$

which is used for inverse interpolation.

Example 4.28: The following table gives the values of x and y

x	30	35	40	45	50
y	15.9	14.9	14.1	13.3	12.5

Find the value of x corresponding to $y = 13.6$.

Solution: Here $y_0 = 15.9$, $y_1 = 14.9$, $y_2 = 14.1$, $y_3 = 13.3$, $y_4 = 12.5$

$$x_0 = 30, x_1 = 35, x_2 = 40, x_3 = 45, x_4 = 50$$

Taking $y = 13.6$ in Eq. (4.21), we have

$$x = \frac{(13.6-14.9)(13.6-14.1)(13.6-13.3)(13.6-12.5)}{(15.9-14.9)(15.9-14.1)(15.9-13.3)(15.9-12.5)} \quad (30)$$

$$+ \frac{(13.6-15.9)(13.6-14.1)(13.6-13.3)(13.6-12.5)}{(14.9-15.9)(14.9-14.1)(14.9-13.3)(14.9-12.5)} \quad (35)$$

$$+ \frac{(13.6-15.9)(13.6-14.9)(13.6-13.3)(13.6-12.5)}{(14.1-15.9)(14.1-14.1)(14.1-13.3)(14.1-12.5)} \quad (40)$$

$$+ \frac{(13.6-15.9)(13.6-14.9)(13.6-14.1)(13.6-12.5)}{(13.3-15.9)(13.3-14.9)(13.3-14.1)(13.3-12.5)} \quad (45)$$

$$+ \frac{(13.6-15.9)(13.6-14.9)(13.6-14.1)(13.6-13.3)}{(12.5-15.9)(12.5-14.9)(12.5-14.1)(12.5-13.3)} \quad (50)$$

$$= 0.4044117 - 4.3237305 + 21.41276 - 2.1470014$$

$$= 43.14185 = 43.1 \text{ (approx.)}$$

4.6.2 Iterative Method

Let us consider Newton's forward interpolation formula

$$y_p = y_0 + p\Delta y_0 + \frac{p(p-1)}{2!}\Delta^2 y_0 + \frac{p(p-1)(p-2)}{3!}\Delta^3 y_0 + \frac{p(p-1)(p-2)(p-3)}{4!}\Delta^4 y_0 + \ldots$$

or

$$p = \frac{1}{\Delta y_0}\left[y_p - y_0 - \frac{p(p-1)}{2!}\Delta^2 y_0 - \frac{p(p-1)(p-2)}{3!}\Delta^3 y_0 - \frac{p(p-1)(p-2)(p-3)}{4!}\Delta^4 y_0 + \ldots\right]$$

$$(4.22)$$

Neglecting second and higher differences for first approximation

$$p^{(1)} = \frac{1}{\Delta y_0}(y_p - y_0) \quad (4.23)$$

Substituting value of Eq. (4.11) in RHS of Eq. (4.22), the second approximation

$$p^{(2)} = \frac{1}{\Delta y_0}\left[y_p - y_0 - \frac{p^{(1)}(p^{(1)}-1)}{2!}\Delta^2 y_0\right]$$

Similarly, third approximation

$$p^{(3)} = \frac{1}{\Delta y_0}\left[y_p - y_0 - \frac{p^{(2)}(p^{(2)}-1)}{2!}\Delta^2 y_0 - \frac{p^{(2)}(p^{(2)}-1)(p^{(2)}-2)}{3!}\Delta^3 y_0\right] \text{ and so on.}$$

Example 4.29: The following values of $y = f(x)$ are given in the following table.

x	10	15	20
y	1754	2648	3564

Find the value of x for $y = 3000$ by iterative method.

Solution: Taking $x_0 = 10$, $h = 5$, the difference table is

x	y	Δy	$\Delta^2 y$
10	1754		
		894	
15	2648		22
		916	
20	3564		

Here $y_p = 3000$, $y_0 = 1754$, $\Delta y_0 = 894$ and $\Delta^2 y_0 = 22$.

The successive approx. to p are

$$p^{(1)} = \frac{1}{894}[3000 - 1754] = 1.39$$

$$p^{(2)} = \frac{1}{894}\left[3000 - 1754 - \frac{(1.39)(1.39-1)}{2} \times 22\right] = 1.387$$

therefore, taking $p = 1.387$ correct to three decimal places, the value of x corresponding to $y = 3000$ is 1.387.

Example 4.30: From the following data

x	1.8	2.0	2.2	2.4	2.6
y	2.9	3.6	4.4	5.5	6.7

find x when $y = 5$, using iterative method.

Solution: The forward difference table:

x	p	Δy	$\Delta^2 y$	$\Delta^3 y$	$\Delta^4 y$
1.8	2.9				
		0.7			
2.0	3.6		0.1		
		0.8		0.2	
2.2	4.4		0.3		-0.4
		1.1		-0.2	
2.4	5.5		0.1		
		1.2			
2.6	6.7				

Here $y_p = 5$, $y_0 = 2.9$, $\Delta y_0 = 0.7$, $\Delta^2 y_0 = 0.1$, $\Delta^3 y_0 = 0.2$, $\Delta^4 y_0 = -0.4$.

Consider Newton's forward interpolation formula

$$y_p = y_0 + p\Delta y_0 + \frac{p(p-1)}{2!}\Delta^2 y_0 + \frac{p(p-1)(p-2)}{3!}\Delta^3 y_0 + \frac{p(p-1)(p-2)(p-3)}{4!}\Delta^4 y_0 + \dots$$

The 1st approximation:

$$p^{(1)} = \frac{1}{\Delta y_0}[y_p - y_0] = \frac{1}{0.7}[5 - 2.9] = 3$$

The second approximation:

$$p^{(2)} = \frac{1}{0.7}[5 - 2.9 - 0.3] = 2.5714, \; p^{(3)} = 2.46143$$

Proceeding in the same way, we get

$p^{(4)} = 2.4724147$	$p^{(5)} = 2.6364724$	$p^{(6)} = 2.5372869$	$p^{(7)} = 2.5985062$
$p^{(8)} = 2.5611579$	$p^{(9)} = 2.5841148$	$p^{(10)} = 2.5700665$	$p^{(11)} = 2.5766447$
$p^{(12)} = 2.573406$	$p^{(13)} = 2.5766447$	$p^{(14)} = 2.5746598$	$p^{(15)} = 2.5758768$

$p = 2.575$ correct to three decimal places.

Hence the value of x corresponding to $y_p = 5 = x_0 + ph = 1.8 + (2.575)(0.2) = 2.315$.

Example 4.31: Find the real root of the equation $x^3 + x - 3 = 0$ which lies between 1.2 and 1.3.

Solution: The forward difference table:

x	p	$y_p = (x^3 + x - 3)$	Δy	$\Delta^2 y$	$\Delta^3 y$
1	-2	-1			
			0.431		
1.1	-1	-0.569		0.066	
			0.497		0.006
1.2	0	-0.072		0.072	
			0.569		0.006
1.3	1	0.497		0.078	
			0.647		
1.4	2	1.144			

In the table $p = \dfrac{x - 1.2}{0.1}$ when $x_0 = 1.2$.

Now using Stirling's formula

$$y_p = y_0 + p\left[\frac{\Delta y_0 + \Delta y_{-1}}{2}\right] + \frac{p^2}{2!}\Delta^2 y_{-1} + \frac{p(p^2 - 1)}{3!}\left[\frac{\Delta^3 y_{-1} + \Delta^3 y_{-2}}{2}\right]$$

$$0 = -0.072 + p\left(\frac{0.569 + 0.497}{2}\right) + \frac{p^2}{2}(0.072) + \frac{p(p^2 - 1)}{6}\left(\frac{0.006 + 0.006}{2}\right)$$

$$0 = -0.072 + 0.532p + 0.036p^2 + 0.001p^3$$

which can be written as

$$p = \frac{0.072}{0.532} - \frac{0.036}{0.532}p^2 - \frac{0.001}{0.532}p^3$$

First approximation

$$p^{(1)} = \frac{0.072}{0.532} = 0.1353$$

$$p^{(1)} = 0.1353 - 0.067(0.1353)^2 - 1.8797 \times (1353)^3 = 0.134$$

Hence the desired root is $1.2 + 0.1 \times 0.134 = 1.2134$.

4.7 SPLINE INTERPOLATION

If we draw a cubic curve, through A_i, A_{i+1}, and another cubic through A_{i+1}, A_{i+2}, such that the slopes and curvatures of the two curves match at A_{i+1}. Such a curve is called *cubic spline*.

Cubic spline: Consider the problem of interpolating between the data points (x_0, y_0), (x_1, y_1) ... (x_n, y_n) by means of spline fitting.

Then the cubic spline function $f(x)$ is such that

1. $f(x)$ is a linear polynomial outside the interval (x_0, x_n).
2. $f(x)$ is a cubic polynomial in each of subintervals.
3. $f'(x), f''(x)$ are continuous in each point in the interval (x_0, x_n).

Before deriving a formula for cubic spline function, we consider linear spline interpolation.

Linear spline interpolation: Straight line is the simplest connection between two points. The first order splines for a group of ordered data points can be defined as a set of linear function.

Let $(x_i, y_i), i = 0, 1, 2, ..., n$ be the given $n + 1$ points.

Now for (x_i, x_{i+1}), $\dfrac{f(x) - f(x_i)}{f(x_{i+1}) - f(x_i)} = \dfrac{x - x_i}{x_{i+1} - x_i}$.

On simplification, we get

$$f(x) = f(x_i) + \frac{f(x_{i+1}) - f(x_i)}{x_{i+1} - x_i}(x - x_i)$$

$$\Rightarrow \qquad = \frac{x_{i+1} - x}{x_{i+1} - x_i} f(x_i) + \frac{x - x_i}{x_{i+1} - x_i} f(x_{i+1}) \quad x_i \le x \le x_{i+1}$$

Taking equal spaced values of x so that $x_{i+1} - x_i = h$

$$f(x) = \frac{x_{i+1} - x}{h} f(x_i) + \frac{x - x_i}{h} f(x_{i+1}) = \frac{1}{h}[(x_{i+1} - x) f(x_i) + (x - x_i) f(x_{i+1})]$$

which is a formula for $f(x)$, here $i = 1, 2, 3, ..., n$ and where $f(x)$ is a linear polynomial.

4.7.1 Cubic Spline Interpolation

Now, we shall derive a formula for $f(x)$, where $f(x)$ is a cubic polynomial in

$$x_i \le x \le x_{i+1}$$

$f''(x)$ is a linear expression in that interval.

By first order Lagrange's interpolation polynomial

$$f''(x) = \frac{1}{h}[(x_{i+1} - x) f''(x_i) - (x - x_i) f''(x_{i+1})] \tag{4.24}$$

Now integrating twice on both sides of Eq. (4.24), w.r.t. x, we get

$$f(x) = \frac{1}{h}\left[\frac{(x_{i+1} - x)^3}{3!} f''(x_i) + \frac{(x - x_i)^3}{3!} f''(x_{i+1})\right] + kx + k' \tag{4.25}$$

where k and k' are constants of integration.

Now we shall find k and k'.

Putting $x = x_i$

$$f(x_i) = \frac{1}{h} \frac{(x_{i+1} - x_i)^3}{3!} f''(x_i) + kx_i + k'$$

$$= \frac{1}{h} \left[\frac{h^3}{3!} f''(x_i) \right] + kx_i + k' \tag{4.26}$$

Putting $x = x_{i+1}$ in Eq. (4.24), we get

$$f(x_{i+1}) = \frac{1}{h} \left[\frac{(x - x_i)^3}{3!} f''(x_{i+1}) \right] + kx_{i+1} + k'$$

$$= \frac{1}{h} \left[\frac{h^3}{3!} f''(x_{i+1}) \right] + kx_{i+1} + k' \tag{4.27}$$

From Eqs (4.25) and (4.26), we have

$$f(x_i) - f(x_{i+1}) = \frac{h^2}{3!} [f''(x_i) - f''(x_{i+1})] + k[x_i - x_{i+1}]$$

$$y_i - y_{i+1} = \frac{h^2}{3!} [f''(x_i) - f''(x_{i+1})] - hk$$

$$k = \frac{h}{3!} [f''(x_i) - f''(x_{i+1})] - (y_i - y_{i+1}) \frac{1}{h}$$

$$= \frac{1}{h} \left[y_{i+1} - \frac{h^2}{3!} f''(x_{i+1}) \right] - \frac{1}{h} \left[y_i - \frac{h^2}{3!} f''(x_i) \right] = b_i - a_i$$

Now from Eq. (4.25)

$$hf(x_i) = \frac{h^3}{3!} f''(x_i) + hkx_i + hk'$$

$$f(x_i) = \frac{h^2}{3!} f''(x_i) + kx_i + k'$$

$$-k' = \frac{h^2}{3!} f''(x_i) - f(x_i) + (b_i - a_i) x_i = (b_i - a_i) x_i - \left(y_i - \frac{h^2}{3!} f''(x_i) \right)$$

$$= (b_i - a_i) x_i - a_i \times h$$

$$= (b_i - a_i) x_i - a_i (x_{i+1} - x_i)$$

$$= b_i x_i - a_i x_i - a_i x_{i+1} + a_i x_i$$

$$= b_i x_i - a_i x_{i+1}, \quad k' = a_i x_{i+1} - b_i x_i$$

$$f(x) = \frac{1}{h} \left[\frac{(x_{i+1} - x)^3}{3!} f''(x_1) + \frac{(x - x_i)^3}{3!} f''(x_{i+1}) \right] + (b_i - a_i) x - b_i x_i + a_i x_{i+1}$$

$$= \frac{1}{h} \left[\frac{(x_{i+1} - x)^3}{3!} f''(x_1) + \frac{(x - x_i)^3}{3!} f''(x_{i+1}) \right] + a_i [x_{i+1} - x] + b_i [x - x_i] \tag{4.27a}$$

where $\quad a_i = \frac{1}{h} \left[y_i - \frac{h^2}{3!} f''(x_i) \right], b_i = \frac{1}{h} \left[y_{i+1} - \frac{h^2}{3!} f''(x_{i+1}) \right]$

Substituting the values of a_i, b_i and writing $f''(x_i) = M_i$, Eq. (4.27a) becomes

$$f(x) = \frac{(x_{i+1} - x)^3}{6h} M_i + \frac{(x - x_i)^3}{6h} M_{i+1}$$

$$+ \frac{(x_{i+1} - x)}{6h}\left(y_i - \frac{h^2}{6} M_i\right) + \frac{(x - x_i)}{h}\left(y_{i+1} - \frac{h^2}{6} M_{i+1}\right) \qquad (4.27b)$$

$$f'(x) = \frac{-(x_{i+1} - x)^2}{2h} M_i + \frac{(x - x_i)^2}{2h} M_{i+1} - \frac{h}{6}(M_{i+1} - M_i) + \frac{1}{h}(y_{i+1} - y_i)$$

Now the cubic polynomial in the interval (x, x_{i+1}) is

$$f(x) = \frac{(x_{i+1} - x)^3}{6h} M_i + \frac{(x - x_i)^3}{6h} M_{i+1}$$

$$+ \frac{x_{i+1} - x}{h}\left(y_i - \frac{h^2}{6} M_i\right) + \frac{x - x_i}{h}\left(y_{i+1} - \frac{h^2}{6} M_{i+1}\right)$$

Differentiating w.r.t. x

$$f'(x) = \frac{-(x_{i+1} - x)^2}{2h} M_i + \frac{(x - x_i)^2}{2h} M_{i+1} - \frac{h}{6}(M_{i+1} - M_i) + \frac{1}{h}(y_{i+1} - y_i)$$

Setting $\qquad x = x_i$

$$f'(x_i) = \frac{-(x_{i+1} - x)^2}{2h} M_i - \frac{h}{6}(M_{i+1} - M_i) + \frac{1}{h}(y_{i+1} - y_i)$$

$$= -\frac{h}{2} M_i - \frac{h}{6}(M_{i+1} - M_i) + \frac{1}{h}(y_{i+1} - y_i)$$

$$= -\frac{2h}{6} M_i - \frac{h}{6} M_{i+1} + \frac{1}{h}(y_{i+1} - y_i)$$

$$h = x_{i+1} - x_i$$

Now, the cubic polynomial in the interval (x_{i-1}, x_i)

$$f(x) = \frac{(x_i - x)^3}{6h} M_{i-1} + \frac{(x - x_{i-1})^3}{6h} M_i$$

$$+ \frac{x_i - x}{h}\left(y_{i-1} - \frac{h^2}{6} M_{i-1}\right) + \frac{(x - x_{i-1})}{h}\left(y_i - \frac{h^2}{6} M_i\right)$$

Differentiating w.r.t. x

$$f'(x) = \frac{-(x_i - x)^2}{h} M_{i-1} + \frac{(x - x_{i-1})^2}{2h} M_i + \frac{h}{6} M_{i-1} - \frac{1}{h} y_{i-1} + \frac{1}{h} y_i - \frac{h}{6} M_i$$

Setting $\qquad x = x_{i-1}$

$$f'(x_i) = \frac{(x_i - x_{i-1})^2}{2h} M_i + \frac{h}{6} M_{i-1} - \frac{h}{6} M_i + \frac{1}{h}[y_i - y_{i-1}]$$

$$= \frac{h}{2} M_i - \frac{h}{6} M_i + \frac{h}{6} M_{i-1} + \frac{1}{h}[y_i - y_{i-1}]$$

$$= \frac{2h}{6} M_i + \frac{h}{6} M_{i-1} + \frac{1}{h}[y_i - y_{i-1}]$$

Apply condition of continuity of $f'(x)$ at $x = x_i$, we get

$$-\frac{2h}{6}M_i - \frac{h}{6}M_{i+1} + \frac{1}{h}(y_{i+1} - y_i) = \frac{h}{6}M_i + \frac{h}{6}M_i + \frac{h}{6}M_{i-1} + \frac{1}{h}(y_i - y_{i-1})$$

$$-\frac{h}{6}(2M_i + M_{i+1}) + \frac{1}{h}(y_{i+1} - y_i) = \frac{h}{6}(2M_i - M_{i-1}) + \frac{1}{h}(y_i - y_{i-1})$$

or
$$M_{i-1} + 4M_i + M_{i+1} = \frac{6}{h^2}(y_{i-1} - 2y_i + y_{i-1}) \qquad (4.27c)$$

$$i = 1 \text{ to } n - 1$$

Now, since the graph is linear for $x < x_0$ and $x > x_n$, we have

$$M = 0, M_n = 0 \qquad (4.27d)$$

Equations (4.27c) and (4.27d) gives $(n + 1)$ equations in $(n + 1)$, unknowns M_i ($i = 0, 1, ..., n$)which can be solved, substituting the value of M_i in Eq. (4.27b) gives the concerned cubic spline.

Example 4.32: Find the cubic splines from the values of x and y given below.

x	1	2	3
y	-8	-1	18

Solution: The points are equispaced with $h = 1$, $n = 2$, $M_0 = M_2 = 0$

$$x_0 = 1, x_1 = 2, x_3 = 0$$
$$y_0 = -8, y_1 = -1, y_2 = 18$$

for the interval $1 \le x \le 2$.

So the recurrence relation

$$M_{i-1} + 4M_i + M_{i+1} = \frac{6}{h^2}(y_{i+1} - 2y_i + y_{i-1}), i = 1, 2, ..., n$$

$$\therefore \qquad M_0 + 4M_1 + M_2 = \frac{6}{1}[18 - 2(-1) + (-8)]$$

$$= 6[18 + 2 - 8] = 6 \times 12 = 72$$
$$4M_1 = 72$$
$$M_1 = 18$$

$$f(x) = \frac{1}{h}\left[\frac{(x_i - x)^3}{6}M_{i-1} + \frac{(x - x_{i-1})^3}{6}M_i + \left(y_{i-1} - \frac{h^2}{6}M_{i-1}\right)(x_i - x)\right.$$

$$\left. + \left(y_i - \frac{h^2}{6}M_i\right)(x_i - x_{i-1})\right]$$

$$= \frac{1}{h}(x_1 - x)^3 M_0 + \frac{(x - x_0)^3}{6}M_1 + \left(y_0 - \frac{1}{h}M_0\right)(x_1 - x)$$

$$+ \left(y_1 - \frac{1}{h}M_1\right)(x - x_0)$$

$$= \left[(1 - x)^3 \times 0 + \frac{(x - 1)^3}{6} \times 18 + \left(-8 - \frac{1}{1} \times 0\right)(2 - x)\right.$$

$$\left. + \left(-1 - \frac{1}{6} \times 18\right)(x - 1)\right]$$

$$= 3(x - 1)^3 - 8(2 - x) + (-1 - 3)(x - 1)$$

$$f(x) = 3(x - 1)^3 - 8(2 - x) - 4(x - 1)$$

$$y(1.5) = -\frac{45}{8}$$

Example 4.33: Consider the function $y = \sin x$ $(0 \le x \le p)$ and interpolate between three of its values, i.e. the points $(0, 0)$, $(\pi/2, 1)$ and $(\pi, 0)$ and find $f(\pi/6)$.

Solution:

x	0	$\pi/2$	π
$f(x)$	0	1	0

Here $n = 2$ and $h = \pi/2$, $M_0 = 0$, $M_2 = 0$.

Recurrence relation gives

$$M_0 + 4M_1 + M_2 = \frac{6 \times 4}{\pi^2}[0 - 2 + 0] = -\frac{48}{\pi^2}$$

$$4M_1 = -\frac{48}{\pi^2}$$

$$M_1 = -\frac{12}{\pi^2}$$

In the interval $(0, \pi/2)$, the spline is given by

$$f(x) = \frac{1}{h}\left[\frac{(x_{i+1} - x)^3}{6}M_i + \frac{(x - x_i)^3}{6}M_{i+1} + \left(y_i - \frac{h^2}{6}M_i\right)(x_{i+1} - x)\right.$$

$$\left. + \left(y_{i+1} - \frac{h^2}{6}M_{i+1}\right)(x - x_i)\right]$$

$$= \frac{1}{\frac{\pi}{2}}\left[\frac{(x - x_0)^3}{6}M_1 + \left(y_0 - \frac{\pi^2}{4}M_0\right)(x_1 - x) + \left(y_1 - \frac{\pi^2}{4 \times 6}M_1\right)(x - x_0)\right]$$

$$= \frac{2}{\pi}\left[\frac{x^3}{6}\left(\frac{12}{\pi^2}\right) + 0 + \left\{1 - \frac{\pi^2}{24}\left(\frac{-12}{\pi^2}\right)\right\}(x - 0)\right]$$

$$= \frac{2}{\pi}\left[-\frac{2x^3}{6} + \left(1 + \frac{1}{2}\right)x\right] = \frac{2}{\pi}\left[-\frac{2x^3}{6} + \frac{3}{2}x\right]$$

Hence $\qquad f(\pi/6) = 0.4815$

Example 4.34: The following values of x and y are given below.

x	1	2	3	4
y	1	2	5	11

Find the cubic spline and evaluate $y(1.5)$ and $y'(3)$.

Solution: Here $n = 3$ and $h = 1$, we obtain the recurrence relations for $i = 1, 2$ as follows.

$$M_{i-1} + 4M_i + M_{i+1} = 6(y_{i-1} - 2y_i + y_{i+1})$$

$$M_0 + 4M_1 + M_2 = 6(y_0 - 2y_1 + y_2) = 12$$

$$M_1 + 4M_2 + M_3 = 6(y_1 - 2y_2 + y_3) = 18$$

Since $M_0 = 0$ and $M_3 = 0$

$$4M_1 + M_2 = 12$$
$$M_1 + 4M_2 = 18$$

we get $M_1 = 2$, $M_2 = 4$ which gives cubic spline by substituting in

$$f(x) = \frac{(x_{i+1} - x)^3}{6h} M_i + \frac{(x - x_i)^3}{6h} M_{i+1} + \frac{x_{i+1} - x}{h}\left(y_i - \frac{h^2}{6} M_i\right) + \frac{x - x_i}{h}\left(y_{i+1} - \frac{h^2}{6} M_{i+1}\right)$$

$$= \begin{cases} \dfrac{1}{3}(x^3 - 3x^2 + 5) & 1 \le x \le 2 \\[2mm] \dfrac{1}{3}(x^3 - 3x^2 + 5) & 2 \le x \le 3 \\[2mm] \dfrac{1}{3}(-2x^3 + 24x^2 - 76x + 81) & 3 \le x \le 4 \end{cases}$$

PROBLEMS 4.4

1. Using cubic spline, find $y(0.5)$ and $y'(1)$ given $M_0 = M_2 = 0$ from the table:

x	0	1	2
y	-5	-4	3

[Ans. $(-81/16), 4$]

2. Find the cubic spline approximation for the function given below:

x	0	1	2	3
$y = f(x)$	1	2	33	244

Assume $M(0) = M(3) = 0$. Also find $y(2.5)$. [Ans. 121.25]

3. Test whether the following functions are cubic splines or not?

   i. $P_1(x) = x^2 - x + 1$ $1 \le x \le 2$

      $P_2(x) = 3x - 3$ $2 \le x \le 3$

  ii. $P_1(x) = -2x^2 + x^3$ $-1 \le x \le 0$

      $P_2(x) = x^2 - 2x^3$ [Ans. i. Not a cubic spline, ii. Not a cubic spline]

4. Find the cubic spline approximate for the function $y = f(x)$ given in the following data. Assume that $y_0'' = y_3'' = 0$.

x	0	1	2	3
y	-5	-4	3	22

$$\text{Ans. } y = \begin{bmatrix} \dfrac{1}{5}(4x^3 + x - 25) & 0 \le x \le 1 \\[2mm] \dfrac{1}{5}(0x^3 - 18x^2 + 19x - 31) & 1 \le x \le 2 \\[2mm] \dfrac{1}{5}(-14x^3 + 126x^2 - 269x + 161) & 2 \le x \le 3 \end{bmatrix}$$

4.8 HERMITE'S INTERPOLATION FORMULA

The Hermitian interpolation is rather similar to the Lagrangian. The interpolation formula so far considered, make use of only a certain number of function values. We now derive an interpolation formula in which both the function and its first derivative values are to be assigned at each point of interpolation. This is referred to as *Hermitian interpolation formula*. The interpolation problem is then defined as given a set of data points (x_p, y_p, y_i'), $i = 0, 1, 2, ..., n$, it is required to determine a polynomial of the least degree, say $H_{2n+1}(x_i)$, such that

$$\left. \begin{aligned} H_{2n+1}(x_i) &= y_i \\ H'_{2n+1}(x_i) &= y_i' \end{aligned} \right\}_{i=0, 1, 2, ..., n} \tag{4.28}$$

where the primes denote differentiation with respect to x. The polynomial $H_{2n+1}(x)$ is called *Hermitian interpolation polynomial*. We have $(2n + 2)$ conditions and therefore, the number of coefficients to be determined is $(2n + 2)$ and the degree of polynomial is $(2n + 1)$. In analogy with the Lagrange's interpolation formula, we seek a representation of the form

$$H_{2n+1}(x) = \sum_{i=0}^{n} u_i(x) y_i + \sum_{i=0}^{n} v_i(x) y_i' \tag{4.29}$$

Here $u_i(x)$ and $v_i(x)$ are polynomial in x of degree $(2n + 1)$.
Using conditions of Eq. (4.27)

$$\left. \begin{aligned} u_i(x_j) &= \begin{cases} 1 \text{ if } i = h, v_i(x) = 0 \\ 0 \text{ if } i \neq j, \text{ for all } i \end{cases} \\ u_i'(x) &= 0, \text{ for all } i; v_i(x_j) = \begin{cases} 1 \text{ if } i = j \\ 0 \text{ if } i \neq j \end{cases} \end{aligned} \right\} \tag{4.30}$$

Since $u_i(x)$ and $v_i(x)$ are polynomials in x of degree $(2n + 1)$, we write

$$\begin{aligned} u_i(x) &= A_i(x) [l_i(x)]^2 \\ \text{and} \qquad v_i(x) &= = B_i(x) [l_i(x)]^2 \end{aligned} \tag{4.31}$$

where $l_i(x)$ are given by Lagrange's interpolation formula. It is easy to see that $A_i(x)$ and $B_i(x)$ are both linear functions in x, we therefore write

$$\begin{aligned} u_i(x) &= (a_i x + b_i) [l_i(x)]^2 \\ \text{and} \qquad v_i(x) &= (c_i x + d_i) [l_i(x)]^2 \end{aligned} \tag{4.32}$$

Using conditions of Eqs (4.31) and (4.32), we obtain

$$\left. \begin{aligned} a_i x_i + b_i &= 1 \\ c_i x_i + d_i &= 0 \\ c_i &= 1 \end{aligned} \right\} \tag{4.33}$$

From Eq. (4.32) we deduce

$$\begin{aligned} a_i &= -2l_i'(x_i) \\ b_i &= 1 + 2x_i l_i'(x_i) \\ c_i &= 1 \\ d_i &= -x_i \end{aligned}$$

Equation (4.30) becomes

$$u_i(x) = [-2x_i l_i'(x_i) + 1 + 2x_i l_i'(x_i)][l_i(x)^2]$$
$$= [1 - 2(x - x)l_i'(x_i)][l_i(x)]^2$$

and $$v_i(x) = (x - x_i)[l_i(x)]^2$$

using the above expression for $u_i(x)$ and $v_i(x)$ in Eq. (4.29)

$$H_{2n+1}(x) = \sum_{i=1}^{n}[1 - 2(x - x_i)\, l_i'(x_i)][l_i(x)]^2\, y_i + \sum_{i=0}^{n}(x - x_i)[l_i(x)]^2\, y_i' \quad (4.34)$$

which is required Hermitian interpolation formula.

Example 4.35: Determine the Hermitian polynomial of fifth degree which fits to the following data and hence find an approximate value of ln 2.7.

x	2.0	2.5	3.0
$y = \ln x$	0.69315	0.91629	1.09861
$y' = 1/x$	0.5	0.4000	0.33333

Solution: $$l_0(x) = \frac{(x - 2.5)(x - 3.0)}{(-0.5)(-1.0)} = 2x^2 - 11x + 15$$

Similarly, we find $$l_1(x) = -(4x^2 - 20x + 24)$$
and $$l_2(x) = 2x^2 - 9x + 10$$
We obtain $$l_0'(x) = 4x - 11$$
$$l_1'(x) = -8x + 20$$
$$l_2'(x) = 4x - 9$$
Hence $$l_0'(x_0) = 4 \times 2 - 11 = -3$$
$$l_1'(x_1) = -8 \times 2.5 + 20 = 0$$
$$l_2'(x_2) = 4 \times 3 - 9 = 3$$
$$u_0(x) = [1 - 2(x - 2)\, l_0'(x_0)][l_0(x)]^2$$
$$= [1 - 2(x - 2)(-3)[2x^2 - 11x + 15]^2$$
$$= (6x - 11)\{2x^2 - 11x + 15\}^2$$
$$u_1(x) = [1 - 2(x - x_1)\, l_1'(x_1)][l_1(x)]^2$$
$$= [1 - 2(x - 2.5) \times 0][-(4x_2 - 20x + 24)]^2$$
$$= (4x^2 - 20x + 24)2$$
$$u_2(x) = [1 - 2(x - x_2)\, l_2'(x_2)][l_2(x)]^2$$
$$= [1 - 2(x - 3)(3)][2x^2 - 9x + 10)]^2$$
$$= (19 - 6x)(2x^3 - 9x + 10)^2$$
$$v_0(x) = (x - x_0)[l_0(x)]^2$$
$$= (x - 2)(2x^2 - 11x + 15)^2$$
$$v_1(x) = (x - x_1)[l_1(x)]^2 = (x - 2.5)(4x^2 - 20x + 24)^2$$
$$v_2(x) = (x - 3)(2x^2 - 9x + 10)^2$$

Substituting these expressions in Hermitian equation, we have

$$H_{2n+1}(x) = \sum_{i=0}^{2} [1 - 2(x - x_i) l_i'(x_i)] [l_i(x)]^2 y_i + \sum_{i=0}^{2} (x - x_i) [l_i(x)]^2 y_i'$$

or

$$H_{2n+1}(x) = \sum_{i=0}^{2} u_i(x) y_i + \sum_{i=0}^{2} v_i(x) y_i'$$

$$\begin{aligned} H_5(x) = {} & (6x - 11)(2x^2 - 11x + 15)^2 (0.69315) + (4x^2 - 20x + 24)^2 (0.91629) \\ & + (19 - 6x)(2x^2 - 9x + 10)^2 (1.09861) + (x - 2)(2x^2 - 11x + 15)^2 (0.5) \\ & + (x - 2.5)(4x^2 - 20x + 24)^2 (0.4) + (x - 3)(2x^2 - 9x + 10)^2 (0.33333) \end{aligned}$$

Putting $x = 2.7$ and simplifying, we obtain

$$H_5(2.7) = 0.993252$$

which is correct to six decimal places. This is therefore, a more accurate result than that obtained by using Lagrange's interpolation formula.

Example 4.36: Apply Hermite interpolation formula to find the value of $\sin(1.05)$ from the following data.

x	1.00	1.10
$y = \sin x$	0.84147	0.89121
$y' = \cos x$	0.54030	0.45360

Solution:

$$l_0(x_0) = \frac{x - 1.10}{1.00 - 1.10} = \frac{x - x_1}{x_0 - x_1}$$

$$l_0(x_0) = -10x + 11$$

$$l_0'(x_0) = -10$$

$$l_1(x_1) = \frac{x - x_0}{x_1 - x_0} = \frac{x - 1.00}{1.10 - 1.00}$$

$$l_1(x) = 10x - 10$$

$$l_1'(x_1) = 10$$

Since $f(x_0) = 0.84147$, $f'(x_0) = 0.54030$, $f(x_1) = 0.89121$, $f'(x_1) = 0.45360$.

Now by Hermite interpolation formula,

substituting these expressions in the equation, we have

$$H_{2n+1}(x) = \sum_{i=0}^{1} [1 - 2(x - x_i) l_i'(x_i)] [l_i(x_i)]^2 f(x) + \sum_{i=0}^{1} (x - x_i) [l_i(x)]^2 f'(x_i)$$

$$\begin{aligned} = {} & [1 - 2l_0'(x_0)(x - x_0)][l_0(x)]^2 f(x_0) + [1 - 2l_1'(x_1)(x - x_1)][l_1(x)]^2 f(x_1) \\ & + (x - x_0)[(l_0(x)]^2 f'(x_0) + (x - x_1)[l_1(x)]^2 f'(x_1) \\ = {} & [1 + 20(x - 1.00)][-10x + 11]^2 (0.84147) + [1 - 20(x - 1.10)] \\ & (10x - 10)^2 (0.89121) + (x - 1.00)(-10x + 11)^2 (0.54030) \\ & + (x - 1.10)(10x - 10)^2 (0.45360) \end{aligned} \tag{i}$$

Putting $x = 1.05$ in Eq. (i), we get

$$H_3(1.05) = \sin(1.05)$$

$$= 1 + 20(1.05 - 1.00)[-10(1.05) + 11]^2(0.84147) + [1 - 20(1.05 - 1.00)]$$
$$[10(1.05) - 10]^2(0.89121) + (1.05 - 100)[-10(1.05) + 11]^2(0.54030)$$
$$+ (1.05 - 1.10)[10(1.05) - 10]^2(0.45360)$$

$$= 0.420735 + 0.445605 + 0.00675375 - 0.00567 = 0.86742$$

Example 4.37: Apply Hermitian formula to find a polynomial from the following data

x	0	1	2
$f(x)$	0	1	0
$f'(x)$	0	0	0

Solution: $x_0 = 0, x_1 = 1, x_2 = 0$

$$f(x_0) = 0, f(x_1) = 1, f(x_2) = 0$$
$$f'(x_0) = 0, f'(x_1) = 0, f'(x_2) = 0$$

$$l_0(x) = \frac{(x - x_1)(x - x_2)}{(x_0 - x_1)(x_0 - x_2)}, l_0(x) = \frac{(x - 1)(x - 2)}{(0 - 1)(0 - 2)} = \frac{1}{2}(x^2 - 3x + 2)$$

$$l_0'(x_0) = \frac{1}{2}(2x_0 - 3) = \frac{1}{2}(-3) = -\frac{3}{2}$$

$$l_1(x) = \frac{(x - x_0)(x - x_1)}{(x_1 - x_0)(x_1 - x_2)} = \frac{x(x - 2)}{(1 - 0)(1 - 2)} = -(x^2 - 2x)$$

$$l_1'(x_1) = -(2x_1 - 2) = -(2 - 2) = 0$$

$$l_2(x) = \frac{(x - x_0)(x - x_1)}{(x_2 - x_0)(x_2 - x_1)} = \frac{x(x - 1)}{(2 - 0)(2 - 0)} = \frac{1}{2}(x^2 - x)$$

$$l_2'(x_2) = \frac{1}{2}(2x_2 - 1) = \frac{1}{2}(4 - 1) = \frac{3}{2}$$

Now by Hermite interpolation formula, we have

$$H_{2n+1}(x) = \sum_{i=0}^{2}[1 - 2l_i'(x_i)(x - x_i)][l_i(x)]^2 f(x_i) + \sum_{i=0}^{2}(x - x_i)[l_i(x)]^2 f'(x_i)$$

$$= [1 - 2l_0(x_0)(x - x_0)][l_0(x)]^2 f(x_0) + [1 - 2l_1'(x_1)(x - x_1)][l_1(x)]^2 f(x_1)$$
$$+ [1 - 2l_2'(x_2)(x - x_2)][l_2(x)]^2 f(x_2) + (x - x_0)[l_0(x)]^2 f'(x_0)$$
$$+ (x - x_1)[l_1(x)]^2 f'(x_1) + (x - x_2)[l_2(x)]^2 f'(x_2)$$

$$= [1 + 3(x - 0)][x^2 - 3x + 2]^2 \cdot \frac{1}{4} \times 0 + (1 - 0)(x^2 - 2x)^2 \cdot 1$$

$$+ [1 - 3(x - 2)]\frac{1}{4}(x^2 - x)^2 \times 0 + (x - 0)\frac{1}{4}(x^2 - 3x + 2)^2 \times 0$$

$$+ (x - 1)(x^2 - 2x)^2 \times 0 + (x - 2)\left[\frac{1}{4}(x^2 - x)^2\right] \times 0$$

$$H_5(x) = (x^2 - 2x)^2$$

This is the required polynomial.

PROBLEMS 4.5

1. Apply Hermitian formula to estimate the value of log 3.2 from the following data.

x	3.0	3.5	4.0
$y = \log x$	1.0986	1.2528	1.3863
$y' = 1/x$	0.3333	0.2857	0.2500

[Ans. 1.1631]

2. Find Hermitian polynomial of third approximating function $y(x)$ such that $y(x_0) = 1, y''(x_0) = 0, y'(x_1) = 0$.

4.9 INTERPOLATION BY ITERATION

Newton's general interpolation formula may be considered as one of a class of methods which generate successively higher order interpolation formula. We describe now another method of this class, due to AC Aitken, which has the advantage of being very easily programmed for a digital computer.

Given $(n + 1)$ points $(x_0, y_0 (x_1, y_1) \dots (x_n, y_n)$

where the values of x need not necessarily be equally spaced, then to find the value of y corresponding to any given value of x, we proceed iteratively as follows. Obtain a first approximati... .· y by considering the first two points only, then obtain its second approximation by considering the first three points and so on. We denote the different interpolation polynomials by $\Delta(x)$ so that at the first stage of approximation, we have

$$\Delta_{01}(x) = y_0 + (x - x_0)[x_0, x_1] = \frac{1}{(x_1 - x_0)}\begin{vmatrix} y_0 & x_0 - x \\ y_1 & x_1 - x \end{vmatrix} \tag{4.35}$$

Similarly, we can form $\Delta_{02}(x)$, $\Delta_{03}(x)$.

Next, we form Δ_{01} by considering the first three poi .ts

$$\Delta_{012}(x) = \frac{1}{(x_2 - x_1)}\begin{vmatrix} \Delta_{01}(x) & x_1 - x \\ \Delta_{02}(x) & x_2 - x \end{vmatrix} \tag{4.36}$$

Similarly, we obtain $\Delta_{013}(x)$, $\Delta_{014}(x)$ etc. At the nth stage of approximation, we obtain

$$\Delta_{012 \dots n}(x) = \frac{1}{x_n - x_{n-1}}\begin{vmatrix} \Delta_{012\dots\overline{n-1}}(x) & x_{n-1} - x \\ \Delta_{012\dots\overline{n-2n}}(x) & x_n - x \end{vmatrix} \tag{4.37}$$

The computation may conveniently be arranged as shown in the table.

		Aitken's Scheme			
x	y				
x_0	y_0				
		$\Delta_{01}(x)$			
x_1	y_1		$\Delta_{012}(x)$		
		$\Delta_{02}(x)$		$\Delta_{0123}(x)$	
x_2	y_2		$\Delta_{013}(x)$		$\Delta_{01234}(x)$
		$\Delta_{03}(x)$		$\Delta_{0124}(x)$	
x_3	y_3		$\Delta_{014}(x)$		
		$\Delta_{04}(x)$			
x_4	y_4				

A modification of this scheme, due to Neville is given in the table below. Neville's scheme is particularly suited for iterated inverse interpolation.

		Neville's scheme			
x	y				
x_0	y_0				
		$\Delta_{01}(x)$			
x_1	y_1		$\Delta_{012}(x)$		
		$\Delta_{12}(x)$		$\Delta_{0123}(x)$	
x_2	y_2		$\Delta_{123}(x)$		$\Delta_{01234}(x)$
		$\Delta_{23}(x)$		$\Delta_{1234}(x)$	
x_3	y_3		$\Delta_{234}(x)$		
		$\Delta_{34}(x)$			
x_4	y_4				

Example 4.38: By using Aitken's scheme, find the value of $\log_{10} 301$.

x	300	304	305	307
$\log_{10} x$	2.4771	2.4829	2.4843	2.4871

Solution:

x	$\log_{10} x$			
300	2.4771			
		2.47855		
304	2.4829		2.47858	
		2.47854		2.47860
305	2.4843		2.47857	
		2.47853		
307	2.4871			

Hence $\log_{10} 301 = 2.4786$.

Empirical Laws and Curve Fitting

5.1 INTRODUCTION

In engineering and science, it is often required to express a given data obtained from observations in the form of law relating to the two variables involved. Such a law is called the *empirical law*.

In order to find the mathematical relation for the given data, we first plot the given data on a sheet of graph paper and then sketch a line that virtually conform to the given data. The curve, thus obtained is called the approximating curve and the algebraic equation of the curve is called *Empirical equation*. We can draw different types of such approximating curves to the given data and thus equations of different types can be obtained to express the given data.

The curve which is most suitable for predicting the values of an unknown is called the curve of the 'best fit'. The process to find the equation of the curve of 'best fit' is known as *curve fitting*.

If it is desired to obtain an equation representing the given data, we can use any of the four methods, i.e. graphical method, method of group averages, method of moments and method of least-squares. The graphical method and method of averages fail to give the values of the unknown constants uniquely and accurately while the other methods can do. The method of least-squares is probably the best to fit a unique curve to a given data.

5.2 GRAPHICAL METHOD

When the curve representing the given data is a linear relation $y = mx + c$, we plot the given points on the graph paper using suitable scale. Then draw a line of 'best fit' such that the points are evenly distributed about this line, some on one side and some on the other side. Taking any two points (x_1, y_1) and (x_2, y_2) on the line, calculate m, the slope of the line, i.e. $m = \dfrac{y_2 - y_1}{x_2 - x_1}$ and c, the intercept on y-axis can be found easily by substituting any point on the line. By substituting the values of m and c, we get the straight line required.

When the points representing the observed values do not approximate to a straight line, a smooth curve is drawn through them and from the shape of the curve, we try to infer the law of the curve and then reduce it to the form $y = mx + c$.

5.3 LAWS REDUCIBLE TO LINEAR LAWS

If the laws of relationship between the variables x and y are not linear laws, reduce them to linear laws by suitable substitutions. Given below are some of the laws of common use to reduce them to linear laws.

1. $y = mx^n + c$, the law which is not linear.

   In order to reduce it to linear law put

   $$x^n = X \text{ and } y = Y, \text{ getting}$$
   $$Y = mX + c, \text{ which is a linear law.}$$

2. $y = ax^n$, the law which is not linear.

   Taking logarithms of both sides, it becomes

   $$\log_{10} y = \log_{10} a + n \log_{10} x$$

   Putting $\log_{10} x = X, \log_{10} a = c \text{ and } \log_{10} y = Y$

   $$Y = c + nx \text{ which is a linear law.}$$

3. $y = ax^n + b \log x$, the law which is not linear,

   writes as $$\dfrac{y}{\log x} = a\dfrac{x^n}{\log x} + b$$

   Taking $$\dfrac{x^n}{\log x} = X \text{ and } \dfrac{y}{\log x} = Y$$

   $$Y = aX + b \text{ which is a linear law.}$$

4. $y = a e^{bx}$, law which is not linear.

   Taking logarithms,

   $$\log_{10} y = (b \log_{10} e) x + \log_{10} a$$
   $$x = X, \log_{10} y = Y$$
   $$Y = mX + c \text{ which is a linear.}$$

5. $xy = ax + by$ which is not linear.

   We have $y = a + by/x$

   Putting $y = Y y/x = X$

   $$Y = bX + a \text{ which is a linear law.}$$

Example 5.1: Fit a curve of the form $y = ax^2 + b$ to the following data.

x	12	16	20	22	24	26	30
y	6.0	7.5	9.0	9.8	10.7	11.5	14.0

Solution: The given relation is

$$y = ax^2 + b$$

Putting $x^2 = X, y = Y$

$$Y = aX + b$$

Table for X and Y is

x	$X = x^2$	$y = Y$
12	144	6.0
16	256	7.5
20	400	9.0
22	484	9.8
24	576	10.7
26	676	11.5
30	900	14.0

Selecting two points (144, 6.0), (900, 14.0)

Slope
$$a = \frac{14 - 6}{900 - 144} = \frac{8}{756} = 0.0106$$

Hence equation becomes
$$Y = (0.0106) X + b$$

Since (144, 6.0) lies on this line
$$6 = (0.0106) (144) + b$$
$$b = 4.48 \text{ (approx.)}$$

Hence the required form is
$$y = (0.0106) x^2 + 4.48$$

Example 5.2: The values of x and y obtained in an experiment are as follows

x	2.30	3.10	4.00	4.92	5.91	7.20
y	33.0	39.1	50.3	67.2	85.6	125.0

The probable law is $y = ae^{bx}$. Test graphically the accuracy of this law and if the law holds good, find the best value of the constants.

Solution: Given law is $y = ae^{bx}$.

Taking the logarithm to base 10, we get
$$\log_{10} y = \log_{10} a + (b \log_{10} e) x$$

Putting $x = X, \log_{10} y = Y$, it becomes
$$Y = (b \log_{10} e) X + \log_{10} a$$
$$Y = A + BX$$

where $B = b \log_{10} e, A = \log_{10} a$

Table for values of X and Y is as follows.

$X = x$	2.30	3.10	4.00	4.92	5.91	7.20
$Y = \log_{10} y$	1.52	1.59	1.71	1.83	1.93	2.1
Points	P_1	P_2	P_3	P_4	P_5	P_6

Scale: 1 small division along x-axis = 0.1

10 small divisions along y-axis = 0.1

Plot these points and draw the line of best fit. As all these points lies along a straight line, the given law is nearly accurate (Fig. 5.1).

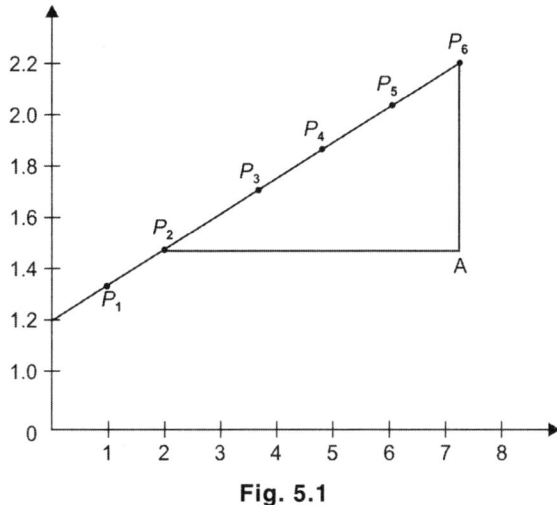

Fig. 5.1

Now slope of this line is

$$B = \frac{P_6 A}{P_2 A} = \frac{2.1 - 1.59}{7.2 - 3.1} = 0.1244$$

Since $P_2(3.1, 1.59)$ lies on (2)

$$1.59 = A + (0.1244)(3.1)$$

$\Rightarrow \qquad A = 1.2044$

But $\qquad b = B/\log_{10}e = 0.1244/0.4343 = 0.2864$

and $\qquad \log_{10}a = A, \quad \therefore a = \text{anti} \log_{10}A = 16.01$

Hence the curve of 'best fit' is $y = 16.01\, e^{0.2864x}$.

5.4 METHOD OF GROUP AVERAGES

Let (x_1, y_1), (x_2, y_2), ..., (x_n, y_n) be given n sets of observations of related data and let the line be

$$y = ax + b \tag{5.1}$$

when $x = x_1$, the observed value of $y = y_1 = M_1 P_1$ and from Eq. (5.1)

$$y = ax_1 + b = M_1 L_1$$

which is known as the expected value or calculated value of y at M_1.

Then $\qquad e_1 = $ observed value at M_1 – expected value at M_1

$$= y_1 - (ax_1 + b) = L_1 P_1$$

$$... = ...$$

$$e_n = y_n - (ax^n + b) = L_n P_n$$

Some of the errors may be +ve and some are –ve.

The method of group averages is based on the assumption that the sum of the residuals at all points is zero. Since there are two unknowns a, b, we require two equations to find them. So we divide the given data into two groups. Let the first group contains k observations (x_1, y_1), (x_2, y_2), ..., (x_n, y_n) and the second group contains the remaining $(n - k)$ observations (x_{k+1}, y_{k+1}), ..., (x_k, y_k) (Fig. 5.2).

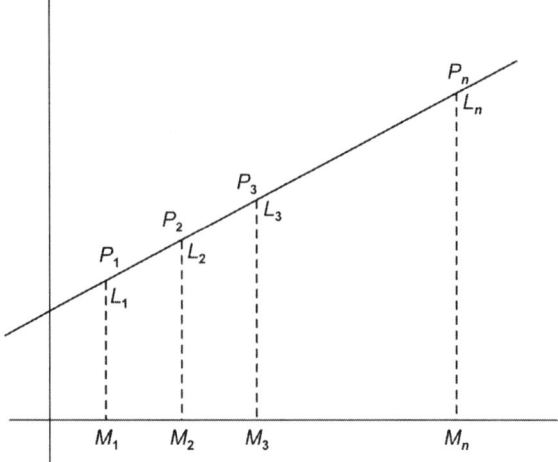

Fig. 5.2

Assuming the sum of the errors in each group to be zero, we get

$$\{y_1 - (ax_1 + b)\} + \{y_2 - (ax_2 + b)\} + \dots + \{y_k - (ax_k + b)\} = 0$$

$$\{y_{k+1} - (ax_{k+1} + b)\} + \{y_{k+2} - (ax_{k+2} + b)\} + \dots + \{y_n - (ax_n + b)\} = 0$$

On simplification, we get

$$\frac{y_1 + y_2 + \dots + y_k}{k} = b + a\,\frac{x_1 + x_2 + \dots + x_k}{k} \tag{5.2}$$

$$\frac{y_{k+1} + y_{k+2} + \dots + y_{k+n}}{n-k} = b + a\,\frac{x_{k+1} + x_{k+2} + \dots + x_n}{n-k} \tag{5.3}$$

In Eq. (5.2), $\frac{1}{k}(x_1 + x_2 + \dots + x_k)$ and $\frac{1}{k}(y_1 + y_2 + \dots + y_k)$ are the average values of x's and y's of the 1st group. Hence the Eqs (5.2) and (5.3) are obtained from Eq. (5.1) by replacing x and y by their respective averages of the two groups. Solving Eqs (5.2) and (5.3), we get a and b.

Note: The main drawback of this method is that a different grouping of observations will give different values of a and b, so in practice, we divide the data in such a way that each group contains an even number of observations.

Example 5.3: Using the method of group averages, fit the law of the form $y = ax^b$ to the following data

x	0.013	0.027	0.042	0.073	0.108	0.151	0.233	0.341
y	1.68	2.45	3.08	4.09	4.97	5.95	7.39	9.00

Solution: Given $\qquad y = ax^b$ $\hfill$ (i)

Taking logarithm on both sides

$$\log_{10} y = \log_{10} a + b\,\log_{10} x$$

$$Y = A + bX \tag{ii}$$

where $\qquad Y = \log_{10} y,\ A = \log_{10} a,\ X = \log_{10} x$

We form the following tables:

Group I			
x	y	$X = \log_{10} x$	$Y = \log_{10} y$
0.013	1.68	$\overline{2}.1139$	0.2253
0.027	2.45	$\overline{2}.4314$	0.3892
0.042	3.08	$\overline{2}.6232$	0.4886
0.073	4.09	$\overline{2}.8633$	0.6117
		$\Sigma X = \overline{6}.0318$	$\Sigma Y = 1.7148$

Group II			
x	y	$X = \log_{10} x$	$Y = \log_{10} y$
0.108	4.97	$\overline{1}.0334$	0.6964
0.151	5.95	$\overline{1}.1790$	0.7745
0.233	7.32	$\overline{1}.3674$	0.8686
0.341	9.00	$\overline{1}.5328$	0.9542
		$\Sigma X = \overline{3}.1126$	$\Sigma Y = 3.2937$

Substituting the averages of x and y of two groups in the given relation in Eq. (ii)

$$\frac{1.7148}{4} = A + b \frac{\overline{6}.0318}{4}$$

$$\frac{3.2937}{4} = A + b \frac{\overline{3}.1126}{4}$$

$$4A + (\overline{6}.0318)b = 1.7148$$
$$4A + (\overline{3}.1126)b = 3.2937$$

Solving, we get $A = 1.1897$, $b = 0.51$

This implies $a = $ antilog $1.1897 = 15.49$

Hence the equation is $y = (15.49) x^{(0.51)}$

Example 5.4: The latent heat of vapourisation of steam r is given in the following table at different temperature t.

t	40	50	60	70	80	90	100	110
r	1069.1	1063.6	1058.2	1052.7	1049.3	1041.8	1036.3	1030.8

For this range of temperature, a relation of the form $r = a + bt$ is known to fit the data. Find the value of a and b by the method of group averages.

Solution: Let us divide the data into two groups each containing four readings, then we have

t	r	t	r
40	1069.1	80	1049.3
50	1063.6	90	1041.8
60	1058.2	100	1036.3
70	1052.7	110	1030.8
$\Sigma t = 220$	$\Sigma r = 4243.6$	$\Sigma t = 380$	$\Sigma r = 4158.2$

Substituting the averages of t's and r's of the two groups in the given relation, $r = a + bt$

$$\frac{4243.6}{4} = a + b\frac{220}{4} \Rightarrow 1060.9 = a + 55b \tag{i}$$

$$\frac{4158.2}{4} = a + b\frac{380}{4} \Rightarrow 1039.55 = a + 95b \tag{ii}$$

Solving Eqs (i) and (ii), we get

$$a = 1090.26$$
$$b = -0.534$$

5.5 LAWS CONTAINING THREE CONSTANTS

So far, we have applied the above methods to fit the data to laws involving two constants only. Now, we shall discuss the laws involving three constants, i.e. $y = a + bx + cx^2$, $y = a + bx^c$ and $a + be^{cx}$. To fit such laws to a set of observations, we proceed as follows to reduce these laws to linear laws.

1. Equation $\qquad y = a + bx + cx^2$ (5.4)

Let (x_1, y_1) be a point on the curve in Eq. (5.4)

$$y_1 = a + bx_1 + cx_1^2 \tag{5.5}$$

From Eqs (5.4) and (5.5), we get

$$y - y_1 = b(x - x_1) + c(x^2 - x_1^2)$$

$$\frac{y - y_1}{x - x_1} = b + c[(x^2 - x_1^2)/(x - x_1)] = b + c(x + x_1)$$

Putting $\qquad \dfrac{y - y_1}{x - x_1} = Y, x + x_1 = X$, we get

$Y = b + cX$. Now b and c can be found by the graphical method.

2. Equation $\qquad y = a + bx^c$ (5.6)

It can be written as $y - a = bx^c$

To find a, let (x_1, y_1), (x_2, y_2), (x_3, y_3) by three points on the curve in Eq. (5.4), such that x_1, x_2, x_3 are in geometrical progression

i.e. $\qquad x_1x_3 = x_2^2$

Then $\qquad (y_1 - a) = bx_1^c$

$\qquad\qquad (y_2 - a) = bx_2^c$

and $\qquad (y_3 - a) = bx_3^c$

$$(y_1 - a)(y_3 - a) = b^2(x_1x_3)^c = b^2x_2^{2c} = (y_2 - a)^2$$

or $\qquad a(y_1 + y_3 - 2y_2) = y_1y_3 - y_2^2 \Rightarrow a = \dfrac{y_1y_3 - y_2^2}{y_1 + y_3 - 2y_2}$

which gives a. Now Eq. (5.4) reduces to law having two constants b and c only.

Taking logarithms, Eq. (5.4) becomes

$$\log_{10}(y - a) = \log_{10}b + c\log_{10}x$$

or $$Y = B + cX \qquad (5.7)$$

where $$X = \log_{10}x, \; Y = \log_{10}(y - a), \; B = \log_{10}b$$

Hence, we can find b and c as before from Eq. (5.7).

3. Equation $\qquad y = a + be^{cx}$

which can be written as

$$y - a = be^{cx}$$

To find a, let (x_1, y_1), (x_2, y_2), (x_3, y_3) be three particular points on the curve in Eq. (5.4), such that x_1, x_2, x_3 are in arithmetical progression, i.e.

$$x_1 + x_3 = 2x_2$$

Thus $$y_1 - a = be^{cx_1}$$

$$y_2 - a = be^{cx_2}$$

$$y_3 - a = be^{cx_3}$$

and $$(y_1 - a)(y_3 - a) = b^2 e^{c(x_1 + x_3)}$$

$$= (be^{cx_2})^2 = (y_2 - a)^2$$

$$a(y_1 + y_3 - 2y_2) = y_1 y_3 - y_2^2$$

which gives a, now reduce Eq. (5.4) to a law containing two constants a and b taking logarithm in Eq. (5.4) becomes

$$\log_{10}(y - a) = \log_{10}b + cx\log_{10}e$$

$$Y = B + Cx \qquad (5.8)$$

where $$Y = \log_{10}(y - a), \; B = \log_{10}b, \; C = c\log_{10}e$$

Hence, we can find b and c as before from Eq. (5.7).

Example 5.5: Using the method of averages, fit a parabola $y = a + bx + cx^2$ to the following data

x	20	40	60	80	100	120
y	5.5	9.1	14.9	22.8	33.3	46.0

Solution: Let the point (20, 5.5) be on the parabola

$$y = a + bx + cx^2 \qquad (i)$$

$$5.5 = a + 20b + 400c \qquad (ii)$$

From Eqs (i) and (ii)

$$\frac{y - 5.5}{x - 20} = b + c(x + 20)$$

$$Y = b + cX$$

where $$Y = \frac{y - 5.5}{x - 20}, \; X = x + 20$$

Choosing first three observations as one group and next three observations as another group, we make the following table for the values of X and Y.

For Group I					
x	y	$x-20$	$y-5.5$	$X = x+20$	$Y = y - 5.5/x - 20$
20	5.5	0	0	–	–
40	9.1	20	3.6	60	0.18
60	14.9	40	9.4	80	0.235
				$\Sigma X = 140$	$\Sigma Y = 0.415$

$$\bar{X}_1 = \frac{140}{2} = 70, \bar{Y}_1 = \frac{0.415}{2} = 0.2075$$

For Group II					
x	y	$x-20$	$y-5.5$	X	Y
80	22.9	60	17.3	100	0.288
100	33.3	80	27.8	120	0.348
120	46	100	40.5	140	0.405
				$\Sigma X = 360$	$\Sigma Y = 1.041$

$$\bar{X}_2 = \frac{360}{3} = 120, \bar{Y}_2 = \frac{1.041}{3} = 0.347$$

Substituting the averages in Eq. (5.7)

$$\bar{Y}_1 = b + c\bar{X}_1$$
$$0.2075 = b + c\,(70)$$
$$\bar{Y}_2 = b + c\bar{X}_2$$
$$0.347 = b + c\,(120)$$

Solving these $\qquad b = 0.0122, c = 0.00279$

Substituting in Eq. (5.7)

$$\frac{y - 5.5}{x - 20} = 0.0122 + 0.00279\,(x + 20)$$

or $\qquad\qquad y = 0.00279x^2 + 0.0122x + 4.14$

which is the required curve.

Example 5.6: In a magnetic arc at constant arc length, voltage v consumed by the arc is observed for values of the current i.

i	0.5	1	2	4	8	12
v	160	120	94	75	62	56

If v and i are connected by the relation of the form $v = ai^b + c$, find a, b and c and hence, the curve by the method of group averages.

Solution: Take $i_1 = 1, i_2 = 2, i_3 = 4$ which are in GP

Then $v_1 = 120, v_2 = 94, v_3 = 75$

$$c = \frac{v_1 v_3 - v_2^2}{v_1 + v_3 - 2v_2} = \frac{(120)\,(75) - (94)^2}{120 + 75 - 2 \times 94} = 23.4$$

$$v = ai^b + c \Rightarrow v - 23.4 = ai^b$$

Taking logarithm on both sides
$$\log_{10}(v - 23.4) = \log_{10}a + b\log_{10}i$$

Putting $\log_{10}(v - 23.4) = y$; $\log_{10}a = A$; $\log_{10}i = X$
$$Y = A + bk \tag{i}$$

which is linear in X and Y.

To fit Eq. (i), we use the method of group averages. So we form the table for X and Y.

			Group I		
i	v	$v - 23.4$	$Y = \log_{10}(v - 23.4)$	$X = \log_{10}i$	
0.5	160	136.6	2.1354		
1	120	96.6	1.9850	0.0000	
2	94	70.6	1.8488	0.3010	
			$\Sigma Y = 5.9692$	$\Sigma X = 0.3010$	

			Group II		
i	v	$v - 23.4$	$Y = \log_{10}(v - 23.4)$	$X = \log_{10}i$	
4	75	51.6	1.7126	0.6021	
8	62	38.6	1.5866	0.9031	
12	56	32.6	1.5132	1.0792	
			$\Sigma Y = 4.8124$	$\Sigma X = 2.5844$	

Substituting the averages of X's and Y's in Eq. (i), we get

$$\frac{5.9692}{3} = A + b\,(0) \implies 3A = 5.9692 \tag{ii}$$

and
$$\frac{4.8124}{3} = A + b \times \frac{2.5844}{3} \tag{iii}$$

$\implies \qquad 3A + 2.5844b = 4.8124$

Solving Eqs (ii) and (iii), we get
$$A = 1.9897, b = -0.4477$$

i.e.
$$a = 10^A = 10^{1.9897} = 97.66$$

Therefore the required equation is
$$v = (97.66)\, i^{-(0.4477)} + 23.4$$

Example 5.7: The variables s and t are connected by the relation $s = a + be^{ni}$ and their corresponding values are given in the following data.

t	1	2	6	8	11
s	12.71	12.46	11.65	11.34	10.99

Fit the curve by the method of group averages.

Solution: The given curve can be written as
$$s - a = b\,e^{nt} \tag{i}$$
$$Y = B + Nt \tag{ii}$$

where $Y = \log_{10}(s - a)$, $B = \log_{10}b$ and $N = n\log_{10}e$

Take $t_1 = 1, t_2 = 6, t_3 = 11$ such that $x_1 + x_3 = 2x_2$.

The corresponding values of $s_1 = 12.71$, $s_2 = 11.65$ and $s_3 = 10.99$

$$a = \frac{s_1 s_3 - s_2^2}{s_1 + s_3 - 2s_2} = \frac{(12.71)(10.99) - (11.65)^2}{12.71 + 10.99 - 2(11.65)}$$

$$= \frac{3.9604}{0.4} = 9.901$$

Group I		
t	s	$Y = \log_{10}(s - 9.901)$
1	12.71	0.4486
2	12.46	0.4081
6	11.65	0.2428
$\Sigma t = 9$		$\Sigma Y = 1.0995$

Group I		
t	s	Y
8	11.34	0.1581
11	10.99	0.0370
$\Sigma t = 19$		$\Sigma Y = 0.1951$

Averages of the first group are

$$\overline{t_1} = \frac{\Sigma t}{3} = \frac{9}{3} = 3 \quad \overline{Y_1} = \frac{\Sigma Y}{3} = \frac{1.0995}{3} = 0.3665$$

and the averages of the second group are

$$\overline{t_2} = \frac{\Sigma t}{2} = \frac{19}{2} = 9.5 \quad \overline{Y_1} = \frac{\Sigma Y}{2} = \frac{0.1951}{2} = 0.0976$$

Substituting the averages of t's and Y's of two groups in Eq. (ii), we get

$$0.3665 = B + 3N \text{ and } 0.0976 = B + 9.5N$$

On solving, we get $B = 0.4907$, $N = -0.0414$

$$b = \text{antilog}_{10} \ B = 3.097 \text{ and } n = \frac{N}{\log_{10} e}$$

Hence the required equation is

$$s = 9.901 + 3.097e^{-0.0953t}$$

PROBLEMS 5.1

1. Fit a curve of the form $y = ae^{bx}$ to the following data:

x	1	2	3	4	5	6
y	14	27	40	55	68	300

[Ans. $y = 7.943\,e^{0.5419x}$]

2. The resistance R of carbon filament lamp was measured at various values of voltage V and the following observations were made

V	62	70	78	84	92
R	73	70.7	69.2	67.8	66.3

Assuming a law of the form, find by graphical method the best value of a and b.

[Ans. $a = 1120, b = 55.1$]

3. R is resistance to motion of a train at speed V. Find a law of the type $R = aV^2 + b$ connecting R and V using the following data

R (kg/ton)	8	10	15	21	30
V (km/h)	10	20	30	40	50

[Ans. $a = 0.0085, b = 7.35$]

4. The following table gives corresponding values of x and y. Obtain an equation of the form $y = a + bx$ using the method of grouping.

x	0	5	10	15	20	24
y	12	15	17	22	24	30

[Ans. $y = 11.1117 + 0.7111x$]

5. Fit a curve of the form $y = ab^x$ using the method of group averages for the following data.

x	2	4	6	8	10	12
y	7.32	8.24	9.20	10.19	11.01	12.05

[Ans. $y = (6.7468)(1.0505)^x$]

6. Convert the equation $y = b/x (x - a)$ to a linear form and hence determine a and b which will fit the following data using the method of group averages.

x	8	10	15	20	30	40
y	13	14	15.4	16.3	17.2	17.8

[Ans. $a = 0.2039, b = 0.051$]

5.6 PRINCIPLE OF LEAST SQUARES

Curve fitting by (i) graphical method, (ii) method of group averages has already been discussed. But these methods have the drawback of being unable to give the unique curve of fit. The values of the constants in the first method depend upon the choice of points on the line and in the second method on grouping the observations. Hence, both the methods fails to give the unique curve of fit. The principle of least squares, however, provides a unique set of values to the constants and hence suggests a curve of best fit to the given data.

Let $(x_i, y_i) i = 1, 2, ..., n$ be n sets of observations and let $y = f(x)$ be the proposed relation between x and y when $x = x_i$, the observed value of y_i is $y_i = P_iM_i$ and expected value, i.e. $L_iM_i = f(x_i)$.

It is obvious that some values of e_i's may be +ve and –ve. We square each of them and form the sum of the squares for giving equal weightage to each residual, i.e.

$$E = \sum_{i=1}^{n} e_i^2 = \sum_{i=1}^{n} [y_i - f(x_i)]^2$$

Now if $E = 0$, then $y_i = f(x_i)$, all points be on the curve. If $E \neq 0$, then the minimum of E results the best fitting curve to the data. Then the curve of best fit is that for which $e's$ are as small as possible, i.e. the sum of the squares of the errors is minimum (Fig. 5.3). This is known as the principle of least squares. Using this principle, we shall fit the following curves:

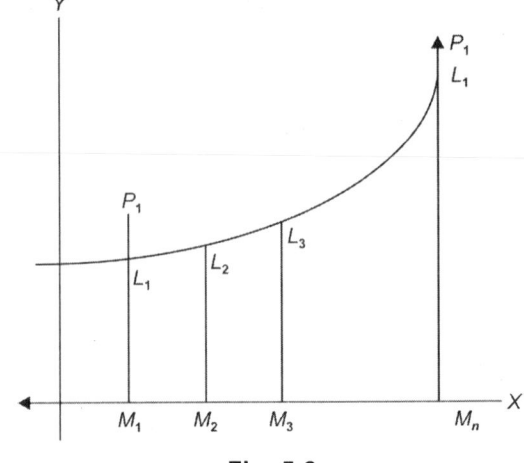

Fig. 5.3

   i. A straight line, $y = ax + b$

   ii. A parabola $y = ax^2 + bx + c$

   iii. The exponential curve $y = ae^{bx}$ and

   iv. The curve $y = ax^b$.

5.7 FITTING A STRAIGHT LINE

Let $(x_i, y_i) i = 1, 2, ..., n$ be set of observations of the related data and $y = ax + b$ be the straight line to be fitted. Now, we have to find a and b when $x = x_i$, the residual as $x = x_i$ is

$$e_i = y_i - f(x_i) = y_i - (ax_i - b); \quad i = 1, 2, ..., n$$

$$E = \sum_{i=1}^{n} e_i^2 = \sum_{i=1}^{n} [y_i - (ax_i + b)]^2$$

By the principle of least squares, E is minimum

$$\frac{\partial E}{\partial a} = 0, \quad \frac{\partial E}{\partial b} = 0$$

$$\sum_{i=1}^{n} 2(y_i - ax_i - b)(-x_i) = 0 \quad \text{or} \quad \sum_{i=1}^{n} y_i(x_i - ax_i^2 - bx_i) = 0$$

and $$\sum_{i=1}^{n} 2(y_i - ax_i - b)(-1) = \sum_{i=1}^{n} (y_i - ax_i - b) = 0$$

$\therefore$
$$\sum_{i=1}^{n} y_i x_i = a \sum_{i=1}^{n} (x_i^2 + b \sum_{i=1}^{n} x_i \tag{5.9}$$

and
$$\sum_{i=1}^{n} y_i = a \sum_{i=1}^{n} x_i + nb \tag{5.9a}$$

Since x_i, y_i are known, on solving Eqs (5.9) and (5.9a), we can find a and b.

Notes: 1. Equations (5.9) and (5.9a) are called the *normal equation* and by dropping the suffixes, they can be written as

$$\Sigma y = a \Sigma x + nb$$
$$\Sigma xy = a \Sigma x^2 + b \Sigma x$$

2. If the values x and y are very large, the calculation can be made easy by using the transformation $X = \dfrac{x-p}{h}, Y = \dfrac{y-q}{k}$ which reduce the linear equation $y = ax + b$ to form $Y = AX + B$.

Example 5.8: If P is the pull required to lift a load W by means of a pulley block, find a linear law of the form $P = mW + c$ connecting P and W, using the following data.

P	12	15	21	25
W	50	70	100	120

when P and W are taken in kg. Compute P, when $W = 150$ kg.

Solution: The law is $\qquad P = mW + c$

The corresponding normal equations are

$$\Sigma P = 4c + m\Sigma W \qquad\qquad\qquad\qquad \text{(i)}$$

Here $n = 4$ $\qquad\qquad \Sigma WP = c\Sigma W + m\Sigma W^2 \qquad\qquad\qquad \text{(ii)}$

The values of ΣW and relative value can be calculated by means of the following table.

W	P	W²	WP
50	12	2500	600
70	15	4900	1050
100	21	10000	2100
120	25	14400	3100
$\Sigma W = 340$	$\Sigma P = 73$	$\Sigma W^2 = 31800$	$\Sigma WP = 6750$

Substituting in Eqs (i) and (ii), we have

$$73 = 4c + 340m \qquad\qquad\qquad\qquad\qquad \text{(iii)}$$

and $\qquad\qquad\qquad 6750 = 340c + 31800 \qquad\qquad\qquad\qquad\qquad \text{(iv)}$

From Eqs (iii) and (iv), we get

$$m = 0.1879, c = 2.2785$$

Hence the line of best fit is

$$P = 2.2785 + 0.1879W$$

When $W = 150$ kg, $\qquad P = 2.2785 + 0.1879 \times 150 = 30.4635$ kg.

Example 5.9: Fit a straight line to the following, using method of least square.

x	1	2	3	4	5
y	14	27	40	55	68

Solution: Method I: Let the straight line of best fit be

$$y = ax + b \qquad\qquad\qquad\qquad\qquad\qquad \text{(i)}$$

The normal equations are

$$\Sigma xy = a\Sigma x^2 + b\Sigma x \qquad\qquad\qquad\qquad\qquad \text{(ii)}$$

$$\Sigma y = a\Sigma x + 5b \qquad\qquad\qquad\qquad\qquad\qquad \text{(iii)}$$

The value of Σx, Σy, Σx^2, Σxy are calculated as below:

x	y	x^2	xy
1	14	1	14
2	27	4	54
3	40	9	120
4	55	16	220
5	68	25	340
$\Sigma x = 15$	$\Sigma y = 204$	$\Sigma x^2 = 55$	$\Sigma xy = 748$

Normal Eqs (ii) and (iii) becomes
$$748 = 55a + 15b$$
$$204 = 15a + 5b$$

Solving, we get $\quad a = 13.6, b = 0$

Putting these values in Eq. (i), we get the line of best fit as $y = 13.6x$

Method II: Let $X = x - 3$, $Y = y - 40$ for making the calculation simpler.

Let the straight line in new variables be
$$Y = AX + B$$

Now the normal equations are
$$\Sigma XY = A\Sigma X^2 + B\Sigma X$$
$$\Sigma Y = A\Sigma X + 5B$$

The value of $2\Sigma X$, ΣY, ΣX^2, ΣXY are calculated as below:

x	y	$X = x - 3$	$Y = y - 40$	X^2	XY
1	14	-2	-26	4	52
2	27	-1	-13	1	13
3	40	0	0	0	0
4	55	1	15	1	15
5	68	2	28	4	56
		$\Sigma X = 0$	$\Sigma Y = 4$	$\Sigma X^2 = 10$	$\Sigma XY = 136$

Substituting these values, we get
$$136 = 10A$$
$$4 = 5B$$

or $\quad A = 13.6, B = 0.8$

The equation is $\quad Y = 13.6X + 0.8$
$$y - 40 = 13.6(x - 3) + 0.8$$
$$y = 13.6x \text{ which is the same as above.}$$

PROBLEMS 5.2

1. Fit a straight line to the following data by the method of least squares.

x	-4	1	2	3
y	4	6	10	8

[Ans. $y = 0.7x + 6.65$]

2. Fit a straight line to the following data by the method of least squares.

x	5	10	15	20	25	
y	16	19	23	26	30	[Ans. $y = 0.7x + 123$]

3. The table below gives the temperature T (in °C) and length l (in mm) of a heated rod. If $l = a_0 T + a_1$, find the best value of a_0, a_1.

T	20	30	40	50	60	70
l	800.3	800.4	800.6	800.7	800.9	801.0

[Ans. $0.0146T + 800$]

4. Fit a straight line to the following data.

x	71	68	73	69	67	65	66	67
y	69	72	70	70	68	67	68	64

Also find y at $x = 70$.   [Ans. $y = 0.4242x + 37.908, y = 67.6030$]

5.8 FITTING A PARABOLA OR A SECOND DEGREE CURVE

Let (x_i, y_i), $[i = 1, 2, 3, ..., n]$ be the given n sets of observation of the related variables x, y and let $y = ax^2 + bx + c$ be the equation which fits the observations best. We have to find a, b, c.

For $x = x_i$, the expected value of y is $ax_i^2 + bx_i + c$ and observed value of y is y_i.

The residual $e = y_i - (ax_i^2 + bx_i + c), i = 1, 2, ..., n$.

Let E denote the sum of the squares of the residuals

$$E = \sum_{i=1}^{n} [y_i - (ax_i^2 + bx_i + c)]^2 \tag{5.10}$$

By the principle of least squares, E should be minimum for which the conditions are

$$\frac{\partial E}{\partial a} = 0, \quad \frac{\partial E}{\partial b} = 0, \quad \frac{\partial E}{\partial c} = 0$$

Therefore, differentiate on both sides of Eq. (5.10) partiallly w.r.t. a, b and c, we get

$$\sum_{i=1}^{n} 2[y_i - (ax_i^2 + bx_i + c)](-x_i^2) = 0$$

$$\sum_{i=1}^{n} 2[y_i - (ax_i^2 + bx_i + c)](-x_i) = 0$$

$$\sum_{i=1}^{n} 2[y_i - (ax_i^2 + bx_i + c)](-1) = 0$$

i.e. $$\sum_{i=1}^{n} (x_i^2 y_i - ax_i^4 - bx_i^3 - cx_i^2) = 0$$

$$\sum_{i=1}^{n} (x_i y_i - ax_i^3 - bx_i^2 - cx_i) = 0$$

$$\sum_{i=1}^{n} (y_i - ax_i^2 - bx_i - c) = 0$$

Simplifying and dropping the suffixes, we get

$$\Sigma y = a\Sigma x^2 + b\Sigma x + nc$$

$$\Sigma xy = a\Sigma x^3 + b\Sigma x^2 + c\Sigma x$$

$$\Sigma x^2 y = a\Sigma x^4 + b\Sigma x^3 + c\Sigma x^2$$

Solving the above normal equations, we get the values of a, b and c. Substituting these values of a, b and c in $y = ax^2 + bx + c$, we get the least fitting parabola.

Example 5.10: Fit a second degree parabola to the data

x	1929	1930	1931	1932	1933	1934	1935
y	352	356	357	358	360	361	361

Solution: Let $\qquad y = ax^2 + bx + c \qquad\qquad$ (i)

be the best fit.

Since the values of x and y are large, we choose the origins for x and y at 1932 and 357 respectively. In other words, we transform x and y.

$$X = x - 1932, \ \ Y = y - 357, \text{ then Eq. (i) can be reduced to}$$

$$Y = aX^2 + bX + c \qquad\qquad \text{(ii)}$$

The normal equations are

$$\Sigma Y = a\Sigma X^2 + b\Sigma X + nc$$

$$\Sigma XY = a\Sigma X^3 + b\Sigma X^2 + c\Sigma X$$

$$\Sigma X^2 Y = a\Sigma X^4 + b\Sigma X^3 + c\Sigma X^2$$

We form the table:

x	y	$X = x - 1932$	$Y = y - 357$	X^2	X^3	X^4	XY	X^2Y
1929	352	-3	-5	9	-27	81	15	-45
1930	356	-2	-1	4	-8	16	2	-4
1931	357	-1	0	1	-1	1	0	0
1932	358	0	1	0	0	0	0	0
1933	360	1	3	1	1	1	3	3
1934	361	2	4	4	8	16	8	16
1935	361	3	4	9	27	81	12	36
		$\Sigma X = 0$	$\Sigma Y = 5$	$\Sigma X^2 = 28$	$\Sigma X^3 = 0$	$\Sigma X^4 = 196$	$\Sigma XY = 40$	$\Sigma X^2Y = 6$

Hence the normal equations became

$$28a + 7c = 6$$

$$28b = 40$$

$$196a + 28c = 6$$

On solving, we get $b = 1.4286$, $a = -0.21429$, $c = 1.7143$.

Equation (ii) becomes

$$Y = -0.21429X^2 + 1.4286X + 1.7143$$

$$y - 357 = -0.21429\,(x - 1932)^2 + 1.4286\,(x - 1932) + 1.7143$$

$$y = -0.21429x^2 + 829.445x - 802265.33 \text{ is the required equation.}$$

Example 5.11: Find a second degree parabola $y = a + bx + x^2$ to the following data:

x	1	3	5	7	9
y	2	7	10	11	9

Solution: Method I: Normal equations are

$$\Sigma y = 5a + b\Sigma x + c\Sigma x^2$$

$$\Sigma xy = a\Sigma x + b\Sigma x^2 + c\Sigma x^3$$

$$\Sigma x^2 y = a\Sigma x^2 + b\Sigma x^3 + c\Sigma x^4$$

We form the table:

x	y	x^2	x^3	x^4	xy	x^2y
1	2	1	1	1	2	2
3	7	9	27	81	21	63
5	10	25	125	625	50	250
7	11	49	343	2401	77	539
9	9	81	729	6561	81	729
$\Sigma x = 25$	$\Sigma y = 39$	$\Sigma x^2 = 165$	$\Sigma x^3 = 1225$	$\Sigma x^4 = 9669$	$\Sigma xy = 231$	$\Sigma x^2 y = 1583$

Putting the values in equations (normal), we obtain a, b, c

$$a = -1.556, b = 3.757, c = -0.2857$$

Hence the fitted equation is

$$y = -1.556 + 3.757x - 0.2857x^2$$

Method II: The above can also be solved by using the change of origin and scale. Thus, we have to take

$$X = \frac{x-5}{2} \text{ and the fitted line will be}$$

$$y = a_1 + b_1 X + c_1 X^2$$

$$= a_1 + b_1 \left(\frac{x-5}{2}\right) + c_1 \left(\frac{x-5}{2}\right)^2$$

$$y = A + BX + CX^2$$

Proceed as above and find A, B and C.

Example 5.12: Fit a second degree parabola to the following data using method of least squares.

x	0	1	2	3	4
y	1	1.8	1.3	2.5	6.3

Solution: Let $y = ax^2 + bx + c$ be the parabola

The normal equations are

$$\Sigma y = a\Sigma x^2 + b\Sigma x + 5c$$

$$\Sigma xy = a\Sigma x^3 + b\Sigma x^2 + c\Sigma x$$

$$\Sigma x^2 y = a\Sigma x^4 + b\Sigma x^3 + c\Sigma x^2$$

The values Σx, Σx^2, ... etc. are calculated by means of the following table:

x	y	x^2	x^3	x^4	xy	x^2y
0	1	0	0	0	0	0
1	1.8	1	1	1	1.8	1.8
2	1.3	4	8	16	2.6	5.2
3	2.5	9	27	81	7.5	22.5
4	6.3	16	64	256	25.2	100.8
$\Sigma x = 10$	$\Sigma y = 12.9$	$\Sigma x^2 = 30$	$\Sigma x^3 = 100$	$\Sigma x^4 = 354$	$\Sigma xy = 37.1$	$\Sigma x^2y = 130.3$

Substituting the obtained values in the normal equations, we get

$$12.9 = 30a + 10b + 5c$$
$$37.1 = 100a + 30b + 10c$$
$$130.3 = 354a + 100b + 30c$$

On solving, we get $a = 0.55, b = -1.07, c = 1.42$.

The required parabola is

$$y = 0.55x^2 - 1.07x + 1.42$$

Example 5.13: Fit a second degree parabola to the following data.

x	1.0	1.5	2.0	2.5	3.0	3.5	4.0
y	1.1	1.3	1.6	2.0	2.7	3.4	4.1

Solution: We shift the origin to $(2.5, 0)$. This results in change of the variable x to X by the relation

$$X = \frac{x - 2.5}{0.5} = \frac{x - \dfrac{5}{2}}{\dfrac{1}{2}} = 2x - 5$$

Let the parabola of fit be $y = a + bX + cX^2$.

The values of ΣX and other corresponding values are calculated as below.

x	X	y	Xy	X^2	X^2y	X^3	X^4
1.0	-3	1.1	-3.3	9	9.9	-27	81
1.5	-2	1.3	-2.6	4	5.2	-8	16
2.0	-1	1.6	-1.6	1	1.6	-1	1
2.5	0	2.0	0.0	0	0.0	0	0
3.0	1	2.7	2.7	1	2.7	1	1
3.5	2	3.4	6.8	4	13.6	8	16
4.0	3	4.1	12.3	9	36.9	27	81
	$\Sigma X = 0$	$\Sigma y = 16.2$	$\Sigma Xy = 14.3$	$\Sigma X^2 = 28$	$\Sigma X^2y = 69.9$	$\Sigma X^3 = 0$	$\Sigma X^4 = 196$

The normal equations are

$$7a + 28c = 16.2$$
$$28b = 14.3$$
$$28a + 196c = 69.9$$

Solving these equations, we get

$$a = 2.07, b = 0.511, c = 0.061$$
$$y = 2.07 + 0.511X + 0.061X^2$$

Replacing X by $2x - 5$, we get

$$y = 2.07 + 0.511 (2x - 5) + 0.061 (2x - 5)^2$$

or $$y = 1.04 - 0.198x + 0.244x^2$$

This is the required parabola of 'best fit'.

5.9 FITTING OF OTHER CURVES

$$y = ax^b$$

Taking logarithm $\log_{10} y = \log_{10} a + b \log_{10} x$

i.e. $Y = A + bX$, where $Y = \log_{10} y$, $A = \log_{10} a$, $X = \log_{10} x$

The normal equations are therefore

$$\Sigma Y = nA + b\Sigma X$$

$$\Sigma XY = A\Sigma X + b\Sigma X^2$$

from which A and b can be determined. Then a can be calculated from $A = \log_{10} a$.

1. $$y = ae^{bx}$$

Taking logarithm $\log_{10} y = \log_{10} a + bx \log_{10} e$

or $Y = A + Bx$

where $Y = \log_{10} y$, $A = \log_{10} a$, $B = b \log_{10} e$

Here the normal equations are

$$\Sigma Y = nA + B\Sigma X$$

$$\Sigma xY = A\Sigma x + B\Sigma x^2$$

from which A and B can be found and consequently a, b can be calculated.

2. $$xy^a = b$$

Taking logarithm, we get

$$\log_{10} x + a \log_{10} y = \log_{10} b$$

or $$\log_{10} y = 1/a \log_{10} b - 1/a \log_{10} x$$

This is of the form $Y = A + BX$

where $Y = \log_{10} y$, $X = \log_{10} x$, $A = 1/a \log_{10} b$, $B = -1/a$

Here the normal equations are

$$\Sigma Y = nA + B\Sigma X$$

$$\Sigma XY = A\Sigma X + B\Sigma X^2$$

from which A and B can be determined and hence a, b are calculated.

Example 5.14: Apply the method of least squares to determine the constants a and b such that $y = ae^{bx}$ fits the following data.

x	0.0	0.5	1.0	1.5	2.0	2.5
y	0.10	0.45	2.15	9.15	40.35	180.75

Solution: The curve to be fitted is $y = ae^{bx}$

or $Y = A + Bx$ when $Y = \log_{10} y$

$A = \log_{10} a$, $B = b \log_{10} e$

The normal equations are

$$\Sigma Y = 6A + B\Sigma x$$
$$\Sigma xY = A\Sigma x + B\Sigma x^2$$

x	y	$Y = \log_{10} y$	x^2	xY
0	0.10	-1	0	0
0.5	0.45	-0.3468	0.25	-0.1734
1.0	2.15	0.3324	1.0	0.3324
1.5	9.15	0.9614	2.25	1.4421
2.0	40.35	1.6058	4.00	3.2116
2.5	180.75	2.2571	6.25	5.6428
$\Sigma x = 7.5$		$\Sigma Y = 3.8099$	$\Sigma x^2 = 13.75$	$\Sigma xY = 10.4555$

Substituting the values in normal equations, we get

$$3.8099 = 6A + 7.5B \quad \text{and} \quad 10.4555 = 7.5A + 13.755B$$

On solving $\qquad A = -0.9916, \quad B = 13013$

$$a = \text{antilog}_{10} A = 0.1019 \quad b = B/\log_{10} e = 2.9963$$

and the curve is $\qquad y = 0.1019 \exp(2.9963x)$

Example 5.15: Fit $y = ab^x$ by the method of least squares to the data given below.

x	0	1	2	3	4	5	6	7
y	10	21	35	59	92	200	400	610

Solution: The curve to be fitted is $y = ab^x$ or $Y = A + Bx$

where $\qquad A = \log_{10} a, \quad B = \log_{10} b$ and $Y = \log_{10} y$

The normal equations are

$$\Sigma Y = 8A + B\Sigma x$$
$$\Sigma xY = A\Sigma x + B\Sigma x^2$$

x	y	$Y = \log_{10} y$	x^2	xY
0	10	1.0000	0	0
1	21	1.3222	1	1.3222
2	35	1.5441	4	3.0882
3	59	1.7708	9	5.3124
4	92	1.9638	16	7.8552
5	200.	2.3010	26	11.5050
6	400	2.6021	36	15.6126
7	610	2.7853	49	19.4971
$\Sigma x = 28$		$\Sigma Y = 15.2893$	$\Sigma x^2 = 140$	$\Sigma xY = 64.1927$

Substituting the values in normal equations, we get

$$15.2893 = 8A + 28B, \quad 64.1927 = 28A + 140B$$

On solving $\qquad A = 1.02115, \quad B = 0.2542892$

$$a = \text{antilog}_{10} A = 10.499 \text{ and } b = \text{antilog } B = 1.7959$$

Hence the required curve is $y = 10.499 (1.7959)^x$.

Example 5.16: The pressure and volume of a gas are related by equation $pV^r = kr$, k being constant. Fit the equation to the following set of observations.

p (kg/cm^2)	0.5	1.0	1.5	2.0	2.5	3.0
V (litres)	1.62	1.00	0.75	0.62	0.52	0.46

Solution: The law is $pV^r = k$

Taking logarithm on both sides, we get

$$\log_{10} p + r \log_{10} V = \log_{10} k$$

$$\log_{10} V = 1/r \log_{10} k - 1/r \log_{10} p \quad \text{or} \quad Y = A + BX$$

where $\qquad X = \log_{10} p,\ Y = \log_{10} V,\ A = 1/r \log_{10} k,\ B = -1/r$

The normal equations are

$$\Sigma Y = 6A + B\Sigma X \qquad \qquad \text{(i)}$$

$$\Sigma XY = A\Sigma X + B\Sigma X^2 \qquad \qquad \text{(ii)}$$

Now ΣX are calculated as follows.

p	V	$X = \log_{10} p$	$Y = \log_{10} V$	XY	X^2
0.5	1.62	-0.3010	0.2095	-0.0640	0.0906
1.0	1.00	0.0000	0.0000	-0.0000	0.0000
1.5	0.75	0.1761	-0.1249	-0.0220	0.0310
2.0	0.62	0.3010	-0.2076	-0.0625	0.0906
2.5	0.52	0.3979	-0.2840	-0.1130	0.1583
3.0	0.46	0.4771	-0.3372	-0.1699	0.2276
		$\Sigma X = 1.0511$	$\Sigma Y = -0.7442$	$\Sigma XY = -0.4214$	$\Sigma X^2 = 0.5981$

Equations (i) and (ii) becomes

$$6A + 1.0511B = -0.7442$$

$$1.0511A + 0.5981B = -0.4214$$

On solving, we get $\qquad A = 0.0132,\ B = -0.7836,\ r = -\dfrac{1}{B} = 1.276$

$$k = \text{antilog } (Ar) = \text{antilog } (0.0168) = 1.039$$

Hence the equation of the best fit is

$$pV^{1.276} = 1.039$$

Example 5.17: Determine the constants a and b by the method of least squares such that $y = ae^{bx}$ fits the following data.

x	2	4	6	8	10
y	4.077	11.084	30.128	81.897	222.62

Solution: The given relation is $y = ae^{bx}$.

Taking logarithm on both sides, we get

$$\log_{10} y = \log_{10} a + bx$$

Setting $\qquad \log_{10} y = Y,\ x = X$

$$\log_{10} a = A \text{ and } b = B$$

We get $Y = A + BX$ which is strong line.

We have the following table.

$X = x$	$Y = \log_{10} y$	X^2	XY
2	1.405	4	2.810
4	2.405	16	9.620
6	3.405	36	20.430
8	4.405	64	35.240
10	5.405	100	54.050
$\Sigma X = 30$	$\Sigma Y = 17.025$	$\Sigma X^2 = 220$	$\Sigma XY = 122.150$

Using normal equations, we get

$$5A + 30B = 17.025$$
$$30A + 220B = 122.150$$

On solving, we get $A = 0.405, \quad B = 0.5$

Hence

$$a = e^A = e^{0.405} = 1.499$$
$$b = B = 0.5$$

Example 5.18: Fit the curve of the form $y = ax^b$ to the data:

X	1	2	3	4	5	6
Y	1200	900	600	200	110	50

Solution: Given $y = ax^b$

Taking logarithm on both sides, we get

$$\log_{10} y = \log_{10} a + \log_{10} x$$

Putting $Y = \log_{10} y, X = \log_{10} x, A = \log_{10} a$, we get

$$Y = A + bX \text{ which is linear in } X \text{ and } Y$$

The normal equations are

$$\Sigma Y = 6A + b\Sigma X \tag{i}$$
$$\Sigma XY = A\Sigma X + b\Sigma X^2 \tag{ii}$$

Now ΣX are calculated as follows:

x	y	$X = \log_{10} x$	$Y = \log_{10} y$	X^2	XY
1	1200	0.0000	3.0792	0.0000	0.0000
2	900	0.3010	2.9542	0.0906	0.8892
3	600	0.4771	2.7782	0.2276	1.3255
4	200	0.6021	2.3010	0.3625	1.3854
5	110	0.6990	2.0414	0.4886	1.4269
6	50	0.7781	1.6990	0.6064	1.3220
		$\Sigma X = 2.8573$	$\Sigma Y = 14.8530$	$\Sigma X^2 = 1.7747$	$\Sigma XY = 6.3490$

The normal equations become

$$14.8530 = 6A + 2.8573b$$
$$6.3490 = 2.8573A + 1.7747b$$

On solving, we get $A = 3.3086, \ b = -1.749$

$$a = \text{antilog } (3.3086) = 2035$$

The required equation is

$$y = (2035) x^{-1.7494}.$$

PROBLEMS 5.3

1. Fit a parabola to the data:

x	0	0.1	0.2	0.3	0.4	0.5	0.6	0.7	0.8	0.9
y	3.1950	3.2299	3.2532	3.2611	3.2516	3.2282	3.1807	3.1266	3.0594	2.9759

[Ans. $y = -0.7653x^2 + 0.4425x + 3.1951$]

2. Fit a parabola to the data:

x	2	4	6	8	10
y	3.07	12.85	31.47	57.38	91.29

[Ans. $y = 0.99x^2 - 83x + 0.61$]

3. Fit a parabola to the data:

x	1	2	3	4	5
y	5	12	26	60	97

[Ans. $y = 5.7x^2 - 11x + 10.3$]

4. Fit a parabola to the data:

x	1	2	3	4	5
y	2	3	5	8	10

[Ans. $y = 0.403 + 0.388x - 2.79x^2$]

5. Fit a parabola to the data:

Year x	1911	1912	1913	1914	1915
Commodity y	10	12	8	10	14

[Ans. $y = 9.4 + 0.6(x - 1913) + 0.7(x - 1913)^2$]

6. Fit a parabola to the data:

Speed v (ft/min)	350	400	500	600
Life t (min.)	61	26	7	2.6

Find the best values of a and b in the law $v = ae^{bt}$ by the method of least squares.

[Ans. $V = (553.4)\, e^{-(0.0083)\, t}$]

7. Fit a law of the type $y = ae^{bx}$ to the data:

x	0	1	2	3
y	1.05	2.10	3.85	8.30

[Ans. $y = 4.64\, e^{0.46x}$]

8. Fit a curve of the form $y = ax^b$ to the given data:

x	61	21	7	2.6
y	350	400	500	600

[Ans. $y = (682.3)\, x^{-(0.1588)}$]

9. Fit a curve of the form $y = ax^b$ to the following data:

x	2	3	4	5	6
y	144	172.8	207.4	248.8	298.5

[Ans. $100\,(1.2)^x$]

5.10 CURVE FITTING BY A SUM OF EXPONENTIALS

A frequently encountered problem in engineering and physics is that of fitting a sum of exponentials of the form

$$y = f(x) = a_1 e^{\lambda_1 x} + a_2 e^{\lambda_2 x} + \dots + a_n e^{\lambda_n x} \tag{5.11}$$

to a set of data points, say $(x_1, y_1), (x_2, y_2), \dots, (x_n, y_n)$. Assuming that n is known and $a_1, a_2, \dots, a_n, \lambda_1, \lambda_2, \dots, \lambda_n$ are unknowns and which are to be determined.

It is clear that $f(x)$ satisfies a differential equation of the type

$$\frac{d^n y}{dx^n} + A_1 \frac{d^{n-1} y}{dx^{n-1}} + A_2 \frac{d^{n-2} y}{dx^{n-2}} + \dots + A_n y = 0 \tag{5.12}$$

where $A_i's$ are all unknown.

To solve the Eq. (5.12), Frogberg suggested a method, in which we numerically evaluate the derivatives $\frac{d^n y}{dx^n}, \frac{d^{n-1} y}{dx^{n-1}}$ at the n data points and substituting them in Eq. (5.12), we therefore, obtain a system of n linear equations for the n unknowns $A_1, A_2, \dots, A_n$ which can then be solved.

Also, it can be verified that $\lambda_1, \lambda_2, \dots, \lambda_n$ are the roots of the algebraic equation

$$\lambda^n + A_1 \lambda^{n-1} + A_2 \lambda^{n-2} + \dots + A_n = 0 \tag{5.13}$$

which when solved enables us to determine $a_1, a_2, \dots, a_n$ from Eq. (5.11) by the method of least squares.

Disadvantage: Due to the numerical evaluations of higher derivatives whose accuracy increases as we evaluate higher order derivatives and leading to unreliable results.

In 1974, Moore described a computational technique which leads to more reliable results. Now demonstrate the method for the case $n = 2$.

Let the function to be fitted to a given data of the form

$$Y = a_1 e^{\lambda_1 x} + a_2 e^{\lambda_2 x} + \dots \tag{5.14}$$

which satisfies a differential equation

$$\frac{d^2 y}{dx^2} = A_1 \frac{dy}{dx} + A_2 y \tag{5.15}$$

where A_1, A_2 are to be determined.

Assuming that x_0 is the initial value of x, we integrating Eq. (5.15) from x_0 to x, we get

$$y'(x) - y'(x_0) = A_1 y(x) - A_1 y(x_0) + A_2 \int_{x_0}^x y(x) dx \tag{5.16}$$

where

$$y' = \frac{dy}{dx}$$

Again integrating Eq. (5.16) from x_0 to x, we get

$$y(x) - y(x_0) - (x - x_0) y'(x_0) = A_1 \int_{x_0}^x y(x) dx - A_1 (x - x_0) y(x_0) + A_2 \int_{x_0}^x \int_{x_0}^x y(x) dx \dots \tag{5.17}$$

Since we know that

$$\int_{x_0 \dots}^x \int_{x_0}^x f(x) dx \dots dx = \frac{1}{(n-1)!} \int_{x_0}^x (x - t)^{n-1} f(t) dt \tag{5.18}$$

Using Eq. (5.18), Eq. (5.17) becomes

$$y(x) - y(x_0) - (x - x_0) y'(x_0) = A_1 \int_{x_0}^x y(x) dx - A_1 (x - x_0) y(x_0) + A_2 \int_{x_0}^x (x - t) y(t) dt \dots \tag{5.19}$$

In order to use the Eq. (5.9) to set up a linear system for A_1 and A_2, we eliminate $y'(x_0)$ in the following way.

Choose two data points x_1 and x_2 such that

$$x_0 - x_1 = x_2 - x_0, \text{ we obtain from Eq. (5.9)}$$

$$y(x_1) - y(x_0) - (x_1 - x_0)\, y'(x_0) = A_1 \int_{x_0}^{x_1} y(x)\, dx - A_1(x - x_0) y(x_0) + A_2 \int_{x_0}^{x_1} (x_1 - t) y(t)\, dt$$

$$(5.20)$$

and $y(x_2) - y(x_2) - (x_2 - x_0)\, y'(x_0) = A_1 \int_{x_0}^{x_2} y(x)\, dx - A_1(x_2 - x_0) y(x_0) + A_2 \int_{x_0}^{x_2} (x_2 - t) y(t)\, dt$

$$(5.21)$$

Adding Eqs (5.20) and (5.21) and satisfying, we get

$$y(x_1) + y(x_2) - 2y(x_0) = A_1 \left[\int_{x_0}^{x_1} y(x)\, dx + \int_{x_0}^{x_2} y(x)\, dx \right] + A_2 \left[\int_{x_0}^{x_1} (x_1 - t) y(t)\, dt + \int_{x_0}^{x_2} (x_2 - t) y(t)\, dt \right]$$

$$(5.22)$$

In Eq. (5.22), we get a linear system of equations for A_1 and A_2 and then we obtain λ_1 and λ_2 from the characteristic equation

$$\lambda^2 = A_1\lambda + A_2 \qquad (5.22a)$$

Finally, a_1 and a_2 can be obtained by the averaging technique.

Example 5.19: Fit a function of the form $y = a_1 e^{\lambda_1 x} + a_2 e^{\lambda_2 x}$ to the data given by

x	1.0	1.1	1.2	1.3	1.4	1.5	1.6	1.7	1.8
y	1.54	1.67	1.81	1.97	2.15	2.35	2.58	2.83	3.11

Solution: Choosing $x_1 = 1.0$, $x_0 = 1.2$, $x_2 = 1.4$, then Eq. (5.22) becomes

$$0.07 = A_1 \left[\int_{1.2}^{1.0} y(x)\, dx + \int_{1.2}^{1.4} y(x)\, dx \right] + A_2 \left[\int_{1.2}^{1.0} (1.0 - t) y(t)\, dt + \int_{1.2}^{1.4} (1.4 - t)\, (t)\, dt \right]$$

Evaluating the integrals by Simpson's rule and simplifying, the above equation becomes

$$1.81A_1 + 2.18A_2 = 2.10 \qquad \text{(i)}$$

Again choosing $x_1 = 1.4$, $x_0 = 1.6$ and $x_2 = 1.8$.
then Eq. (i) after simplifying, we get

$$2.88A_1 + 3.10A_2 = 3 \qquad \text{(ii)}$$

Solving Eqs (i) and (ii), we get

$$A_1 = 0.03204, \ A_2 = 0.9364$$

Equation (5.22a) becomes

$$\lambda^2 - 0.03204\,\lambda - 0.9364 = 0 \qquad \text{(iii)}$$

Solving Eq. (iii), we get

$$\lambda_1 = 0.99, \ \lambda_2 = -0.96$$

Using least squares method, we get

$$a_1 = 0.499, \ a_2 = 0.491$$

Hence

$$y = 0.499\, e^{0.99x} + 0.491\, e^{-0.96x}$$

6 Solution to Numerical, Algebraic and Transcendental Equations

6.1 INTRODUCTION

The following equation

$$f(x) = a_0 x^n + a_1 x^{n-1} + \dots + a_{n-1} x + a_n$$

where a_is are constants ($a_0 \neq 0$) and n is a positive integer called a polynomial in x of degree n, and the equation $f(x) = 0$ is called an algebraic equation of degree n.

If $f(x)$ contains some other functions like exponential, trigonometric, logarithmic, etc. then $f(x) = 0$ is called a transcendental equation, e.g.

$$x^3 - 3x + 6 = 0, \, x^5 - 7x^4 + 3x^2 + 36x + 7 = 0$$

are algebraic equations of third and fifth degree, whereas $x^2 - 3\cos x + 1 = 0$, $x^2 e^x - 2 = 0$, $x \log_{10} x = 1.2$, etc. are transcendental equations.

In both the cases, if the coefficients are of pure numbers, they are called numerical equations. For the algebraic equations of degree two, three or four, methods are available to solve them. But the need often arises to solve higher degree or transcendental equation for which no direct method exists. Such equations can best be solved by approximate methods.

Intermediate Value Property

If $f(x)$ is a continuous function in $[a, b]$ and $f(a), f(b)$ have opposite signs, then the equation $f(x) = 0$ must have at least one root between $x = a$ and $x = b$.

Note: Every equation of odd degree has at least one real root.

Descartes' Rule of Signs

The equation, $f(x) = 0$ cannot have more positive roots than change of signs in $f(x) = 0$, and more –ve roots than the change of signs in $f(-x)$, e.g.

$$f(x) = 2x^5 - 3x^4 + 5x^3 - 8x^2 + 1 = 0 \qquad \text{(signs of } f(x) = + - + - +)$$

As $f(x)$ has four changes of signs from (+ve to –ve or –ve to +ve), above equation cannot have more than four positive roots.

Also
$$f(x) = 2(-x)^5 - 3(-x)^4 + 5(-x)^3 - 8(-x)^2 + 1$$
$$= -2x^5 - 3x^4 - 5x^3 - 8x^2 + 1$$

This shows that $f(x)$ has one change of sign. It cannot have more than one negative root.

Existence of Imaginary Roots

If an equation of nth degree has (at the most) p positive roots and q negative roots, then it follows that the equation has at least $n - (p + q)$ imaginary roots.

E.g. $\qquad f(x) = x^7 - 3x^4 + 2x^3 - 1 = 0$ has at least four imaginary roots.

$\qquad\qquad f(x) = x^7 - 3x^4 + 2x^3 - 1$ has three possible changes or at the most

has three positive roots and no negative root

$$f(-x) = -x^7 - 3x^4 - 2x^3 - 1$$

Degree of equation $= 7 \quad \therefore$ at least 4 imaginary roots

6.2 BISECTION METHOD (BOLZONO METHOD OR INTERVAL HALVING METHOD)

It is based on repeated application of intermediate value property for root to exist in the given interval $[a, b]$.

Let $f(x)$ be a continuous function in $[a, b]$

and $\qquad\qquad f(a) < 0$ and $f(b) > 0$

or $\qquad\qquad f(a)\, f(b) < 0$

Then first approximation to the root is

$$x_1 = \frac{a + b}{2}$$

If $f(x_1) = 0$, then x_1 is a root of $f(x)$, otherwise root lies between a and x_1 or x_1 and b as $f(x_1) > 0$ or $f(x_1) < 0$ (Fig. 6.1).

Then we proceed as before and continue till the desired accuracy is achieved.

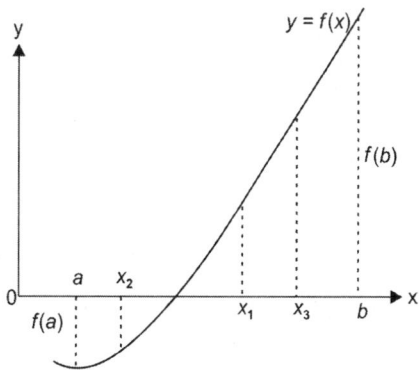

Fig. 6.1

Advantages

1. The main advantage of this method is that if $f(x)$ is continuous in $[a, b]$ and this interval actually contains a root, then this method is guaranteed to work.
2. The number of iterations required to achieve a specific accuracy is almost known in advance. Since the interval $[a, b]$ is halved each time, the last value of x differ from the true root by less than $1/2$ the last interval or we can say error after the nth

$$\text{iteration} < \left| \frac{b - a}{2^n} \right|.$$

In other words, after 'n' bisection, the length of the subinterval which contain x^n is $\dfrac{b-a}{2^n}$. If the error is to be made less than a small quantity $\in$ say

$$\frac{b-a}{2^n} < \in, \text{ i.e. } 2^n > \frac{b-a}{\in}$$

The number of iterations should be greater than $\dfrac{\log \dfrac{b-a}{\in}}{\log 2}$.

Disadvantages

1. Its rate of convergence is slow.
2. When there are multiple root, interval halving (bisection method) may not be applicable.

Example 6.1: Find a root of the equation $x^3 - x - 11 = 0$ correct to four decimal points using bisection method.

Solution: Let
$$f(x) = x^3 - x - 11$$
$$f(2) = -5 < 0$$
$$f(3) = 13 > 0$$

$\therefore$ root lies between 2 and 3, therefore, first approximation to the root is

$$x_1 = \frac{(2+3)}{2} = 2.5$$

Now
$$f(2.5) = (2.5)^3 - 2.5 - 11 = 2.215 \text{ (+ve)} > 0$$
$\therefore$ root lies between 2 and 2.5.

2nd approximation

$$x_2 = \frac{2 + 2.5}{2} = 2.25$$

Now
$$f(2.25) = (2.25)^3 - 2.25 - 11 = -1.859375 < 0$$
$\therefore$ root lies between 2.5 and 2.25.

3rd approximation

$$x_3 = \frac{x_1 + x_2}{2} = \frac{2.5 + 2.25}{2} = 2.375$$

Now
$$f(2.375) = (2.375)^3 - 2.375 - 11 = 0.0214843 > 0$$
$\therefore$ root lies between x_2 and x_3, i.e. 2.25 and 2.375.

4th approximation

$$x_4 = \frac{2.25 + 2.375}{2} = 2.3125$$

$$f(2.3125) = (2.3125)^3 - 2.3125 - 11 = -0.9460449 < 0$$
$\therefore$ root lies between 2.375 and 2.3125.

5th approximation

$$x_5 = \frac{x_3 + x_4}{2} = \frac{2.375 + 2.3125}{2} = 2.34375$$

$$= -0.469137 < 0$$
$\therefore$ root lies between x_3 and x_5.

6th approximation

$$x_6 = \frac{x_3 + x_5}{2} = \frac{2.375 + 2.34375}{2} = 2.359375$$

Now $f(2.359375) = (2.359375)^3 - 2.359375 - 11 = -0.225592 < 0$

∴ root lies between x_3 and x_6.

7th approximation

$$x_7 = \frac{x_3 + x_6}{2} = \frac{2.375 + 2.359375}{2} = 2.3671875$$

$$f(2.3671875) = (2.3671875)^3 - 2.3671875 - 11 = -0.1024708 < 0$$

∴ root lies between x_3 and x_7.

8th approximation

$$x_8 = \frac{2.375 + 2.3671875}{2} = 2.3710938$$

$$f(2.3710938) = (2.3710938)^3 - 2.3710938 - 11 = -0.040601 < 0$$

∴ root lies between 2.375 and 2.3710938.

9th approximation

$$x_9 = \frac{2.375 + 2.3710938}{2} = 2.3730469$$

$$f(2.3730469) = (2.3730469)^3 - 2.3730469 - 11 = -9.585468 \times 10^{-3} < 0$$

∴ root lies between x_3 and x_9.

10th approximation

$$x_{10} = \frac{2.375 + 2.3730469}{2} = 2.3740235$$

$$f(2.3740235) = (2.3740235)^3 - 2.3790235 - 11 = 5.942661 \times 10^{-3} > 0$$

∴ root lies between x_9 and x_{10}.

11th approximation

$$x_{11} = \frac{2.3730469 + 2.3740235}{2} = 2.3735352$$

$$f(2.3735352) = (2.3735352)^3 - 2.3735352 - 11 = -1.823101 \times 10^{-3} < 0$$

∴ root lies between x_{10} and x_{11}.

12th approximation

$$x_{12} = \frac{x_{10} + x_{11}}{2} = \frac{2.3740235 + 2.3735352}{2} = 2.3737794$$

$$f(2.3737794) = (2.3737794)^3 - 2.3737794 - 11 = 2.059952 \times 10^{-3} > 0$$

∴ root lies between x_{11} and x_{12}.

Following the same procedure, we have

$$x_{13} = 2.3736573$$
$$x_{14} = 2.3735963$$
$$x_{15} = 2.3736268 \qquad \therefore x_{14} = x_{15}$$

Hence, the root correct to four decimal places is 2.3736.

Example 6.2: Using bisection method, find the negative root of $x^3 - x + 11 = 0$.

Solution: Let
$$f(x) = x^3 - x + 11$$
$$f(-x) = -x^3 + x + 11$$

The negative root of $f(x) = 0$ is positive root of $f(-x) = 0$.

Therefore, we will find the positive root of $f(-x) = 0$

or
$$\phi(x) = x^3 - x - 11 = 0$$

In last example, we got $x = 2.3736$, hence, the negative root is -2.3736.

Example 6.3: Find a positive root of the equation $xe^x = 1$, which lies between 0 and 1.

Solution:
$$f(x) = xe^x - 1$$
$$f(0) = -1 < 0$$
$$f(1) = 1.718 > 1$$

∴ root lies between 0 and 1.

n	a	b	x	$f(x)$
1	0	1	0.5	−ve
2	0.5	1	0.75	+ve
3	0.5	0.75	0.625	+ve
4	0.5	0.625	0.5625	−ve
5	0.5625	0.625	0.59375	+ve
6	0.5625	0.59375	0.5781	+ve
7	0.5625	0.5781	0.5703	+ve
8	0.5625	0.5703	0.5664	−ve
9	0.5664	0.5703	0.5684	+ve
10	0.5664	0.5684	0.5674	+ve
11	0.5664	0.5674	0.5669	
			≈ 0.567	−ve
12	0.5669	0.5674	0.5671	
			≈ 0.567	+ve

The required root is 0.567, which is correct to three decimal places.

Example 6.4: Find one root of $e^x - 3x = 0$ correct to two decimal places.

Solution:
$$f(x) = e^x - 3x$$
$$f(0) = 1 = \text{positive}$$
$$f(1) = e - 3 = \text{negative}$$
$$f(1.5) = e^{1.5} - 4.5 = \text{negative}$$
$$f(2) = e^2 - 6 = \text{positive}$$
$$f(1.6) = e^{1.6} - 4.8 = \text{positive}$$

∴ root lies between 1.5 and 1.6, therefore, first approximation to the root is:

$$x_1 = \frac{1.5 + 1.6}{2} = 1.55$$

$$f(x_1) = 0.06 = +\text{ve}$$

∴ root lies between 1.5 and 1.55.

2nd approximation

$$x_2 = \frac{1.5 + 1.55}{2} = 1.525$$

$$f(x_2) = 0.02 = +ve$$

∴ root lies between 1.5 and 1.525.

3rd approximation

$$x_3 = \frac{1.5 + 1.525}{2} = 1.5125$$

$$f(1.5125) = 0.00056 = +ve$$

∴ root lies between 1.5 and 1.5125.

4th approximation

$$x_4 = \frac{1.5 + 1.5125}{2} = 1.5062$$

$$f(1.5062) = -0.00904 = -ve$$

∴ root lies between 1.5062 and 1.55.

$$x_5 = \frac{1.5062 + 1.5125}{2} = 1.50935$$

The required root is 1.50 which is correct to two decimal places.

Example 6.5: Find the root of $x^3 - 4x - 9 = 0$ using bisection method in four stages.

Solution:
$$f(x) = x^3 - 4x - 9 = 0$$
$$f(1) = \text{negative}$$
$$f(2) = \text{negative}$$
$$f(3) = \text{positive}$$

∴ root lies between 2 and 3, therefore, first approximation to the root is:

$$x_1 = \frac{2 + 3}{2} = 2.5$$

$$f(2.5) = (2.53)^2 - 4(2.5) - 9 = -3.375 \ (-ve)$$

∴ root lies between 2.5 and 3.

2nd approximation

$$x_2 = \frac{3 + 2.5}{2} = 2.75$$

$$f(2.75) = (2.75)^2 - 4(2.75) - 9 = 0.7969 > 0$$

∴ root lies between 2.75 and 2.5.

3rd approximation

$$x_3 = \frac{2.75 + 2.5}{2} = 2.625$$

$$f(2.625) = (2.625)^3 - 4(2.625) - 9 = -1.4121 \ (-ve)$$

∴ root lies between 2.625 and 2.75.

4th approximation

$$x_4 = \frac{2.625 + 2.75}{2} = 2.6875$$

Hence fourth approximation = 2.6875

Example 6.6: Find the root of $\tan x + x = 0$ up to two decimal places which lies between 2 and 2.1.

Solution:
$$f(x) = \tan x + x$$
$$f(2) = -0.18 \ (-ve)$$
$$f(2.1) = 0.39 \ (+ve)$$

∴ root lies between 2.0 and 2.1, therefore, first approximation to the root is:

$$x_1 = \frac{2 + 2.1}{2} = 2.05$$

$$f(2.05) = 0.12 \ (+ve)$$

∴ root lies between 2 and 2.05.

2nd approximation

$$x_2 = \frac{2 + 2.05}{2} = 2.025$$

$$f(2.025) = -0.023$$

∴ root lies between 2.025 and 2.05.

3rd approximation

$$x_3 = \frac{2.025 + 2.05}{2} = 2.0375$$

$$f(2.0375) = 0.054 = +ve$$

∴ root lies between 2.025 and 2.0375.

4th approximation

$$x_4 = \frac{2.025 + 2.0375}{2} = 2.03125$$

The required root is 2.03 which is correct to two decimal places.

6.3 METHOD OF FALSE POSITION (REGULA-FALSI METHOD)

This method (also known as Regula–Falsi method) is used for finding the real root of an equation $f(x) = 0$ and is somewhat similar to the bisection method.

Consider the equation $f(x) = 0$. Let a and b $(a < b)$ be two values of x such that $f(a)$ and $f(b)$ are of opposite signs. Then the graph of $y = f(x)$ crosses the x-axis at some point between a and b (Fig. 6.2).

Therefore, the equation of chord joining the two points A $[a, f(a)]$ and B $[b, f(b)]$ is

$$y - f(a) = \frac{f(b) - f(a)}{b - a}(x - a) \tag{6.1}$$

Let in the interval (a, b) the graph of the function be considered as a straight line. So the intersection of line given by Eq. (6.1) with x-axis will give an approximate value of the root. Putting $y = 0$ in Eq. (6.1), we get

$$f(a) = \frac{f(b) - f(a)}{b - a}(x - a)$$

$$x = \frac{af(b) - bf(a)}{f(b) - f(a)}$$

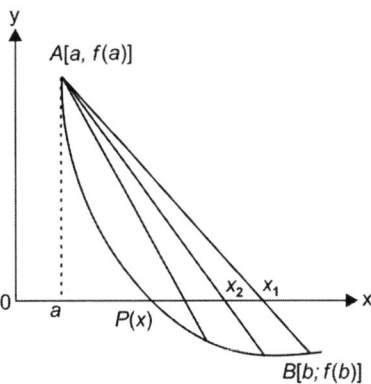

Fig. 6.2

Hence the first approximation to the root is given by

$$x_1 = \frac{af(b) - bf(a)}{f(b) - f(a)} \tag{6.2}$$

Now if $f(x_1)$ and $f(a)$ are of opposite sign, then the root lies in between a and x_1. So, we replace b by x_1 in Eq. (6.2) and get the next approximation x_2. But if $f(x_1)$ and $f(a)$ are of same sign, then $f(x_1)$ and $f(b)$ will be of opposite sign, therefore, the root lies between x_1 and b. Hence we replace a by x_1 in Eq. (6.2) and get the next approximate x_2. The process is to be repeated till the root is found to be of desired accuracy.

Example 6.7: Find root of the equation $x \log_{10} x - 1.2 = 0$ correct to three decimal places.

Solution: $f(x) = x \log_{10} x - 1.2$
$f(1) = -1.2 = -\text{ve}$
$f(2) = -0.5979 = -\text{ve}$
$f(3) = 0.23136 = +\text{ve}$

∴ root lies between 2 and 3.

Taking $a = 2, b = 3$
$f(a) = f(2) = -0.59795$
$f(b) = f(3) = 0.23136$

By method of false position

1st iteration $x_1 = \dfrac{af(b) - bf(a)}{f(b) - f(a)} = \dfrac{2(0.23136) - 3(-0.59794)}{0.23136 + 0.59794} = 2.72102$

$f(2.72102) = -0.0170863$ is the root lies between 2.72102 and 3.

Now $a = 2.72102, b = 3$
$f(2.72102) = -0.01709$
$f(3) = 0.23136$

2nd iteration $x_2 = \dfrac{af(b) - bf(a)}{f(b) - f(a)} = \dfrac{2.72101 \times 0.23136 - 3 \times -0.01709}{0.23136 + 0.01709} = 2.74021$

$f(2.74021) = -0.0003802$

The root lies between 2.74021 and 3.

Proceeding in the same way, we get 3rd and 4th iterations
$$x_3 = 2.74024, \, x_4 = 2.74063$$
∴ root is 2.740 correct to three decimal places.

Example 6.8: Find the real root of equation $x^3 - 2x - 5 = 0$ by the method of false position correct to three decimal places.

Solution:
$$f(x) = x^3 - 2x - 5$$
$$f(1) = -ve$$
$$f(2) = -ve$$
$$f(3) = +ve$$
∴ the root of $f(x) = 0$ lies between 2 and 3.

Taking $a = 2, b = 3$
$$f(a) = f(2) = -1 \, (-ve)$$
$$f(b) = f(3) = 16 \, (+ve)$$

1st approximation $\quad x_1 = \dfrac{af(b) - bf(a)}{f(b) - f(a)} = \dfrac{2 \times 16 - 3(-1)}{16 + 1} = 2.0588$

$$f(2.0588) = (2.055)^3 - 2(2.0588) - 5 = -0.391 \, (-ve)$$
∴ the root lies between 2.0588 and 3.

Now $\quad a = 2.0588, f(a) = -0.391, b = 3, f(b) = 16$

2nd approximation $\quad x_2 = \dfrac{2.0588(16) - 3(-0.391)}{16 - (-0.391)} = 2.08125$

$$f(2.08125) = -0.147 \, (-ve)$$
∴ the root lies between 2.08125 and 3.

Now $\quad a = 2.08125, f(a) = -0.147, b = 3, f(b) = 16$

3rd approximation $\quad x_3 = \dfrac{2.08125(16) - 3(-0.147)}{16 + 0.147} = 2.0896$

$$f(2.0896) = -0.0551 \, (-ve)$$
∴ the root lies between 2.0896 and 3.

4th approximation $\quad x_4 = \dfrac{2.0896(16) - 3(-0.0551)}{16 + 0.0551} = 2.0927$

$$f(2.0927) = -0.0206 \, (-ve)$$
∴ the root lies between 2.0927 and 3.

5th approximation $\quad x_5 = \dfrac{2.0927(16) - 3(-0.0206)}{16 + 0.0206} = 2.0939$

$$f(2.0939) = -0.00726 \, (-ve)$$
∴ the root lies between 2.0939 and 3

6th approximation $\quad x_6 = \dfrac{2.0939(16) - 3(-0.00726)}{16 + 0.00726} = 2.0943$

$$f(x_6) = -0.0028$$
∴ the root lies between 2.0943 and 3.

7th approximation $\quad x_7 = \dfrac{2.0943\,(16) - 3\,(-0.0028)}{16 + 0.0028} = 2.0944$

Therefore 2.094 is correct to three decimal places.

Example 6.9: Find real root of $xe^x = \cos x$ correct up to four decimal places.

Solution: $\qquad\qquad f(x) = \cos x - xe^x = 0$

$\qquad\qquad\qquad\qquad f(0) = 1$

$\qquad\qquad\qquad\qquad f(1) = -2.17798$

$\therefore$ the root lies between 0 and 1.

$\qquad\qquad\qquad\qquad a = 0, b = 1$

$\qquad\qquad\qquad f(0) = f(1), f(b) = -2.17798$

1st approximation $\quad x_1 = \dfrac{af(b) - bf(a)}{f(b) - f(a)} = \dfrac{0(-2.17798) - 1}{-2.17798 - 1} = 0.31467$

$\qquad\qquad f(0.31467) = 0.51987 \ (+\text{ve})$

$\therefore$ the root lies between 0.31467 and 1.

$\qquad\qquad\qquad f(a) = f(0.31467) = 0.51987$

$\qquad\qquad\qquad f(b) = f(1) = -2.17798$

2nd approximation $\quad x_2 = \dfrac{0.31467\,(-2.17798) - 1(0.51987)}{-2.17798 - 0.51987} = 0.44673$

$\qquad\qquad f(x_2) = 0.20356 \ (+\text{ve})$

$f(0.44673)$ and $f(1)$ are of opposite sign.

$\qquad\qquad\qquad f(0.44673) = 0.20356$

$\qquad\qquad\qquad f(1) = -2.17798$

3rd approximation $\quad x_3 = \dfrac{0.446733\,(-2.17798) - 1 \times (0.20356)}{-2.17798 - 0.20356} = 0.49402$

$\qquad\qquad f(x_3) = 0.070788$

$\therefore$ the root lies between 0.49402 and 1.

Proceeding in the same way, we have

$\qquad\qquad\qquad\qquad x_4 = 0.50995$

$\qquad\qquad\qquad\qquad x_5 = 0.51520$

$\qquad\qquad\qquad\qquad x_6 = 0.51692$

$\qquad\qquad\qquad\qquad x_7 = 0.51748$

$\qquad\qquad\qquad\qquad x_8 = 0.51767$

$\qquad\qquad\qquad\qquad x_9 = 0.51775$

Hence, the root is 0.5177 correct to four decimal places.

Example 6.10: Find the real root of $xe^x - 3 = 0$ correct to three decimal places.

Solution: $\qquad\qquad f(x) = xe^x - 3, f(0) = -3 \ (-\text{ve})$

$\qquad\qquad\qquad f(1) = e - 3 = -0.28172 \ (-\text{ve})$

$\qquad\qquad\qquad f(1.5) = 1.5e^{1.5} - 3 = 3.72253 \ (+\text{ve})$

∴ the root lies between 1 and 1.5.

$$a = 1, b = 1.5$$
$$f(a) = f(1) = -0.28172$$
$$f(b) = f(1.5) = 3.72253$$

1st approximation $\quad x_1 = \dfrac{af(b) - bf(a)}{f(b) - f(a)} = \dfrac{1(3.72253) - 1.5(-0.28172)}{3.72253 - (-0.28172)} = 1.035$

$$f(x_1) = -0.0864 \ (-ve)$$

∴ the root lies between 1.5 and 1.035.

$$a = 1.035, b = 1.5$$
$$f(a) = -0.028648$$
$$f(b) = 3.72253$$

2nd approximation $\quad x_2 = \dfrac{af(b) - bf(a)}{f(b) - f(a)} = \dfrac{(1.035)(3.72253) - 1.5(-0.028648)}{3.72253 - (-0.0286485)} = 1.045$

$$f(x_2) = -0.0286485 \ (-ve)$$

∴ the root lies between 1.045 and 1.5.

$$f(1.045) = -0.0286485$$
$$f(1.5) = 3.72253$$

3rd approximation $\quad x_3 = \dfrac{(1.045)(3.72253) - 1.5(-0.0286485)}{3.72253 + 0.0286485} = 1.048$

$$f(1.0480) = -0.0111652 \ (-ve)$$

∴ the root lies between 1.0480 and 1.5.

$$f(1.0480) = -0.0111652$$
$$f(1.5) = 3.72253$$

4th approximation $\quad x_4 = \dfrac{1.0480(3.72253) - 1.0480(-0.0111652)}{3.72253 + 0.0111652} = 1.049$

$$f(1.049) = -5.320155 \times 10^{-3} \ (-ve)$$

∴ the root lies between 1.049 and 1.5.

Similarly $\qquad\qquad x_5 = 1.0496$
$$x_6 = 1.0498$$

Hence, the root is 1.05.

Example 6.11: Find the real root of $x^3 - 18 = 0$.

Solution: Given $\qquad x^3 - 18 = 0 = f(x)$
$$f(2) = -10 \ (-ve)$$
$$f(3) = 9 \ (+ve)$$

∴ the root lies between 2 and 3.

$$a = 2, b = 3$$

1st approximation $\quad x_1 = \dfrac{af(b) - bf(a)}{f(b) - f(a)} = \dfrac{2 \times 9 - 3 \times (-10)}{9 + 10} = 2.5263$

$$f(2.5263) = -1.87666$$

∴ the root lies between 2.5263 and 3.

2nd approximation $x_2 = \dfrac{2.5263(9) - 3(-1.87666)}{9 + 1.87666} = 2.6080$

$$f(2.6080) = -2.61260$$

∴ the root lies between 2.6080 and 3.

3rd approximation $x_3 = \dfrac{2.6080(9) - 3(-2.61260)}{9 + 2.61260} = 2.6961$

$$f(2.6961) = 1.5998$$

∴ the root lies between 2.6080 and 2.6961.

$$f(a) = -2.61260, f(b) = 1.5998$$

4th approximation $x_4 = \dfrac{2.6080(1.5998) - 2.6961(-2.61260)}{1.5998 + 2.61260} = 2.6626$

$$f(2.6626) = 8.7784$$

∴ the root lies between 2.6080 and 2.6626.

5th approximation $x_5 = \dfrac{2.6080(8.7784) - 2.6626(2.61260)}{8.7784 + 2.6126} = 2.6148$

$$f(x_5) = -12.1580$$

∴ the root lies between 2.648 and 2.6626.

$$f(2.6148) = -12.1580, \ f_2(2.6626) = 8.7784$$

6th approximation $x_6 = \dfrac{2.6148(8.7784) + 2.6626(12.1580)}{20.9364} = 2.642$

Hence, the root is 2.6.

6.4 NEWTON–RAPHSON METHOD

Let x_0 be an approximate root of the equation $f(x) = 0$. If $x_1 = x_0 + h$ be the exact root, and h being a small quantity, then $f(x_1) = 0$.

Expanding it by Taylor's theorem, we get

$$f(x_1) = f(x_0 + h) = 0$$

$$f(x_0) + hf'(x_0) + \frac{h^2}{21} f''(x_0) + \ldots = 0$$

Since h is small, we can neglect h^2 and higher power of h,

Hence, $f(x_0) + hf'(x_0) = 0$

or $h = -\dfrac{f(x_0)}{f'(x_0)}, \ f'(x_0) \neq 0$

Hence $x_1 = x_0 + h = x_0 - \dfrac{f(x_0)}{f'(x_0)}$

$$x_2 = x_1 - \frac{f(x_1)}{f'(x_1)}$$

$$\ldots = \ldots$$

$$x_{n+1} = x_n - \frac{f(x_n)}{f'(x_n)} \quad (n = 0, 1, 2, \ldots)$$

Notes: 1. This method is useful in case of large values of $f'(x)$, i.e. when the graph of $f(x)$ which crosses x-axis is almost vertical, then h will be small and hence, the root can be calculated in less time or in short time.

2. Newton's formula provides quick root if the initial approximation x_0 is choosen sufficiently close to the root.

6.5 CONVERGENCE OF NEWTON–RAPHSON METHOD

Newton–Raphson formula

$$x_{n+1} = x_n - \frac{f(x_n)}{f'(x_n)}$$

or

$$x_{n+1} = \phi(x_n) = x_n - \frac{f(x_n)}{f'(x_n)}$$

In general

$$\phi(x) = x_n - \frac{f(x)}{f'(x)}$$

which gives

$$\phi'(x) = 1 - \frac{[f'(x)]^2 - f(x)f''(x)}{[f'(x)]^2} = 1 - 1 + \left[\frac{f(x)f''(x)}{[f'(x)]^2}\right] = \frac{f(x)f''(x)}{[f(x)]^2}$$

Since iteration method converges if $|\phi'(x)| < 1$.

Newton's method will converge if

$$\left|\frac{f(x)f''(x)}{[f(x)f''(x)]^2}\right| < 1$$

$$|f(x)f''(x)| < [f'(x)]^2$$

in the interval considered.

The interval containing the root 'a' of $f(x) = 0$ should be selected in which the above condition is satisfied.

Notes: 1. The process will fail if $f'(x) = 0$ is in neighbourhood of the root. In such case, Regula-Falsi method should be used.

2. If the initial approximation to the root is not given, choose two values of x, say a and b such that $f(a)$ and $f(b)$ are of opposite sign. If $|f(a)| < |f(b)|$, then take 'a' as the initial approximation.

3. Newton-Raphson method is also referred to as method of tangents.

Example 6.12: Find the negative root of $x^3 - 21x + 3500 = 0$ correct to two decimal places.

Solution: Let

$$f(x) = x^3 - 21x + 3500$$

$$f(-15) = (-15)^3 + 21 \times 15 + 3500$$

$$= -3375 + 315 + 3500 = 440 > 0$$

$$f(-16) = -260 < 0$$

since $f(-16) < f(-15)$ (numerically)

$\therefore$ root lies near -16.

$$x_0 = -15.8$$

$$x_{n+1} = x_n - \frac{f(x_n)}{f'(x_n)} = x_n - \frac{x_n^3 - 21x_n + 35}{3x_n^2 - 21} = \frac{2x_n^3 - 3500}{3x_n^2 - 21}$$

$$x_1 = \frac{2(-15.8)^3 - 3500}{3(-15.8)^2 - 21} = -15.645$$

$$x_2 = \frac{2(-15.64)^3 - 3500}{3(-15.64)^2 - 21} = -15.643$$

$$x_3 = \frac{2(-15.643)^3 - 3500}{3(-15.643)^2 - 21} = -15.643$$

Hence, the root is -15.64 correct to two decimal places.

Example 6.13: Using Newton–Raphson, solve $x \sin x + \cos x = 0$ which is near to $x = \pi$.

Solution: Let $f(x) = x \sin x + x \cos x$

$$f'(x) = x \cos x + \sin x - \sin x = x \cos x$$

$$x_{n+1} = x_n - \frac{f(x_n)}{f'(x_n)} = \frac{x_n^2 \cos x_n - x_n \sin x_n - \cos x_n}{x_n \cos x_n}$$

$$x_1 = \frac{\pi^2 \cos \pi - \pi \sin \pi - \cos \pi}{\pi \cos \pi} = \frac{\pi^2 + 1}{-\pi} = 2.8232$$

$$x_2 = \frac{2.8232 \cos (2.8232) - 2.8232 \sin (2.8232) - \cos (2.8232)}{2.8232 \cos (2.8232)} = 2.79819$$

$$x_3 = \frac{2.79819 \cos(2.79819) - 2.79819 \sin (2.79819) - \cos (2.79819)}{2.79819 \cos(2.79819)}$$

$$= 2.79819$$

Hence, the required root is 2.79819.

Example 6.14: Find a real root of equation $x = e^{-x}$ by Newton–Raphson method.

Solution: $\qquad\qquad f(x) = xe^x - 1 = 0$

$$f'(x) = e^x + xe^x = (1 + x) e^x, \ f(0) = -1, f(1) = e - 1 = +ve$$

$\therefore$ the root lies between 0 and 1.

Let $\qquad\qquad x_0 = 1$

Thus $\qquad\qquad x_1 = 1 - \dfrac{e-1}{2e} = \dfrac{1}{2}\left(1 + \dfrac{1}{e}\right) = 0.6839397$

$$f(x_1) = 0.3553424$$
$$f'(x_1) = 3.337012$$

so that $\qquad\qquad x_2 = 0.6839397 - \dfrac{0.3553424}{3.337012} = 0.577545$

$$x_3 = 0.5672297$$
$$x_4 = 0.56714323$$

The required root is 0.567.

Example 6.15: Solve by Newton's method, the root of equation $e^x = 4x$ which is approximately correct to three decimal places.

Solution: $\qquad\qquad f(x) = e^x - 4x$

$$f(2) = e^2 - 8 = -0.610944 \ (-ve)$$
$$f(3) = e^3 - 12 = 8.085537 \ (+ve)$$
$$f(2) f(3) < 0$$

The root of $f(x) = 0$ lies between 2 and 3; $|f(2)| < |f(3)|$

∴ the root lies nearer to 2.

Let us take
$$x_0 = 2.1$$
$$f(x) = e^x - 4x$$
$$f'(x) = e^x - 4$$
$$f(x_0) = f(2.1) = e^{2.1} - 4(2.1) = 8.16617 - 8.4 = -0.23383$$
$$f'(x_0) = f'(2.1) = e^{2.1} - 4 = 4.16617$$

1st approximation $\quad x_1 = x_0 - \dfrac{f(x_0)}{f'(x_0)} = 2.1 - \dfrac{(-0.23383)}{4.16617} = 2.1 + 0.0561258 = 2.1561$

2nd approximation $\quad x_2 = x_1 - \dfrac{f(x_1)}{f'(x_1)} = 2.1561 - \dfrac{f(2.1561)}{f'(2.1561)} = 2.1561 - \dfrac{0.0129861}{4.6373861}$

$$= 2.1533$$

3rd approximation $\quad x_3 = x_2 - \dfrac{f(x_2)}{f'(x_2)} = 0.1533 - \dfrac{(0.0013484)}{4.6106516} = 2.1532$

The value of root correct to three decimal places = 2.1532.

Example 6.16: Solve $x^4 - 5x^3 + 20x^2 - 40x + 60 = 0$ by Newton–Raphson method given that all the roots are complex.

Solution:
$$f(x) = x^4 - 5x^3 + 20x^2 - 40x + 60$$
$$f'(x) = 4x^3 - 15x^2 + 40x - 40$$

Using Newton–Raphson formula, we have

$$x_{n+1} = x_n - \frac{f(x_n)}{f'(x_n)} = x_n - \frac{x_n^4 - 5x_n^3 + 20x_n^2 - 40x_n + 60}{4x_n^3 - 15x_n^2 + 40x_n - 40}$$

$$= \frac{3x_n^4 - 10x_n^3 + 20x_n^2 - 60}{4x_n^3 - 15x_n^2 + 40x_n - 40}$$

Putting $n = 0$ and taking $x_0 = 2(1 + i)$ by trial, we get

$$x_1 = \frac{3(2+2i)^4 - 10(2+2i)^3 + 20(2+2i)^2 - 60}{4(2+2i)^3 - 15(2+2i)^2 + 40(2+2i) - 40} = 1.92(1+i)$$

$$x_2 = \frac{3(1.92+1.92i)^4 - 10(1.92+1.92i)^3 + 20(1.92+1.92i)^2 - 60}{4(1.92+1.92i)^3 - 15(1.92+1.92i)^2 + 40(1.92+1.92i) - 40} = 1.915 + 1.908i \quad \text{(i)}$$

Since imaginary roots occur in conjugate pairs so the root of i are $1.915 \pm 1.908i$ up to three decimal places. Assuming that the other pair of roots of i is $\alpha + i\beta$, we have

sum of the roots $= (\alpha + i\beta) + (\alpha - i\beta)$ or $(1.915 + 1.908i) + (1.915 - 1.908i) = 5$

$2\alpha + 3.83 = 5$ or $\alpha = 0.585$

Also $\quad$ product of roots $= (\alpha^2 + \beta^2)$ or $\{(1.915)^2 + (1.908)^2\} = 60$

which gives $\quad\quad \beta = 2.805$

Hence the other roots are $0.585 \pm 2.805i$.

Example 6.17: Using Newton–Raphson method, establish iteration $x_{n+1} = \dfrac{1}{2}\left[x_n + \dfrac{N}{x_n}\right]$, where N is a positive number.

Solution: Let

$$\sqrt{N} = x$$
$$x^2 = N$$
$$x^2 - N = 0$$

Let

$$f(x) = x^2 - N$$
$$f'(x) = 2x$$

Now, Newton–Raphson formula

$$x_{n+1} = x_n - \frac{f(x_n)}{f'(x_n)} = x_n - \frac{x_n^2 - N}{2x_n} = \frac{1}{2}\left[x_n + \frac{N}{x_n}\right], \quad N = 0, 1, 2, 3, \ldots$$

Example 6.18: Using Newton-Raphson method, establish the formula

$$x_{n+1} = \frac{1}{3}\left[2x_n + \frac{N}{x_n^2}\right],$$

where N is a positive number.

Solution: Let

$$x = N^{1/3}$$
$$x^3 = N$$
$$x^3 - N = 0$$
$$f(x) = x^3 - N$$
$$= 3x^2$$

Newton–Raphson formula is

$$x_{n+1} = x_n - \frac{f(x)}{f'(x)} = x_n - \frac{x_n^3 - N}{3x_n^2} = \frac{3x_n^2 - x_n^3 + N}{3x_n^2} = \frac{2x_n^3 + N}{3x_n^2}$$

$$= \frac{1}{3}\left[2x_n + \frac{N}{x_n^2}\right] \quad N = 0, 1, 2, \ldots$$

Example 6.19: Find cube root of 12.

Solution: Let

$$x = 12^{1/3}$$
$$x^3 - 12 = 0$$
$$f(x) = x^3 - 12 = 0 \qquad\qquad (\therefore x = 12^{1/3})$$
$$f(1) = -\text{ve}$$
$$f(2) = -\text{ve} = 8 - 12 = -4$$
$$f(3) = +\text{ve} = 21 - 12 = 9$$

$\therefore$ the root lies between 2 and 3 $|f(2)| < |f(3)|$

$\therefore$ the root lies near to 2.

Let us take $x_0 = 2.1$

By Newton–Raphson method

$$x_{n+1} = \frac{1}{3}\left[2x_n + \frac{N}{x_n^2}\right]$$

$$x_1 = \frac{1}{3}\left[2x_0 + \frac{N}{x_0^2}\right] = \frac{1}{3}\left[2 \times 2.1 + \frac{12}{(2.1)^2}\right] = 2.307$$

$$x_2 = 2.2901.$$
$$x_3 = 2.29$$

Hence, the root is 2.29 correct to two decimal places.

Example 6.20: Find the smallest root of equation $e^{-x} = \sin x$ correct to four decimal places.

Solution:
$$f(x) = e^{-x} - \sin x, \quad f'(x) = -e^{-x} - \cos x$$

$$x_{n+1} = x_n + \frac{e^{-x_n} - \sin x_n}{e^{-x_n} + \cos x_n}$$

$$f(0) = 1 - \sin 0 = 1$$
$$f(1) = e^{-1} - \sin 1 = -0.47359$$
$$|f(0)| > |f(1)|, \text{ root is near to } 1$$

Let us take $\quad x_0 = 0.6$

$$x_1 = 0.6 + \frac{e^{-0.6} - \sin 0.6}{e^{-0.6} + \cos 0.6} = 0.6 - \frac{0.01583}{1.3745} = 0.6 - 0.01151 = 0.588548$$

$$x_2 = 0.58848 + \frac{e^{-0.58848} - \sin(0.58848)}{e^{-0.58848} + \cos(0.58848)} = 0.588559287$$

Hence, the desired root is 0.5885.

Example 6.21: Find by Newton–Raphson method, the real root of equation $3x = \cos x + 1$ correct to four decimal places.

Solution: Let $\quad f(x) = 3x - \cos x - 1 = 0$
$$f(0) = -1 - 1 = -2 \,(-ve)$$
$$f(1) = 3 - 0.5403 - 1 = 1.4597 = +ve$$

Root lies between 0 and 1, $|f(0)| > |f(1)|$.

Let us take $x_0 = 0.6$

Newton–Raphson method

$$x_{n+1} = x_n - \frac{f(x_n)}{f'(x_n)} = x_n - \frac{3x_n - \cos x_n - 1}{3 + \sin x_n}$$

$$= \frac{x_n \sin x_n + \cos x_n + 1}{3 + \sin x_n}$$

$n = 0, 1, 2, 3, \ldots$, we get successive approximation

1st approximation $\quad x_1 = \frac{x_0 \sin x_0 + \cos x_0 + 1}{3 + \sin x_0} = \frac{(0.6)\sin(0.6) + \cos(0.6) + 1}{3 + \sin(0.6)} = 0.6071$

2nd approximation $\quad x_2 = \frac{0.6071 \times 0.57049 + 0.831311}{3 + 0.57049} = 0.6071$

Hence, the approximate correct root to four decimal places is 0.6071.

Example 6.22: Find the real root of $xe^x = 2$ correct to two decimal places.

Solution: Given $\quad f(x) = xe^x - 2$
$$f(0) = -2$$
$$f(1) = e - 2 = 0.71828 = +ve$$

Root lies between 0 and 1

$$|f(0)| > |f(1)| \quad \therefore \text{ root lies nearest to 1}$$

Let us take $x_0 = 1$

$$x_{n+1} = x_n - \frac{f(x_n)}{f'(x_n)}$$

$$x_1 = x_0 - \frac{f(x_0)}{f'(x_0)} = \frac{e+2}{2e} = 0.867879$$

$$x_2 = 0.867879 - \frac{0.06716}{4.44902} = 0.85278$$

$$x_3 = 0.853$$

Hence, the root is 0.853 correct to two decimal places.

Example 6.23: Find a root of equation $x \sin x + \cos x = 0$.

Solution:

$$f(x) = x \sin x + \cos x$$
$$f'(x) = x \cos x$$
$$f(0) = 1 \ (+\text{ve})$$
$$f\left(\frac{\pi}{2}\right) = \frac{\pi}{2} \ (+\text{ve})$$
$$f(\pi) = -1 \ (-\text{ve})$$

$$\therefore \text{ root lies between } = \frac{\pi}{2}, \pi \left|f\left(\frac{\pi}{2}\right)\right| > |f(\pi)|$$

Root lies near to π.

Taking $x_0 = \pi$, we have

$$x_{n+1} = x_n - \frac{x_n \sin x_n + \cos x_n}{x_n \cos x_n} = 3.1416 - \frac{3.1416 \sin(3.1416) + \cos(3.1416)}{3.1416 \cos(3.1416)} = 2.8233$$

$$x_2 = 2.7986$$
$$x_3 = 2.7984$$

Hence, the root is 2.798 correct to three decimal places.

Example 6.24: Using Newton–Raphson method, find the root of equation $x \log_{10} x = 1.2$.

Solution: Let

$$f(x) = x \log_{10} x - 1.2$$

$$f'(x) = \log_{10} x + x (\log_{10} e / x) \qquad \left[\frac{d}{dx} \log_{10} x = \frac{1}{x} \log_a e\right]$$

$$= \log_{10} x + 0.4343$$

$$f(2.5) = -0.2051499 < 0$$

$$f(3) = 0.2313636 > 0$$

$\therefore$ root lies between 2.5 and 3.

Root lies nearer to 3.

Let us take $x_0 = 27$

By Raphson's method

$$x_{n+1} = x_n - \frac{x_n \log_{10} x_n - 1.2}{\log_{10} x_n + 0.4343} = \frac{0.4343 x_n + 1.2}{\log_{10} x_n + 0.4343}$$

$$x_1 = \frac{0.4343(2.7) + 1.2}{\log_{10} 2.7 + 0.4343} = 2.7407986$$

$$x_2 = 2.7406461$$
$$x_3 = 2.7406461$$

Hence, the root is 2.7406461.

Example 6.25: Solve $\sin x = 1 + x^3$, find the −ve root using Newton–Raphson method.

Solution: Let
$$f(x) = \sin x - 1 - x^3, f'(x) = \cos x - 3x^2$$
$$f(1) = \sin 1 - 1 - 1 = -1.1585$$
$$f(2) = -9.09070$$
$$f(-1) = \sin(-1) - 1 + 1 = -0.8414709 < 0$$
$$f(-2) = 6.0907026 > 0$$

∴ root lies between −1 and −2, $|f(-1)| < |f(-2)|$

$$x_{n+1} = \frac{x_n \cos x_n - \sin x_n - 2x_n^3 + 1}{\cos x_n - 3x_n^2}$$

Root lies near to −1, take $x_0 = -1$

$$x_1 = \frac{x_0 \cos x_0 - \sin x_0 - 2x_0^3 + 1}{\cos x_0 - 3x_0^2} = -1.2763653$$

$$x_2 = -1.2497465$$
$$x_3 = -1.2490526$$
$$x_4 = -1.2490522$$
$$x_5 = -1.2490521$$
$$x_6 = -1.2490522$$

Hence, the root is −1.249052 correct to six decimal places.

6.6 GENERALIZED NEWTON'S METHOD FOR MULTIPLE ROOTS

If a be a root of $f(x) = 0$ which is repeated m times, then

$$x_{n+1} = x_n - m\frac{f(x_n)}{f'(x_n)}$$

which is called generalized Newton's formula.

Notes: 1. If a is a root of $f(x) = 0$ with multiplicity m, then it is also a root of $f'(x) = 0$ with multiplicity $m - 1$ of $f''(x) = 0$ with multiplicity $m - 2$ and so on. Hence, if initial approximation x_0 is sufficiently close to root, then the expression

$$x_0 - m\frac{f(x_0)}{f'(x_0)}, x_0 - (m-1)\frac{f'(x_0)}{f''(x_0)}, x_0 - (m-2)\frac{f''(x_0)}{f'''(x_0)}$$

will have the same value.

2. Generalized Newton's formula has a second order of convergence for determining multiple root.

Example 6.26: Find the double root of the equation $x^3 - x^2 - x + 1 = 0$.

Solution: Let $x^3 - x^2 - x + 1 = f(x)$

$f(x) = 0$ and $f'(x) = 0$ will have the same root

$$f(x) = x^3 - x^2 - x + 1$$
$$f'(x) = 3x^2 - 2x - 1 \quad f(x) = 0 \text{ and } f'(x) = 0 \text{ will have the same root}$$
$$f''(x) = 6x - 2$$

Starting with $x_0 = 0.9$, we have

1st approximation $x_1 = x_0 - m\dfrac{f(x_0)}{f'(x_0)} = 0.9 - \dfrac{2 \times 0.19}{-0.37} = 1.003$

and $x_1 = x_0 - (m-1)\dfrac{f'(x_0)}{f''(x_0)} = 0.9 - (2-1)\dfrac{(-0.37)}{3.4} = 1.009$

The closeness of these values implies that there is a double root near $x = 1$. Choosing $x_1 = 1.01$ for the next approximation, we get

$$x_2 = x_1 - m\dfrac{f(x_1)}{f'(x_1)} = 1.01 - \dfrac{2 \times 0.0002}{0.0403} = 1.0001$$

$$= x_1 - (m-1)\dfrac{f'(x_1)}{f''(x_1)} = 1.01 - \dfrac{0.0403}{4.06} = 1.0001$$

This show that there is a double root at $x = 1.000$ which is quite near the actual root $x = 1$.

Example 6.27: Find the multiple root of $x^4 - 6x^3 + 12x^2 - 10x + 3 = 0$ with multiplicity 3, by modified Newton–Raphson method with an initial guess of $x_0 = 0$.

Solution: $f(x) = x^4 - 6x^3 + 12x^2 - 10x + 3$

$f'(x) = 4x^3 - 18x^2 + 24x - 10$

Given the initial value $x_0 = 0$ and $m = 3$

$$x_1 = x_0 - m\dfrac{f(x_0)}{f'(x_0)} = 0 - 3\dfrac{3}{(-10)} = 0.9$$

$$x_2 = 0.9 - 3\dfrac{f(0.9)}{f'(0.9)} = 0.9$$

$$= 0.9 - 3\left[\dfrac{(0.9)^4 - 6(0.9)^3 + 12(0.9)^2 - 10(0.9) + 3}{4(0.9)^3 - 18(0.9)^2 + 24(0.9) - 10}\right] = 0.998$$

$$x_3 = 0.998 - 3\left[\dfrac{f(0.998)}{f'(0.998)}\right] = 1.096$$

$$x_4 = 1.096 - 3\left[\dfrac{f(1.096)}{f'(1.096)}\right] = 1.0965$$

Hence, $x = 1.0965$ is the required root.

Example 6.28: Find the double root of $x^3 - 5.4x^2 + 9.24x - 5.096 = 0$ given that it is nearer to 1.5.

Solution: Let

$$f(x) = x^3 - 5.4x^2 + 9.24x - 5.096$$

$$f'(x) = 3x^2 - 10.8x + 9.24$$

$$x_{n+1} = x_n - m\frac{f(x_n)}{f'(x_n)} = x_n - (2)\frac{f(x_n)}{f'(x_n)}$$

$$= x_n - (2)\frac{[x_n^3 - 5.4x_n^2 + 9.24x_n - 5.096]}{[3x_n^2 - 10.8x_n + 9.24]} = \frac{x_n^3 - 9.24x_n + 10.192}{3x_m^2 - 10.8x + 9.24}$$

Given the initial value $x_0 = 1.5$

$$x_1 = \frac{(1.5)^3 - (9.24 \times 1.5) + 10.192}{3 \times (1.5)^2 - (10.8 \times 1.5) + 9.24} = 1.3952$$

$$x_2 = \frac{(1.3952)^3 - (9.24 \times 1.3952) + 10.192}{3 \times (1.3952)^2 - (10.8 \times 1.3952) + 9.24} = 1.3966$$

$$x_3 = \frac{(1.3966)^3 - (9.24 \times 1.3966) + 10.192}{3 \times (1.3966)^2 - (10.8 \times 1.3966) + 9.24} = 1.4024$$

$$x_4 = \frac{(1.4024)^3 - (9.24 \times 1.4024) + 10.192}{3 \times (1.4024)^2 - (10.8 \times 1.4024) + 9.24} = 1.4211$$

Hence, $x = 1.4$ correct to one decimal place.

6.7 SECANT METHOD

Although the Newton-Raphson method is often very efficient, there are situations where it performs poorly, and it is not convenient for polynomials and many other functions, there are certain functions whose derivative may be extremely difficult or inconvenient to evaluate. For these cases, the derivative can be approximated by a finite divided difference as shown in Fig. 6.3.

$$f'(x) = \frac{f(x_n) - f(x_{n-1})}{x_n - x_{n-1}}$$

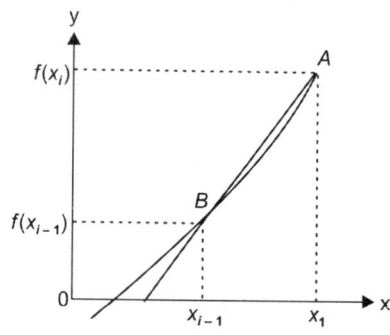

Fig. 6.3

The approximation can be substituted in the Newton–Raphson method to yield the following iterative equation

$$x_{n+1} = x_n - \frac{x_n - x_{n-1}}{f(x_n) - f(x_{n-1})} f(x_n), n \geq 1$$

which is called the formula for the secant method.

This method is an improvement over the method of false position as to condition $f(x_0) f(x_1) < 0$ is not required. At each iteration that the interval must obtain the root.

Note: If somehow $f(x_n) = f(x_{n-1})$, then this method fails and hence doesn't converge. This is the drawback of secant method over the method of position, which always converges. But if secant method converges, its rate of convergence is faster than Regula–Falsi method.

Example 6.29: Solve $e^{-x} - x = 0$ by secant method.

Solution:
$$f(x) = e^{-x} - x$$
$$f(0) = e^0 - 0 = 1 \ (+ve)$$
$$f(1) = e^{-1} - 1 = -0.6321 \ (-ve)$$

∴ A root lies between 0 and 1.
$$x_0 = 0, x_1 = 1$$

$$x_2 = x_1 - \frac{x_1 - x_0}{f(x_1) - f(x_0)} f(x_1) = 1 - \frac{(1-0)(-0.6321)}{(1) - (0.6321)} = 0.61270$$

$$f(x_2) = f(0.61270) = -0.07081 \ (-ve)$$

$$x_3 = x_2 - \frac{x_2 - x_1}{f(x_2) - f(x_1)} f(x_2)$$

$$= 0.61270 - \frac{(0.61270 - 1)}{(-0.07081) + 0.63212}(-0.07081) = 0.56384$$

$$f(x_3) = f(0.56384) = 0.00518 \ (+ve)$$

$$x_4 = 0.56384 - \frac{(0.00518)(0.56384 - 0.61270)}{0.00518 + 0.07081} = 0.56717$$

Hence, $x = 0.56$ is correct to two decimal places.

Example 6.30: Find a root of $xe^x = \cos x$ correct to four decimal places using secant method.

Solution: Let
$$f(x) = \cos x - xe^x = 0$$

Let
$$x_0 = 0, x_1 = 1$$
$$f(x_0) = 1, f(x_1) = \cos 1 - e = -2.17798$$

By Secant method

$$x_2 = x_1 - \frac{x_1 - x_0}{f(x_1) - f(x_0)} f(x_1)$$

$$f(x_1) = 1 - \frac{1-0}{-2.17798 - 1}(-2.17798)) = 0.31467$$

$$f(x_2) = 0.51987$$

$$x_3 = x_2 - \frac{(x_2 - x_1)}{f(x_2) - f(x_1)} f(x_2)$$

$$= 0.31467 - \frac{(0.31467 - 1)}{0.51987 + 2.17798}(0.51987) = 0.44673$$

$$f(x_3) = 0.20354$$

$$x_4 = x_3 - \frac{x_3 - x_2}{f(x_3) - f(x_2)} f(x_3) = 0.53171$$

$$x_5 = 0.51690$$

$$x_6 = 0.51775, \; x_7 = 0.51776 \text{ etc.}$$

Hence 0.5177 is the required root.

6.8 FIXED-POINT ITERATION METHOD

In this method, we express $f(x) = 0$ in the form $x = \phi(x)$.

Let $x = x_0$ be an initial approximation of the desired root. Then the first approximation to the root is given by

$$x_1 = \phi(x_0)$$
$$x_2 = \phi(x_1)$$
$$\dots = \dots$$
$$\dots = \dots$$
$$x_n = \phi(x_{n-1})$$

If the sequences of approximation $x_1, x_2, \dots, x_n$ converges to α, it is taken as root of the equation $f(x) = 0$. For convergence purpose, the initial approximation x_0 is to be done carefully. The choice of x_0 is determined according to sufficient condition.

6.8.1 Sufficient Conditions for Convergence

i. If α be a root of $f(x) = 0$ which is equivalent to $x = \phi(x)$
ii. I be any interval containing the point $x = \alpha$.
iii. $|\phi'(x)| < 1 \; (x \in I)$

then the sequence of approximation $x_0, x_1, \dots, x_n$ will converge to α provided the initial approximation x_0 is chosen in I.

Proof: Since α is a root of $x = f(x)$, we have $\alpha = \phi(\alpha)$.

If x_{n-1} and x_n be two successive approximation to α, we have

$$x_n = \phi(x_{n-1}) \tag{6.3}$$
$$x_n - \alpha = \phi(x_{n-1}) - \phi(\alpha)$$

By mean value theorem

$$\frac{\phi(x_{n-1}) - \phi(\alpha)}{x_{n-1} - \alpha} = \phi'(\xi), \quad \text{where } x_{n-1} < \xi < \alpha$$

Hence Eq. (6.3) becomes $x_n - \alpha = (x_{n-1} - \alpha) \, \phi'(\xi)$
If $|\phi'(x_i)| \leq K \leq 1$ for all i, then

$$|x_n - \alpha| \leq K |x_{n-1} - \alpha| \tag{6.4}$$

Similarly

$$|x_{n-1} - \alpha| \le K |x_{n-2} - \alpha|$$
$$|x_n - \alpha| \le K^2 |x_{n-2} - \alpha|$$

Proceeding in this way, we have

$$|x_n - \alpha| \le K^n |x_0 - \alpha|$$

As $n \to \infty$, the RHS tends to zero, therefore, the sequence of approximation converges to the root α.

Notes: 1. The smaller the value of $\phi'(x)$, the more rapid will be the convergence.

2. This method of iteration is particularly useful for finding the real roots of an equation in the form of an infinite series.

Example 6.31: Solve $e^{-x} = 10x$ using iteration method correct to four decimal points.

Solution: Let $\qquad\qquad f(x) = e^{-x} - 10x = 0$

Then $\qquad\qquad f(0) = 1, f(1) = -9.6321$

∴ the root lies between 0 and 1.

Rewriting $\qquad\qquad x = \dfrac{1}{10} e^{-x} = \phi(x)$

$$\phi'(x) = \dfrac{-1}{10} e^{-x}$$

$$|\phi'(x)| = \dfrac{e^{-x}}{10} = \dfrac{1}{e^x \cdot 10} < 1 \ \forall \ x \in (0, 1)$$

Hence, this method can be applied.

Let $\qquad\qquad x_0 = 0,$

then $\qquad\qquad x_1 = \phi(x_0) = 1/10 = 0.1$

$$x_2 = \dfrac{e^{-0.1}}{10} = 0.09048$$

$$x_3 = 0.091349$$

$$x_4 = 0.091274$$

Hence required root is 0.0913.

Example 6.32: Find the root of the equation $f(x) = x^3 + x^2 - 1 = 0$ by iteration method.

Solution: $\qquad\qquad f(0) = -1, f(1) = 1$

∴ root lies between 0 and 1.

Rewriting the given equation

$$x^2(x + 1) = 1$$

$$x = \dfrac{1}{\sqrt{x+1}} = \phi(x)$$

$$\phi'(x) = \dfrac{-1}{2(x+1)^{3/2}}$$

as $|\phi'(x)| < 1$, for all $x \in (0, 1)$.

Hence, method of iteration is applicable here.

Taking initial approximation as $x_0 = 1$

$$x_1 = \phi(x_0) = \frac{1}{\sqrt{1 + x_0}} = \frac{1}{\sqrt{2}} = 0.70711, \; f(x_1) = -0.14644$$

$$x_2 = \phi(x_1) = \frac{1}{\sqrt{1 + 0.70711}} = 0.76537, \; f(x_2) = 0.3414$$

$$x_3 = 0.75263, \qquad f(x_3) = 7.2213 \times 10^{-3}$$
$$x_4 = 0.75536 \qquad f(x_4) = 1.5565 \times 10^{-3}$$
$$x_5 = 0.75477 \qquad f(x_5) = -3.44323 \times 10^{-4}$$
$$x_6 = 0.7549 \qquad f(x_6) = 7.38295 \times 10^{-5}$$

Hence, required root is 0.7549.

Note: The given equation can be written in many ways $x^2 = 1 - x^3$

or

$$x = \sqrt{(1 - x^3)} = \phi(x)$$

$$\phi'(x) = \frac{3x^2}{2(1 - x^3)^{3/2}} > 1 \text{ in } (0, 1)$$

The condition $|\phi'(x)| < 1$ is not satisfied.

Example 6.33: Solve $x^3 + x^2 - 100 = 0$ using iteration method.

Solution: Let $\quad x^2(x + 1) = 100$

$$f(4) = -20$$
$$f(5) = 125 + 25 - 100 = 50$$

$\therefore$ root lies between 4 and 5.

$$x = \frac{10}{\sqrt{x + 1}} = \phi(x)$$

$$\phi'(x) = \frac{-10}{2(x + 1)^{3/2}}$$

$$|\phi'(x)| < 1, \text{ for all } x \in (4, 5)$$

Let $\quad\quad x_0 = 4$

$$\frac{10}{\sqrt{x_0 + 1}} = x_1 = \frac{10}{\sqrt{5}} = 4.472$$

$$x_2 = \frac{10}{\sqrt{5.472}} = 4.2749$$

$$x_3 = \frac{10}{\sqrt{4.2749 + 1}} = 4.354$$

$$x_4 = \frac{10}{\sqrt{4.354 + 1}} = 4.3217$$

$$x_5 = \frac{10}{\sqrt{4.3217 + 1}} = 4.3348$$

$$x_6 = \frac{10}{\sqrt{4.3348 + 1}} = 4.3295$$

$$x_7 = \frac{10}{\sqrt{4.3295 + 1}} = 4.33168$$

$$x_8 = 4.33079$$

$$x_9 = 4.331159$$

$$x_{10} = 4.33100$$

Hence, required root is 4.331 correct to three decimal places.

Example 6.34: Find the root of equation $x = \dfrac{1}{2} + \sin x = f(x)$ by iteration method.

Solution: $\qquad f(x) = \dfrac{1}{2} + \sin x - x = 0$

$$f(0) = \frac{1}{2}$$

$$f(1) = \frac{1}{2} + \sin 1 - 1 = 0.3414 > 0$$

$$f(2) = -0.09070 < 0$$

∴ root lies between 1 and 2.

Rewriting the above equation

$$\phi(x) = \frac{1}{2} + \sin x$$

$$\phi'(x) = \cos x$$

$$|\phi'(x)| < 1 \text{ for all } x \in (1, 2)$$

Let $\qquad\qquad x_0 = 1$

$$x_1 = \frac{1}{2} + \sin 1 = 1.3414$$

$$x_2 = \frac{1}{2} + 0.4738 = 1.4738, \; x_3 = 1.4952$$

$$x_4 = 1.49715, \; x_5 = 1.4972$$

Hence, the required root is 1.497.

Example 6.35: Find $\sqrt{30}$ by iteration method.

Solution: Let $\qquad\qquad x = \sqrt{30}$

$$f(x) = x^2 - 30$$

$$f(5) = -5$$

$$f(6) = -6$$

∴ root lies between 5 and 6.

$$x^2 - 30 = 0$$

$$x^2 + x - x - 30 = 0$$

$$x(x + 1) = x + 30$$

$$x = \frac{x+30}{x+1} = 1 + \frac{29}{x+1} = \phi(x)$$

$$\phi'(x) = \frac{-29}{(x+1)^2}$$

$$|\phi'(x)| = \frac{29}{(x+1)^2} < 1; \quad 5 < x < 6$$

Hence, iteration method is applicable taking $x_0 = 5$.

$$x_1 = \phi(x_0) = 1 + \frac{29}{6} = 5.83$$

$$x_2 = \phi(x_1) = 1 + \frac{29}{6.83} = 5.2439$$

$$x_3 = \phi(x_2) = 1 + \frac{29}{6.245} = 5.64453$$

$x_4 = 5.36449$	$x_5 = 5.55653$	$x_6 = 5.42307$
$x_7 = 5.514974$	$x_8 = 5.451284073$	$x_9 = 5.495229$
$x_{10} = 5.464815$	$x_{11} = 5.48582$	$x_{12} = 5.47129$
$x_{13} = 5.48133$	$x_{14} = 5.47439$	$x_{15} = 5.47918$
$x_{16} = 5.47587$	$x_{17} = 5.47816$	$x_{18} = 5.47658$
$x_{19} = 5.47767$	$x_{20} = 5.47691$	$x_{21} = 5.47744$
$x_{22} = 5.477077$		

Hence, the root is 5.477.

Example 6.36: Find a real root of the equation $x - \frac{x^3}{3} + \frac{x^5}{5} - \frac{x^7}{42} + \frac{x^9}{216} - \frac{x^{11}}{1320} + \ldots = 0.433$ by iteration method.

Solution: $\quad x = 0.433 + \frac{x^3}{3} - \frac{x^5}{5} + \frac{x^7}{42} - \frac{x^9}{216} + \frac{x^{11}}{1320}$

Let $\quad x_0 = 0.433$

$$x_1 = 0.433 + \frac{(0.433)^3}{3} - \frac{(0.433)^5}{5} + \frac{(0.433)^7}{42} - \frac{(0.433)^9}{216} + \frac{(0.433)^{11}}{1320} = 0.45708$$

$$x_2 = 0.433 + \frac{(0.45708)^3}{3} - \frac{(0.45708)^5}{5} + \frac{(0.45708)^7}{42} = 0.460939$$

$$x_3 = 0.433 + \frac{0.46094}{3} - \frac{(0.46094)^5}{5} + \frac{(0.46094)^7}{42} = 0.46158$$

$$x_4 = 0.461695$$

Hence, the root is 0.461 correct to three decimal places.

Example 6.37: Solve by iteration method $x^2 + 1 = x^3$.

Solution: $\quad\quad f(x) = x^3 - x^2 - 1$

Root lies between 1 and 2.

$$f(1.4) = 1.784$$

$$f(2) = 3$$

Numerically $\quad f(1.4) < f(2)$

Root lies near to 1.4

$$x = (x^2 + 1)^{1/3} = \phi(x)$$

$$|\phi'(x)| = |\phi'(1.4)| = 0.388 < 1 \text{ and } |\phi'(x)| = |\phi'(2)| = 0.455 < 1.$$

We shall consider the interval (1.4, 2)

Let
$$x_0 = 1.4$$

$$x_1 = \phi(x_0) = [1 + (1.4)^2]^{1/3} = 1.435$$

$$x_2 = [1 + (1.435)^2]^{1/3} = 1.4252$$

$$x_3 = [1 + (1.422)^2]^{1/3} = 1.459$$

$$x_4 = [1 + (1.459)^2]^{1/3} = 1.462$$

$$x_5 = [1 + (1.462)^2]^{1/3} = 1.464$$

$$x_6 = [1 + (1.464)^2]^{1/3} = 1.464$$

Hence, required root is 1.464.

Example 6.38: Find a real root of $2x - \log_{10} x = 7$ using iteration method.

Solution:
$$f(x) = 2x - \log_{10} x - 7 = 0$$

$$f(3) = 6 - \log_{10} 3 - 7 = -1.4471$$

$$f(4) = 8 - 0.602 - 7 = 0.398$$

The root lies between 3 and 4.

$$x = \frac{1}{2}(\log_{10} x + 7) = \phi(x)$$

$$\phi'(x) = \frac{1}{2}\left(\frac{1}{x}\log_{10} e\right)$$

$$|\phi'(x)| < 1 \text{ when } 3 < x < 4$$

Iteration can be applied

$$x_0 = 3.6 \text{ (root near to 4)}$$

$$x_1 = \frac{1}{2}(\log_{10} 3.6 + 7) = 3.77815$$

$$x_2 = \frac{1}{2}(\log_{10} 3.77815 + 7) = 3.78863$$

$$x_3 = \frac{1}{2}(\log_{10} 3.78863 + 7) = 3.78924$$

$$x_4 = \frac{1}{2}(\log_{10} 3.78924 + 7) = 3.78927$$

x_3 and x_4 are correct to four decimal places.

Hence, the root is 3.7892 correct to four decimal places.

Example 6.39: Find the root of $xe^x = 1$ by iteration method if the root lies between 0 and 1.

Solution:
$$x = e^{-x} = \phi(x)$$

$$\phi'(x) = -e^{-x}$$

$$|\phi'(x)| = |e^{-x}| < 1 \ \forall \ 0 < x < 1$$

Let us take $x_0 = 1$

$x_1 = 0.3678794$	$x_2 = 0.6922006$	$x_3 = 0.5004735$
$x_4 = 0.6062435$	$x_5 = 0.5453957$	$x_6 = 0.5796123$
$x_7 = 0.5601154$	$x_8 = 0.5711431$	$x_9 = 0.5648793$
$x_{10} = 0.5684287$	$x_{11} = 0.5664147$	$x_{12} = 0.5675566$
$x_{13} = 0.5669087$	$x_{14} = 0.5672762$	$x_{15} = 0.5670679$
$x_{16} = 0.567186$	$x_{17} = 0.567119$	$x_{18} = 0.567157$
$x_{19} = 0.5671354$	$x_{20} = 0.5671477$	

Hence the root is 0.5671.

6.9 MULLER'S METHOD

Method I: In this method, $f(x)$ is approximated by a second degree curve in the vicinity of a root. The roots of the quadratic are then assumed to be the approximations to the root of the equation $f(x) = 0$.

Let x_{i-2}, x_{i-1}, x_i be three distinct approximations to a root of $f(x) = 0$ and let y_{i-2}, y_{i-1}, y_i be the corresponding value of $y = f(x)$. Assuming that

$$P(x) = A(x - x_i)^2 + B(x - x_i) + y_i \tag{6.5}$$

is the parabola passing through the points (x_{i-2}, y_{i-2}), (x_{i-1}, y_{i-1}) and (x_i, y_i), we have

$$y_{i-1} = A(x_{i-1} - x_i)^2 + B(x_{i-1} - x_i) + y_i \tag{6.5a}$$
$$y_{i-2} = A(x_{i-2} - x_i)^2 + B(x_{i-2} - x_i) + y_i \tag{6.5b}$$

from Eqs (6.5a) and (6.5b), we get

$$y_{i-1} - y_i = A(x_{i-1} - x_i)^2 + B(x_{i-1} - x_i)$$
$$y_{i-2} - y_i = A(x_{i-2} - x_i)^2 + B(x_{i-2} - x_i)$$

Solving the above equations, we get

$$A = \frac{(x_{i-2} - x_i)(y_{i-1} - y_i) - (x_{i-1} - x_i)(y_{i-2} - y_i)}{(x_{i-2} - x_{i-1})(x_{i-1} - x_i)(x_{i-2} - x_i)}$$

$$B = \frac{(x_{i-2} - x_i)^2(y_{i-1} - y_i) - (x_{i-1} - x_i)^2(y_{i-2} - y_i)}{(x_{i-2} - x_{i-1})(x_{i-1} - x_i)(x_{i-2} - x_i)}$$

With the values of A and B given above, the quadratic equation

$P(x) = A(x - x_i)^2 + B(x - x_i) + y_i = 0$ give the next approximation x_{i+1}

$$x_{i+1} - x_i = \frac{\left(-B \pm \sqrt{B^2 - 4Ay_i}\right)}{2} \tag{6.6}$$

A direct solution from Eq. (6.6) gives inaccurate result, so we can rewrite Eq. (6.6) as

$$A + \frac{B}{x - x_i} + \frac{y_i}{(x - x_i)^2} = 0$$

$$\frac{1}{x - x_i} = \frac{-B \pm \sqrt{B^2 - 4Ay_i}}{2y_i}$$

$$x - x_i = \frac{-2y_i}{\left(B \mp \sqrt{B^2 - 4Ay_i}\right)}$$

$$x_{i+1} = x_i - \frac{2y_i}{\left(B \mp \sqrt{B^2 - 4Ay_i}\right)} \tag{6.7}$$

The sign in the denominator should be chosen so that it will be the largest in magnitude (i.e. if $B > 0$, choose plus, if $B < 0$, choose minus, if $B = 0$, choose either).

Note: Muller's method like Newton-Raphson's method will find a complex. Of course, the computation must use complex arithmetic root if given complex starting values. This method converges at the same rate as Newton–Raphson's method.

Method II: This method is developed by writing a quadratic equation that fits through three points in the vicinity of root, in the form $av^2 + bv + c$.

Let $$h_1 = x_1 - x_0$$

and $$h_2 = x_0 - x_2$$

We find the coefficient by evaluating $P_2(v)$ at three points

$$v = 0,\ a\,(0)^2 + b(0) + c = f_0$$

$$= h_1 \Rightarrow ah_1^2 + bh_1 + c = f_1$$

$$= -h_2 \Rightarrow ah_2^2 - bh_2 + c = f_2$$

From the first equation $c = f_0$. Let $h_2/h_1 = v$, we can solve the other equations for a and b.

$$a = \frac{vf_1 - f_0(1 + v) + f_2}{vh_1^2(1 + v)},\ b = \frac{f_1 - f_0 - ah_1^2}{h_1}$$

After computing a, b and c, we solve for the root of $av^2 + bv + c = 0$ by quadratic formula, chosing the root nearest to middle point x_0, putting this value

$$\text{root} = x_0 - \frac{2c}{b \mp \sqrt{b^2 - 4ac}}$$

With the sign in the denominator taken to give the largest absolute value of the denominator (i.e. $b > 0$, choose plus, if $b < 0$, choose minus, if $b = 0$, choose either).

Example 6.40: Find a root of $\cos x = xe^x$ which lies between 0 and 1.

Solution: Let $$y = \cos x - xe^x$$

Let $$x_{i-2} = -1,\ x_{i-1} = 0,\ x_i = 1$$

$$y_{i-2} = 0.9082,\ y_{i-1} = 1,\ y_i = -2.178$$

$$A = \frac{(x_{i-2} - x_i)(y_{i-1} - y_i) - (x_{i-1} - x_i)(y_{i-2} - y_i)}{(x_{i-2} - x_{i-1})(x_{i-1} - x_i)(x_{i-2} - x_i)}$$

$$B = \frac{(x_{i-2} - x_i)^2(y_{i-1} - y_i) - (x_{i-1} - x_i)^2(y_{i-2} - y_i)}{(x_{i-2} - x_{i-1})(x_{i-1} - x_i)(x_{i-2} - x_i)}$$

Substituting the value of $x_{i-2}, x_{i-1}, x_i, y_{i-2}, y_{i-1}, y_i$ in above formula, we have

$$A = \frac{(-1-1)(1+2.178) - (0-1)(0.9082+2.178)}{(0+1)(0-1)(-1-1)} = -1.6349$$

$$B = \frac{(-1-1)^2(1+2.178) - (0-1)^2(0.9082+2.178)}{(-1-0)(0-1)(-1-1)} = -4.8129$$

The quadratic equation is given by

$$-1.6349(x-1)^2 - 4.8129(x-1) - 2.178 = 0$$

The next approximation to the desired root is

$$x_{i+1} = x_i - \frac{2y_i}{B \pm \sqrt{B^2 - 4Ay_i}}$$

$$= 1 - \frac{2(-2.178)}{-4.8129 - \sqrt{(-4.8129)^2 - 4(1.6349)(-2.178)}} = 0.4415$$

For the next iteration, we have

$$x_{i-2} = 0, y_{i-2} = 1$$
$$x_{i-1} = 0.4415, y_{i-1} = 0.21756$$
$$x_i = 1, y_i = -2.178$$

$$A = \frac{(0-1)(0.21756+2.178) - (0.4415-1)(1+2.178)}{(0.4415-0)(0.4415-1)(0-1)}$$

$$= \frac{-2.3956 + 1.775}{(0.4415)(-1)(-0.5585)} = \frac{-0.6206}{0.2465} = -2.517$$

$$B = \frac{(1)^2(0.21756+2.178) - (-0.5585)^2(1+2.178)}{(-0.4415)(-0.5585)(-1)}$$

$$= \frac{2.3956 - (0.31192)(3.178)}{-0.2466} = \frac{1.4043}{-0.2466} = -5.695$$

Next approximation

$$x_{i+1} = x_i - \frac{2y_i}{\left(B \pm \sqrt{B^2 - 4Ay_i}\right)} = 1 - \frac{2(-2.178)}{-5.695 - \sqrt{10 - 5.0}}$$

$$= 1 - \frac{(-4.356)}{-5.695 - 3.24} = 1 - \frac{(-4.356)}{(-8.925)}$$

$$= 1 - 0.4875 = 0.5124$$

After a third iteration, we get 0.5177 as the value for the root.

PROBLEMS 6.1

1. Find the positive root of $x - \cos x = 0$ by bisection method correct to three decimal places. [Ans. $x = 0.738$]

2. Find the positive root of the following equations by bisection method.

a. $x^3 - x - 1 = 0$ [Ans. 1.32475]

b. $3x - \cos x - 1 = 0$ [Ans. 66664]

c. $x^4 - x^2 - 2x^2 - 6x = 4$ [Ans. 2.7065]

d. $x \log_{10} x = 1.2$ which lies between 2 and 3 [Ans. 2.74]

e. $3x + \sin x = e^x$ [Ans. 0.34375]

3. Find a real root of the equation $x^3 - 2x - 5 = 0$ by the Regula-Falsi method correct to three decimal places. [Ans. 2.0944]

4. Find the real root of $xe^x = 2$ by Regula-Falsi method correct to three decimal places. [Ans. 0.85261]

5. Find the real root of the following equation correct to four decimal places using Regula-Falsi method.

a. $x^3 - 4x - 1 = 0$ [Ans. 0.2541]

b. $x = \cos x$ [Ans. 0.7391]

c. $e^x \sin x = 1$ [Ans. 0.5885]

d. $2x - \log_{10} x = 7$ [Ans. 3.7892]

e. $x^3 + 1 = 9x$ [Ans. 2.943]

f. $2x - 3 \sin x = 5$ [Ans. 1.5236]

g. $x - e^{-x} = 0$ [Ans. 0.567]

6. Using Newton–Raphson method, find a root correct to three decimal places of the following equations.

a. $x^x = 1000$ [Ans. 3.592]

b. $x(1 - \log_e x) = 0.5$ [Ans. 0.187]

c. $2x \tan x = 1$ [Ans. 0.653]

d. $\sin x = 1 - x$ [Ans. 0.511]

e. $x^3 - 3x^2 + 7x - 8 = 0$ [Ans. 1.675]

f. $2x - 3 \sin x = 5$ [Ans. 1.5236]

7. Use Muller's method to find a root of the following equations.

a. $x^3 - 3x - 5 = 0$ [Ans. 2.26]

b. $x^3 - x^2 - x - 1$ [Ans. 1.839]

8. Use Secant method to find a root of the following equations

a. $2x^3 - 3x - 6 = 0$ [Ans. 1.78376]

b. $x^3 - 6x + 4 = 0$ [Ans. 0.73205]

c. $x = \cos x$ [Ans. 0.73908]

d. $3x = \cos x + 1$ [Ans. 0.60710]

7 Numerical Differentiation and Integration

7.1 INTRODUCTION

Consider a set of values (x_i, y_i) of a function. The process of computing the derivative or derivatives of that function at some value of x from the given set of values is called *numerical differentiation*. This may be done by first approximating the function by a suitable interpolation formula and then differentiating it as many times as desired.

If the value of x are equispaced and the derivative is required near the beginning of the table, we employ *Newton forward interpolation formula*. If it is required near the end of the table, we used *Newton backward interpolation formula*.

For values near the middle of the table, dy/dx is calculated by means of Stirling's or Bessel's formula.

If the value of x are not equispaced, we use Newton's divided differences or Lagrange's interpolation formula to get the derivative value.

Note: While using these formula, it must be observed that the table of values defines the function at these points only and does not completely defines the function and the function may not be differentiable at all.

As such the process of numerical differentiation should be used only if the tabulated values are such that differences of some order are constant otherwise, errors are bound to creep in and go on increasing as derivatives of higher order are found. This is due to the fact that the difference between $f(x)$ and the approximating polynomial $\phi(x)$ may be small at the data point $f'(x) - \phi'(x)$ may be large.

7.2 FORMULA FOR DERIVATIVES

Consider the function $y = f(x)$ which is tabulated for the values
$$x_i = x_0 + ih, i = 0, 1, 2, ..., n$$

7.2.1 Derivatives Using Newton's Forward Difference Formula

Newton's forward interpolation formula
$$y = y_0 + p\Delta y_0 + \frac{p(p-1)}{2!}\Delta^2 y_0 + \frac{p(p-1)(p-2)}{3!}\Delta^3 y_0 + ...$$

Method I: Differentiating both sides w.r.t. p, we have
$$\frac{dy}{dp} = \Delta y_0 + \frac{2p-1}{2!}\Delta^2 y_0 + \frac{3p^2 - 6p + 2}{3!}\Delta^3 y_0 + ...$$

Since $p = \dfrac{x - x_0}{h}$, therefore $\dfrac{dp}{dx} = \dfrac{1}{h}$

Now $\dfrac{dy}{dx} = \dfrac{dy}{dp} \cdot \dfrac{dp}{dx}$

$$= \frac{1}{h}\left[\Delta y_0 + \frac{2p-1}{2!}\Delta^2 y_0 + \frac{3p^2 - 6p + 2}{3!}\Delta^3 y_0 + \frac{4p^3 - 18p^2 + 22p - 6}{4!}\Delta^4 y_0 + \ldots\right]$$

(7.1)

At $x = x_0$, $p = 0$, hence putting $p = 0$, we have

$$\left(\frac{dy}{dx}\right)_{x_0} = \frac{1}{h}\left[\Delta y_0 - \frac{1}{2}\Delta^2 y_0 + \frac{1}{3}\Delta^3 y_0 - \frac{1}{4}\Delta^4 y_0 + \frac{1}{5}\Delta^5 y_0 - \frac{1}{6}\Delta^6 y_0 + \ldots\right]$$

(7.2)

Again differentiating Eq. (7.1) w.r.t. x, we get

$$\frac{d^2 y}{dx^2} = \frac{d}{dp}\left(\frac{dy}{dx}\right)\frac{dp}{dx}$$

$$= \frac{1}{h^2}\left[\frac{2}{2!}\Delta^2 y_0 + \frac{6p-6}{3!}\Delta^3 y_0 + \frac{12p^2 - 36p + 22}{4!}\Delta^4 y_0 + \ldots\right]$$

Putting $p = 0$, we obtain

$$\left(\frac{d^2 y}{dx^2}\right) = \frac{1}{h^2}\left[\Delta^2 y_0 - \Delta^3 y_0 + \frac{11}{12}\Delta^4 y_0 - \frac{5}{6}\Delta^5 y_0 + \frac{137}{180}\Delta^6 y_0 + \ldots\right]$$

(7.3)

Similarly $\qquad \left(\dfrac{d^3 y}{dx^3}\right) = \dfrac{1}{h^3}\left[\Delta^2 y_0 - \dfrac{3}{2}\Delta^4 y_0 + \ldots\right]$ and so on.

Method II: We know that

$$1 + \Delta = E = e^{hD}$$

$$hD = \log(1 + \Delta) = \Delta - \frac{1}{2}\Delta^2 + \frac{1}{3}\Delta^3 - \frac{1}{4}\Delta^4 + \ldots$$

$$D = \frac{1}{h}\left[\Delta - \frac{1}{2}\Delta^2 + \frac{1}{3}\Delta^3 - \frac{1}{4}\Delta^4 + \ldots\right]$$

$$D^2 = \frac{1}{h^2}\left[\Delta - \frac{1}{2}\Delta^2 + \frac{1}{3}\Delta^3 - \frac{1}{4}\Delta^4 + \ldots\right]^2$$

$$= \frac{1}{h^2}\left[\Delta^2 - \Delta^3 + \frac{11}{12}\Delta^4 - \frac{5}{6}\Delta^5 + \ldots\right]$$

$$D^3 = \frac{1}{h^3}\left[\Delta^3 - \frac{3}{2}\Delta^4 + \frac{7}{4}\Delta^5 + \ldots\right]$$

Now applying the above identities to y_0, we get

$$Dy_0 = \left(\frac{dy}{dx}\right)_{x_0} = \frac{1}{h}\left[\Delta y_0 - \frac{1}{2}\Delta^2 y_0 + \frac{1}{3}\Delta^3 y_0 - \frac{1}{4}\Delta^4 y_0 + \frac{1}{5}\Delta^5 y_0 - \frac{1}{6}\Delta^6 y_0 + \ldots\right]$$

$$D^2 y_0 = \left(\frac{d^2 y}{dx^2}\right)_{x_0} = \frac{1}{h^2}\left[\Delta^2 y_0 - \Delta^3 y_0 + \frac{11}{12}\Delta^4 y_0 - \frac{5}{6}\Delta^5 y_0 + \frac{137}{180}\Delta^6 y_0 + \ldots\right]$$

$$\left(\frac{d^3 y}{dx^3}\right)_{x_0} = \frac{1}{h^2}\left[\Delta^3 y_0 - \frac{3}{2}\Delta^4 y_0 + \ldots\right]$$

7.2.2 Derivatives Using Backward Difference Formula

Newton's backward interpolation formula

$$y = y_n + p\nabla y_n + \frac{p(p+1)}{2!}\nabla^2 y_n + \frac{p(p+1)(p+2)}{3!}\nabla^3 y_n + \ldots$$

Method I: Differentiating both sides w.r.t. p, we get

$$\frac{dy}{dp} = \nabla y_n + \frac{2p+1}{2!}\nabla^2 y_n + \frac{3p^2 + 6p + 2}{3!}\nabla^3 y_n + \ldots$$

Since $p = \dfrac{x - x_n}{h}$, therefore $\dfrac{dp}{dx} = \dfrac{1}{h}$

$$\frac{dy}{dx} = \frac{dy}{dp}\cdot\frac{dp}{dx}$$

$$= \frac{1}{h}\left[\nabla y_n + \frac{2p+1}{2!}\nabla^2 y_n + \frac{3p^2 + 6p + 2}{3!}\nabla^3 y_n + \ldots\right] \qquad (7.4)$$

Put $x = x_n$, $p = 0$, here putting $p = 0$ in Eq. (7.4), we get

$$\left(\frac{dy}{dx}\right)_{x_n} = \frac{1}{h}\left[\nabla y_n + \frac{1}{2}\nabla^2 y_n + \frac{1}{3}\nabla^3 y_n + \frac{1}{4}\nabla^4 y_n + \ldots\right] \qquad (7.5)$$

Again differentiating Eq. (7.5) w.r.t. x, we have

$$\frac{d^2 y}{dx^2} = \frac{d}{dp}\left(\frac{dy}{dx}\right)\frac{dp}{dx}$$

$$= \frac{1}{h}\left[\nabla^2 y_n + \frac{6p+6}{3!}\nabla^3 y_n + \frac{6p^2 + 18p + 11}{12}\nabla^2 y_n + \ldots\right]$$

Putting $p = 0$, we obtain

$$\left(\frac{d^2 y}{dx^2}\right)_{x_n} = \frac{1}{h^2}\left[\nabla^2 y_n + \nabla^3 y_n + \frac{11}{12}\nabla^4 y_n + \frac{5}{6}\nabla^5 y_n + \frac{137}{180}\nabla^6 y_n + \ldots\right]$$

and so on.

Method II: We know that $1 - \nabla = E^{-1} = e^{-hD}$

$$-hD = \log(1 - \nabla) = -\left[\nabla + \frac{1}{2}\nabla^2 + \frac{1}{3}\nabla^3 + \frac{1}{4}\nabla^4 + \ldots\right]$$

$$D = \frac{1}{h}\left[\nabla + \frac{1}{2}\nabla^2 + \frac{1}{3}\nabla^3 + \frac{1}{4}\nabla^4 + \ldots\right]$$

$$D^2 = \frac{1}{h^2}\left[\nabla + \frac{1}{2}\nabla^2 + \frac{1}{3}\nabla^3 + \ldots\right]^2 = \frac{1}{h^2}\left[\nabla^2 + \nabla^3 + \frac{11}{12}\nabla^4 + \ldots\right]$$

Similarly

$$D^3 = \frac{1}{h^3}\left[\nabla^3 + \frac{3}{2}\nabla^4 + \frac{7}{4}\nabla^5 + \ldots\right]$$

Applying these identities to y_n, we get

$$Dy_n = \left(\frac{dy}{dx}\right)_{x_n} = \frac{1}{h}\left[\nabla y_n + \frac{1}{2}\nabla^2 y_n + \frac{1}{3}\nabla^3 y_n + \frac{1}{4}\nabla^4 y_n + \ldots\right]$$

$$D^2 y_n = \left(\frac{d^2 y}{dx^2}\right)_{x_n} = \frac{1}{h^2}\left[\nabla^2 y_n + \nabla^3 y_n + \frac{11}{12}\nabla^4 y_n + \ldots\right]$$

$$\left(\frac{d^3 y}{dx^3}\right)_{x_n} = \frac{1}{h^3}\left[\nabla^3 y_n + \frac{3}{2}\nabla^4 y_n + \ldots\right]$$

7.2.3 Derivatives Using Central Difference Formula

$$y_p = y_0 + \frac{p}{1!}\left[\frac{\Delta y_0 + \Delta y_{-1}}{2}\right] + \frac{p^2}{2!}\Delta^2 y_{-1} + \frac{p(p^2-1)}{3!}\left[\frac{(\Delta^3 y_{-1} + \Delta^3 y_{-2})}{2}\right] + \frac{p^2(p^2-1)}{4!}\Delta^4 y_{-2} + \ldots$$

(7.6)

Differentiating Eq. (7.6), both sides w.r.t. p, we get

$$\frac{dy}{dp} = \left[\frac{\Delta y_0 + \Delta y_{-1}}{2}\right] + \frac{2p}{2!}\Delta^2 y_{-1} + \frac{3p^2-1}{3!}\left[\frac{(\Delta^3 y_{-1} + \Delta^3 y_{-2})}{2}\right] + \frac{4p^3 - 2p}{4!}\Delta^4 y_{-2} + \ldots$$

Since

$$p = \frac{x - x_0}{h}, \frac{dp}{dx} = \frac{1}{h}, \frac{dy}{dx} = \frac{dy}{dp}\cdot\frac{dp}{dx}$$

$$\frac{dy}{dx} = \frac{1}{h}\left[\left(\frac{y_0 + \Delta y_{-1}}{2}\right) + p\Delta^2 y_{-1} + \frac{3p^2-1}{6}\left(\frac{\Delta^3 y_{-1} + \Delta^3 y_{-2}}{2}\right) + \frac{2p^3 - p}{12}\Delta^4 y_{-2} + \ldots\right]$$ (7.7)

At $x = x_0$, $p = 0$, hence putting $p = 0$ in Eq. (7.7), we get

$$\left(\frac{dy}{dx}\right)_{x_0} = \frac{1}{h}\left[\frac{\Delta y_0 + \Delta y_{-1}}{2} - \frac{1}{6}\left(\frac{\Delta^3 y_{-1} + \Delta^3 y_{-2}}{2}\right) + \frac{1}{30}\left(\frac{\Delta^5 y_{-2} + \Delta^5 y_{-3}}{2}\right)\right]$$

Similarly

$$\left(\frac{d^2 y}{dx^2}\right) = \frac{1}{h^2}\left[\Delta^2 y_{-1} - \frac{1}{12}\Delta^4 y_{-2} + \frac{1}{90}\Delta^6 y_{-3} + \ldots\right] \text{ and so on.}$$

7.3 MAXIMA AND MINIMA OF A TABULATED FUNCTION

Newton's forward interpolation formula:

$$y = y_0 + p\Delta y_0 + \frac{p(p-1)}{2!}\Delta^2 y_0 + \frac{p(p-1)(p-2)}{3!}\Delta^3 y_0 + \ldots$$

Differentiate w.r.t. p, we get

$$\frac{dy}{dp} = \Delta y_0 + \frac{2p-1}{2}\Delta^2 y_0 + \frac{3p^2 - 6p + 2}{6}\Delta^3 y_0 + \dots \tag{7.8}$$

For maxima or minima $\dfrac{dy}{dp} = 0$. Hence, equating the right hand side of Eq. (7.8) to zero, retaining only up to third differences, we obtain

$$\Delta y_0 + \frac{2p-1}{2}\Delta^2 y_0 + \frac{3p^2 - 6p + 2}{6}\Delta^3 y_0 = 0 \tag{7.9}$$

or $\left(\dfrac{1}{2}\Delta^3 y_0\right)p^2 + (\Delta^2 y_0 - \Delta^3 y_0)\,p + \left(\Delta y_0 - \dfrac{1}{2}\Delta^2 y_0 + \dfrac{1}{3}\Delta^3 y_0\right) = 0$

Substituting the value of $\Delta y_0, \Delta^2 y_0, \Delta^3 y_0$ from the difference table, we solve this quadratic for p. Then the corresponding value of x are given by $x = x_0 + ph$ at which y is maximum or minimum.

Example 7.1: Find the first, second and third derivatives of $f(x)$ at $x = 1.5$ if

x	1.5	2.0	2.5	3.0	3.5	4.0
$f(x)$	3.375	7.000	13.625	24.000	38.875	59.000

Solution: We have to find the derivative at point $x = 1.5$ which is at the beginning of the given data. Therefore, we use here the derivatives of Newton's forward interpolation formula.

x	$y = f(x)$	Δy	$\Delta^2 y$	$\Delta^3 y$	$\Delta^4 y$
1.5	3.375				
		3.625			
2.0	7.000		3.000		
		6.625		0.750	
2.5	13.625		3.750		0
		10.375		0.750	
3.0	24.000		4.500		0
		14.875		0.750	
3.5	38.875		5.250		
		20.125			
4.0	59.000				

Here $x_0 = 1.5$, $y_0 = 3.375$, $\Delta y_0 = 3.625$, $\Delta^2 y_0 = 3$, $\Delta^3 y_0 = 0.75$, $h = 0.5$

$$\left.\frac{dy}{dx}\right)_{x = x_0} = f'(x) = \frac{1}{h}\left[\Delta y_0 - \frac{1}{2}\Delta^2 y_0 + \frac{1}{3}\Delta^3 y_0 - \frac{1}{4}\Delta^4 y_0 + \dots\right]$$

$$f'(1.5) = \frac{1}{0.5}\left[3.625 - \frac{1}{2}(3) + \frac{1}{3}(0.75)\right] = 4.75$$

$$\left.\frac{d^2 y}{dx^2}\right)_{x = x_0} = f''(x_0) = \frac{1}{h^2}\left[\Delta^2 y_0 - \Delta^3 y_0 + \frac{11}{12}\Delta^4 y_0 + \dots\right]$$

$$f''(1.5) = \frac{1}{(0.5)^2}\,[3 - 0.75] = 9$$

$$\left.\frac{d^3y}{dx^3}\right)_{x=y_0} = f'''(x) = \frac{1}{h^3}\left[\Delta^3 y_0 - \frac{3}{2}\Delta^4 y_0 + \dots\right]$$

$$f'''(1.5) = \frac{1}{(0.5)^3}[0.75] = 6$$

Example 7.2: The population of a certain town is shown in the following table.

Year	1951	1961	1971	1981	1991
Population (in thousands)	19.96	36.65	58.81	77.21	94.61

Find the rate of growth of population in the year 1981.

Solution: We have to find the derivative at point $x = 1981$ which is near the end of the given data. Therefore, we use here the derivatives of Newton's backward interpolation formula. The table of differences is as follows:

x	$y = f(x)$	Δy	$\Delta^2 y$	$\Delta^3 y$	$\Delta^4 y$
1951	19.96				
		16.69			
1961	36.65		5.47		
		22.16		−9.23	
1971	58.81		−3.76		11.99
		18.40		2.76	
1981	77.21		−1		
		17.40			
1991	94.61				

Here $h = 10$, $x_n = 1991$, $\nabla y_n = 17.4$, $\nabla^2 y_n = -1$, $\nabla^3 y_n = 2.76$, $\nabla^4 y_n = 11.99$

$$\left.\frac{dy}{dx}\right)_{x=x_0} = \frac{1}{h}\left[\nabla y_n + \frac{2p+1}{3}\nabla^2 y_n + \frac{3p^2+6p+2}{6}\nabla^3 y_n + \dots\right]$$

We have to find out the rate of growth of population in the year 1981, i.e. $\left.\dfrac{dy}{dx}\right)_{x=x_n}$,

now $x_n + ph = 1981$

$$p = \frac{1981 - 1991}{10} = -1$$

But $\qquad p = -1, h = 10$

$$y'(1981) = \frac{1}{10}\left[17.4 + \frac{2(-1)+1}{2}(-1) + \frac{3(-1)^2 + 6(-1) + 2}{6}(2.76) + \right.$$

$$\left. \frac{2(-1)^3 + 9(-1)^2 + 11(-1) + 3}{12}(11.99)\right] = 1.6440833$$

Example 7.3: Obtain the value of $f'(90)$ using Stirling's formula to the following data.

x	60	75	90	105	120
$f(x)$	28.2	38.2	43.2	40.9	37.7

Also find the maximum value of function from the data.

Solution: Since $x = 90$ is in the middle of the table, we use central difference formula. We use Stirling's formula

x	$y = f(x)$	Δy	$\Delta^2 y$	$\Delta^3 y$	$\Delta^4 y$
60	28.2				
		10			
75	38.2		−5		
		5		−2.3	
90	43.2		−7.3		8.7
		−2.3		6.4	
105	40.9		−0.9		
		−3.2			
120	37.7				

Here $x_0 = 90$, $y_0 = 43.2$, $\nabla y_0 = -2.3$, $\nabla y_{-1} = 5$, $\nabla^3 y_{-1} = -2.3$, $\nabla^3 y_{-2} = 6.4$ and $h = 15$.

$$\left.\frac{dy}{dx}\right)_{x=y_0} = \frac{1}{h}\left[\left\{\frac{\Delta y_0 + \Delta_{y-1}}{2}\right\} - \frac{1}{6}\left\{\frac{\Delta^3 y_{-1} + \Delta^3_{y-2}}{2}\right\} + \ldots\right]$$

$$f'(90) = \frac{1}{15}\left[\left(\frac{-2.3+5}{2}\right) - \frac{1}{6}\left(\frac{-2.3+6.4}{2}\right)\right] = 0.0672222$$

To find the maximum value of tabular function by Stirling's formula

$$y = y(x_0 + ph) = y_0 + \frac{p}{2}(\Delta y_0 + \Delta y_{-1}) + \frac{p^2}{2!}\Delta^2 y_{-1}$$

$$+ \frac{p(p^2 - 1)}{3!}\left[\frac{(\Delta^3 y_{-1} + \Delta^3 y_{-2})}{2}\right] + \frac{p^2(p^2 - 1)}{4!}\Delta^4 y_{-2} + \ldots$$

Substituting the values from table, we get after simplification
$$y = 43.2 + 1.35p - 3.65p^2 + 0.3417(p^3 - p)$$
$$= 0.3417p^3 - 3.65p^2 + 1.0083p + 43.2$$

If y is maximum, $\frac{dy}{dp} = 0$,

i.e. $1.0251p^2 - 7.3p + 1.0083 = 0$

$$p = \frac{7.3 + \sqrt{(7.3)^2 - 4(1.0251)(1.0083)}}{2(1.0251)}$$

$$= 6.9803 \text{ or } 0.1409$$
$$p = 6.9803 \text{ is out of range}$$
$\therefore p = 0.1409$ because p lies between $-1/4$ and $1/4$.

Hence $\qquad x = x_0 + ph = 90 + 15(0.1409) = 92.1135$

and maximum value of $y = 0.3417(0.1409)^3 - 3.65(0.1409)^2 + 1.0083(0.1409) + 43.2$

$$= 43.27$$

Example 7.4: Using Bessel's formula, find $f'(7.5)$ from the following data:

x	7.47	7.48	7.49	7.50	7.51	7.52	7.53
$f(x)$	0.193	0.195	0.198	0.201	0.203	0.206	0.208

Solution: Taking $x_0 = 7.50$, $h = 0.01$, we have

p	x	y_p	Δy	$\Delta^2 y$	$\Delta^3 y$	$\Delta^4 y$	$\Delta^5 y$	$\Delta^6 y$
-3	7.47	0.193						
			0.002					
-2	7.48	0.195		0.001				
			0.003		-0.001			
-1	7.49	0.198		0.000		0.000		
			0.003		-0.001		0.003	
0	7.50	0.201		-0.001		0.003		-0.010
			0.002		0.002		-0.007	
1	7.51	0.203		0.001		-0.004		
			0.003		-0.002			
2	7.52	0.206		-0.001				
			0.002					
3	7.53	0.208						

Bessel's formula is

$$y_p = y_0 + p\Delta y_0 + \frac{p(p-1)}{2!}\left[\frac{\Delta^2 y_{-1} + \Delta^2 y_0}{2}\right] + \frac{\left(p-\frac{1}{2}\right)p(p-1)}{3!}\Delta^3 y_{-1}$$

$$+ \frac{(p+1)p(p-1)(p-2)}{4!}\left[\frac{\Delta^4 y_{-2} + \Delta^4 y_{-1}}{2}\right] + \frac{\left(p-\frac{1}{2}\right)(p+1)p(p-1)(p-2)}{5!}\Delta^5 y_{-2}$$

$$+ \frac{(p+2)(p+1)p(p-1)(p-2)(p-3)}{6!}\left[\frac{\Delta^6 y_{-3} + \Delta^6 y_{-2}}{2}\right] \qquad \text{(i)}$$

Since
$$p = \frac{x - x_0}{h}, \frac{dp}{dx} = \frac{1}{h}$$

$$\frac{dy}{dx} = \frac{dy}{dp}\cdot\frac{dp}{dx} = \frac{1}{h}\frac{dy}{dp}$$

$$\frac{dy}{dx}\bigg]_{x = x_0} = \frac{1}{h}\cdot\left[\frac{dy}{dp}\right]_{p=0}$$

Differentiate Eq. (i) w.r.t. p and put $p = 0$

$$\frac{dy}{dx} = \frac{1}{h}\left[\Delta y_0 - \frac{1}{4}(\Delta^2 y_{-1} + \Delta^2 y_0) + \frac{1}{12}\Delta^3 y_{-1} + \frac{1}{24}(\Delta^4 y_{-2} + \Delta^4 y_{-1}) - \frac{1}{120}\Delta^5 y_{-2}\right.$$

$$\left. - \frac{1}{240}(\Delta^6 y_{-3} + \Delta^6 y_{-2})\right]$$

$$\left(\frac{dy}{dx}\right)_{7.5} = \frac{1}{0.01}\left[0.002 - \frac{1}{4}(-0.001 + 0.001) + \frac{1}{12}(0.002) + \frac{1}{24}(-0.004 + 0.003)\right.$$

$$\left. - \frac{1}{120}(-0.007) - \frac{1}{240}(-0.010 + 0)\right]$$

$$= 0.223$$

Example 7.5: A slider in a machine moves along a straight rod. Its distance x cm along the rod is given below for various values of the time t seconds. Find the velocity of the slider and its acceleration when $t = 0.3$ seconds.

t	0	0.1	0.2	0.3	0.4	0.5	0.6
x	30.13	31.62	32.87	33.64	33.95	33.81	33.24

Solution: Taking $x_0 = 7.50$, $h = 0.01$, we have

t	x	Δ	Δ^2	Δ^3	Δ^4	Δ^5	Δ^6
0	30.13						
		1.49					
0.1	31.62		−0.24				
		1.25		−0.24			
0.2	32.87		−0.48		0.26		
		0.77		0.02		−0.27	
0.3	33.64		−0.46		−0.01		0.29
		0.31		0.01		0.02	
0.4	33.95		−0.45		0.01		
		−0.14		0.02			
0.5	33.81		−0.43				
		−0.57					
0.6	33.24						

As the derivatives are required near the middle of the table, we use Stirling's formula

$$\left(\frac{dx}{dt}\right)_{t_0} = \frac{1}{h}\left[\left(\frac{\Delta x_0 + \Delta x_{-1}}{2}\right) - \frac{1}{6}\left(\frac{\Delta^3 x_{-1} + \Delta^3 x_{-2}}{2}\right) + \frac{1}{30}\left(\frac{\Delta^5 x_{-2} + \Delta^5 x_{-3}}{2}\right) + \ldots\right]$$

$$\left(\frac{d^2x}{dt^2}\right)_{t_0} = \frac{1}{h^2}\left[\Delta^2 x_{-1} - \frac{1}{12}\Delta^4 x_{-2} + \frac{1}{90}\Delta^6 x_{-3} + \ldots\right]$$

Here $h = 0.1$, $t_0 = 0.3$, $\Delta x_0 = 0.31$, $\Delta x_{-1} = 0.77$, $\Delta^2 x_{-1} = -0.46$ etc.

Put the values

$$\left(\frac{dx}{dt}\right)_{0.3} = \frac{1}{0.1}\left[\left(\frac{0.31 + 0.77}{2}\right) - \frac{1}{6}\left(\frac{0.01 + 0.02}{2}\right) + \frac{1}{30}\left(\frac{0.02 - 0.27}{2}\right) + \ldots\right]$$

$$\left(\frac{d^2x}{dt^2}\right)_{0.3} = \frac{1}{(0.1)^2}\left[-0.46 - \frac{1}{12}(-0.01) + \frac{1}{90}(0.29) + \ldots\right]$$

$$= -45.6$$

velocity is 5.33 cm/sec and acceleration is −45.6 cm/sec^2.

Example 7.6: Find $f'(10)$ from the following data:

x	3	5	11	27	34
$f(x)$	−13	23	899	17315	35606

Solution: As the value of x are not equispaced, we shall use Newton's divided difference formula.

The divided difference table:

x	f(x)	1st divided difference	2nd divided difference	3rd divided difference	4th divided difference
3	−13				
		18			
5	23		16		
		146		0.998	
11	899		29.96		0.0002
		1025		1.003	
27	17315		69.04		
		2613			
34	35606				

Newton's divided difference formula:

$$f(x) = f(x_0) + (x - x_0) f(x_0, x_1) + (x - x_0)(x - x_1) f(x_0, x_1, x_2)$$
$$+ (x - x_0)(x - x_1)(x - x_2) f(x_0, x_1, x_2, x_3)$$
$$+ (x - x_0)(x - x_1)(x - x_2)(x - x_3) f(x_0, x_1, x_2, x_3, x_4)$$

Differentiating w.r.t. x, we get

$$f'(x) = f(x_0, x_1) + (2x - x_0 - x_1) f(x_0, x_1, x_2) + [3x^2 - 2x(x_0 + x_1 + x_2)$$
$$+ x_0 x_1 + x_1 x_2 + x_2 x_0] f(x_0, x_1, x_2, x_3) + [4x^2 - 3x^2 (x_0 + x_1 + x_2 + x_3)$$
$$+ 2x(x_0 x_1 + x_1 x_2 + x_2 x_3 + x_2 x_0 + x_1 x_3 + x_0 x_2)$$
$$- (x_0 x_1 x_2 + x_1 x_2 x_3 + x_2 x_3 x_0 + x_0 x_1 x_3)] f(x_0, x_1, x_2, x_3, x_4)$$

Put $x_0 = 3$, $x_1 = 5$, $x_2 = 11$, $x_3 = 27$ and $x = 10$

$$f'(10) = 18 + 12 \times 16 + 23 \times 0.998 - 426 \times 0.0002$$
$$= 232.869$$

Example 7.7: Given the following data, find the maximum value of y:

x	−1	1	2	3
f(x)	−21	15	12	3

Solution: Since the arguments are not equispaced, we will form the divided difference table.

x	y	Δy	$\Delta^2 y$	$\Delta^3 y$
−1	−21			
		18		
1	15		−7	
		−3		1
2	12		−3	
		−9		
3	3			

Using Newton's divided difference formula

$$y = y_0 + (x - x_0) \Delta y_0 + (x - x_0)(x - x_1) \Delta^2 y_0 + (x - x_0)(x - x_1)(x - x_2) \Delta^3 y_0$$
$$= -21 + (x + 1)(18) + (x + 1)(x - 1)(-7) + (x + 1)(x - 1)(x - 2)(1)$$
$$= x^3 - 9x^2 + 17x + 6$$

for maximum value, $\dfrac{dy}{dx} = 0 \Rightarrow 3x^2 - 18x + 17 = 0$

$$x = \dfrac{+18 \pm \sqrt{(-18)^2 - 4(3)(17)}}{6}$$

$$= 4.8257 \text{ or } 1.1743$$

But $\qquad x = 4.8257$ is out of range

$\therefore$ $\qquad = 1.1743$ is the value giving the maximum of y

Maximum of y (at $x = 1.1743$) $= (1.743)^3 - 9(1.1743)^2 + 17(1.1743) + 6$

$$= 15.171612$$

PROBLEMS 7.1

1. From the following table of values, estimate $y'(1.05)$ and $y''(1.25)$.

x	1.00	1.05	1.10	1.15	1.20	1.25	1.30
y	1.0000	1.0247	1.0488	1.0724	1.0954	1.1180	1.1402

[Ans. $y'(1.05) = 0.4876$, $y''(1.25) = -0.1583$]

2. Find $f'(8)$ given $f(6) = 1.556$, $f(7) = 1.908$, $f(12) = 2.158$. \qquad [Ans. $f'(8) = 0.109$]

3. Find the values of $f'(0.5)$ using Stirling's formula from the following data.

x	0.35	0.40	0.45	0.50	0.55	0.60	0.65
$f(x)$	1.521	1.506	1.488	1.467	1.444	1.418	1.389

[Ans. -0.44]

4. Find the value of $\cos(1.74)$ from the following table.

x	1.7	1.74	1.78	1.82	1.86
$\sin x$	0.9916	0.9857	0.9857	0.9691	0.9584

[Ans. -0.17125]

5. Find $y'(1.2)$ using Bessel's formula from the following data.

x	1.0	1.1	1.2	1.3	1.4
y	43.1	47.7	52.1	56.4	60.8

[Ans. 43.1666]

6. The following data gives corresponding value of pressure and specific volume of a superheated steam:

v	1.0	1.2	1.4	1.6	1.8	2.0
p	0	0.128	0.544	1.296	2.432	4.00

  i. Find the rate of change of pressure with respect to volume when $V = 2$

  ii. Find the rate of change of volume with respect to pressure

Hint: $\left(\dfrac{dv}{dp} = \dfrac{1}{\dfrac{dp}{dv}} \right)$ \qquad [Ans. i. -52.4, ii. -0.01908]

7. If $y = f(x)$ and y_n denotes $f(x_0 + nh)$. Prove that if powers of h above h^6 be neglected

$$\left(\frac{dy}{dx} \right)_{x=x_0} = \frac{3}{4h} \left[(y_1 - y_{-1}) - \frac{1}{5}(y_2 - y_{-2}) + \frac{1}{45}(y_3 - y_{-3}) \right]$$

8. From the following table, find the maximum value of y:

x	0	2	4	6
y	3	3	11	27

[Ans. $x = 1, y = 2$]

7.4 CUBIC SPLINE METHOD

The cubic spline derived in interpolation with cubic spline can conveniently be used to compute the first and second derivatives of a function. For a natural cubic spline, the recurrence formula are given in Eqs (7.1) and (7.2).

$$\frac{h_i}{6} M_{i-1} + \frac{1}{3}(h_i + h_{i+1}) M_i + \frac{h_{i+1}}{6} M_{i+1} = \frac{y_{i+1} - y_i}{h_{i+1}} - \frac{y_i - y_{i-1}}{h_i} \quad (i = 1, 2, 3, ..., N-1) \quad (7.10)$$

For equal intervals, we have $h_i = h_{i+1} = h$ and equation simplifies to

$$M_{i-1} + 4M_i + M_{i+1} = \frac{6}{h^2}(y_{i+1} - 2y_i + y_{i-1})(i = 1, 2, 3, ..., N-1) \quad (7.11)$$

may be used to compute the spline second derivatives depending upon the choice of subdivisions.

Thus,
$$S_{i+1}(x) = \frac{1}{h_{i+1}}\left[\frac{(x_{i+1} - x)^3}{6} M_i + \frac{(x - x_i)^3}{6} M_{i+1}\right.$$

$$\left. + \left(y_i - \frac{h_{i+1}^2}{6} M_i\right)(x_{i+1} - x) + \left(y_{i+1} - \frac{h_{i+1}^2}{6} M_{i+1}\right)(x - x_i)\right] \quad (7.12)$$

gives the spline in the interval of interest from which the first derivatives can be computed.

For the first derivatives at the tabular points, it would of course, be easier to use the formula given below directly if on the other hand, end conditions involving the first derivatives are given

$$S_i'(x_i) = \frac{h}{2} M_i - \frac{1}{h_i}\left(y_{i-1} - \frac{h_i^2}{6} M_{i-1}\right) + \frac{1}{h_i}\left(y_i - \frac{h_i^2}{6} M_i\right)$$

$$= \frac{1}{h_i}(y_i - y_{i-1}) + \frac{h_i}{6} M_{i-1} + \frac{h_i}{3} M_i \ (i = 1, 2, ..., N)$$

$$S_{i+1}'(x_{i+}) = \frac{1}{h_{i+1}}(y_{i+1} - y_i) - \frac{h_{i+1}}{3} M_i - \frac{h_{i+1}}{6} M_{i+1} \ (i = 1, 2, ..., N-1)$$

Thus the recurrence formula is

$$\frac{1}{h} m_{i-1} + 2\left(\frac{1}{h_i} + \frac{1}{h_{i+1}}\right) m_i + \left(\frac{1}{h_{i+1}}\right) m_{i+1} = \frac{3}{h_{i+1}^2}(y_{i+1} - y_i) + \frac{3}{h_i^2}(y_i - y_{i-1})(i = 1, 2, ..., N-1)$$

for equally spaced knots,

i.e. $\quad m_{i-1} + 4m_i + m_{i+1} = \frac{3}{h}(y_{i+1} - y_{i-1})(i = 1, 2, ..., N-1)$

may be used to compute the remaining first derivatives.

Example 7.8: We consider the function $f(x) = \sin x$ in $[0, \pi]$.

Solution: $\qquad M_0 = M_N = 0$

Let $N = 2$, i.e. $h = \pi/2$

The

x	0	$\pi/2$	π
y	0	1	0

$$M_0 = M_2 = 0$$

Using formula of Eq. (7.11), we obtain

$$M_0 + 4M_1 + M_2 = \frac{6}{h^2}(y_0 - 2y_1 + y_2)$$

$$M_1 = \frac{-12}{\pi^2}$$

Equation (7.12) now gives the spline in each interval thus, for $0 \le x \le \pi/2$, we obtain

$$S(x) = \frac{2}{\pi}\left[0 + \frac{(x-0)^3}{6} \times \left(\frac{-12}{\pi^2}\right) + \frac{3x}{2}\right] = \frac{2}{\pi}\left[-\frac{2x^3}{\pi^2} + \frac{3x}{2}\right]$$

which gives

$$S'(x) = \frac{2}{\pi}\left[\frac{-2}{\pi^2}(3x^2) + \frac{3}{2}\right]$$

$$S'\left(\frac{\pi}{4}\right) = \frac{2}{\pi}\left[\frac{-6}{\pi^2}\left(\frac{\pi^2}{16}\right) + \frac{3}{2}\right] = \frac{9}{4\pi} = 0.71619725$$

Exact value of $S'\left(\frac{\pi}{4}\right) = \cos\frac{\pi}{4} = \frac{i}{\sqrt{2}} = 0.70710681$

The percentage error in the computed value of $S'\left(\frac{\pi}{4}\right)$ is 1.28%.

From (i)

$$S'(x) = \frac{-24}{\pi^3}x$$

$$S''\left(\frac{\pi}{4}\right) = \frac{-24}{\pi^3}\cdot\frac{\pi}{4} = \frac{-6}{\pi^2} = -0.60792710$$

Since the exact value is $\frac{-1}{\sqrt{2}}$ the percentage error in this result is 14.03%.

We now consider value of $y = \sin x$ in interval of $10°$ from $x = 0$ to π. To obtain the spline second derivatives, we have used a computer and the results are given in the table below (up to $x = 90°$).

x in degrees	$y''(x)$	
	Exact	Cubic spline
10	−0.173648178	−0.174089426
20	−0.342020143	−0.342889233
30	−0.500000000	−0.501270524
40	−0.642787610	−0.644420964
50	−0.766044443	−0.767990999
60	−0.866025404	−0.868226016
70	−0.939692621	−0.942080425
80	−0.984807753	−0.987310197
90	−1.000000000	−1.002541048

It is seen that there is a greatest accuracy in the values of spline second derivatives.

7.5 NUMERICAL INTEGRATION

The process of computing $\int_a^b y\,dx$, where $y = f(x)$ is given by a set of tabulated values $[x_i, y_i]$, $i = 1, 2, ..., n$; $a = x_0$ and $b = x_n$ is called *numerical integration*.

This process when applied to a function of a single variable $y = f(x)$ is called *quadrature*.

7.6 QUADRATURE FORMULA

Let us consider an interval (a, b) divided into n equal parts of width h (Fig. 7.1), so that

$$x_0 = a, x_0 + h = x_1$$
$$x_0 + 2h = x_2, ...; x_0 + nh = x_n = b$$

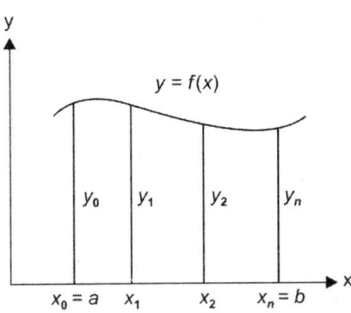

Fig. 7.1

Now
$$I = \int_{x_0}^{x_n + nh} f(x)\,dx$$

Put
$$x = x_0 + ph, \text{ so that } dx = hdp$$

or
$$I = h\int_0^n f(x_0 + ph)\,dp$$

Replacing $f(x_0 + ph)$ by Newton's forward interpolation formula, we get

$$I = h\int_0^n \left(y_0 + p\Delta y_0 + \frac{p(p-1)}{2!}\Delta^2 y_0 + ... \right)dp$$

Integrating each term by term, we have after substituting the limit

$$\int_{x_0}^{x_0 + nh} f(x)\,dx = h\left[ny_0 + \frac{n^2}{2}\Delta y_0 + \frac{1}{2}\left(\frac{n^3}{3} - \frac{n^2}{2} \right)\Delta^2 y_0 + \frac{1}{6}\left(\frac{n^4}{4} - n^3 + n^2 \right)\Delta^3 y_0 \right.$$

$$\left. + \frac{1}{24}\left(\frac{n^5}{5} - \frac{3n^4}{2} + \frac{11n^3}{3} - 3n^2 \right)\Delta^4 y_0 + ... \right] \qquad (7.13)$$

This is known as Newton-Cotes' quadrature formula, from the general formula, we deduce the following important quadrature rules by taking $n = 1, 2, 3, ...$

7.6.1 Trapezoidal Rule

Put $n = 1$ in Eq. (7.13) and neglecting second and higher order differences, we get

$$\int_{x_0}^{x_0 + h} f(x)\,dx = h\left[y_0 + \frac{1}{2}\Delta y_0 \right]$$

$$= h\left[y_0 + \frac{1}{2}(y_1 - y_0) \right] = \frac{h}{2}[y_0 + y_1]$$

$$\int_{x_0+h}^{x_0+2h} f(x)\,dx = \frac{h}{2}[y_1 + y_2] \tag{7.14}$$

$$\int_{x_0+2h}^{x_0+3h} f(x)\,dx = \frac{h}{2}[y_2 + y_3] \tag{7.15}$$

$$\int_{x_0+(n-1)h}^{x_0+nh} f(x)\,dx = \frac{h}{2}[y_{n-1} + y_n] \tag{7.16}$$

Adding $$\int_{x_0}^{x_0+nh} f(x)\,dx = \frac{h}{2}[y_0 + y_n + 2(y_1 + y_2 + \dots + y_{n-1})] \tag{7.17}$$

Note: The area of each strip (trapezium) is found separately. Then the area under the curve and the ordinates x_0 and $x_0 + nh$ is approximately equal to sum of the area of n trapeziums.

7.6.2 Simpson's 1/3 Rule

Put $n = 2$ in Eq. (7.13) above and taking the curve through (x_0, y_0), (x_1, y_1) and (x_2, y_2) as a parabola, i.e. a polynomial of second order so that differences of order higher than second order will vanish, we get

$$\int_{x_0}^{x_0+2h} f(x)\,dx = h\left[2y_0 + 2\Delta y_0 + \frac{1}{2}\left(\frac{8}{3} - 2\right)\Delta^2 y_0\right]$$

$$= h\left[2y_0 + 2(y_1 - y_0) + \frac{1}{3}(y_2 - 2y_1 + y_0)\right] = \frac{h}{3}[y_0 + 4y_1 + y_2]$$

$$\int_{x_0+2h}^{x_0+4h} f(x)\,dx = \frac{h}{3}(y_2 + 4y_3 + y_4)$$

$$\dots = \dots$$

$$\int_{x_0+(n-2)h}^{x_0+nh} f(x)\,dx = \frac{h}{3}(y_{n-2} + 4y_{n-1} + y_n), n \text{ being even}$$

Adding $$\int_{x_0}^{x_0+nh} f(x)\,dx = \frac{h}{3}[(y_0 + y_n) + 4(y_1 + y_3 + \dots + y_{n-1}) + 2(y_2 + y_4 + \dots + y_{n-2})]$$

This is known as Simpson's 1/3 rule or Simpson's rule and is most commonly used.

Note: While applying Simpson's 1/3rd rule, the given interval must be divided into even number of equal subintervals, since we find the area of two strips at a time.

7.6.3 Simpson's 3/8 Rule

Putting $n = 3$ and in Eq. (7.13) neglecting all differences above the third order, we have

$$\int_{x_0}^{x_0+3h} f(x)\,dx = h\left[3y_0 + \frac{9}{2}\Delta y_0 + \frac{1}{2}\left(\frac{27}{3} - \frac{9}{2}\right)\Delta^2 y_0 + \dots\right]$$

$$= h\left[3y_0 + \frac{9}{2}(y_1 - y_0) + \frac{9}{4}(y_2 - 2y_1 + y_0) + \frac{3}{8}(y_3 - 3y_2 + 3y_1 - y_0)\right]$$

$$= \frac{3h}{8}[y_0 + 3y_1 + 3y_2 + y_3]$$

$$\int_{x_0 + 3h}^{x_0 + 6h} f(x)dx = \frac{3h}{8}(y_3 + 3y_4 + 3y_5 + y_6)$$

$$\dots = \dots$$

$$\int_{x_n + (n-3)h}^{x_0 + nh} f(x)dx = \frac{3h}{8}(y_{n-3} + 3y_{n-2} + 3y_{n-1} + y_n) \qquad \text{(if } n \text{ is a multiple of 3)}$$

$$\int_{x_0}^{x_0 + nh} f(x)dx = \frac{3h}{8}[y_0 + y_n + 3(y_1 + y_2 + y_4 + y_5 + \dots + y_{n-1}) + 2(y_3 + y_6 + y_9 + \dots + y_{n-3})$$

This is known as Simpson's 3/8 rule.

Note: Application of Simpson's rule: If various ordinates in a table represents equispaced cross-sectional area then Simpson's rule gives the volume of the solid. If ordinates denote the velocities at equal intervals of time, then Simpson's rule gives the distance travelled.

7.6.4 Boole's Rule

Put $n = 4$ in Eq. (7.13) and neglecting all differences above fourth order, we have

$$\int_{x_0}^{x_0 + 4h} f(x)dx = 4h\left[y_0 + 2\Delta y_0 + \frac{5}{3}\Delta^2 y_0 + \frac{2}{3}\Delta^3 y_0 + \frac{7}{90}\Delta^4 y_0\right]$$

$$= \frac{2h}{5}[7y_0 + 32y_1 + 12y_2 + 32y_3 + 7y_4]$$

$$\int_{x_0 + 4h}^{x_0 + 8h} f(x)dx = \frac{2h}{45}[7y_4 + 32y_5 + 12y_6 + 32y_7 + 7y_8]$$

Adding all these integrals from x_0 to $x_0 + nh$, where n is multiple of 4, we get

$$\int_{x_0}^{x_0 + nh} f(x)dx = \frac{2h}{45}[7y_0 + 32y_1 + 12y_2 + 32y_3 + 14y_4 + 32y_5 + 12y_6 + 32y_7 + 14y_8 + \dots]$$

This is known as Boole's rule.

Note: The number of subintervals should be taken as a multiple of 4.

7.6.5 Weddle's Rule

Put $n = 6$ in Eq. (7.13) and neglecting all differences above sixth order, we obtain

$$\int_{x_0}^{x_0 + 6h} f(x)dx = h\left[6y_0 + 18\Delta y_0 + 27\Delta^2 y_0 + 24\Delta^3 y_0 + \frac{123}{10}\Delta^4 y_0 + \frac{33}{10}\Delta^5 y_0 + \frac{41}{140}\Delta^6 y_0\right]$$

If we replace the coefficient of $\Delta^6 y_0$ by $\dfrac{42}{140}$ (the error made will be negligible) and the difference in term of $y's$, we get

$$\int_{x_0+3h}^{x_0+6h} f(x)dx = \frac{3h}{10}[y_0 + 5y_1 + y_2 + 6y_3 + y_4 + 5y_5 + y_6]$$

Similarly

$$\int_{x_0+6h}^{x_0+12h} f(x)dx = \frac{3h}{10}[y_6 + 5y_7 + y_8 + 6y_9 + y_{10} + 5y_{11} + y_{12}]$$

Adding all these integrals and rearranging them in an order

$$\int_{x_0}^{x_0+nh} f(x)dx = \frac{3h}{10}[(y_0 + y_n) + (5y_1 + y_2 + 6y_3 + y_4 + 5y_5) + (2y_6 + 5y_7 + y_8 + 6y_9$$
$$+ y_{10} + 5y_{11}) + (2y_{n-6} + 5y_{n-5} + y_{n-4} + 6y_{n-3} + y_{n-2} + 5y_{n-1})]$$

This formula is used when number of subintervals is a multiple of 6.

Note: Weddle's rule is generally more accurate than any others. Of the two Simpson rules, the 1/3 rule is better.

Simpson rule is useful to civil engineers for calculating the amount of earth that must be moved to fill a depression or make a dam.

In the above formula, the coefficients may be remembered in groups of six.

First group: Coefficients:	1, 5	1, 6	1, 5
All interior group: Coefficients:	2, 5	1, 6	1, 5
Last group: Coefficients:	2, 5	1, 6	1, 5

Note: If there are only 7 ordinates, the coefficients are 1, 5, 1, 6, 1, 5, 1.

1. In trapezoidal rule $f(x)$ is a linear function of x, i.e. of the form $f(x) = ax + b$. It is the simplest rule but least accurate.
2. In Simpson's 1/3 rule, $f(x)$ is polynomial of second degree, i.e. $f(x) = ax^2 + bx + c$. To apply this rule, the number of intervals n must be even, in other words, the number of ordinates must be odd.
3. In Simpsopn's 3/8 rule, $f(x)$ is a polynomial of third degree, i.e. $f(x) = ax^3 + bx^2 + cx + d$. To apply this rule, the number of intervals n must be a multiple of 3.
4. In Weddle's rule, $f(x)$ is a polynomial of degree 6.

To apply this rule, the number of intervals n must be a multiple of 6, but it requires at least seven consecutive values of the function.

7.7 NEWTON–COTES' FORMULA

As the Lagrange's formula is based on the same assumption as an approximate integration formula, viz. $f(x)$ is a polynomial, it can be used in finding an approximate formula.

Let y_0, y_n be values of the function $y = f(x)$ corresponding to the argument $x_0, x_1, ..., x_n$. Let the values of the arguments be at equal intervals, say h,

i.e. $$x_i = x_0 + ih$$

$$x_1 = x_0 + h, x_2 = x_0 + 2h, ... \text{ i.e. } x_n = x_0 + nh$$

Now by Lagrange's formula

$$f(x) \approx P(x) = \sum_{K=1}^{n} L_K(x) y_k$$

where $$L_k(x) = \frac{(x-x_0)(x-x_1)(x-x_2)\dots(x-x_{k-1})(x-x_{k+1})(x-x_n)}{(x_k-x_0)(x_k-x_1)(x_k-x_2)\dots(x_k-x_{k-1})(x_k-x_{k+1})(x_k-x_n)}$$

Thus $$I = \int_{x_0}^{x_0+nh} f(x)dx = \int_{x_0}^{x_0+nh} P(x)dx$$

$$= \int_{x_0}^{x_0+nh} \sum_{k=0}^{n} L_k(x)y_k dx = \int_{x_0}^{x_0+nh} [L_0(x)y_0 + \dots L_n(x)y_n]dx$$

$$= y_0 \int_{x_0}^{x_0+nh} L_0(x)dx + \dots + y_n \int_{x_0}^{x_0+nh} L_n(x)dx = \sum_{K=0}^{n} y_k \int_{x_0}^{x_0+nh} L_k(x)dx$$

Substituting $x = x_0 + uh \Rightarrow dx = hdu$, we get

$$I = h \sum_{K=0}^{n} y_k \int_0^n L_k du = nh \sum y_k \frac{1}{n} \int_0^n L_k du \tag{7.18}$$

where $L_k = L_k(x_0 + uh)$

$$= \frac{(x_0 + uh - x_0)(x_0 + uh - x_1)\dots(x_0 + uh - x_{k-1})(x_0 + uh - x_{k+1})\dots(x_0 + uh - x_n)}{(x_k - x_0)(x_k - x_1)\dots(x_k - x_{k-1})(x_k - x_{k+1})\dots(x_k - x_n)}$$

Putting $x_k = x_0 + kh$, we get

$$L_k = \frac{u(u-1)\dots(u-k+1)(u-k-1)\dots(u-n)}{k(k-1)\dots(+1)(-1)\dots(k-n)} \tag{7.19}$$

Now putting $\frac{1}{n}\int_0^n L_k \, du = C_k^n$, we get

$$I = nh \sum_{k=0}^{n} y_k C_k^n \tag{7.20}$$

This is the required formula, it is known as Cotes' formula.

The numbers $C_k^n, 0 \le k \le n$ are called Cotes' numbers. We can evaluate I from Eq. (7.20) in case we know the values of Cotes' numbers.

Properties of Cotes' numbers:

1. $C_k^n = C_{n-k}^n$

2. $\sum_{k=0}^{n} C_k^n = 1$

Proof: We have shown that

$$C_k^n = \frac{1}{n}\int_0^n L_k \, du$$

$$= \frac{1}{n} \int_0^n \frac{u(u-1)\dots[u-(k-1)][u-(k+1)]\dots(u-n)}{k(k-1)\dots(+1)\cdot(-1)(k-n)}\,du$$

$$= \frac{1}{n} \int_0^n \frac{u(u-1)\dots[u-(k-1)][u-(k+1)]\dots(u-n)}{k!(-1)^{n-k}(n-k)!}\,du$$

Again $\qquad C_k^n = \dfrac{1}{n} \displaystyle\int_0^n \dfrac{u(u-1)\dots[u-(n-k-1)][u-(n-k+1)]\dots(u-n)}{(n-k)!k!(-1)^k}\,du$

Putting $u - n = -t$ or $u = n - t$, $du = -dt$, we get

$$C_k^n = \frac{1}{n} \int_0^n \frac{(n-t)(n-t-1)\dots(-t+k-1)(-t+k-1)(-t)}{k!(n-k)!(-1)^k}(-dt)$$

$$= \frac{1}{n} \int_0^n \frac{(-1)^n(t-1)\dots(t-k+1)(t-k-1)(t-n)}{k!(n-k)!(-1)^k}(-dt)$$

$$= \frac{1}{n} \int_0^n \frac{(-1)^{n-k} t(t-1)\dots[t-(k-1)][(t-(k+1)\dots(t-n)]}{k!(n-k)!}(dt)$$

$$= \frac{1}{n} \int_0^n \frac{t(t-1)\dots[t-(k-1)][(t-(k+1)\dots(t-n)]}{k!(n-k)!(-1)^{n-k}}(dt)$$

[Multiplying numerator and denominator by $(-1)^{n-k}$]

$$= C_{n-k}^n$$

$$\int_a^b f(x)dx = \int_a^b f(t)dt, \text{ by the properties of definite integrals}$$

Some deduction from Cotes' formula: On putting $n = 1$ in Eq. (7.20), we get

$$I = \int_{x_o}^{x_1} f(x)dx = 1\cdot h \sum_{k=0}^{n} C_k^1 y_k = h[C_0^1 y_0 + C_1^1 y_1]$$

Again by $\qquad C_k^n = \dfrac{1}{n} \displaystyle\int_0^n L_k\,du$, we have

$$C_0^1 = \frac{1}{1}\int_0^1 L_0\,du = \int_0^1 \frac{(u-1)}{0-1}\,du$$

$$= (-1)\int_0^1 (u-1)du = (-1)\left(\frac{u^2}{2} - u\right)_0^1 = \frac{1}{2}$$

$$C_1^1 = \frac{1}{1}\int_0^1 L_1 du = \int_0^1 \frac{(u-0)}{(1-0)}\,du = \int_0^1 u\,du = \frac{1}{2}$$

Thus $I = \dfrac{h}{2}[y_0 + y_1]$. This is trapezoidal rule.

Again putting $n = 2$ in Eq. (7.20), we get

$$I = \int_{x_0}^{x_0+nh} f(x)dx = 2h \sum_{k=0}^{2} C_k^2 y_k = 2h[y_0 C_0^2 + y_1 C_1^2 + y_2 C_2^2]$$

Now

$$C_0^2 = \frac{1}{2}\int_0^2 L_0 \, du = \frac{1}{2}\int_0^2 \frac{(u-1)(u-2)}{(0-1)(0-2)} \, du = \frac{1}{6}$$

$$C_1^2 = \frac{1}{2}\int_0^2 L_1 \, du = \frac{1}{2}\int_0^2 \frac{u(u-2)}{(1-0)(1-2)} \, du = \frac{2}{3}$$

$$C_2^2 = \frac{1}{6} \text{ by using } (C_k^n = C_{n-k}^n)$$

Hence

$$I = 2h\left(\frac{1}{6}y_0 + \frac{2}{3}y_1 + \frac{1}{6}y_2\right) = \frac{h}{3}(y_0 + 4y_1 + y_2)$$

This is Simpson's 1/3 rule.

Similarly putting $n = 3$ in Eq. (7.20), we can find the Simpson's 3/8 rule.

Putting $n = 4$ in Eq. (7.20), we can find the Boole's rule and putting $n = 6$ in Eq. (7.20), we can find the Weddle's rule.

Below, we give the Cotes' numbers for some value of n.

	C_0^n	C_1^n	C_2^n	C_3^n	C_4^n	C_5^n	C_6^n
$n = 1$	1/2	1/2					
$n = 2$	1/6	4/6	1/6				
$n = 3$	1/8	3/8	3/8	1/8			
$n = 4$	7/90	32/90	12/90	32/90	7/90		
$n = 5$	19/208	75/288	50/288	50/288	75/288	19/288	
$n = 6$	41/840	216/840	27/840	278/840	27/840	216/840	41/800

Example 7.9: Evaluate $\int_0^4 e^x dx$ by Simpson's rule using $e^1 = 2.72$, $e^2 = 7.39$, $e^3 = 20.09$, $e^4 = 54.60$.

Solution: Here $y = f(x) = e^x$. Divide the interval (0.4) into four equal intervals $(0, 1)$, $(1, 2)$, $(2, 3)$, $(3, 4)$ which are even and $h = 1$, we apply 1.3 Simpson's rule.

x	$y = f(x)$	Odd ordinates	Even ordinates	Extreme ordinates
$x_0 = 0$	$y_0 = e^0 = 1$			1
$x_1 = 1$	$y_1 = e^1 = 2.72$	2.72		
$x_2 = 2$	$y_2 = e^2 = 7.39$		7.39	
$x_3 = 3$	$y_3 = e^3 = 20.09$	20.09		
$x_4 = 4$	$y_4 = e^4 = 54.60$			54.60
	Total	22.81	7.397	55.60

By 1/3 Simpson's rule

$$\int_0^4 e^x dx = \frac{h}{3}[\text{sum of extreme ordinates} + 4 \text{ (odd ordinate)} + 2 \text{ (even ordinate)}]$$

$$= \frac{1}{3}[55.60 + 4(22.81) + 2(7.397)] = 53.872$$

Example 7.10: Calculate $\int_0^{\frac{\pi}{2}} e^{\sin x} dx$ correct to four decimal places, using Simpson's 3/8th rule.

Solution: Let us divide the interval into three equal subintervals by the ordinates $x_0 = 0$, $x_1 = \pi/6$, $x_2 = 2\pi/6$, $x_3 = 3\pi/6$.

x	$y = f(x)$	
$x_0 = 0$	$y_0 = e^0 = 1$	Hence $h = \pi/6$
$x_1 = \pi/6$	$y_1 = e^{\sin \pi/6} = e^{1/2} = 1.64872$	
$x_2 = \pi/3$	$y_2 = e^{\sin \pi/3} = e^{\sqrt{3}/2} = 2.3632$	
$x_3 = \pi/2$	$y_3 = e^{\sin \pi/2} = e = 2.71828$	

By Simpson's 3/8th rule

$$\int_0^{\frac{\pi}{2}} e^{\sin x} dx = \frac{3h}{8}[(y_0 + y_3) + 3(y_1 + y_2)]$$

$$= \frac{3 \times \frac{\pi}{2}}{8}[(1 + 2.71828) + 3(1.64872 + 2.3622)] = 0.091121$$

Example 7.11: Evaluate $\int_0^1 e^{-x^2} dx$ by Simpson's rule.

Solution: Divide the interval $(0, 1)$ into ten equal subintervals.

x	x^2	$y = e^{-x^2}$	Odd ordinates	Even ordinates	Extreme ordinates
$x_0 = 0.0$	0.0	$y_0 = e^0 = 1$			1.00000
$x_1 = 0.1$	0.01	$y_1 = e^{-0.01} = 0.99005$	0.99005		
$x_2 = 0.2$	0.04	$y_2 = e^{-0.04} = 0.96080$		0.96080	
$x_3 = 0.3$	0.09	$y_3 = e^{-0.09} = 0.91391$	0.91391		
$x_4 = 0.4$	0.016	$y_4 = e^{-0.16} = 0.85214$		0.85214	
$x_5 = 0.5$	0.25	$y_5 = e^{-0.25} = 0.77880$	0.77880		
$x_6 = 0.6$	0.36	$y_6 = e^{-0.36} = 0.69768$		0.69768	
$x_7 = 0.7$	0.49	$y_7 = e^{-0.49} = 0.61263$	0.61263		
$x_8 = 0.8$	0.64	$y_8 = e^{-0.64} = 0.5273$		0.52730	
$x_9 = 0.9$	0.81	$y_9 = e^{-0.81} = 0.4449$	0.44492		
$x_{10} = 1.0$	1.0	$y_{10} = e^{-1} = 0.36788$			0.36788
		Total	3.74031	3.03792	1.36788

By Simpson's 1/3rd rule, here $h = 0.1$

$$\int_0^1 e^{-x^2} dx = \frac{h}{3}[1.36788 + 4(3.74031) + 2(3.0379)]$$

$$= \frac{1}{3}[1.36788 + 14.96124 + 6.07584] = 0.746832$$

Example 7.12: The velocity v (km/min) of a moped which starts from rest at fixed intervals of time t (minute) as follows.

t	2	4	6	8	10	12	14	16	18	20
v	10	18	25	29	32	20	11	5	2	0

Find the distance travelled in 20 minutes.

Solution: We have $\dfrac{ds}{dt} = v$

$$[s]_0^{20} = \int_0^{20} v\, dt$$

Let us divide the interval (0, 20) into 10 equal subintervals.

t	v	Extreme ordinate	Odd ordinates	Even ordinates
$t_0 = 0$	$v_0 = 0$	0		
$t_1 = 2$	$v_1 = 10$		10	
$t_2 = 4$	$v_2 = 18$			18
$t_3 = 6$	$v_3 = 25$		25	
$t_4 = 8$	$v_4 = 29$			29
$t_5 = 10$	$v_5 = 32$		32	
$t_6 = 12$	$v_6 = 20$			20
$t_7 = 14$	$v_7 = 11$		11	
$t_8 = 16$	$v_8 = 5$			5
$t_9 = 18$	$v_9 = 2$		2	
$t_{10} = 20$	$v_{10} = 0$	0		
	Total	0	80	72

By Simpson's 1/3rd rule, here $h = 2$

$$[s]_0^{20} = \int_0^{20} v\, dt = \frac{h}{3}[0 + 4 \times 80 + 2 \times 72] = \frac{2}{3}[320 + 144] = \frac{2}{3}[464] = \frac{928}{3}$$

Distance = 309.3 km

Example 7.13: A curve is drawn to pass through the following points:

x	1	1.5	2	2.5	3	3.5	4
y	2	2.4	2.7	2.8	3	2.6	2.1

Estimate the area bounded by the curve, x-axis and the lines $x = 1$, $x = 4$. Also find the volume of solid generating by revolving this area using Weddle's rules.

Solution: The area bounded by the curve, x-axis and the lines $x = 1$, $x = 4$ is given by

$$A = \int_1^4 y\, dx = \frac{3h}{10}[y_0 + 5y_1 + y_2 + 6y_3 + y_4 + 5y_5 + 2y_6]$$

$$= 3 \times \frac{0.5}{10}[2 + 5(2.4) + 2.7 + 6(2.8) + 3 + 5(2.6) + 2(2.1)]$$

$$= 8.055 \text{ square units}$$

Volume of the solid generated

$$V = \pi \int_1^4 y^2 dx = \pi \frac{3h}{10} [y_0^2 + 5y_1^2 + y_2^2 + 6y_3^2 + y_4^2 + 5y_5^2 + 2y_6^2]$$

$$= \pi \cdot \frac{3(0.5)}{10} [4 + 5(2.4)^2 + (2.7)^2 + 6(2.8)^2 + 9 + 5(2.6)^2 + 2(2.1)^2]$$

$$= \pi \times 20.8125 = 63.384397 \text{ cubic units}$$

Example 7.14: Evaluate $\int_0^{10} \frac{dx}{1+x^2}$ by using

1. Trapezoidal rule   2. Simpson's 1/3rd rule
3. Simpson's 3/8th rule   4. Weddle's rule
Compare the result with the actual values.

Solution: Taking $h = 1$, divide the whole range of integration [0, 10] into ten equal parts by 10 points $x_0 = 0$, $x_1 = 1$, $x_2 = 2$, $x_3 = 3$, $x_4 = 4$, $x_5 = 5$, $x_6 = 6$, $x_7 = 7$, $x_8 = 8$, $x_9 = 9$, $x_{10} = 10$.

x	$y = \dfrac{1}{1+x^2}$
0	$y_0 = 1$
1	$y_1 = 0.5$
2	$y_2 = 0.2$
3	$y_3 = 0.1$
4	$y_4 = 0.0588235$
5	$y_5 = 0.0384615$
6	$y_6 = 0.027027$
7	$y_7 = 0.02$
8	$y_8 = 0.0153846$
9	$y_9 = 0.0121951$
10	$y_{10} = 0.0099009901$

1. Trapezoidal rule

$$\int_0^{10} \frac{1}{1+x^2} dx = \frac{h}{2} [(y_0 + y_{10}) + 2(y_1 + y_2 + y_3 + y_4 + y_5 + y_6 + y_7 + y_8 + y_9)]$$

$$= \frac{1}{2} [(1 + 0.009900990) + 2(0.5 + 0.2 + 0.1 + 0.0588235 + 0.0384615$$

$$+ 0.027027 + 0.02 + 0.0153846 + 0.0121951)]$$

$$= 1.4768422$$

2. Simpson's 1/3rd rule

$$\int_0^{10} \frac{1}{1+x^2} dx = \frac{h}{2} [(y_0 + y_{10}) + 4(y_1 + y_3 + y_5 + y_7 + y_9) + 2(y_2 + y_4 + y_6 + y_8)]$$

$$= \frac{1}{3} [(1 + 0.009900990) + 4(0.5 + 0.1 + 0.0384615 + 0.02 + 0.0121951)$$

$$+ 2(0.2 + 0.0588235 + 0.027027 + 0.0153846]$$

$$= 1.4316659$$

3. **Simpson's 3/8th rule**

$$\int_0^{10} \frac{1}{1+x^2} dx = \frac{3h}{8}[(y_0 + y_{10}) + 3(y_1 + y_2 + y_4 + y_5 + y_7 + y_8) + 2(y_3 + y_6 + y_9)]$$

$$= \frac{3}{8}[(1 + 0.009900990) + 3(0.5 + 0.2 + 0.0588235 + 0.0384615$$

$$+ 0.02 + 0.0153846) + 2(0.1 + 0.027027 + 0.0121951]$$

$$= 1.4198828$$

4. **Weddle's rule**

$$\int_0^{10} \frac{1}{1+x^2} dx = \frac{3h}{8}[(y_0 + y_{10}) + (5y_1 + y_2 + 6y_3 + y_4 + 5y_5 + 2y_6 + 5y_7 + y_8 + 6y_9)]$$

$$= \frac{3}{10}[(1 + 0.009900990) + 5(0.5) + 0.2 + 6(0.1) + 0.0588235 + 5(0.0384615)$$

$$+ 2(0.027027) + 5(0.02) + 0.0153846 + 6(0.0121951)]$$

$$= 1.4410924$$

Now $\qquad \int_0^{10} \frac{1}{1+x^2} dx = [\tan^{-1} x]_0^{10} = \tan^{-1} 10 = 1.4711237$

which show that the integral found by Weddle's rule is the nearer to the actual value than other.

Example 7.15 Evaluate $\int_0^{10} e^x dx$ by Weddle's rule given that

$$e^0 = 1, e^1 = 2.72, e^2 = 7.39, e^3 = 20.09, e^4 = 54.60, e^5 = 148.41$$
$$e^6 = 403.43, e^7 = 1096.63, e^8 = 2980.96, e^9 = 8103.08, e^{10} = 22026.47$$

Solution: Here $h = 1$.

By Weddle's rule

$$\int_0^{10} e^x dx = \frac{3h}{10}[(y_0 + y_{10}) + 5y_1 + y_2 + 6y_3 + y_4 + 5y_5 + 2y_6 + 5y_7 + y_8 + 6y_9]$$

$$= \frac{3}{10}[(1 + 22026.47) + 5(2.72) + 7.39 + 6(20.09) + 54.60$$

$$+ 5(148.41) + 2(403.43) + 5(1096.63) + 2980.96 + 6(8103.08)]$$

$$= 24256.53$$

Example 7.16: Evaluate $\int_{0.2}^{1.4} (\sin x - \log_e x + e^x) dx$ by

1. Trapezoidal rule
2. Simpson's 1/3rd rule
3. Simpson's 3/8th rule
4. Weddle's rule

Solution: Let $y = f(x) = \sin x - \log_e x + e^x$. Divide the interval (0.2, 1.4) into six equal parts by the ordinates at point.

$$x_0 = 0.2, x_1 = 0.4, x_2 = 0.6, x_3 = 0.8,$$
$$x_4 = 1.00, x_5 = 1.2, x_6 = 1.4$$

so that

x	$\sin x$	$-\log_e x$	e^x	y	Trapezoidal rule	Simpson's 1/3rd rule	Simpson's 1/3rd rule	Weddle's 3/8th rule
$x_0 = 0.2$	0.1987	−1.6095	1.2214	$y_0 = 3.0296$	$y_0 = 3.0296$	$y_0 = 3.0296$	$y_0 = 3.0296$	$y_0 = 3.0296$
$x_1 = 0.4$	0.3894	−0.9163	1.4918	$y_1 = 2.79752$	$y_1 = 5.5950$	$4y_1 = 11.1900$	$3y_1 = 8.3925$	$5y_1 = 13.9875$
$x_2 = 0.6$	0.5646	−0.5108	1.8221	$y_2 = 2.8975$	$2y_2 = 5.7950$	$2y_2 = 5.7889$	$3y_2 = 8.6925$	$y_2 = 2.8975$
$x_3 = 0.8$	0.7174	−0.2232	2.2255	$y_3 = 3.1661$	$2y_3 = 6.3322$	$4y_3 = 12.6644$	$2y_3 = 6.3322$	$6y_3 = 18.9966$
$x_4 = 1.0$	0.8415	−0.000	2.7183	$y_4 = 3.5598$	$2y_4 = 7.1196$	$2y_4 = 7.1196$	$2y_4 = 10.6794$	$y_4 = = 3.5598$
$x_5 = 1.2$	0.9320	−0.1824	3.3201	$y_5 = 4.0698$	$2y_5 = 8.1396$	$4y_5 = 16.2792$	$3y_5 = 12.2074$	$5y_5 = 10.3490$
$x_6 = 1.4$	0.9855	0.3365	4.0552	$y_6 = 4.7042$	$y_6 = 4.7042$	$y_6 = 4.7042$	$y_6 = 4.7042$	$y_6 = 4.7042$
					40.7152	60.7820	54.0398	67.5242

1. **By trapezoidal rule**

$$\int_{0.2}^{1.4} f(x)\,dx = \frac{h}{2}[40.7152] = \frac{0.2}{2}[40.7152] = 4.0715$$

2. **By Simpson's 1/3 rule**

$$\int_{0.2}^{1.4} f(x)\,dx = \frac{h}{3}[64.7820] = \frac{0.2}{3}[64.7820] = 4.0521$$

3. **By Simpson's 3/8 rule**

$$\int_{0.2}^{1.4} f(x)\,dx = \frac{3h}{8}[54.0398] = \frac{3}{8}(0.2)[54.0398] = 4.0529$$

4. **By Weddle's rule**

$$\int_{0.2}^{1.4} f(x)\,dx = \frac{3h}{10}[67.5242] = \frac{3}{10} \times 0.2\,[67.5242] = 4.0515$$

Example 7.17: A rocket is launched from the ground. Its acceleration is registered during the first 80 seconds and is given in the table below. Using Simpson's 1/3rd rule, find the velocity of the rocket at $t = 80$.

$t(s)$	0	10	20	30	40	50	60	70	80
$f(\text{cm/sec}^2)$	30	31.63	33.64	35.47	37.75	40.33	43.50	46.69	50.67

Solution: Now $\dfrac{dv}{dt} = f, h = 10$

$$[v]_0^{80} = \int_0^{80} f\,dt$$

$$= \frac{h}{3}[(f_0 + f_8) + 4(f_1 + f_3 + f_5 + f_7) + 2(f_2 + f_4 + f_6)]$$

$$= \frac{10}{3}[(30 + 50.67) + 4(31.63 + 35.47 + 40.33 + 46.69) + 2(33.64 + 37.75 + 43.50)]$$

$$= \frac{10}{3}[80.67 + 616.76 + 228.68] = \frac{926.11}{3}$$

$$= 3087.3 \text{ cm/sec}^2 = 30.87 \text{ m/sec}^2$$

Example 7.18: A reservoir discharging water through sluices at a depth h feet below the water surface has a surface area A for various values of h as given below.

$h\,(f)$	10	11	12	13	14
A (in sq ft)	950	1070	1200	1350	1530

If t denotes time in minutes, the rate of the fall of the surface is given by $\dfrac{dh}{dt} = \dfrac{-48\sqrt{h}}{A}$.

Estimate the time taken for the water level to fall from 14 to 10 feet above the sluices.

Solution:
$$\frac{A}{\sqrt{h}} = \frac{950}{\sqrt{10}}, \frac{1070}{\sqrt{11}}, \frac{1200}{\sqrt{12}}, \frac{1350}{\sqrt{13}}, \frac{1530}{\sqrt{14}}$$

$$= 300.41, 322.61, 346.41, 374.47, 408.90$$

Now
$$t = \int_{14}^{10} \frac{A}{-48\sqrt{h}} dh$$

By Simpson's 1/3rd rule, we have
$$h = -1$$

$$t = \int_{14}^{10} \frac{A}{-48\sqrt{h}} dh$$

$$= -\frac{1}{48} \cdot \frac{-1}{3} [(300.41 + 408.90) + 4(322.61 + 374.41) + 2(346.41)]$$

$$= \frac{1}{144} [709.31 + 2788.08 + 692.82] = \frac{4190.21}{144} = 29.09$$

$$= 29 \text{ minutes (approx)}$$

Time taken for water to fall from 14 to 10 feet above the sluices = 29 minutes (approx.)

Example 7.19: Integrate numerically $\int_0^{\frac{\pi}{2}} \sqrt{\cos\theta}\, d\theta$.

Solution:
$$x = 0, y = \sqrt{\cos\theta}$$

x	y
$x_0 = 0$	$y_0 = 1$
$x_1 = \dfrac{\pi}{6}$	$y_1 = 0.930$
$x_2 = \dfrac{2\pi}{6}$	$y_2 = 0.707$
$x_3 = \dfrac{3\pi}{6}$	$y_3 = 0$

$$\int_0^{\frac{\pi}{2}} \sqrt{\cos\theta}\, d\theta = \frac{3}{8}(h)[(y_0 + y_3) + 3(y_1 + y_2)]$$

$$= \frac{3}{8}\left(\frac{\pi}{6}\right)[1 + 3(1.637)] = \frac{\pi}{16}[5.911] = 1.1600$$

Example 7.20:

$x = 1$	1.5	2	2.5	3	3.5	4
$y = 2$	2.4	2.7	2.8	3.0	2.6	2.1

Estimate the area bounded by the curve, x-axis to the lines $x = 1, x = 4$

Solution:
$$\text{Area} = \int_1^4 y\, dx = \frac{3}{8}h[(y_0 + y_6) + 3(y_1 + y_2 + y_4 + y_5) + 2 \times y_3]$$

$$= \frac{3}{8}(0.5)[(2 + 2.1) + 3(2.4 + 2.7 + 3 + 2.67) + 2(2.8)]$$

$$= \frac{3}{8}(0.5)[4.1 + 3(10.7) + 5.6] = \frac{3}{8}(0.5)[41.8] = 7.8375$$

Example 7.21: A curve is given by

x	0	1	2	3	4	5	6
y	0	2	2.5	2.3	2	1.7	1.5

The x-coordinate of CG of the area bounded by the curve, the end coordinates and x-axis is given by $A\bar{x} = \int_0^6 xy\,dx$, where A is the area. Find $\bar{x}$ by using Simpson's rule.

Solution: Now the area $A = \int_0^6 y\,dx$

By Simpson's 1/3rd rule

$$= \frac{1}{3}[(y_0 + y_6) + 4(y_1 + y_3 + y_5) + 2(y_2 + y_4)]$$

$$= \frac{1}{3}[(0.0 + 1.5) + 4(2 + 2.3 + 1.7) + 2(2.5 + 2)]$$

$$= \frac{1}{3}[1.5 + 24 + 9] = \frac{34.5}{3} = 11.5$$

Now

x	0	1	2	3	4	5	6
xy	0	2	6	6.9	8	8.5	9

By Simpson's 1/3rd rule

$$\int_0^6 xy\,dy = \frac{1}{3}[(0 + 9) + 4(17.4) + 2(13)] = \frac{1}{3}[140.6] = 34.866$$

$$A\bar{x} = \int_0^6 xy\,dx$$

$$(11.5)\bar{x} = 34.886$$

$$\bar{x} = \frac{34.866}{11.5} = 3.0318$$

Example 7.22: A body is in the form of a solid of revolution. The diameter D in cm of its section at distance x cm from one end are given below. Estimate volume of the solid.

x	0	2.5	5.0	7.4	10.0	12.5	15.0
D	5	5.5	6.0	6.75	6.25	15.5	4.0
y	25	30.25	36	45.56	39.06	36.25	16.0

Solution: Volume of revolution $= \pi \int_0^{15} r^2\,dx$

$$\pi \int_0^{15} \left(\frac{D}{2}\right)^2 dx = \frac{\pi}{4} \int_0^{15} D^2\,dx \quad h = 2.5$$

By Simpson's 1/3rd rule

$$\text{Volume} = \frac{\pi}{4}\left(\frac{2.5}{3}\right)[(25+16)+4(30.25+45.56+36.25)+2(36+39.06)]$$

$$= \frac{\pi}{12}(2.5)[41+424.24+150.12] = \frac{\pi}{12} \times 2.5[615.36] = 402.5$$

PROBLEMS 7.2

1. Find the integral $\int_{1.0}^{1.8} \frac{e^x + e^{-x}}{2} dx$ using Simpson's 1/3 rule by taking $h = 0.2$

[Ans. 1.7670]

2. Find the integral $\int_0^6 [f(x)]^2 dx$ using Simpson's 1/3 rule, given that

x	0	1	2	3	4	5	6
f(x)	1	0	1	4	9	16	25

[Ans. 626]

3. Compute the integral using Simpson's 3/8 rule $\int_0^{\frac{\pi}{2}} \frac{dx}{\sin^2 x + \frac{1}{4}\cos^2 x}$. [Ans. 3.1416]

4. Compute the integral $I = \sqrt{\frac{2}{\pi}} \int_0^1 e^{-\frac{x^2}{2}} dx$ using Simpson's 1/3 rule taking $h = 0.125$.

[Ans. 0.6827]

5. A river is 80 feet wide. Depth d in feet at a distance of x feet from one bank is given by the following table.

x	0	10	20	30	40	50	60	70	80
y = d	0	4	7	9	12	15	14	8	3

Find the approximate area of the cross-section. [Ans. 710 sq feet]

6. When a train is moving at 30 minutes an hour, stream is shut off and brakes are applied. The speed of the train in miles per hour after t seconds is given below.

t	0	5	10	15	20	24	30	35	40
v	30	24	19.5	16	13.6	11.7	10.0	8.5	7.0

Determine how far the train has moved in 40 seconds. [Ans. 296.7 yards]

7. A curve passes through the points (1, 0.2) (2, 0.7) (3, 1) (4, 1.3) (5, 1.5) (6, 1.7) (7, 1.9) (8, 2.1) (9, 2.3) using Weddle's rule, estimate the volume generated by revolving the area between the curve, x-axis and the ordinates $x = 1$ and $x = 9$ about the x-axis.

[Ans. 59.68 cu units]

8. The following table gives the velocity v of a particle at time t.

t (seconds)	0	2	4	6	8	10	12
v (m/sec)	4	6	16	34	60	94	136

Find the distance moved by the particle in 12 seconds and also the acceleration at $t = 2$ sec. [Ans. 552 m, 4 m/sec²]

9. Evaluate by (i) Trapezoidal rule (ii) Simpson's 1/3 rule (iii) Simpson's 3/8 rule, (iv) Weddle's rule $\int_4^{5.2} \log x \, dx$. Given that

x	4.0	4.2	4.4	4.6	4.8	5.0	5.2
Log x	1.3863	1.4351	1.4816	1.5261	1.5686	1.6094	1.6484

[Ans. 1.8276551, 1.8278472, 1.8278470, 1.8278474]

7.8 ERRORS IN QUADRATURE FORMULA

$P(x)$ is a polynomial representing $y = f(x)$ in the interval $[a, b]$.

7.8.1 Errors in Trapezoidal Rule

Divide the interval $[a, b]$ into n subintervals $(x_0, x_1), (x_1, x_2), ..., (x_{n-1}, x_n)$, where $a = x_0$ and $b = x_n$.

Error in the interval $(x_0, x_1) = \int_{x_0}^{x_1} f(x)dx - \int_{x_0}^{x_1} P(x)dx$

Expanding $y = f(x)$ by Taylor's series around $x = x_0$, we have

$$y = f(x) = f(0) + (x - x_0) f'(0) + \frac{(x - x_0)^2}{2!} f''(0) + \frac{(x - x_0)^3}{3!} f'''(0) + ...$$

$$= y_0 + (x - x_0) y'(0) + \frac{(x - x_0)^2}{2!} y''(0) + \frac{(x - x_0)^3}{3!} y'''(0) + ...$$

$$\int_{x_0}^{x_1} f(x)dx = \int_{x_0}^{x_0+h} \left[y_0 + (x - x_0)y_0' + \frac{(x - x_0)^2}{2!} y_0'' + \frac{(x - x_0)^3}{3!} y_0''' + ... \right] dx$$

$$= \left[y_0 + \frac{(x - x_0)^2}{2!} y_0' + \frac{(x - x_0)^3}{3!} y_0'' + \frac{(x - x_0)^4}{4!} y_0''' + ... \right]_{x_0}^{x_0+h}$$

$$= y_0 h + \frac{h^2}{2!} y_0' + \frac{h^3}{3!} y_0'' + \frac{h^4}{4!} y_0''' + ...$$

Also $\int_{x_0}^{x_1} P(x)dx = \frac{h}{2}[y_0 + y_1]$ by trapezoidal rule

$$= \frac{h}{2}[(f(x_0) + f(x_0 + h)] = \frac{h}{2}\left[y_0 + y_0 + hy_0' + \frac{h^2}{2!} y_0'' + \frac{h^3}{3!} y_0''' + ... \right]$$

$$= \frac{h}{2}\left[2y_0 + hy_0' + \frac{h^2}{2!} y_0'' + \frac{h^3}{3!} y_0''' + ... \right]$$

$$= hy_0 + \frac{h^2}{2!} y_0' + \frac{h^3}{2.2!} y_0'' + \frac{h^4}{2.3!} y_0''' + ...$$

Error in $(x_0, x_1) = \int_{x_0}^{x_1} f(x)dx - \int_{x_0}^{x_1} P(x)dx$

$$= h^3 \left(\frac{1}{3!} - \frac{1}{2.2!} \right) y_0'' + ... = -\frac{1}{12} h^2 y_0'' + ...$$

Principal part of the error in $[x_0, x_1] = -\frac{1}{12} h^3 y_0''$

Principal part of the error in $[x_1, x_2] = -\frac{1}{12} h^3 y_1''$

Principal part of the error in $[x_2, x_3] = -\dfrac{1}{12}h^3 y_2''$

Principal part of the error in $[x_{n-1}, x_n] = -\dfrac{h^3}{12}[y_{n-1}'']$

Let $y''(x)$ be the largest of $y_0'' + y_1'' + \dots + y_{n-1}''$

Total error $\qquad Z < -\dfrac{1}{12}ny''(x)h^3 = -\dfrac{1}{12}(b-a)h^2 y''(x) \quad [\because nh = b - a]$

Hence, the error in the trapezoidal rule is of the order h^2.

7.8.2 Errors in Simpson's 1/3 Rule

Error in the interval (x_0, x_2) is $\displaystyle\int_{x_0}^{x_0+2h} f(x)\,dx - \int_{x_0}^{x_0+2h} P(x)\,dx$.

Now by Taylor's series

$$y = f(x) = y_0 + (x - x_0)y_0' + \frac{(x-x_0)^2}{2!} y_0'' + \frac{(x-x_0)^3}{3!} y_0''' + \dots$$

$$\int_{x_0}^{x_2} f(x)\,dx = \int_{x_0}^{x_0+2h} \left[y_0 + (x-x_0)y_0' + \frac{(x-x_0)^2}{2!} y_0'' + \frac{(x-x_0)^3}{3!} y_0''' + \frac{(x-x_0)^4}{4!} y_0'''' + \dots \right] dx$$

$$= 2hy_0 + \frac{4h^2}{2!} y_0' + \frac{8h^3}{3!} y_0'' + \frac{16h^4}{4!} y_0''' + \frac{32h^5}{5!} y_0'''' + \dots$$

Also

$$\int_{x_0}^{x_0+2h} P(x)\,dx = \frac{h}{3}[y_0 + 4y_1 + y_2] \text{ by Simpson's 1/3rd rule}$$

$$= \frac{h}{3}[y_0 + 4(f(x_0 + h)) + f(x_0 + 2h)]$$

$$= \frac{h}{3}\left[y_0 + 4\left(y_0 + hy_0' + \frac{h^2}{2!} y_0'' + \frac{h^3}{3!} y_0''' + \frac{h^4}{4!} y_0'''' \right) \right.$$

$$\left. + \left(y_0 + 2hy_0' + \frac{4h^2}{2!} y_0'' + \frac{8h^3}{3!} y_0''' + \frac{16h^4}{4!} y_0'''' \right) \right]$$

$$= \frac{h}{3}\left[6y_0 + y_0'h(4+2) + y_0'' \frac{h^2}{2!}(4+4) + y_0''' \frac{h^3}{3!}(4+8) + y_0'''' \frac{h^4}{4!}(4+16) + \dots \right]$$

$$= \frac{h}{3}\left[6y_0 + 6y_0'h + 4y_0''h^2 + 2y_0'''h^3 + \frac{20}{24} y_0''''h^4 + \dots \right]$$

$$= 2y_0h + 2y_0'h^2 + \frac{4}{3}y_0''h^3 + \frac{2}{3}y_0'''h^4 + \frac{5}{18}y_0''''h^5 + \dots$$

$$E_1 = \int_{x_0}^{x_0+2h} f(x)dx - \int_{x_0}^{x_0+2h} P(x)dx = \left(\frac{32}{5!} - \frac{5}{18}\right) y_0'''' h^5$$

$$= \left(\frac{4}{15} - \frac{5}{18}\right) h^5 y_0'''' + \ldots = -\frac{h^5}{90} y_0'''' + \ldots$$

Principle part of the error in (x_0, x_2) $= -\dfrac{h^5}{90} y_0''''$

Principle part of the error in (x_2, x_4) $= -\dfrac{h^5}{90} y_0''''$

Principle part of the error in (x_{n-2}, x_n) $= -\dfrac{h^5}{90} y_{n-2}''''$

$$\text{Total error} = -\frac{h^5}{90}[y_0'''' + y_2'''' \ldots y_{n-2}''''] \cong -\frac{h^5}{90} n y''''(x)$$

7.8.3 Errors in Simpson's 3/8 Rule

$$\int_{x_0}^{x_0+3h} f(x)dx - \int_{x_0}^{\tilde{x}_0+3h} P(x)dx = \left[xy_0 + \frac{(x-x_0)^2}{2!} y_0' + \frac{(x-x_0)^3}{3!} y_0'' + \frac{(x-x_0)^4}{4!} y_0''' + \ldots \right]_{x_0}^{x_0+3h}$$

$$- \frac{3h}{8}(y_0 + 3y_1 + 3y_2 + y_3)$$

$$= 3hy_0 + \frac{(3h)^2}{2!} y_0' + \frac{(3h)^3}{3!} y_0'' + \frac{(3h)^4}{4!} y_0''' + \ldots$$

$$- \frac{3h}{8}[y_0 + 3f(x_0+h) + 3f(x_0+2h) + f(x_0+3h)]$$

$$= 3hy_0 + \frac{(3h)^2}{2!} y_0' + \frac{(3h)^3}{3!} y_0'' + \frac{(3h)^4}{4!} y_0''' + \ldots$$

$$- \frac{3h}{8}\left[y_0 + 3\left(y_0 + hy_0' + \frac{h^2}{2!} y_0'' + \frac{h^3}{3!} y_0''' + \ldots \right) \right.$$

$$+ 3\left(y_0 + 2hy_0' + \frac{(2h)^2}{2!} y_0'' \right)$$

$$- \frac{3h}{8}\left[y_0 + 3\left(y_0 + hy_0' + \frac{h^2}{2!} y_0'' + \frac{h^3}{3!} y_0''' + \ldots \right) \right.$$

$$+ 3\left(y_o + 2hy_0' + \frac{(2h)^2}{2!} y_0'' + \frac{(2h)^3}{3!} y_0''' + \ldots \right)$$

$$\left. + y_0 + 3hy_0' + \frac{(3h)^2}{2!} y_0'' + \frac{(3h)^3}{3!} y_0''' + \ldots \right]$$

On simplifying, we get

$$\int_{x_0}^{x_0+3h} f(x)\,dx - \int_{x_0}^{x_0+3h} P(x)\,dx = -\frac{3h^5}{80}\, y_0'''' + ...$$

$$\int_{x_0+3h}^{x_0+6h} f(x)\,dx - \int_{x_0+3h}^{x_0+6h} P(x)\,dx = -\frac{3h^5}{80}\, y_3'''' + ...$$

Adding, we get total error $Z < -\dfrac{3h^5}{80}\, y''''(x)$

Similarly, Boole's and Weddle's rule can be established.

7.8.4 Error in Boole's Rule

In this case, the principal part of the error in the interval

$$[x_0,\ x_4] = \frac{-8h^7}{945}\, y''''''$$

7.8.5 Error in Weddle's Rule

In this case, principal part of the error in error in the interval

$$[x_0,\ x_6] = \frac{-h^7}{140}\, y''''''$$

7.9 ROMBERG'S METHOD OF INTEGRATION

This method provides a simple modification to the quadrature formula for finding their better approximation. Let us improve upon the value of integral

$$I = \int_a^b f(x)\,dx$$

By trapezoidal rule: If I_1, I_2 be the value of I with subintervals of width h_1 and h_2 and E_1 and E_2 be their corresponding errors respectively. Then

$$E_1 = \frac{-(b-a)}{12}\, h_1^2 y''(\bar{X})$$

$$E_2 = \frac{-(b-a)}{12}\, h_2^2 y''(\bar{X})$$

where $y''(\bar{X})$ is also the largest value as $y''(X)$, we can assume that $y''(X)$ and $y''(\bar{X})$ are nearly equal.

$\therefore$

$$\frac{E_1}{E_2} = \frac{h_1^2}{h_2^2} \text{ or } \frac{E_1}{E_2 - E_1} = \frac{h_1^2}{h_2^2 - h_1^2} \qquad (7.21)$$

$$\frac{E_2}{E_1} - 1 = \frac{h_2^2}{h_1^2} - 1$$

As

$$I = I_1 + E_1 = I_2 + E_2 \text{ and } E_2 - E_1 = I_1 - I_2 \qquad (7.22)$$

From Eqs (7.21) and (7.22), we have

$$E_1 = \frac{h_1^2}{h_2^2 - h_1^2}(l_1 - l_2)$$

Hence

$$I = I_1 + E_1 = I_1 + \frac{h_1^2}{h_2^2 - h_1^2}(I_1 - I_2)$$

$$= \frac{I_1 h_2^2 - I_2 h_1^2}{h_2^2 - h_1^2} \tag{7.23}$$

which is a better approximation of I.

To find I systematically, we take

$$h_1 = h \text{ and } h_2 = \frac{1}{2}h.$$

So that Eq. (7.23) gives $I = \dfrac{I_1\left(\dfrac{h}{2}\right)^2 - I_2 h_2^2}{\left(\dfrac{h}{2}\right)^2 - h^2} = \dfrac{4I_2 - I_1}{3}$

i.e.

$$I\left(h, \frac{h}{2}\right) = \frac{1}{3}\left[4I\left(\frac{h}{2}\right) - I(h)\right] \tag{7.24}$$

We got the result by applying trapezoidal rule.

Now we use the trapezoidal rule several times successively halving h and apply Eq. (7.24) to each pair of values as per the following scheme:

$I(h)$			
	$I(h, h/2)$		
$I(h/2)$		$I(h, h/2, h/4)$	
	$I(h/2, h/4)$		$I(h, h/2, h/4, h/8)$
$I(h/4)$		$I(h/2, h/4, h/8)$	
	$I(h/4, h/8)$		
$I(h/8)$			

The computation is continued till successive values are close to each other. This method is called Richardson's deferred approach to the limit and its systematic refinement is called Romberg's method.

Example 7.23: Use Romberg's method to compute $\int_0^1 \dfrac{dx}{1+x}$. Hence find $\log_e 2$ correct to two decimal places.

Solution: We take $h = 0.5, 0.25, 0.125$ successively and evaluate the given integral using trapezoidal rule.

i. When $h = 0.5$, the value of $y = (1+x)^{-1}$ are

x	0	0.5	1
$1/1+x = y$	1	0.666	0.5

Trapezoidal rule

$$I(0.5) = \int_0^1 \frac{dx}{1+x} = \frac{0.5}{2}[1+5+2(0.666)] = 0.708$$

ii. **W hen** $h = 0.25$, the value of $y = (1+x)^{-1}$ are

x	0	0.25	0.5	0.75	1
y	1	0.8	0.666	0.571	0.5

$$I(0.25) = \frac{0.25}{2}[(1+0.5)+2(0.8+0.666+0.571)] = 0.69675$$

iii. When $h = 0.125$, the value of $y = (1+x)^{-1}$ are

x	0	0.125	0.25	0.375	0.5	0.625	0.75	0.875	1
y	1	0.88	0.8	0.7272	0.66	0.615	0.571	0.53	0.5

$$I(0.125) = \frac{0.125}{2}[(1+0.5)+2(0.88+0.8+0.7272+0.66+0.615+0.571+0.53)]$$

$$= 0.69165$$

Thus, we have

$$I(h) = 0.708, \ I(h/2) = 0.69675, \ I(h/4) = 0.69165$$

$$I\left(h, \frac{h}{2}\right) = \frac{1}{3}\left[4I\left(\frac{h}{2}\right) - I(h)\right] = \frac{1}{3}[4 \times 0.69675 - 0.708] = 0.693$$

$$I\left(\frac{h}{2}, \frac{h}{4}\right) = \frac{1}{3}\left[4I\left(\frac{h}{4}\right) - I\left(\frac{h}{2}\right)\right] = \frac{1}{3}[4 \times 0.69165 - 0.69675] = 0.68995$$

$$I\left(h, \frac{h}{2}, \frac{h}{4}\right) = \frac{1}{3}\left[4I\left(\frac{h}{2}, \frac{h}{4}\right) - I\left(h, \frac{h}{2}\right)\right] = \frac{1}{3}[4 \times 0.68995 - 0.693] = 0.68893$$

The value of the integral = 0.6893.

Example 7.24: Use Romberg's method to find $\int_0^1 \frac{dx}{1+x^2}$ correct to four decimal places.

Solution: We take $h = 0.5, 0.25$ and 0.125 successively and find the given integral using trapezoidal rule.

1. When $h = 0.5$, the value of $y = (1+x^2)^{-1}$ are

x	0	0.5	1.0
y	1	0.8	0.5

$$I = \frac{0.5}{2}[1+0.5+2 \times 0.8] = 0.775$$

2. When $h = 0.25$, the value of $y = (1+x^2)^{-1}$ are

x	0	0.25	0.5	0.75	1.0
y	1	0.9412	0.8	0.64	0.5

$$I = \frac{0.25}{2}[1+0.5+2(0.9412+0.8+0.64)] = 0.7828$$

3. When $h = 0.125$, we find that $I = 0.7848$

Thus, we have $I(h) = 0.7750$, $I(h/2) = 0.7828$, $I(h/4) = 0.7848$

$$I\left(h, \frac{h}{2}\right) = \frac{1}{3}\left[4I\left(\frac{h}{2}\right) - I(h)\right] = \frac{1}{3}[3.1312 - 0.775] = 0.7854$$

$$I\left(\frac{h}{2}, \frac{h}{4}\right) = \frac{1}{3}\left[4I\left(\frac{h}{4}\right) - I\left(\frac{h}{2}\right)\right] = \frac{1}{3}[3.1392 - 0.7828] = 0.7855$$

$$I\left(h, \frac{h}{2}, \frac{h}{4}\right) = \frac{1}{3}\left[4I\left(\frac{h}{2}, \frac{h}{4}\right) - I\left(h, \frac{h}{2}\right)\right] = \frac{1}{3}[3.142 - 0.7534] = 0.7855$$

The value of integral $= 0.7855$

PROBLEMS 7.3

Use Romberg's method to evaluate the following, up to four decimal places.

1. $\int_0^1 e^x dx$ [Ans. 1.7183]

2. $\int_0^1 \cos x\, dx$ [Ans. 0.8415]

7.10 EULER-MACLAURIN FORMULA

Taking $\Delta F(x) = f(x)$, we define the inverse operator Δ^{-1} as

$$F(x) = \Delta^{-1} f(x) \qquad (7.25)$$

Now $\qquad F(x_1) - F(x_0) = \Delta F(x_0) = f(x_0)$

Similarly $\quad F(x_2) - F(x_1) = \Delta F(x_1) = f(x_1)$

$$... = ...$$
$$... = ...$$
$$... = ...$$

$$F(x_n) - F(x_{n-1}) = \Delta F(x_{n-1}) = f(x_{n-1})$$

Adding all these, we get

$$F(x_n) - F(x_0) = \sum_{i=0}^{n-1} f(x_i) \qquad (7.26)$$

where $x_0, x_1, ..., x_n$ are the $(n + 1)$ equispaced value of x with difference h.

From Eq. (7.25), we have

$$F(x) = \Delta^{-1} f(x) = (E - 1)^{-1} f(x) = (e^{hD} - 1)^{-1} f(x) \quad [\because E = e^{hD}]$$

$$= \left[\left(1 + hD + \frac{h^2 D^2}{2!} + \frac{h^3 D^3}{3!} + ...\right) - 1\right]^{-1} f(x)$$

$$= (hD)^{-1}\left[1 + \frac{hD}{2!} + \frac{h^2 D^2}{3!} + ...\right]^{-1} f(x)$$

$$= \frac{1}{h} D^{-1}\left[1 - \frac{hD}{2} + \frac{h^2 D^2}{12} - \frac{h^4 D^4}{720} + ...\right] f(x)$$

$$= \frac{1}{h} \int f(x)\,dx - \frac{1}{2} f(x) + \frac{h}{12} f'(x) - \frac{h^3}{720} f'''(x) + \dots \qquad (7.27)$$

Putting $x = x_n$ and $x = x_0$ in Eq. (7.27) and then subtracting, we get

$$F(x_n) - F(x_0) = \frac{1}{h} \int_{x_0}^{x_n} f(x)\,dx - \frac{1}{2}[f(x_n) - f(x_0)]$$

$$+ \frac{h}{12}[f'(x_n) - f'(x_0)] - \frac{h^3}{720}[f'''(x_n) - f'''(x_0)] + \dots \qquad (7.28)$$

From Eqs (7.25) and (7.28), we have

$$\sum_{i=0}^{n-1} f(x_i) = \frac{1}{h} \int_{x_0}^{x} f(x)\,dx - \frac{1}{2}[f(x_n) - f(x_0)]$$

$$+ \frac{h}{12}[f'(x_n) - f'(x_0)] - \frac{h^3}{720}[f'''(x_n) - f'''(x_0)] + \dots$$

$$\frac{1}{h} \int_{x_0}^{x_n} f(x)\,dx = \sum_{i=0}^{n-1} f(x_1) + \frac{1}{2}[f(x_n) - f(x_0)] - \frac{h}{12}[f'(x_n) - f'(x_0)]$$

$$+ \frac{h^3}{720}[f'''(x_n) - f'''(x_0)] + \dots$$

$$= \frac{1}{2}[f(x_0) + 2f(x_1) + 2f(x_2) + \dots + 2f(x_{n-1}) + f(x_n)]$$

$$- \frac{h}{12}[f'(x_n) - f'(x_0)] + \frac{h^3}{720}[f'''(x_n) - f'''(x_0)] + \dots$$

We can write as

$$\int_{x_0}^{x_0 + nh} y\,dx = \frac{h}{2}[y_0 + 2y_1 + 2y_2 + \dots + 2y_{n-1} + y_n]$$

$$- \frac{h^2}{12}(y'_n - y'_0) + \frac{h^4}{720}(y'''_n - y''') + \dots$$

which is called Euler–Maclaurin formula.

Note: This formula is often used to find the sum of a series of the form

$$y(x_0) + y(x_0 + h) + \dots + y(x_0 + nh)$$

Example 7.25: Apply Euler–Maclaurin formula to find $\dfrac{1}{51^2} + \dfrac{1}{53^2} + \dfrac{1}{55^2} + \dots + \dfrac{1}{99^2}$.

Solution: Taking $y = \dfrac{1}{x^2}$, $x_0 = 51$, $h = 2$, $n = 24$, we have $y' = \dfrac{-2}{x^3}$, $y''' = \dfrac{-24}{x^5}$.

Then Euler–Maclaurin formula gives

$$\int_{51}^{99} \frac{dx}{x^2} = \frac{2}{5}\left[\frac{1}{51^2} + \frac{2}{53^2} + \frac{2}{55^2} + \dots + \frac{2}{97^2} + \frac{2}{99^2}\right]$$

$$- \frac{(2)^2}{12}\left[\frac{-2}{99^3} - \frac{(-2)}{51^3}\right] + \frac{(2)^4}{720}\left(\frac{-24}{99^5} - \frac{(-24)}{51^5}\right)$$

$$\frac{1}{51^2} + \frac{1}{53^2} + \frac{1}{55^2} + ... + \frac{1}{99^2} = \frac{1}{2} \int_{51}^{99} \frac{dx}{x^2} + \left(\frac{1}{51^2} + \frac{1}{99^2} \right) + \frac{1}{3} \left(\frac{1}{51^3} - \frac{1}{99^3} \right)$$

$$- \frac{8}{30} \left(\frac{1}{51^5} - \frac{1}{99^5} \right) + ...$$

$$= \frac{1}{2} \left| -\frac{1}{x} \right|_{51}^{99} + 0.000243 + 0.0000022$$

$$= 0.00499 \text{ approx.}$$

Example 7.26: Use Euler-Maclaurin's formula to prove that

$$\sum_{1}^{n} x^2 = \frac{n(n+1)(2n+1)}{6}$$

Solution: Euler–Maclaurin's formula is given by

$$\frac{1}{h} \int_{x_0}^{x_0 + nh} f(x) \, dx = \sum_{i=0}^{n} f(x_i) - \frac{1}{2} [f(x_n) + f(x_0)]$$

$$- \frac{h}{12} [f'(x_0 + nh) - f'(x_0)] + \frac{h^3}{720} [f'''(x_0 + nh) - f'''(x_0)] + ...$$

Now putting $f(x) = x^2$, $f'(x) = 2x$, $x_0 = 0$, $h = 1$, $x_0 + nh = n$, $x_i = x_0 + ih = i$, we easily obtain

$$\int_{0}^{n} x^2 dx = \sum_{i=0}^{n} i^2 - \frac{1}{2} [n^2 + 0] - \frac{1}{12} [2n - 0]$$

[$\because$ third and higher order derivatives of x^2 are zero]

$$\frac{n^3}{3} = \sum_{i=0}^{n} i^2 - \frac{n^2}{2} - \frac{n}{6}$$

$$\sum_{i=0}^{n} i^2 = \frac{n^3}{3} + \frac{n^2}{2} + \frac{n}{6} = \frac{n(2n^2 + 3n + 1)}{6} = \frac{n(n+1)(2n+1)}{6}$$

$$\sum_{x=1}^{n} x^2 = \frac{n(n+1)(2n+1)}{6}$$

Example 7.27: Find the sum of the fourth powers of the first n natural numbers by means of the Euler–Maclaurin formula.

Solution: The general form of the Euler–Maclaurin formula is

$$\frac{1}{h} \int_{x_0}^{x_0 + nh} f(x) \, dx = \sum_{0}^{n} f(x + nh) - \frac{1}{2} [f(x_0 + nh) + f(x_0)]$$

$$- \frac{h}{12} [f'(x_0 + nh) - f'(x_0)] + \frac{h^3}{720} [f'''(x_0 + nh) - f'''(x_0)]$$

Putting $x_0 = 0, h = 1, f(x) = x_i$, we get

$$\int_0^n x^4 dx = \sum_0^n x^4 - \frac{1}{12}[n^4 + 0^4] - \frac{1}{12}[5n^3 - 4(0)^3] + \frac{1}{720}[24n - 24(0)]$$

$$\frac{n^5}{5} = \sum_0^n x^4 - \frac{n^4}{2} - \frac{n^3}{3} + \frac{n}{40}$$

$$\sum x^4 = \frac{n^5}{5} + \frac{n^4}{2} + \frac{n^3}{3} - \frac{n}{30}$$

PROBLEMS 7.4

1. Evaluate $\int_0^{\frac{\pi}{2}} \sin x dx$ using the Euler–Maclaurin formula.

   i. $h = \pi/8$ [Ans. 1.000003]

   ii. $h = \pi/4$ [Ans. 0.837336]

2. Sum the series by using Euler–Maclaurin summation formula.

   i. $\dfrac{1}{400} + \dfrac{1}{402} + \ldots + \dfrac{1}{498} + \dfrac{1}{505}$ [Ans. 0.11382114]

   ii. $\dfrac{1}{100} + \dfrac{1}{101} + \dfrac{1}{102} + \dfrac{1}{103} + \dfrac{1}{104}$ [Ans. 0.0490291]

   iii. $\dfrac{1}{201^2} + \dfrac{1}{203^2} + \ldots + \dfrac{1}{299^2}$ [Ans. 0.0008333290]

3. Find the sum of the fifth power of the first n natural numbers by means of Euler–Maclaurin formula.

$$\left[\text{Ans. } \frac{n^6}{6} + \frac{n^5}{2} + \frac{5n^4}{12} - \frac{n^2}{12}\right]$$

4. If $f(x)$ and its derivative $f'(x)$ are continuously increasing in the range $0 < x < n$, show that $\int_0^n f(x) dx < \frac{1}{2}[f(0) + f(n)] + \sum_{x=0}^{n-1} f(x).$

[**Hint:** In Euler–Maclaurin formula, put $a = 0, h = 1$]

5. Obtain the approximate quadrature formula $\int_0^1 \dfrac{dx}{1 + x^2} = \dfrac{1}{4a}\left(1 + \dfrac{1}{6a}\right) + \sum_{x=1}^a \dfrac{a}{a^2 + x^2}.$

[**Hint:** Use Euler–Maclaurin formula with $f(x) = \dfrac{a}{a^2 + x^2}$ and range $(0, a)$

$$\int_0^a \frac{a}{a^2 + x^2} dx = \sum_{x=0}^a \frac{a}{a^2 + x^2} - \frac{1}{2}\left(\frac{1}{a} + \frac{1}{2a}\right) + \frac{1}{2a} \cdot \frac{1}{2a^2}$$

put $x = ay$ in the integral]

6. Evaluate $\int_0^1 \dfrac{dx}{1 + x}$ correct to five decimal places by using Euler–Maclaurin formula.

                                                         [Ans. 0.69315]

7.11 GAUSSIAN INTEGRATION

Earlier, some integration formula were derived to evaluate the integral $I = \int_a^b f(x)\,dx$ when values of the function were known as equispaced points of the intervals. Gauss derived a formula which uses the same number of functional values but with different spacing and yields better results.

Gauss's formula is expressed in the form

$$\int_{-1}^{+1} f(u)\,du = w_1 F(u_1) + w_2 F(u_2) + \ldots + w_n F(u_n)$$

$$= \sum_{i=0}^{n} w_i F(u_i) \tag{7.29}$$

where the w_i and u_i are called weights and abscissae respectively. An advantage of this formula is that the abscissae and weights are symmetrical with respect to the middle point of the interval. In Eq. (7.29) there exist $2n$ unknowns.

Thus $2n$ relations between them are necessary which can be obtained such that the formula is exact for all polynomials of degree not exceeding $(2n - 1)$. Thus, we may consider

$$F(u) = C_0 + C_1 u + C_2 u^2 + C_3 u^3 + \ldots + C_{2n-1} u^{2n-1} \tag{7.30}$$

Now from the left hand side of Eq. (7.29), we readily get

$$\int_{-1}^{+1} F(u)\,du = \int_{-1}^{1} [C_0 + C_1 u + C_2 u^2 + C_2 u^3 + \ldots + C_{2n-1} u^{2n-1}]\,du$$

$$= 2C_0 + \frac{2}{3}C_2 + \frac{2}{5}C_3 + \ldots \tag{7.31}$$

By setting $u = u_i$ in Eq. (7.30), we obtain

$$F(u_i) = C_0 + C_1 u_i + C_2 u_i^2 + C_3 u_i^3 + \ldots + C_{2n-1} u_i^{2n-1} \tag{7.32}$$

Now substituting these values on the right hand side of Eq. (7.29), we get

$$\int_{-1}^{+1} F(u_i)\,du = w_1 [C_0 + C_1 u_1 + C_2 u_1^2 + C_3 u_1^3 + \ldots + C_{2n-1} u_1^{2n-1}]$$

$$+ w_2 [C_0 + C_1 u_2 + C_2 u_2^2 + C_3 u_2^3 + \ldots + c_{2n-1} u_2^{2n-1}] + w_3 [C_0 + C_1 u_3 + C_2 u_3^3$$

$$+ \ldots + C_{2n-1} u_3^{2n-1}] + \ldots + w_n [C_0 + C_1 u_n + C_2 y_n^3 + \ldots + C_{2n-1} u_n^{2n-1}]$$

$$\int_{-1}^{+1} F(u_i)\,du = C_0 [w_1 + w_2 + w_3 + \ldots + w_n] + C_1 [w_1 u_1 + w_2 u_2 + w_3 u_3 + \ldots + w_n u_n]$$

$$+ C_1 [w_1 u_1^2 + w_2 u_2^2 + w_3 u_3^2 + \ldots + w_n u_n^2]$$

$$+ C_{2n-1} [w_1 u_1^{2n-1} + w_2 u_2^{2n-1} + w_3 u_3^{2n-1} + \ldots + w_n u_n^{2n-1}] \tag{7.32a}$$

But the Eqs (7.31) and (7.32a) are identical for all C_i, hence comparing coefficients of C_i, we obtain $2n$ equations in $2n$ unknown w_i and u_i ($i = 1, 2, 3, ..., n$), yielding

$$\left. \begin{aligned} w_1 + w_2 + w_3 + ... + w_n &= 2 \\ w_1 u_1 + w_2 u_2 + w_3 u_3 + ... + w_n u_n &= 0 \\ w_1 u_1^2 + w_2 u_2^2 + w_3 u_3^2 = ... + w_n u_n^2 &= \frac{2}{3} \\ ... &= ... \\ w_1 u_1^{2n-1} + w_2 u_2^{2n-1} + w_3 u_3^{2n-1} + ... + w_n u_n^{2n-1} &= 0 \end{aligned} \right\}$$ (7.33)

Theoretically, we can obtain n unknown w_is and u_is by solving the above of non-linear equations by the usual algebraic methods but practically it is very difficult to solve even for small values of n.

As an illustration, we consider the case $n = 2$, then the formula is

$$\int_{-1}^{1} F(u)\,du = w_1 F(u_1) + w_2 F(u_2)$$

Since this formula is exact when $F(u)$ is a polynomial of degree not exceeding 3, we put successively $F(u) = 1, u, u^2$ and u^3. Then Eq. (7.32) gives (four equations)

$$\left. \begin{aligned} w_1 + w_2 &= 2 \\ w_1 u_1^2 + w_2 u_2^2 &= \frac{2}{3} \\ w_1 u_1 + w_2 u_2 &= 0 \\ w_1 u_1^3 + w_2 u_2^3 &= 0 \end{aligned} \right\}$$ (7.34)

Solving the above equations, we have

$$\left. \begin{aligned} w_1 + w_2 &= 1 \\ u_2 = -u_1 &= \frac{1}{\sqrt{3}} \end{aligned} \right\}$$ (7.35)

This method when applied to the general system above, will be extremely complicated and difficult, and an alternate method must be chosen to solve the nonlinear system.

It can be shown that u_i are the zeros of the $(n + 1)^{th}$ Legendre polynomials $P_{n+1}(u)$ which can be generated using the recurrence relation

$$(n + 1) P_{n+1}(u) = (2n + 1) u P_n(u) - nP_{n-1}(u)$$

where $P_0(u) = 1$ and $P_1(u) = u$. The first five Legendre polynomials are given by

$$\left. \begin{aligned} P_0(u) &= 1 \\ P_1(u) &= u \\ P_2(u) &= \frac{1}{2}(3u^2 - 1) \\ P_3(u) &= \frac{1}{2}(5u^3 - 3u) \\ P_4(u) &= \frac{1}{8}(35u^4 - 30u^2 + 3) \end{aligned} \right\}$$ (7.36)

It can also be shown that the corresponding weights w_i are given by

$$w_i = \int_{-1}^{1} p^n_{\substack{j=0 \\ j \pm i}} \left(\frac{u - u_j}{u_i - u_j} \right) du \tag{7.37}$$

where the u_1 are the abscissae.

As an example, when $n = 1$, we solve $P_2(u) = 0$, i.e.

$$\frac{1}{2}(3u^2 - 1) = 0$$

which gives the two abscissae

$$u_0 = -\frac{1}{\sqrt{3}} = -\frac{\sqrt{3}}{3} \text{ and } u_1 = \frac{1}{\sqrt{3}} = \frac{\sqrt{3}}{3}$$

The corresponding weights are given by

$$w_0 = \int_{-1}^{1} \frac{u - u_1}{u_0 - u_1} du = \frac{1}{u_0 - u_1} \left[\frac{u^2}{2} - u_1 u \right]_{-1}^{1} = 1$$

$$w_1 = \int_{-1}^{1} \frac{u - u_0}{u_1 - u_0} du = \frac{1}{u_1 - u_0} \left[\frac{u^2}{2} - u_0 u \right]_{-1}^{1} = 1$$

Similarly for $n = 3$, we solve $P_4(u) = 0$

$$\frac{1}{8}(35u^4 - 30u^2 + 3) = 0$$

which give the four abscissae

$$u_1 = \pm \left(\frac{15 \pm 2\sqrt{30}}{35} \right)^{\frac{1}{2}}$$

The weight w_i can then be found from Eq. (7.37). It should be noted, however, that the abscissae u_i and weight w_i are extensively tabulated for different value of n. We list below, in table, the abscissae and weight for values of n up to $n = 6$.

In general case, the limits of the integral in $\int_a^b f(x)dx$ have to be changed to those in Eq. (7.29) by means of the transformation

$$x = \frac{1}{2}u(b - a) + \frac{1}{2}(a + b)$$

Here, we tabulate the abscissae u_i and the weight w_i for the different value of n ($n = 2$, 3, 4, 5, 6).

Abscissae and weight for Gaussian integration is given below.

n	$\pm ui$		wi	
2	0.57735	0.2692	1.00000000	
3		0.0		0.8888888889
	0.77459	6692	0.55555	55556
4	0.33998	10436	0.65214	51549
	0.86113	63116	0.34785	48451

5		0.0	0.56888	88889
	0.53846	93101	0.47862	86705
	0.90617	98459	0.23692	68851
6	0.23861	91861	0.46791	39346
	0.66120	93865	0.36076	15730
	0.93246	95142	0.17132	44924

Example 7.28: Find $I = \int_0^1 x\,dx$ by Gauss's formula with $n = 4$ up to five decimal places.

Solution: First we change the limits of integration with the help of $x = \dfrac{1}{2}(u + 1)$.

This gives
$$I = \frac{1}{4}\int_{-1}^{1}(u + 1)dx = \frac{1}{4}\sum_{i=1}^{n} w_i F(u_i)$$

where
$$F(u_i) = u_i + 1$$

Now we choose $n = 4$ and use the abscissae and weights corresponding $n = 4$ in standard table and obtain

$$I \approx \frac{1}{4}[(-0.86114 + 1)(0.34785) + (-0.33998 + 1)(0.65214)(0.3398 + 1)(0.65214)$$
$$+ (0.86114 + 1)(0.34785)]$$
$$= 0.49999$$

where the abscissae and weights have been rounded to five decimal places.

7.12 NUMERICAL DOUBLE INTEGRATION

7.12.1 Trapezoidal Rule

Suppose we have $I = \int_a^b\int_c^d f(x, y)\,dx\,dy$, the region of integration for which the area is bounded by the lines $x = c$, $x = d$, $y = a$, $y = b$ in the xy-plane.

Now we consider the double integral

$$I = \int_{y_j}^{y_{j+1}} \int_{x_i}^{x_{i+1}} f(x, y)\,dx\,dy \tag{7.38}$$

where $x_{i+1} = x_i + h$ and $y_{k+1} = y_j + k$.

Also here $x_{i+1} - x_i = h$ and $y_{j+1} - y_j = k$ for all i, j.

Here I denotes the area of the region ABCD as shown in Fig. 7.2.

Now let us apply trapezoidal rule repeated to get the value of I in Eq. (7.38), we get

$$I = \int_{y_i}^{y_{j+1}} \frac{h}{2}[(f(x_i + y) + f(x_{i+1}, y)]\,dy$$

$$= \frac{h}{2}\left[\int_{y_i}^{y_{i+1}} f(x_i, y)\,dy + \int_{y_i}^{y_{j+1}} f(x_{i+1}, y)\,dy\right]$$

$$= \frac{h}{2}\left[\frac{k}{2}[(f(x_i, y_{j+1}) + f(x_i, y_j)] + \frac{k}{2}[f(x_{i+1}, y_{j+1}) + f(x_{i+1}, y_j)]\right]$$

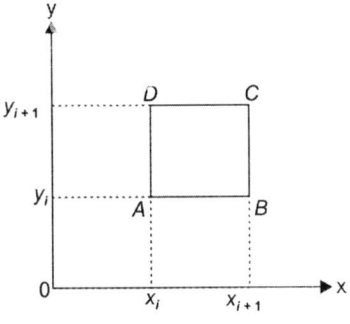

Fig. 7.2

i.e.
$$I = \frac{hk}{4} \left[(f(x_i, y_{j+1}) + f(x_i, y_j)\, f(x_{i+1}, y_{j+1}) + f(x_{i+1}, y_j)) \right]$$

$$= \frac{hk}{4} \left[(f_{i,j+1} + f_{i,j} + f_{i+1,j+1} + f_{i+1,j}) \right] \tag{7.38a}$$

where $f_{i,j} \equiv f(x_i, y_j)$.

Using Fig. 7.2, we have

$$I = \frac{hk}{4} \text{ [sum the value of } f(x, y) \text{ at the four corner points]}$$

i.e.
$$= \frac{hk}{4} [f_A + f_B + f_C + f_D] \tag{7.39}$$

Suppose we require $\int_c^d \int_a^b f(x, y)\, dx\, dy$, where a, b, c, d are constants. The double integral denotes the area of the region bounded by the line $x = a$, $x = b$, $y = x$ and $y = d$ as shown in Fig. 7.3.

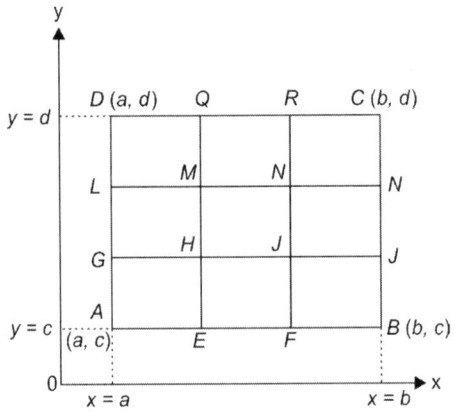

Fig. 7.3

First divide the region of integration (rectangle) $ABCD$ into meshes (or series of subsquares) by dividing AB into three equal parts (in general into equal parts), each part length being h, and AD into three equal parts (in general n equal parts) each part length being k. Here, the integral over the whole rectangle is equal to sum of the integrals over each mesh.

$$I = \int_c^d \int_a^b f(x, y)\, dx\, dy$$

$$= \iint_{AEHG} + \iint_{EFJH} + \iint_{FBKJ} + \iint_{GHML} + \iint_{HJNM} + \iint_{JKPN} + \iint_{LMQD} + \iint_{MNRQ} + \iint_{NPCR}$$

$$= \frac{hk}{4}[(f_A + f_E + f_H + f_G) + (f_E + f_F + f_J + f_H) + ...]\,[\text{using Eq. (7.39)}]$$

$$= \frac{hk}{4}[(f_A + f_B + f_C + f_D) + 2(f_E + f_K + f_F + f_P$$

$$+ f_R + f_Q + f_L + f_G) + 4(f_H + f_J + f_M + f_N)]\quad [\text{on simplification}]$$

$I = \dfrac{hk}{4}$ [(sum of the values of f at the four corners) + 2 (sum of the value of f at the remaining nodes on the boundary) + 4 (sum of the value of f at the interior nodes)].

This equation is known as trapezoidal rule for double integration.

7.12.2 Simpson's Rule for Double Integration

Apply the Simpson's 1/3 rule, the number of intervals must be even, i.e. number of ordinates must be odd.

Now we consider the integral

$$I = \int_{y_j}^{y_{j+1}} \int_{x_i}^{x_{i+2}} f(x, y)\, dx\, dy \tag{7.40}$$

where $y_{j+1} = y_j + k$, $x_{i+1} = x_i + h$.

The region of integration is shown in Fig. 7.4.

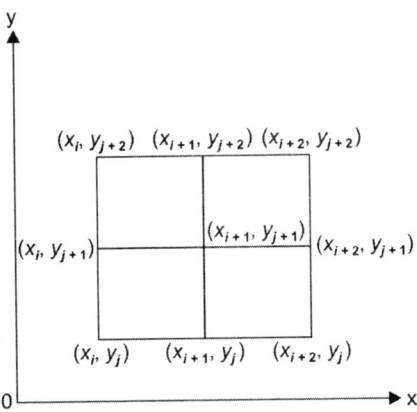

Fig. 7.4

Let us apply Simpson's 1/3 rule to the integral in Eq. (7.40), we get

$$I = \int_{y_j}^{y_{j+2}} \left[\int_{x_i}^{x_{i+2}} f(x, y)\, dx \right] dy$$

$$= \int_{y_j}^{y_{j+2}} \frac{h}{3}[f(x_i, y_j) + f(x_{i+2}, y_j) + 4f(x_{i+1}, y_j)]\, dy$$

Again apply Simpson's 1/3 rule for each integral in y direction

$$= \frac{h}{3}\left[\frac{k}{3} f(x_i, y_j) + f(x_i, y_{j+2}) + 4f(x_i, y_{j+1})\right]$$

$$+ \frac{k}{3} f(x_{i+2}, y_j) + f(x_{i+2}, y_{j+2}) + 4f(x_{i+2}, y_{j+1})\Big]$$

$$+ \frac{4k}{3} [f(x_{i+1}, y_j) + f(x_{i+1}, y_{j+2}) + 4f(x_{i+1}, y_{j+1})]$$

$$= \frac{hk}{9} [(f_{i,j} + f_{i,j+2} + f_{i+2,j} + f_{i+2,j+2}) + 4(f_{i,j+1} + f_{i+1,j}$$

$$+ f_{i+1,j+2} + f_{i+2,j+1}) + 16 f_{i+1,j+1}] \tag{7.41}$$

i.e. $I = \dfrac{hk}{9}$ (sum of the value of f at four corners) $+ 4$ (sum of the values of f at the remaining nodes in the boundary) $+ 16$ (value of f at the central point). $\tag{7.42}$

Suppose we require $I = \int_c^d \int_a^b f(x, y)\, dx\, dy$, where the limits are constant. Here the region of integral is bounded by $x = a$, $x = b$, $y = c$ and $y = d$ as shown in Fig. 7.5.

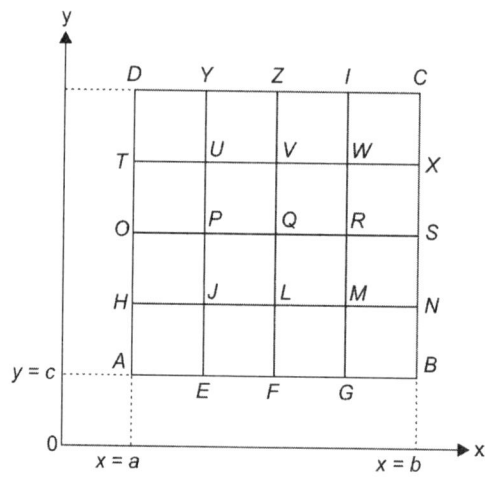

Fig. 7.5

Firstly divide the interval (a, b) into four (in general even number) equal parts and the interval (c, d) into four (in general even number) equal parts. Here length of each interval in (a, b) is h and in (c, d) is k.

Apply Simpson's 1/3 rule given by Eq. (7.41) to the four rectangles AFQO, FBSQ, SCZQ and ZDOQ. Hence we have

$$I = \frac{hk}{9} [(f_A + f_F + f_Q + f_O) + 4(f_E + f_L + f_P + f_H) + 16 f_J]$$

$$+ [(f_F + f_B + f_S + f_Q) + 4(f_G + f_N + f_R + f_L) + 16 f_M]$$

$$+ [(f_S + f_C + f_Z + f_Q) + 4(f_X + f_I + f_V + f_R) + 16 f_W]$$

$$+ [(f_Z + f_D + f_O + f_Q) + 4(f_Y + f_T + f_P + f_V) + 16 f_U]$$

$$= \frac{hk}{9}[(f_A + f_B + f_C + f_D) + 2(f_F + f_S + f_Z + f_O) + 4(f_E + f_G + f_N + f_X$$
$$+ f_T + f_H) + 4(f_Q + 8(f_P + f_R) + 16(f_J + f_M + f_W + f_V)]$$

$I = \frac{hk}{9}$ [(sum of the value of f at four corners) + 2 (sum of the value of f at the odd positions on the boundary except the corners) + 4 (sum of the values of f at the even positions on the boundary) + 4 (the values of f at the centre point) + 8 (sum of the value of f at the even positions on the horizontal central line) + 8 (sum of the value of f at even positions on the vertical central line) + 16 (sum of values of f at the even position on even rows of the matrix)].

The above equation is known as Simpson's rule for double integrals.

Example 7.29: Using (i) Simpson's 1/3 rule (ii) Trapezoidal rule, evaluate
$$\int_0^1 \int_0^1 \frac{1}{(1 + x + y)} dx\, dy,$$

taking $h = k = 0.5$.

Solution: Here $\qquad f(x, y) = \dfrac{1}{1 + x + y}$

We prepare the following table

x / y	0	0.5	1
0	1	0.66	0.50
0.5	0.66	0.50	0.40
1	0.50	0.40	0.33

i. Using Simpson's 1/3 rule

$$\int_0^1 \int_0^1 = \frac{(0.5)(0.5)}{9}[(1 + 0.5 + 0.5 + 0.33) + 4(0.66 + 0.4 + 0.4 + 0.66) + 16(0.5)]$$
$$= (0.0278)[2.33 + 8.48 + 8] = 0.5229$$

ii. Using trapezoidal rule

$$\int_0^1 \int_0^1 \frac{1}{(1 + x = y)} dx\, dy = \frac{(0.5)(0.5)}{9}[(1 + 0.5 + 0.5 + 0.33)$$
$$= + 2(0.66 + 0.4 + 0.4 + 0.66) + 4(0.5)]$$
$$= (0.0625)[2.33 + 4.24 + 2] = 0.5356$$

Example 7.30: Evaluate $I = \int_0^1 \int_0^1 e^{x+y} dx\, dy$ using the trapezoidal and Simpson's rules with $h = k = 0.5$.

Solution:

We prepare the following table

x / y	0	0.5	1
0	1	1.6487	2.7183
0.5	1.6487	2.7183	4.4817
1	2.7183	4.4817	7.3891

i. Using the trapezoidal rule, we obtain

$$I = \frac{0.25}{4}[1.0 + 4(1.6487) + 6(2.7183) + 4(4.4817) + 7.389] = 3.0762$$

ii. Using Simpson's rule, we obtain

$$I = \frac{0.25}{4}[1.0 + 2.7183 + 7.3891 + 2.7183 + 4(1.6387 + 4.4817 + 4.4817 + 1.6487)$$

$$+ 16(2.7183)]$$

$$= \frac{26.59042}{9} = 2.9545$$

The exact value of the double integral is 2.9525 and therefore, it can be verified that the result given by Simpson's rule is about sixty times more accurate than that given by trapezoidal rule.

Example 7.31: Evaluate $\int_2^{2.4}\int_4^{4.4} xy\, dx\, dy$ using (i) trapezoidal rule (ii) Simpson's rule. Verify your result by actual integration.

Solution: To evaluate $\int_2^{2.4}\int_4^{4.4} xy\, dx\, dy$

Here $\qquad\qquad f(x, y) = xy$

i. To use trapezoidal rule, prepare the following table [i.e. value of $f(x, y)$]

y \ x	4	4.4
2	8	8.8
2.4	9.6	10.56

Using trapezoidal rule

$$\int_2^{2.4}\int_4^{4.4} xy\, dx\, dy = \frac{hk}{4}[\text{sum of the value of } f \text{ at the four corner points}]$$

$$= \frac{(0.4)(0.4)}{4}[9.6 + 10.56 + 8.8 + 8] = (0.04)(36.96) = 1.4784$$

ii. To apply Simpson's rule, the range of x and y in two equal parts

$$h = \frac{4.4 - 4}{2} = 0.2, \quad k = \frac{2.4 - 2}{2} = 0.2$$

We prepare the following table for the value of $f(x, y)$

y \ x	4	4.2	4.4
2	8	8.4	8.8
2.2	8.8	9.24	9.68
2.4	9.6	10.08	10.56

Using Simpson's rule

$$\int_2^{2.4}\int_4^{4.4} xy\,dx\,dy = \frac{(0.2)\,(0.2)}{9}[(8+9.6+10.56+8.8)+4(8.8+10.08+9.68+8.4)$$

$$+16(9.24)]$$

$$= (0.004)\,[36.96+147.84+147.84] = 1.33056$$

By actual integration

$$\int_2^{2.4}\int_4^{4.4} xy\,dx\,dy = \int_2^{2.4} y\,dy\int_4^{4.4} x\,dx$$

$$= \left[\frac{y^2}{2}\right]_2^{2.4}\left[\frac{x^2}{2}\right]_4^{4.4} = \frac{1}{4}[(2.4)^2-2^2][(4.4)^2-4^2]$$

$$= \frac{1}{4}(3.36)\times(1.76) = 1.4784$$

The actual value and the value of trapezoidal rule are equal.

PROBLEMS 7.5

1. Using trapezoidal rule, evaluate $\int_1^2\int_1^2\dfrac{dy\,dx}{x+y}$ taking four subintervals.   [Ans. 0.3407]

2. Evaluate $\int_0^1\int_0^1 xe^y\,dx\,dy$ using trapezoidal rule $(h=k=0.5)$ [Ans. 0.876]

3. Apply trapezoidal rule to find $\int_1^5\int_1^5\dfrac{dx\,dy}{\sqrt{x^2+y^2}}$ taking two subintervals.

   [Ans. 4.134]

4. Evaluate $\int_1^{2.6}\int_2^{3.2}\dfrac{dx\,dy}{x+y}$ using Simpson's rule.   [Ans. 0.49]

5. Find the value of double integral $\int_1^{1.2}\int_1^{1.4}\dfrac{dx\,dy}{x+y}$ by trapezoidal rule.   [Ans. 0.0349]

6. Evaluate $\int_0^{\pi}\int_0^{\pi} \sin(x+y)\,dx\,dy$ using Simpson's rule.   [Ans. 2.0080]

Using both trapezoidal and Simpson's rules, evaluate the following integrals from 7 to 10. Check your answer by direct integration.

7. Find $\int_0^1\int_0^1 e^{2x+3y}\,dx\,dy$.

8. Find $\int_3^{3.8}\int_4^{4.8}\dfrac{1}{x^2+y^2}\,dx\,dy$ taking $h=k=0.2$.

9. Find $\int_1^{1.8}\int_2^{2.8}(x^2+y^2)\,dx\,dy$ using $h=k=0.2$.

10. Find $\int_0^1\int_0^1 e^{-(x+y)}\,dx\,dy$.

11. Find the double integral $I = \int_0^1 \int_0^2 \dfrac{2xy}{(1+x^2)(1+y^2)} \, dy \, dx$ by using trapezoidal rule with step length $h = k = 0.25$. [Ans. 0.5582]

12. Compute numerically $\iint\limits_D \dfrac{dx \, dy}{x^2 + y^2}$, where D is the square with corners at $(1, 1)$, $(2, 1)$ $(2, 2)$, $(1, 2)$. [Ans. 0.2313]

8

Numerical Solutions of Ordinary Differential Equations

8.1 INTRODUCTION

The analytical methods for solution with which we are familiar can be applied to a limited class of differential equations. If differential equations appearing in the physical problem do not belong to any of these familiar type, we resort to numerical methods. A variety of numerical methods have been evolved for solving differential equation of order one or more than one.

To describe various numerical methods of solving ordinary differential equations, we consider the first order differential equation

$$dy/dx = f(x, y); \text{ given } y(x_0) = y_0$$

In most of these methods, we replace the differential equation by a difference equation and solve it. These solutions yield solutions in one of the two forms:

i. A series of y in terms of power of x, from which the value of y can be obtained by direct substitution.

ii. A set of tabulated values of x and y.

The methods of Taylor's and Picard belongs to class (i) of solutions, whereas those of Euler, Runge-Kutta, Adams-Bashforth, etc. belong to class (ii). These methods are called step-by-step methods or marching methods because the values of y are computed by steps ahead for equal intervals 'h' of the independent variables. In methods of Euler and Runge-Kutta, the interval 'h' should be kept small, and therefore, these methods are used for computing y over a limited range of values of x, whereas Milne's Adams–Bashforth methods are applied for finding y over a wide range of values of x. Milne's Adams–Bashforth methods require starting values which are usually obtained by Picard's, Taylor's series or Runge–Kutta methods.

a. **Initial value problems:** Problems involving first order differential equations in which all the conditions are specified at the initial point only are called initial value problems such as $dy/dx = f(x, y), y(x_0) = y_0$.

b. **Boundary value problems:** Problems involving second and higher order differential equations in which all the conditions are specified at two or more points are called boundary value problems.

8.2 PICARD'S METHOD

Consider the first order differential equation with initial condition

$$\frac{dy}{dx} = f(x, y) \ y(x_0) = y_0 \tag{8.1}$$

Integrating Eq. (8.1), we have

$$\int_{y_0}^{y} dy = \int_{x_0}^{x} f(x, y) dx$$

$$y = y_0 + \int_{x_0}^{x} f(x, y) dx \tag{8.2}$$

As a first approximation y_1, we put $y = y_0$ in $f(x, y)$ and then integrate, we get

$$y_1 = y_0 + \int_{x_0}^{x} f(x, y_0) dx$$

As a second approximation y_2, we put $y = y_1$ in $f(x, y)$ and then integrate, we get

$$y_2 = y_0 + \int_{x_0}^{x} f(x, y_1) dx \text{ and so on}$$

to get a sequence giving better approximation and hence

$$y_n = y_0 + \int_{x_0}^{x} f(x, y_{n-1}) dx$$

This is known as Picard's method.

Example 8.1: Employ Picard's method to find the solution of $\dfrac{dy}{dx} = x^2 + y^2$ for $x = 0.4$ given that $y = 0$ when $x = 0$ up to four decimal places.

Solution: Here $x_0 = 0$, $y_0 = 0$, $f(x, y) = x^2 + y^2$

By Picard's method

$$y = y_0 + \int_{x_0}^{x} f(x, y) dx$$

$$= 0 + \int_{0}^{x} (x^2 + y^2) dx$$

For the first approximation y_1, put $y = 0$ in $f(x, y)$

$$y_1 = \int_{0}^{x} x^2 dx = \frac{x^3}{3}$$

For the second approximation y_2, put $y = \dfrac{x^3}{3}$

$$y_2 = \int_{0}^{x} \left[x^2 + \left(\frac{x^3}{3} \right)^2 \right] dx = \int_{0}^{x} \left(x^2 + \frac{x^6}{9} \right) dx = \frac{x^3}{3} + \frac{x^7}{63}$$

For the third approximation y_3, put $y = \dfrac{x^3}{3} + \dfrac{x^7}{63}$ in $f(x, y)$

$$y_3 = \int_{0}^{x} \left[x^2 + \left(\frac{x^3}{3} + \frac{x^7}{63} \right)^2 \right] dx$$

$$= \int_{0}^{x} \left(x^2 + \frac{x^6}{9} + \frac{x^{14}}{(63)^2} + \frac{2x^{10}}{3 \times 63} \right) dx = \frac{x^3}{3} + \frac{x^7}{63} + \frac{x^{15}}{59535} + \frac{2x^{11}}{2079}$$

For next approximation, the integral becomes very difficult to evaluate, so we stop up to y_3.

Now when $x = 0.4$

$$y_1 = \frac{(0.4)^3}{3} = 0.021$$

$$y_2 = \frac{(0.4)^3}{3} + \frac{(0.4)^7}{63} = 0.02133333 + 0.00006260 = 0.02139593$$

$$y_3 = \frac{(0.4)^3}{3} + \frac{(0.4)^7}{63} + \frac{(0.4)^{15}}{59535} + \frac{2(0.4)^{11}}{2079}$$

$$= 0.21333333 + 0.00000107 + 0.0000839 + 0.0000000403$$

$$= 0.021418341$$

Hence $(y_2) \cong (y_3)$, we have $y(0.4) = 0.0214$.

Example 8.2: Use Picard's method to approximate y when $x = 0.1$ and 0.2 for $dy/dx = x + y^2$, where $y = 0$ when $x = 0$.

Solution: Method I: Here $y = 0$ when $x = 0$, $x_0 = 0$, $y_0 = 0$.

$$f(x, y) = x + y^2$$

By Picard's method

$$y = 0 + \int_0^x (x + y^2) dx \qquad (i)$$

For first approximation y_1, put $y = 0$ in RHS of Eq. (i)

$$y_1 = \int_0^x x\, dx = \frac{x^2}{2}$$

For second approximation y_2, put $y = x^2/2$ in RHS of Eq. (i)

$$y_2 = \int_0^x \left(x + \frac{x^4}{4} \right) dx = \frac{x^2}{2} + \frac{x^5}{20}$$

For third approximation y_3, put $y = x^2/2 + x^5/20$ in RHS of Eq. (i)

$$y_3 = \int_0^x \left[x + \left(\frac{x^2}{2} + \frac{x^5}{20} \right)^2 \right] dx = \int_0^x x + \left(\frac{x^4}{4} + \frac{1}{400} x^{10} + \frac{1}{20} x^7 \right) dx$$

$$= \frac{x^2}{2} + \frac{1}{20} x^5 + \frac{1}{4400} x^{11} + \frac{1}{1600} x^8$$

For $x = 0.1$

$$y_1 = \frac{(0.1)^2}{2} = \frac{0.01}{2} = 0.005$$

$$y_2 = \frac{0.01}{2} + \frac{0.00001}{20} = 0.005 + 0.0000005 = 0.00500$$

$$y_3 = \frac{0.01}{2} + \frac{0.00001}{20} + \frac{0.00000001}{160} + \dots = 0.00500$$

There is no difference between y_2 and y_3

$\therefore$ $\qquad\qquad y(0.1) = 0.00500$

For $x = 0.2$, we may take $x_0 = 0.1$, $y_0 = 0.005$, we first approximate y_1 and then y_2

$$y_1 = 0.005 + \int_{0.1}^{x}(x + 0.000025)\,dx = 0.005 + \left[\frac{x^2}{2} + 0.000025x\right]_{0.1}^{x}$$

$$= 0.005 + \frac{x^2}{2} + 0.000025x - \frac{(0.1)^2}{2} - 0.000025(0.1)$$

$$= 0.005 - 0.005 - 0.0000025 + \frac{1}{2}x^2 + \frac{25}{10^6}x$$

$$= \frac{0.04}{2} + \frac{25}{10^6}(0.2) - 0.0000025 = 0.02 + 0.0000025 = 0.0200$$

$$y_2 = 0.005 + \int_{0.1}^{x}(x + 0.0004)\,dx$$

$$= 0.005 + \left[\frac{x^2}{2} + 0.0004x\right]_{0.1}^{x} = \frac{x^2}{2} + \frac{4}{10^4}x - \frac{4}{10^5}$$

For $x = 0.2$ $\qquad\qquad y_2 = 0.02 + 0.00008 - 0.00004 = 0.02004$

y_1 and y_2 are approximately the same up to four decimal places

$\therefore$ $\qquad\qquad y(0.2) = 0.02004$

Method II: Put $x = 0.2$, then

$$y_2 = \frac{x^2}{2} + \frac{x^5}{20}$$

$$y_3 = \frac{x^2}{2} + \frac{x^5}{20} + \frac{x^{11}}{4400} + \frac{x^8}{1600},\text{ we get}$$

$$y_2 = 0.0200 \quad y_3 = 0.02004$$

Hence $y(0.2) = 0.0200$ correct to four decimal places.

Example 8.3: Use Picard's method to solve $\dfrac{dy}{dx} = y$, $y(0) = 1$ and compare it with the exact solution.

Solution: Here $x_0 = 0$, $y_0 = 1$, $f(x, y) = y$

First approximation

$$y_1 = 1 + \int_0^x y_0\,dx = 1 + x$$

Second approximation

$$y_2 = 1 + \int_0^x (1 + x)\,dx = 1 + x + \frac{x^2}{2!}$$

Third approximation

$$y_3 = 1 + \int_0^x \left(1 + x + \frac{x^2}{2!}\right)dx = 1 + x + \frac{x^2}{2!} + \frac{x^3}{3!}$$

Fourth approximation

$$y_4 = 1 + \int_0^x \left(1 + x + \frac{x^2}{2!} + \frac{x^3}{3!}\right) dx = 1 + x + \frac{x^2}{2!} + \frac{x^3}{3!} + \frac{x^4}{4!}$$

$$y_5 = 1 + \int_0^x \left(1 + x + \frac{x^2}{2!} + \frac{x^3}{3!} + \frac{x^4}{4!}\right) dx = 1 + x + \frac{x^2}{2!} + \frac{x^3}{3!} + \frac{x^4}{4!} + \frac{x^5}{5!}$$

Similarly

$$y_6 = 1 + x + \frac{x^2}{2!} + \frac{x^3}{3!} + \frac{x^4}{4!} + \frac{x^5}{5!} + \frac{x^6}{6!} \text{ and so on.}$$

Exact solution of $\dfrac{dy}{dx} = y, x = 0, y = 1$

$$dy = y\, dx$$

$$\frac{dy}{y} = dx$$

$$y = e^{x + h}$$

when $x = 0, y = 1$

$$1 = e^h$$

$$y = e^x = 1 + x + \frac{x^2}{2!} + \frac{x^3}{3!} + \frac{x^4}{4!}$$

which is the same as Picard's solution.

Example 8.4: Using Picard's method, find the solution of the equation up to third approximation $\dfrac{dy}{dx} = 1 + xy, y(0) = 0.$

Solution: We have $\qquad y = y_0 + \int_{x_0}^x f(x, y) dx$

Here $y_0 = 0, x_0 = 0$

$$y = 0 + \int_0^x f(x, y)\, dx = 0 + \int_0^x (1 + xy)\, dx \tag{i}$$

First approximation putting $y = 0$

$$y_1 = 0 + \int_0^x (1 + x \cdot 0)\, dx = x$$

Second approximation putting $y = x$ in Eq. (i)

$$y_2 = 0 + \int_0^x (1 + x \cdot x)\, dx = x + \frac{x^3}{3}$$

Third approximation putting $y = x + x^3/3$ in Eq. (i)

$$y_3 = 0 + \int_0^x 1 + x\left(x + \frac{x^3}{3}\right) dx = x + \frac{x^3}{3} + \frac{x^5}{15}$$

Example 8.5: Solve the equation $\dfrac{dy}{dx} = \dfrac{x^2}{y^2 + 1}$ with initial condition $y = 0$ when $x = 0$. Use Picard's method to obtain y for $x = 0.25, 0.5, 1.0$ correct to three decimal places.

Solution: We have
$$y = y_0 + \int_{x_0}^x f(x, y)dx$$

$$= 0 + \int_0^x \frac{x^2}{y^2 + 1} dx \tag{i}$$

Putting $y = 0$ in Eq. (i) to get first approximation y_1

First approximation

$$y_1 = 0 + \int_0^x x^2 dx = \frac{1}{3}x^3$$

Second approximation putting $y = \frac{1}{3}x^3$ in Eq. (i), we have

$$y_2 = 0 + \int_0^x \frac{x^2}{\frac{1}{9}x^6 + 1} dx = \tan^{-1}\left(\frac{1}{3}x^3\right) = \frac{1}{3}x^3 - \frac{1}{81}x^9 + \dots$$

Neglecting $\frac{1}{81}x^9$ and higher powers as 0.25, 0.5, 1 are small.

Hence
$$y(0.25) = \frac{1}{3}(0.25)^3 = 0.005$$

$$y(0.5) = \frac{1}{3}(0.5)^3 = 0.042$$

$$y(1.0) = \frac{1}{3} - \frac{1}{81} = 0.321 \quad \left(\text{taking into consideration } \frac{1}{81}x^9\right)$$

Example 8.6: Use Picard's method to approximate the value of y when $x = 0.1$ given that $y = 1$ when $x = 0$ and $\dfrac{dy}{dx} = \dfrac{y - x}{y + x}$.

Solution:
$$\frac{dy}{dx} = f(x, y) = \frac{y - x}{y + x} \quad (\because y_0 = 1, x_0 = 0) \tag{i}$$

First approximation $\quad y_1 = y_0 + \int_{x_0}^x f(x, y_0)dx = y_0 + \int_{x_0}^x \frac{y - x}{y + x} dx$

$$= 1 + \int_0^x \frac{1 - x}{1 + x} dx = 1 + \int_0^x \left(\frac{2}{1 + x} - 1\right) dx$$

$$= 1 + [2\log(1 + x) - x]_0^x = 1 + 2\log(1 + x) - x$$

$$= 1 - x + 2\log(1 + x)$$

Second approximation

$$y_2 = 1 + \int_0^x \frac{y_1 - x}{y_1 + x} dx$$

$$= 1 + \int_0^x \frac{1 - x + 2\log(1 + x) - x}{1 - x + 2\log(1 + x) + x} dx \tag{ii}$$

$$= 1 - 2\int_0^x \frac{x}{1 + 2\log(1 + x)} dx + x$$

which is quite difficult to integrate putting $x = 0.1$ in Eq. (ii), we get

First approximation $\quad y_1 = 1 - 0.1 + 2\log(1 + 0.1) = 0.9828$.

Example 8.7: Use Picard's method to solve the equations

$$\frac{dx}{dt} = -y$$

$$\frac{dy}{dt} = x$$

Given that $x = 1$, $y = 0$ when $t = 0$.

Solution: $\quad\quad\quad\quad\frac{dx}{dt} = -y$ \hfill (ii)

$$\frac{dy}{dt} = x$$

Integrating Eq. (i) w.r.t. t between $t = 0$ to t

$$[x]_0^x = -\int_0^t y\, dt$$

$$x - 1 = -\int_0^t y\, dt$$

$$x = 1 - \int_0^t y\, dt \hfill \text{(iii)}$$

Integrating Eq. (ii) w.r.t. t from $t = 0$ to t, we get

$$[y]_1^y = \int_0^t x\, dt$$

$$y - 0 = \int_0^t x\, dt$$

$$y = \int_0^t x\, dt \hfill \text{(iv)}$$

Replacing y by 0 in Eq. (iii) and x by 1 in Eq. (iv), we get

$$x = 1 - \int_0^t 0\, dt \quad \text{and} \quad y = \int_0^t 1\, dt = t$$

$$= 1 - \int_0^t t\, dt = 1 - \frac{t^2}{2} \quad \text{and} \quad y = \int_0^t \left(1 - \frac{t^2}{2}\right) dt = t - \frac{t^3}{6}$$

$$= 1 - \int_0^t \left(t - \frac{t^3}{6}\right) dt = 1 - \frac{t^2}{2} + \frac{t^4}{24}$$

$$y = \int_0^t \left(1 - \frac{t^2}{2} + \frac{t^4}{24}\right) dt = t - \frac{t^3}{6} + \frac{t^5}{120}$$

PROBLEMS 8.1

1. Using Picard's method, find $y(0.2)$ given that $y' = x - y$; $y(0) = 1$.

[Ans. $y(0.2) = 0.837$]

2. Using Picard's method, obtain a solution up to fifth approximation to the equation $y' = y + x$, such that $y(0) = 1$. Check your answer by finding the exact particular solution. Also find $y(0.1)$ and $y(0.2)$.

[Ans. $y(0.1) = 1.1103$, $y(0.2) = 1.2428$]

3. Using Picard's method, find $y(0.2)$ and $y(0.4)$ given that $y' = 1 + y^2$ and $y(0) = 0$.

[Ans. $y(0.2) = 0.2027, y(0.4) = 0.4227$]

4. Using Picard's method to approximate the value of y when $x = 0.1$ given that $y(0) = 1$ and $y' = 3x + y^2$.

[Ans. $y(0.1) = 1.127$]

5. Using Picard's method obtain the second approximation of the solution to the given equation $y'' - x^3 y' - x^3 y = 0$ so that $y(0) = 1, y'(0) = 0.5$.

$$\left[\text{Ans. } 1 + \frac{1}{2}x + \frac{3}{40}x^5 \right]$$

8.3 TAYLOR'S SERIES METHOD

Consider the equation

$$\frac{dy}{dx} = f(x, y) \tag{8.3}$$

$$y' = f(x, y)$$

$$\frac{d^2 y}{dx^2} = \frac{df}{dx} = \frac{\partial f}{\partial x} + \frac{\partial f}{\partial y}\frac{dy}{dx} \tag{8.4}$$

$$y'' = f_x + f_y \frac{dy}{dx} = f_x + f_y f$$

Differentiating this successively, we get y''', y'''' so on.

Now the Taylor's series about $x = x_0$ is

$$y = y_0 + (x - x_0)y_0' + \frac{(x - x_0)^2}{2!}y_0'' + \frac{(x - x_0)^3}{3!}y_0''' + \dots \tag{8.5}$$

Putting $x = x_0$ and $y = y_0$ in y', y'', y''', we get the values of y_0', y_0'', y_0'''.

Substituting these values in Taylor's series, we get the value of y for every x for which Eq. (8.5) converges. On finding the value y_1 for $x = x_1$ from Eq. (8.5), y', y'', y''' can be obtained at $x = x_1$ by means of Eqs (8.3), (8.4) etc. Then y can be expanded about $x = x_1$. In this way, the solution can be extended beyond the range of convergence of series in Eq. (8.5).

Example 8.8: From Taylor's series for $y(x)$, find $y(0.1)$ correct to four decimal places if $y(x)$ satisfies $\frac{dy}{dx} = x - y^2$, $y(0) = 1$.

Solution: The Taylor's series for $y(x)$ is given by

$$y = y_0 + (x - x_0)y_0' + \frac{(x - x_0)^2}{2!}y_0'' + \frac{(x - x_0)^3}{2!}y_0''' + \dots$$

Now $x_0 = 0, y_0 = 1$

Then

$$y = 1 + xy_0' + \frac{x^2}{2}y_0'' + \frac{x^3}{6}y_0''' + \frac{x^4}{24}y_0'''' + \dots$$

Now

$$\frac{dy}{dx} = x - y^2$$

$$y' = x - y^2 \qquad\qquad y_0' = -1$$

$$y'' = 1 - 2yy'$$
$$y_0'' = 1 - 2(-1) = 3$$

$$y''' = 0 - 2yy'' - 2y'^2$$
$$y_0''' = -8$$

$$y'''' = -2yy''' - 2y'y''$$
$$y_0'''' = 34$$

$$y(0.1) = 1 - 0.1 + \frac{3}{2}(0.1)^2 - \frac{4}{3}(0.1)^3 - \frac{17}{12}(0.1)^4 \ldots$$

$$= 0.9138$$

Example 8.9: Find by Taylor's series method, the values of y at $x = 0.1$, $x = 0.2$ to five decimal places from $\dfrac{dy}{dx} = x^2 y - 1$, $y(0) = 1$.

Solution: Here $x = 0$, $y = 1$

$$y' = x^2 y - 1$$
$$y_0' = -1$$

$$y'' = 2xy + x^2 \frac{dy}{dx}$$
$$y_0'' = 0$$

$$y''' = 2y + 4xy' + x^2 y''$$
$$y_0''' = 2$$

$$y'''' = 6y' + 6xy'' + x^2 y''$$
$$y_0'''' = -6 \text{ etc.}$$

Putting these values in Taylor's series, we have

$$y = 1 + x(-1) + \frac{x^2}{2!}(0) + \frac{x^3}{3!}(2) + \frac{x^4}{4!}(-6) + \ldots$$

$$= 1 - x + \frac{x^3}{3} - \frac{x^4}{4} + \ldots$$

$$y(0.1) = 1 - (0.1) + \frac{(0.1)^3}{3} - \frac{(0.1)^4}{4} + \ldots = 0.90033$$

$$y(0.2) = 1 - (0.2) + \frac{(0.2)^3}{3} - \frac{(0.2)^4}{4} + \ldots = 0.80227$$

Example 8.10: Apply Taylor's series method to integrate

$$y' = 2t + 3y, \, y(0) = 1, \, t \in [0, 0.1], \, h = 0.1$$

Solution:

$$\frac{dy}{dt} = 2t + 3y \qquad \left(\frac{dy}{dt}\right)_{t=0} = 3, \text{ when } t = 0, y = 1$$

$$\frac{d^2 y}{dt^2} = 2 + 3\frac{dy}{dt} \qquad \left(\frac{d^2 y}{dt^2}\right)_{t=0} = 2 + 3 \times 3 = 11$$

$$\frac{d^3 y}{dt^3} = 3\frac{d^2 y}{dt^2} \qquad \left(\frac{d^3 y}{dt^3}\right)_{t=0} = 3 \times 11 = 33$$

$$\frac{d^4 y}{dt^4} = 3\frac{d^3 y}{dt^3} \qquad \left(\frac{d^4 y}{dt^4}\right)_{t=0} = 3 \times 33 = 99$$

$$\frac{d^5 y}{dt^5} = 3\frac{d^4 y}{dt^4} \qquad \left(\frac{d^5 y}{dt^5}\right)_{t=0} = 3 \times 99 = 297$$

Taylor's series about $t = t_0 = 0$

$$y = y_0 + (t - t_0)\left(\frac{dy}{dt}\right)_{t=0} + \frac{(t - t_0)^2}{2!}\left(\frac{d^2y}{dt^2}\right)_{t=0} + \frac{(t - t_0)^3}{3!}\left(\frac{d^3y}{dt^3}\right)_{t=0} + \frac{(t - t_0)^4}{4!}\left(\frac{d^4y}{dt^4}\right)_{t=0}$$

$$= 1 + 3t + \frac{11}{2!}t^2 + \frac{33}{3!}t^3 + \frac{99}{4!}t^4 + \frac{297}{5!}t^5 + \dots$$

$$y(0, 0.1) = 1 + 3(0.1) + \frac{11}{2}(0.1)^2 + \frac{33}{6}(0.1)^3 + \frac{99}{24}(0.1)^4 + \frac{297}{120}(0.1)^5 + \dots$$

$$= 1 + 0.3 + 0.055 + 0.0055 + 0.0004125 + 0.00002475$$

$$= 1.36093725$$

Example 8.11: Find the solutions $y(0.1)$ and $y(0.2)$ of the initial value problem $y' = x(1 - 2y^2)$, $y(0) = 1$, using the first two nonzero terms of the Taylor's series and $h = 0.1$.

Solution:
$$y' = x(1 - 2y^2)$$
$$y'' = (1 - 2y^2) + x(-4yy')$$
$$y''' = -4yy' - 4yy' + x(-4y'^2 - 4yy'')$$
$$y'''' = 8y'^2 - 8yy'' + (-4y'^2 - 4yy'') + x(-8y'y'' - 4y'y'' - 4yy''')$$

Putting
$$x = 0, y = 1$$
$$y' = 0$$
$$y_0'' = 1 - 2(1)^2 = -1$$
$$y_0''' = 0 + 0 = 0$$
$$y_0'''' = 0 - 8(1)(-1) + [0 - 4(-1)] + 0 = 12$$

Now the Taylor's series about $x = x_0$ is

$$y = y_0 + xy_0' + \frac{x^2}{2!}y_0'' + \frac{x^3}{3!}y_0''' + \frac{x^4}{4!}y_0'''' + \dots$$

$$y(0.1) = 1 + 0 + \frac{(0.1)^2}{2!}(-1) + \frac{(0.1)^3}{3!} \times 0 + \frac{(0.1)^4}{4!} \times 12$$

$$= 1 + -0.005 + 0.00005 = 0.99505$$

for
$$x_1 = 0.1, y_1 = 0.99505$$

Now Taylor's series about $x = x_1$

$$y = y_1 + (x - x_1)y_1' + \frac{(x - x_1)^2}{2!}y_1'' + \frac{(x - x_1)^3}{3!}y_1''' + \frac{(x - x_1)^4}{4!}y_1'''' + \dots$$

$$y_2 = y_1 + hy_1' + \frac{h^2}{2!}y_1'' + \frac{h^3}{3!}y_1''' + \frac{h^4}{4!}y_1'''' + \dots$$

Now
$$y_1' = 0.1[1 - 2(0.99505)^2] = -0.098024$$
$$y_1'' = (1 - 2(0.99505)^2 + 0.1[-4 \times 0.99505 \times -0.098024] = -0.94122$$

$$y(0.2) = 0.99505 + 0.1[-0.098024] + \frac{0.01}{2}[-0.94122]$$

$$= 0.99505 - 0.0098024 - 0.005 \times 0.04122 + \dots$$

$$= 0.9805415$$

Example 8.12: Employ Taylor's series method to obtain the approximate value of y at $x = 0.2$ for $y' = 2y + 3e^x$, $y(0) = 0$. Compare the approximate value with the exact value.

Solution: Here

$x_0 = 0$	$y_0 = 1$
$y' = 2y + 3e^x$	$y_0' = 3$
$y'' = 2y' + 3e^x$	$y_0'' = 6 + 3 = 9$
$y''' = 2y'' + 3e^x$	$y_0''' = 18 + 3 = 21$
$y'''' = 2y''' + 3e^x$	$y_0'''' = 42 + 3 = 45$

Now by Taylor's series expansion of $y = f(x)$ about $x = x_0 = 0$.

We have
$$y = y_0 + (x - 0)y_0' + \frac{(x-0)^2}{2!}y_0'' + \frac{(x-0)^3}{3!}y_0''' + \frac{(x-0)^4}{4!}y_0'''' + \dots$$

$$= y_0 + xy_0' + \frac{x^2}{2!}y_0'' + \frac{x^3}{3!}y_0''' + \frac{x^4}{4!}y_0'''' + \dots$$

$$= 3x + \frac{x^2}{2!} \times 9 + \frac{x^3}{3!} \times 21 + \frac{x^4}{4!} \times 45 + \dots$$

$$= 3x + \frac{9x^2}{2} + \frac{7x^3}{2} + \frac{15x^4}{8} + \dots$$

when $x = 0.2$

$$y(0.2) = 3(0.2) + \frac{9(0.2)^3}{2} + \frac{7(0.2)^3}{2} + \frac{15(0.2)^4}{8} + \dots$$

$$= 0.6 + 0.18 + 0.028 + 0.0030 = 0.811$$

Exact value
$$\frac{dy}{dx} = 2y + 3e^x$$

$$\frac{dy}{dx} - 2y = 3e^x$$

$$\text{I.F.} = e^{-2\int dx} = e^{-2x}$$

$$e^{-2x}y = \int 3e^{-x}dx + C = -3e^{-x} + C$$

$$y = -3e^x + Ce^{+2x}$$

when
$$x = 0, y = 0$$

$$0 = -3 + C, C = 3$$

$$y = -3e^x + 3e^{2x} = 3e^x(e^x - 1)]$$

when
$$x = 0.2, y = 3e^2(e^2 - 1)]$$

Hence, the value of y at $x = 0.2$ is 0.811.

Example 8.13: Use Taylor's series method to solve $\dfrac{dy}{dx} = x^2 - y$, $x = 0$, $y = 1$ at $x = 0.1$, $x = 0.2$, $x = 0.3$, $h = 0.1$.

Solution: Here

$$x_0 = 0 \qquad\qquad y_0 = 1$$
$$y' = x^2 - y \qquad\qquad y_0' = x_0^2 - y_0 = -1$$
$$y'' = 2x - y' \qquad\qquad y_0'' = 2x_0 - y_0' = 1$$
$$y''' = 2 - y'' \qquad\qquad y_0''' = 2 - y_0'' = 1$$
$$y'''' = -y''' \qquad\qquad y_0'''' = -y_0''' = -1$$

Now by Taylor's series expand $y = f(x)$ about $x = x_0$, we have

$$y = y_0 + (x - x_0)y_0' + \frac{(x - x_0)^2}{2!} y_0'' + \frac{(x - x_0)^3}{3!} y_0''' + \dots$$

For $x_1 = 0.1$

$$y_1 = y_0 + (x_1 - x_0)y_0' + \frac{(x_1 - x_0)^2}{2!} y_0'' + \frac{(x_1 - x_0)^3}{3!} y_0''' + \dots$$

$$= y_0 + hy_0' + \frac{h^2}{2!} y_0'' + \frac{h^3}{3!} y_0''' + \dots$$

$$= 1 + 0.1\,(-1) + \frac{(0.1)^2}{2!}\,(1) + \frac{(0.1)^3}{3!}\,(1) + \dots = 0.905125$$

$$x_1 = 0.1, \; y_1 = 0.905125$$

Now
$$y_1' = x_1^2 - y_1 = (0.1)^2 - 0.905125 = -0.895125$$
$$y_1'' = 2x_1 - y_1' = 2(0.1) + (0.895725) = 1.095125$$
$$y_1''' = 2 - y_1'' = 2 - (1.095125) = 0.904875$$
$$y_1'''' - y_1''' = -0.904875$$

Expanding by Taylor's series about $x = x_1$, we have

$$y = y_1 + (x - x_1)\,y_1' + \frac{(x - x_1)^2}{2!}\,y_1'' + \dots$$

For $x_2 = 0.2$ $\quad y_2 = y_1 + hy_1' + \dfrac{h^2}{2!}\,y_1'' + \dfrac{h^3}{3!}\,y_1'''+ \dots$

$$= (0.905125) + (0.1)\,(0.895125) + \frac{(0.1)^2}{2!}\,(1.095125) + \frac{(0.1)^3}{3!}\,(0.904875) + \dots$$

$$= 0.8212354$$

$$x_2 = 0.2, \quad y_2 = 0.8212352$$
$$y_2' = -0.7812352$$
$$y_2'' = 1.1812332$$
$$y_3''' = 0.8187648$$

Expanding by Taylor's series about $x = x_2$

$$y = y_2 + (x - x_2)y_2' + \frac{(x - x_2)^2}{2!}\, y_2'' + \dots$$

For $x_3 = 0.3$

$$y_3 = y_2 + hy_2' + \frac{h^2}{2!}y_2'' + \frac{h^3}{3!}y_2''' + \dots$$

$$= (0.8212352) + (0.1)(-0.7812352) + \frac{(0.1)^2}{2!}(1.1812332)$$

$$+ \frac{(0.1)^3}{3!}(0.8187645) + \dots$$

$$= 0.7491509$$

PROBLEMS 8.2

1. Using Taylor's series method, solve $y' = xy + y^2$, $y(0) = 1$ at $x = 0.1, 0.2, 0.3$.
   [Ans. $y(0.1) = 1.1167$, $y(0.2) = 1.2767$, $y(0.3) = 1.5023$]

2. Use Taylor's series method to compute $y(0.2)$ to three decimal places given
   $\frac{dy}{dx} = 1 - 2xy$ that $y(0) = 0$.   [Ans. 2.0206]

3. Solve $y' = y^2 + x$, $y(0) = 1$, using Taylor's series method and compute $y(0.1)$ and $y(0.2)$.   [Ans. 1.1164, 1.2725]

4. Evaluate $y(0.1)$ correct to six decimal places by Taylor's series method if $y(x)$ satisfies $y' = xy + 1$, $y(0) = 1$.   [Ans. 1.053425]

5. Using Taylor's series method, find $y(1.1)$, $y(1.2)$, given $\frac{dy}{dx} = xy^{1/3}$ and $y(1) = 1$.
   [Ans. 1.106808, 0.809333]

6. Solve $\frac{dy}{dx} = x^2 + y^2$, $y(0) = 1$. Determine the first three nonzero terms and hence obtain $y(1)$.
   $\left[\text{Ans. } \frac{1}{3}x^3 + \frac{1}{63}x^7 + \frac{2}{2079}x^{11} + \dots, y(1) = 0.3502\right]$

8.4 EULER'S METHOD

Consider the equation

$$\frac{dy}{dx} = f(x, y) \tag{8.6}$$

given that $y(x_0) = y_0$. Its curve of solution through $P(x_0, y_0)$ is shown dotted in Fig. 8.1. Now we have to find the ordinate of any other point Q on this curve.

Let us divide ZM into n subintervals each of width h at $Z_1, Z_2, \dots$ so that h is quite small.

In the interval ZZ_1, we approximate the curve by the tangent at P. If the ordinate through Z_1 meets the tangent in $P_1(x_0 + h, y_1)$, then

$$y_1 = Z_1 P_1 = ZP + R_1 P_1 = y_0 + PR_1 \tan \theta$$

$$= y_0 + h\left(\frac{dy}{dx}\right)_P = y_0 + hf(x_0, y_0) \tag{8.7}$$

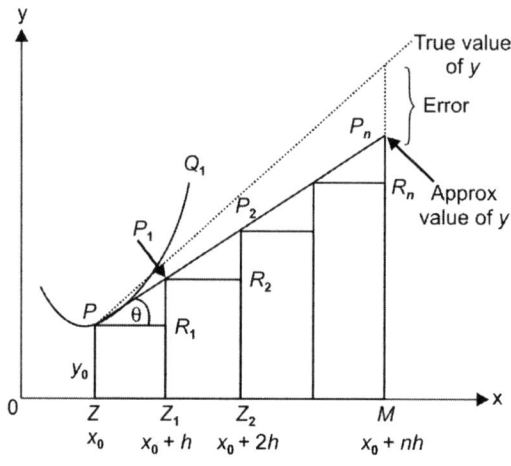

Fig. 8.1

Let P_1Q_1 be the curve of solution of Eq. (8.6) through P_1 and its tangent at P_1 meet the ordinate through Z_2 in $P_2(x_0 + 2h, y_2)$. Then

$$y_2 = y_1 + hf(x_0 + h, y_1) \tag{8.7a}$$

Repeating this process n times, we finally reach on an approximation MP_n of MQ given by

$$y_n = y_{n-1} + hf(x_0 + \overline{(n-1)}h, y_{n-1})$$

This is Euler's method of finding an approximate solution of Eq. (8.5).

Note: In Euler's method, we approximate the curve of the solution by the tangent in each interval, i.e. by a sequence of short lines. Unless h is small, the error is bound to be quite significant. This sequence of lines may also deviate considerably from the curve of solution. As such, the method is very slow and hence there is a modification in this method.

8.5 MODIFIED EULER'S METHOD

In the Euler's method, the curve of solution in the interval ZZ_1 is approximated by the tangent at P (Fig. 8.1), such that at P_1, we have

$$y_1 = y_0 + hf(x_0, y_0) \tag{8.8}$$

Then the slope of the curve of solution through P_1, i.e.

$$\left(\frac{dy}{dx}\right)_{P_1} = f(x_0 + h, y_1)$$

is computed and the tangent at P_1 and P_1Q is drawn meeting the ordinate through Z_2 in $P_2(x_0 + 2h, y_2)$, we find better approximation y_1 of $y(x_0 + h)$ by taking the curve as the mean of slopes of the tangents at P and P_1, etc.

$$y_1^{(1)} = y_0 + \frac{h}{2}[f(x_0, y_0) + f(x_0 + h, y_1)] \tag{8.9}$$

Similarly a better value of y_1 can be obtained

$$y_1^{(2)} = y_0 + \frac{h}{2}[f(x_0, y_0) + f(x_0 + h, y_1^{(1)})]$$

we repeat this step till two consecutive values of y agree.

This is then taken as the starting point for the next interval $Z_1 Z_2$.

Once y_1 is obtained to desired degree of accuracy y corresponding to Z_2 found from Euler's

$$y_2 = y_1 + hf(x_0 + h, y_1)$$

and a better approximation $y_2^{(1)}$ can be obtained by

$$y_2^{(1)} = y_1 + \frac{h}{2}[f(x_0 + h, y_1) + f(x_0 + 2h, y_2)]$$

We repeat this step till y_2 became stationary. Similarly, we can calculate y_3 and better approximate.

Example 8.14: Use Euler's method to find the approximate value of y when $x = 0.8$, if $\frac{dy}{dx} = 1 - 2xy$, given that $y = 0$ when $x = 0$ and $h = 0.2$.

Solution: $y_0 = 0$, $x_0 = 0$.

For $x_1 = 0.2$ $y_1 = y_0 + hf(x_0, y_0)$ by Euler's formula
$$= 0 + 0.2[1 - 2(0)] = 0.2$$

For $x_2 = 0.4$ $y_2 = y_1 + hf(x_1, y_1) = 0.2 + 0.2[1 - 2(0.2)(0.2)] = 0.384$

For $x_3 = 0.6$ $y_3 = y_2 + hf(x_2, y_2) = 0.384 + 0.2[1 - 2(0.4)(0.384)] = 0.52256$

For $x_4 = 0.8$ $y_4 = y_3 + hf(x_3, y_3) = 0.52256 + 0.2[1 - 2(0.6)(0.52256)]$
$$= 0.5971456$$

when $x = 0.8$, $y = 0.5971456$.

Example 8.15: Use Euler's method to solve for y at $x = 0.1$ from the equation $\frac{dy}{dx} = x + y + xy$ and $y(0) = 1$, taking step size $= 0.025$.

Solution: By Euler's method $x_0 = 0$, $x_1 = 0.025$, $x_2 = 0.05$, $x_3 = 0.075$, $x_4 = 0.1$.

For $x_1 = 0.025$ $y_1 = y_0 + hf(x_0, y_0)$
$$= 1 + 0.025(0 + 1 + 0)) = 1.025$$

For $x_2 = 0.05$ $y_2 = 1.025 + 0.025[0.025 + 1.025 + 0.025 \times 1.025]$
$$= 1.025 + 0.025[1.05 + 0.025625] = 1.025 + 0.02689 = 1.05189$$

For $x_3 = 0.075$ $y_3 = 1.05189 + 0.025[0.05 + 1.05189 + 0.05 \times 1.05189]$
$$= 1.05189 + 0.025[1.10189 + 0.05259]$$
$$= 1.05189 + 0.025 \times 1.15455$$
$$= 1.05189 + 0.0287125 = 1.0806$$

Similarly $x_4 = 0.1$ $y_4 = 1.0806 + 0.025[0.075 + 1.0806 + 0.075 + 1.0806]$
$$= 1.0806 + 0.025[1.1556 + 0.08105]$$
$$= 1.0806 + 0.025 \times 1.23665 = 1.0806 + 0.0309 = 1.1115$$

Example 8.16: Using modified Euler's method, find the approximate value of y at $x = 2$ in steps of 0.2 from $\dfrac{dy}{dx} = 2 + \sqrt{xy}$, $y(1) = 1$.

Solution: $\dfrac{dy}{dx} = 2 + \sqrt{xy}$, $y(1) = 1$, $h = 0.2$, $y_0 = 1$, $x_0 = 1$, $x_1 = 1.2$, $x_2 = 1.4$, $x_3 = 1.6$, $x_4 = 1.8$, $x_5 = 2$.

By Euler's method $\quad y_1 = y_0 + hf(x_0, y_0)$

$$= 1 + 0.2[2 + \sqrt{1}] = 1 + 0.2(3) = 1.6$$

By modified Euler's method

For $x_1 = 1.2$ $\qquad y_1^{(1)} = y_0 + \dfrac{h}{2}[f(x_0, y_0) + f(x_1, y_1)]$

$$= 1 + 0.1[3 + 2 + \sqrt{(1.2)(1.6)}] = 1.63856$$

$$y_1^{(2)} = y_0 + \dfrac{h}{2}[f(x_0, y_0) + f(x_1, y_1^{(1)})]$$

$$= 1 + 0.1[3 + 2 + \sqrt{(1.2)(1.63856)}] = 1.6402$$

$$y_1^{(3)} = y_0 + \dfrac{h}{2}[f(x_0, y_0) + f(x_1, y_1^{(2)})]$$

$$= 1.6402$$

Hence, when $x_1 = 1.2$, $y_1 = 1.6402$.

By Euler's method $\quad y_2 = y_1 + hf(x_1, y_1)$

$$= 1.6402 + 0.2[2 + \sqrt{(1.2)(1.6402)}]$$

$$= 1.402 + 0.2[2 + 1.4025] = 2.32078$$

Apply modified Euler's method

$$y_2^{(1)} = y_1 + \left[\dfrac{h}{2} f(x_1, y_1) + f(x_2, y_2)\right]$$

$$= 1.6402 + 0.1[3.4029 + 2\sqrt{(1.4)(2.32078)} = 2.36074$$

$$y_2^{(2)} = y_1 + \dfrac{h}{2}\left[f(x_1, y_1) + f(x_2, y_2^{(1)})\right]$$

$$= 1.6402 + 0.1[3.4029 + 2\sqrt{(1.4)(2.36074)} = 2.36228$$

$$y_2^{(3)} = y_1 + \dfrac{h}{2}\left[f(x_1, y_1) + f(x_2, y_2^{(2)})\right]$$

$$= 1.6402 + 0.1[3.4029 + 2\sqrt{(1.4)(2.36228)} = 2.36234$$

$$y_2^{(4)} = 1.6402 + 0.1[3.4029 + 2\sqrt{(1.4)(2.36234)} = 2.36234$$

Hence $\qquad x_2 = 1.4$, $y_2 = 2.36234$

$x_3 = 1.6$ by Euler's method $y_3 = y_2 + hf(x_2, y_2)$

$$= 2.36234 + 0.2[2 + \sqrt{(1.4)(2.36234)}] = 3.126058$$

Apply modified Euler's method

$$y_3^{(1)} = y_2 + \frac{h}{2}[f(x_2, y_2) + f(x_3, y_3)]$$

$$= 2.36234 + 0.1[3.81859 + 2 + \sqrt{(1.6)\,(3.126058)}] = 3.16784$$

$$y_3^{(2)} = y_2 + \frac{h}{2}[f(x_2, y_2) + f(x_3, y_3^{(1)})]$$

$$= 2.36234 + 0.1[3.81859 + 2 + \sqrt{(1.6)\,(3.16784)}] = 3.16933$$

$$y_3^{(3)} = 2.36234 + 0.1[3.81859 + 2 + \sqrt{(1.6)\,(3.16933)}] = 3.1693$$

when $x_3 = 1.6$, $y_3 = 3.1693$

$x_4 = 1.8$ by Euler's method $y_4 = y_3 + hf(x_3, y_3)$

$$= 3.1693 + 0.2[2 + \sqrt{3.1693)\,(1.6)}]$$

$$= 3.1693 + 0.2\,(4.25186) = 4.0196$$

Apply modified Euler's method

$$y_4^{(1)} = 3.1693 + 0.1[4.25186 + 2 + \sqrt{(1.8)\,(4.0196)}] = 4.06347$$

$$y_4^{(2)} = 3.1693 + 0.1[4.25186 + 2 + \sqrt{(1.8)\,(4.06347)}] = 4.06493$$

$$y_4^{(3)} = 3.1693 + 0.1[4.25186 + 2 + \sqrt{(1.8)\,(4.0693)}] = 4.06493$$

Hence, $\qquad x_4 = 1.8$, $y_4 = 4.06493$

$x_5 = 2.00$ by Euler's method $y_5 = y_4 + hf(x_4, y_4)$

$$= 4.06493 + 0.2[2 + \sqrt{(1.9)\,(4.06491)}] = 5.00589$$

By modified Euler's method

$$y_5^{(1)} = 4.0649 + 0.1[4.7049 + 2 + \sqrt{2 \times 5.00589}] = 5.051803$$

$$y_5^{(2)} = 4.0649 + 0.1[4.7049 + 2 + \sqrt{2 \times 5.051803}] = 5.05325$$

$$y_5^{(3)} = 5.05329$$

when $x_5 = 2$, $y_5 = 5.0532$,

hence $x = 2$, $y = 5$.

Example 8.17: Solve $\dfrac{dy}{dx} = 1 - y$, $y(0) = 0$ in the range $0 \le x \le 0.2$ by taking $h = 01$ by using Euler's and modified Euler's method.

Solution: By Euler's method

$$y_1 = y_0 + hf(x_0, y_0) = 0 + 0.1\,[1 - 0] = 0.1$$

By modified Euler's method

$$y_1^{(1)} = y_0 + \frac{h}{2}[f(x_0, y_0) + f(x_1, y_1)] = 0 + \frac{0.1}{2}[1 + (1 - 0.1)] = 0.095$$

$$y_1^{(2)} = 0 + \frac{0.1}{2}[1 + (1 - 0.095)] = 0.0952$$

$$y_1^{(3)} = 0 + \frac{0.1}{2}[1 + (1 - 0.095)] = 0.09524$$

Hence, $y_1 = 0.09524$ when $x_1 = 0.01$

By Euler's method $\qquad y_2 = y_1 + hf(x_1, y_1) = + 0.09524 + 0.1\,[1 - 0.09524] = 0.185716$

By modified Euler's method

$$y_2^{(1)} = y_1 + \frac{h}{2}[f(x_1, y_1) + g(x_2, y_2]$$

$$= 0.09524 + \frac{0.1}{2}[(1 - 0.09524) + (1 - 0.185716)]$$

$$= 0.09524 + 0.0859522 = 0.1855$$

$$y_2^{(2)} = 0.09524 + \frac{0.1}{2}[(1 - 0.09524) + (1 - 0.1811922)] = 0.18141839$$

Hence $y_2 = 0.1855$ when $x_2 = 0.2$.

Example 8.18: Using modified Euler's method, obtain a solution of $\dfrac{dy}{dx} = x + \sqrt{y} = f(x, y)$

with initial condition $y = 1$ at $x = 0$ for the range $0 \le x \le 0.6$ in steps of 0.2.

Solution: Here $x_0 = 1$, $y_0 = 1$, $h = 0.2$.

By Euler's method $\qquad y_1 = y_0 + hf(x_0, y_0) = 1 + 0.2[1 + 0] = 1.2$

By modified Euler's method

$$y_1^{(1)} = y_0 + \frac{h}{2}[f(x_0, y_0) + f(x_1, y_1)]$$

$$= 1 + \frac{0.2}{2}[1 + 0.2 + \sqrt{1.2}] = 1 + 0.1[1.2 + 1.095] = 1.2295$$

$$y_1^{(2)} = 1 + 0.1[1.2 + \sqrt{1.2295}] = 1.2309$$

$$y_1^{(3)} = 1 + 0.1[1.2 + \sqrt{1.2309}] = 1.2309$$

Hence we have $y_1 = 1.2309$ when $x_1 = 0.2$.

By Euler's method $\qquad y_2 = y_1 + hf(x_1, y_1) = 1.2309 + 0.2[0.2 + \sqrt{1.2309}]$

$$= 1.2309 + 0.2[0.2 + 1.109] = 1.4590$$

By modified Euler's method

$$y_2^{(1)} = 1.2309 + \frac{0.2}{2}[0.2 + \sqrt{(1.2309)} + 0.4 + \sqrt{(1.4590)}]$$

$$= 1.2309 + 0.1[0.6 + 1.109 + 1.222] = 1.5240$$

$$y_2^{(2)} = 1.2309 + 0.1[(0.2 + \sqrt{1.2309}) + (0.4 + \sqrt{1.5240})]$$

$$= 1.2309 + 0.1[0.6 + 1.109 + 1.235] = 1.5253$$

$$y_2^{(3)} = 1.2309 + 0.1[(0.2 + \sqrt{1.2309}) + (0.4 + \sqrt{1.5253})] = 1.5253$$

Hence, we have $y_2 = 1.5253$, $x_2 = 0.4$.

Again by Euler's method

$$y_3 = y_2 + 0.2(x_2 + \sqrt{y_2})$$
$$= 1.5253 + 0.2[0.4 + \sqrt{1.5253}]$$
$$= 1.5253 + 0.2[0.4 + \sqrt{1.235}] = 1.78057$$

By modified Euler's method

$$y_3^{(1)} = y_2 + \frac{0.2}{2}[f(x_2, y_2) + f(x_3, y_3)]$$
$$= 1.5253 + 0.1[(0.4 + \sqrt{1.5253}) + (0.6 + \sqrt{1.78057})]$$
$$= 1.5253 + 0.1[1 + 1.235 + 1.361] = 1.8849$$

$$y_3^{(2)} = 1.5253 + 0.1[1 + \sqrt{1.5253}) + \sqrt{1.8849}]$$
$$= 1.5253 + 0.1[1 + 1.235 + 1.373] = 1.8851$$

$$y_3^{(3)} = 1.8851$$

Hence $x = 0.6$, when $y = 1.8851$.

Example 8.19: Using Euler's method, obtain the solution of the equation $\dfrac{dy}{dx} = x + y$ with initial condition for the range $0 \le x \le 0.3$ in step of 0.1, $y_0 = 1$, $x_0 = 0$.

Solution: For $x_1 = 0.1$ by Euler's method,

$$y_1 = y_0 + hf(x_0, y_0)$$
$$= 1 + 0.1\,[1] = 1.1$$

Apply modified Euler's method

$$y_1^{(1)} = y_0 + \frac{1}{2}h\,[f(x_0, y_0) + f(x_1, y_1)]$$

$$= 1 + \frac{0.1}{2}[f(0.1) + f(0.1, 1.1)] = 1 + 0.05[(0 + 1) + f(0.1, 1.1)]$$
$$= 1 + 0.05(2.2) = 1 + 0.1[1.1] = 1 + 0.11 = 1.11$$

$$y_1^{(2)} = 1 + \frac{0.1}{2}[f(0.1) + f(0.1, 1.11)] = 1 + \frac{0.1}{2}[1 + 0.1 + 1.11]$$

$$= 1 + \frac{0.1}{2}[2.2] = 1 + 0.1[1.105] = 1 + 0.1105 = 1.1105$$

$$y_1^{(3)} = 1 + \frac{0.1}{2}[1 + 0.1 + 1.1105] = 1 + \frac{0.1}{2}[2.2105]$$
$$= 1 + 0.1[1.105] = 1 + 0.1105 = 1.1105$$

Hence $y_1 = 1.1105$ when $x_1 = 0.1$
For $x_2 = 0.2$ by Euler's method

$$y_2 = y_1 + hf(x_1, y_1)$$
$$= 1.1105 + 0.1\,[x_1 + y_1] = 1.1105 + 0.1\,[0.1 + 1.1105]$$
$$= 1.1105 + 0.1\,[1.2105]$$
$$= 1.1105 + 0.12105 = 1.23155 = 1.2316$$

First better approximation to y_2

$$y_2^{(1)} = y_1 + \frac{h}{2}[f(x_1, y_1) + f(x_2, y_2)]$$

$$= 1.1105 + \frac{0.1}{2}[1.2105 + 1.4316]$$

$$= 1.1105 + \frac{0.1}{2}[2.6421] = 1.1105 + 0.1[1.32105]$$

$$= 1.1105 + 0.1321 = 1.2426$$

$$y_2^{(2)} = y_1 + \frac{h}{2}[f(x_1, y_1) + f(x_2, y_2^{(1)})]$$

$$= 1.1105 + \frac{0.1}{2}[1.2105 + 1.4426]$$

$$= 1.1105 + \frac{0.1}{2}[2.6531] = 1.1105 + 0.1[1.3266]$$

$$= 1.1105 + 0.13266 = 1.24316$$

$$y_2^{(3)} = 1.1105 + \frac{0.1}{2}[1.2105 + 1.44316]$$

$$= 1.1105 + \frac{0.1}{2}[2.65366] = 1.1105 + 0.1[1.3268]$$

$$= 1.1105 + 0.13268 = 2.4318$$

Hence $y_2 = 2.43216$, when $x = 0.2$

For $x = 0.3$, $\qquad y_3 = y_2 + hf(x_2, y_2)$

$$= 2.4316 + (0.1)[0.2 + 2.4316] = 2.4316 + (0.1)[2.6316]$$

$$= 2.4316 + 0.26316 = 2.69476$$

$$y_3^{(1)} = y_2 + \frac{h}{2}[f(x_2, y_2) + f(x_3, y_3)]$$

$$= 2.4316 + \frac{0.1}{2}[2.6316 + 0.3 + 2.6958]$$

$$= 2.4316 + \frac{0.1}{2}[5.5264] = 2.4316 + 0.1[2.7632]$$

$$= 2.4316 + 0.2763 = 2.6679$$

Hence $y_3 = 2.6679$, when $x = 0.3$.

Example 8.20: Using Euler's method, find an approximate value of y when $x = 0.1$ given that $\frac{dy}{dx} = x + y$ and $y = 1$ when $x = 0$.

Solution: The problem can be solved as shown in the table. Taking $h = 0.1$, the various calculations are arranged as follows, here h is very small.

x	y	$x + y = \dfrac{dy}{dx}$	old $y + 0.1\left(\dfrac{dy}{dx}\right) = $ new y
0.0	1.00	1.00	$1.00 + 0.1\,(1.00) = 1.10 = y_1$
0.1	1.10	1.20	$1.10 + 0.1\,(1.20) = 1.22 = y_2$
0.2	1.22	1.42	$1.22 + 0.1\,(1.42) = 1.36 = y_3$
0.3	1.36	1.66	$1.36 + 0.1\,(1.66) = 1.53 = y_4$
0.4	1.53	1.93	$1.53 + 0.1\,(1.93) = 1.72 = y_5$
0.5	1.72	2.22	$1.72 + 0.1\,(2.22) = 1.94 = y_6$
0.6	1.94	1.54	$1.94 + 0.1\,(2.54) = 2.19 = y_7$
0.7	2.19	2.89	$2.19 + 0.1\,(2.89) = 2.48 = y_8$
0.8	2.48	3.26	$2.48 + 0.1\,(3.26) = 2.81 = y_9$
0.9	2.81	3.71	$2.81 + 0.1\,(3.71) = 3.18 = y_{10}$
1.0	3.18	–	–

Thus, the required approximate value of $y = 3.18$.

Example 8.21: Solve the following by Euler's modified method $\dfrac{dy}{dx} = \log(x + y),\ y(0) = 2$ at $x = 1.2$ and 1.4 with $h = 0.2$.

Solution: The problem can also be solved by Euler's modified method as shown in the following table.

x	$\log(x + y) = y'$	Mean slope	Old $y + 0.2$ (mean slope) = new y
0.0	$\log(0 + 2)$	–	$2 + 0.2\,(0.301) = 2.0602$
0.2	$\log(0.2 + 2.0602)$	½ (0.301 + 0.3541)	$2 + 0.2\,(0.3276) = 2.0655$
0.2	$\log(0.2 + 2.0655)$	½ (0.301 + 0.3552)	$2 + 0.2\,(0.3281) = 2.0656$
0.2	0.3552	–	$2.066 + 0.2\,(0.3552) = 2.1366$
0.4	$\log(0.4 + 2.1366)$	½ (0.3552 + 0.4042)	$2.0656 + 0.2\,(0.3797) = 2.1415$
0.4	$\log(0.4 + 2.1415)$	½ (0.3552 + 0.4051)	$2.0656 + 0.2\,(0.3801) = 2.1416$
0.4	0.4051	–	$2.1416 + 0.2\,(0.4051) = 2.2226$
0.6	$\log(0.6 + 2.2226)$	½ (0.4051 + 0.4506)	$2.1416 + 0.2\,(0.4279) = 2.2272$
0.6	$\log(0.6 + 2.2272)$	½ (0.4051 + 0.4514)	$2.1416 + 0.2\,(0.4282) = 2.2272$
0.6	0.4514	–	$2.2272 + 0.2\,(0.4514) = 2.3175$
0.8	$\log(0.8 + 2.3175)$	½ (0.4514 + 0.4938)	$2.2272 + 0.2\,(0.4726) = 2.3217$
0.8	$\log(0.8 + 2.3217)$	½ (0.4514 + 0.4943)	$2.2272 + 0.2\,(0.4727) = 2.3217$
0.8	0.4943	–	$2.3217 + 0.2\,(0.4943) = 2.4206$
1.0	$\log(1 + 2.4206)$	½ (0.4943 + 0.5341)	$2.3217 + 0.2\,(0.5142) = 2.4245$
1.0	$\log(1 + 2.4245)$	½ (0.4943 + 0.5346)	$2.3217 + 0.2\,(0.5144) = 2.4245$
1.0	0.5346	–	$2.4245 + 0.2\,(0.5346) = 2.5314$
1.2	$\log(1.2 + 2.5314)$	½ (0.5346 + 0.5719)	$2.4245 + 0.2\,(0.5532) = 2.5351$
1.2	$\log(1.2 + 2.5351)$	½ (0.5346 + 0.5723)	$2.4245 + 0.2\,(0.5534) = 2.5351$
1.2	0.5723	–	$2.5351 + 0.2\,(0.5723) = 2.6496$
1.4	$\log(1.4 + 2.6496)$	½ (0.5723 + 0.6074)	$2.5354 + 0.2\,(0.5898) = 2.6531$
1.4	$\log(1.4 + 2.6531)$	½ (0.5723 + 0.6078)	$2.5351 + 0.2\,(0.5900) = 2.6531$

Hence $y(1.2) = 2.5351$, $y(1.4) = 2.6531$ approximately.

Notes: 1. We obtain the exact value of y as 3.44, whereas by Euler's method $y = 3.18$ in example 8.20. If the interval is divided in small intervals, the accuracy would be increased considerably but at the cost of double labour. The Euler's method is though simple, not the best.

2. The modified Euler's method is an improvement over the Euler's method and the result is very near to the exact result.

8.6 RUNGE-KUTTA (RK) METHODS

This method has an advantage over Taylor's series method where we have to calculate higher order derivatives. In this method, we are not required to calculate higher order derivatives and give greater accuracy. These methods agree with Taylor series solution up to term in h^r where r differs from method to method and is called the order of that method.

1. First Order RK Method

By Euler's method, we have

$$y_1 = y_0 + hf(x_0, y_0) = y_0 + hy_0' \quad [\because y' = f(x, y)]$$

Expanding by Taylor's series

$$y_1 = y(x_0 + h) = y_0 + hy_0' + \frac{h^2}{2!} y_0'' + \dots$$

It shows that Euler's method agrees with Taylor's series solution up to terms in h. Hence Euler's method is RK method of first order.

2. Second Order RK Method

The modified Euler's method gives

$$y_1 = y_0 + \frac{h}{2} [f(x_0, y_0) + f(x_0 + h, y_1^{(1)})] \tag{8.10}$$

where

$$y_1^{(1)} = y_0 + hf(x_0, y_0)$$

i.e.

$$y_1 = y_0 + \frac{h}{2} f(x_0, y_0) + f(x_0 + h, y_0 + hf(x_0, y_0)) \tag{8.11}$$

Now, by Taylor's series

$$y_1 = y(x_0 + h) = y_0 + hy_0' + \frac{h^2}{2!} y_0'' + \dots \quad \text{(by Taylor's series)} \tag{8.12}$$

Now expanding $f[x_0 + h, y_0 + hf(x_0, y_0)]$ by Taylor's series for a function of two variables.

$$y_1 = y_0 + \left[\frac{h}{2} f(x_0, y_0) + f \left\{ (x_0, y_0) + g \left(\frac{\partial f}{\partial x} \right)_{(x_0, y_0)} + hf(x_0, y_0) \left(\frac{\partial f}{\partial x} \right)_{(x_0, y_0)} \right\} \right.$$

$$\left. + \text{ term containing second and higher power of } h \right]$$

$$= y_0 + \frac{1}{2}\left[hf(x_0, y_0) + hf(x_0, y_0) + h^2 \left\{ \left(\frac{\partial f}{\partial x}\right)_{(x_0, y_0)} + \left(\frac{\partial f}{\partial y}\right)_{(x_0, y_0)} f(x_0, y_0) \right\} \right.$$

$$\left. + O(h^3)\,(\text{term containing } h^3 \text{ onwards}) \right]$$

$$= y_0 + \frac{1}{2}\left[2hf(x_0, y_0) + h^2 \left\{ \left(\frac{\partial f}{\partial x}\right)_{(x_0, y_0)} + f(x_0, y_0)\left(\frac{\partial f}{\partial y}\right)_{(x_0, y_0)} \right\} \right] + O(h^3)$$

$$= y_0 + hf_0 + \frac{h^2}{2} f_0' + O(h^3) \qquad \left[\because \frac{df}{dx} = \frac{\partial f}{\partial x} + f \frac{\partial f}{y} \right]$$

$$= y_0 + hy_0' + \frac{h^2}{2} y_0'' + O(h^3) \qquad\qquad\qquad\qquad (8.13)$$

Comparing Eqs (8.12) and (8.13), it follows that modified Euler's method is RK method of second order. These agree up to third term.

Working rule:
$$y_1 = y_0 + \frac{1}{2}(k_1 + k_2)$$
$$k_1 = hf(x_0, y_0)$$
$$k_2 = h[f(x_0 + h, y_0 + k_1)]$$

3. Third Order RK Method

The third order RK formula is

$$y_1 = y_0 + \frac{1}{6}(k_1 + 4k_2 + k_3)$$

where $k_1 = hf(x_0, y_0)$
$$k_2 = hf(x_0 + h, y_0 + k_1)$$
$$k_3 = hf(x_0 + h, y_0 + k_2)$$
$$k_4 = hf\left(x_0 + \frac{h}{2}, y_0 + \frac{k_1}{2} \right)$$
$$k = \frac{1}{6}(k_1 + 4k_4 + k_3)$$
$$y_1 = y_0 + k$$

4. Fourth Order RK Method

This method is commonly used.

Working rule: Consider

$$\frac{dy}{dx} = f(x, y),\ y(x_0) = y_0$$

$$k_1 = hf(x_0, y_0)$$

$$k_2 = hf\left(x_0 + \frac{h}{2}, y_0 + \frac{k_1}{2}\right)$$

$$k_3 = hf\left(x_0 + \frac{h}{2}, y_0 + \frac{k_2}{2}\right)$$

$$k_4 = hf(x_0 + h, y_0 + k_3)$$

$$k = \frac{1}{6}(k_1 + 2k_2 + 2k_3 + k_4)$$

k is weighted mean of k_1, k_2, k_3, k_4

Then $\qquad\qquad y_1 = y_0 + k$

and $\qquad\qquad x_1 = x_0 + h$

The increment in y in second interval is computed in a similar manner by means of the following formula.

$$k_1 = hf(x_1, y_1)$$

$$k_2 = hf\left(x_1 + \frac{h}{2}, y_1 + \frac{k_1}{2}\right)$$

$$k_3 = hf\left(x_1 + \frac{h}{2}, y_1 + \frac{k_1}{2}\right)$$

$$k_4 = hf(x_1 + h, y_1 + k_3)$$

$$k = \frac{1}{6}(k_1 + 2k_2 + 2k_3 + k_4)$$

$$y_2 = y_1 + k \text{ and } x_2 = x_1 + h$$

and so on for succeeding intervals.

Note: One of the advantage of these methods is that the operation is identical whether the differential equation is linear or nonlinear.

Example 8.22: Using RK method of fourth order, find $y\,(0.2)$ for the equation $\dfrac{dy}{dx} = \dfrac{y - x}{y + x}$, $y\,(0)=1$. Take $h = 0.2$.

Solution: Given $y' = f(x, y) = \dfrac{y - x}{y + x}$, $x_0 = 0, y_0 = 1, h = 0.2$.

$$k_1 = hf(x_0, y_0) = 0.2\left[\frac{1-0}{1+0}\right] = 0.2$$

$$k_2 = h\left[f\left(x_0 + \frac{h}{2}, y_0 + \frac{k_1}{2}\right)\right] = 0.2\left[\frac{\left(y_0 + \frac{k_1}{2}\right) - \left(x_0 + \frac{h}{2}\right)}{\left(y_0 + \frac{k_1}{2}\right) + \left(x_0 + \frac{h}{2}\right)}\right]$$

$$= 0.2\left[\frac{1 + \frac{0.2}{2} - 0 - \frac{0.2}{2}}{1 + \frac{0.2}{2} + 0 - \frac{0.2}{2}}\right] = 0.166$$

$$k_3 = hf\left(x_0 + \frac{h}{2}, y_0 + \frac{k_2}{2}\right) = 0.2\left[\frac{\left(y_0 + \frac{k_1}{2}\right) - \left(x_0 + \frac{h}{2}\right)}{\left(y_0 + \frac{k_1}{2}\right) + \left(x_0 + \frac{h}{2}\right)}\right]$$

$$= 0.2\left[\frac{\left(1 + \frac{0.1666}{2}\right) - \left(0 + \frac{0.2}{2}\right)}{\left(1 + \frac{0.1666}{2}\right) + \left(0 + \frac{0.2}{2}\right)}\right] = 0.16619$$

$$k_4 = h[f(x_0 + h, y_0 + k_3)] = 0.2\left[\frac{y_0 + k_3 - (x_0 + h)}{y_0 + k_3 + (x_0 + h)}\right]$$

$$= 0.2\left[\frac{1 + 0.16619 - 0.2}{1 + 0.16619 + 0.2}\right] = 0.2 \times \frac{0.96619}{1.36619} = 0.14144$$

$$k = \frac{1}{6}(0.2 + 0.1666 + 0.16619 + 0.14144] = 0.16783$$

$$y(0.2) = y_0 + k = 1 + 0.16783 = 1.16783$$

Example 8.23: Using RK method of order four, compute $y(0.2)$ and $y(0.4)$ from

$10\frac{dy}{dx} = x^2 + y^2$, $y(0) = 1$, taking $h = 0.1$.

Solution: Given $y' = f(x, y) = x^2 + y^2$, $x_0 = 0$, $y_0 = 1$, $h = 0.1$

Here
$$k_1 = (0.1)\left(\frac{x_0^2 + y_0^2}{10}\right) = 0.1$$

$$k_2 = \frac{0.1}{10}\left(x_0 + \frac{h}{2}\right)^2 + \left(y_0 + \frac{k_1}{2}\right)^2 = 0.01012525$$

$$k_3 = \frac{0.1}{10}\left(x_0 + \frac{h}{2}\right)^2 + \left(y_0 + \frac{k_2}{2}\right)^2 = 0.0101265$$

$$k_4 = \frac{0.1}{10}[(x_0 + h)^2 + (y_0 + k_3)^2] = 0.010303$$

$$k = \frac{1}{6}[0.01 + 2(0.01012525) + 2(0.0101265) + 0.0101344] = 0.0101344$$

$$y_1 = y_0 + 0.101344 = 1.010344$$

For $y(0.2)$, $x_1 = x_0 + h = 0 + 0.1 = 0.1$

$y_1 = 1.010344$, $h = 0.1$

$$k_1 = \frac{0.1}{10}[(0.1)^2 + (1.010134)^2] = 0.010303$$

$$k_2 = hf\left(x_1 + \frac{h}{2}, y_1 + \frac{k_1}{2}\right)$$

$$= 0.1f\left(0.1 + \frac{0.1}{2}, 1.01034 + \frac{0.010303}{2}\right) = 0.010533$$

$$k_3 = hf\left(x_1 + \frac{h}{3}, y_1 + \frac{k_2}{2}\right) = 0.1f\left(0.1 + \frac{0.1}{2}, 1.01034 + \frac{0.010533}{2}\right)$$

$$= \frac{0.1}{10}\left[\left(0.1 + \frac{0.1}{2}\right)^2 + \left(1.010134 + \frac{0.010533}{2}\right)^2\right] = 0.0105354$$

$$k_4 = \frac{0.1}{10}\left[(0.1 + 0.1)^2 + (1.010134 + 0.0105354)^2\right] = 0.010817$$

$$k = \frac{1}{6}\left[0.010303 + 2(0.010533 + 0.0105354) + 0.010817\right] = 0.01054$$

$$y_2 = y_1 + k = 1.010134 + 0.01054 = 1.0206066$$

For $y(0.3)$,
$$x_2 = x_1 + h = 0.1 + 0.1 = 0.2$$
$$y_2 = 1.0206066, h = 0.1$$

$$k_1 = hf(x_2, y_2) = 0.1\left(\frac{x_2^2 + y_2^2}{10}\right) = 0.010816$$

$$k_2 = hf\left(x_2 + \frac{h}{2}, y_2 + \frac{k_1}{2}\right) = \frac{0.1}{10}\left[\left(x_2 + \frac{h}{2}\right)^2 + \left(y_2 + \frac{k_1}{2}\right)^2\right] = 0.011152$$

$$k_3 = hf\left(x_2 + \frac{h}{2}, y_2 + \frac{k_2}{2}\right) = \frac{0.1}{10}\left[\left(x_2 + \frac{h}{2}\right)^2 + \left(y_2 + \frac{k_1}{2}\right)^2\right] = 0.0111553$$

$$k_4 = hf(x_2 + h, y_2 + k_3) = \frac{1.0}{10}\left[(x_2 + h)^2 + (y_2 + k_3)^2\right] = 0.011545$$

$$k = \frac{1}{6}(k_1 + 2k_2 + 2k_3 + k_4) = 0.0111626$$

$$y_2 = 1.0206066 + 0.0111626 = 1.03176$$

For $y(0.4)$,
$$x_3 = x_2 + h = 0.2 + 0.1 = 0.3, y_3 = 1.03176, h = 0.1$$

$$k_1 = hf(x_3, y_3) = 0.1\left(\frac{x_3^2 + y_3^2}{10}\right) = 0.0115453$$

$$k_2 = hf\left(x_3 + \frac{h}{2}, y_3 + \frac{k_1}{2}\right) = \frac{0.1}{10}\left[\left(x_3 + \frac{h}{2}\right)^2 + \left(y_3 + \frac{k_1}{2}\right)^2\right] = 0.011989$$

$$k_3 = hf\left(x_3 + \frac{h}{2}, y_3 + \frac{k_2}{2}\right) = \frac{0.1}{10}\left[\left(x_3 + \frac{h}{2}\right)^2 + \left(y_3 + \frac{k_2}{2}\right)^2\right] = 0.0119943$$

$$k_4 = hf(x_3 + h, y_3 + k) = \frac{0.1}{10}\left[(x_3 + h)^2 + (y_3 + k_3)^2\right] = 0.012494$$

$$k = \frac{1}{6}(k_1 + 2k_2 + 2k_3 + k_4) = 0.01200093$$

$$y_4 = y_3 + k = 1.03176 + 0.01200093 = 1.04376$$

Example 8.24: Using RK method of fourth order find $y(0.2)$ given that $\dfrac{dy}{dx} = 3x + \dfrac{y}{2}$, $y(0) = 1$ taking $h = 0.1$.

Solution: For $y(0.1)$ $\qquad k_1 = 0.1\,(3x_0 + y_0/2) = 0.1 \times 0.5 = 0.05$.

$$k_2 = hf\left(x_0 + \frac{h}{2}, y_0 + \frac{k_1}{2}\right)$$

$$= 0.1\left[\frac{3h}{2} + \frac{y_0 + \dfrac{k_1}{2}}{2}\right] = 0.1\left[\frac{3(0.1)}{2} + \frac{1 + \dfrac{0.05}{2}}{2}\right] = 0.06625$$

$$k_3 = hf\left(x_0 + \frac{h}{2}, y_0 + \frac{k_2}{2}\right) = 0.1\left[\frac{3h}{2} + \frac{y_0 + \dfrac{k_2}{2}}{2}\right] = 0.06665625$$

$$k_4 = hf\,(x_0 + h, y_0 + k_3) = 0.1\left[3h + \frac{y_0 + k_3}{2}\right] = 0.0833328$$

$$k = \frac{1}{6}[0.05 + 0.06625 + 0.06665625 + 0.833328] = 0.0665242$$

$$y_1 = 1 + 0.0665242 = 1.0665242$$

For $y(0.2)$ $\qquad x_1 = 0 + 0.1 = 0.1,\ y_1 = 1.0665242,\ h = 0.1$

$$k_1 = 0.1\left(3x_1 + \frac{y_1}{2}\right) = 0.1\left[3(1) + \frac{1.0665242}{2}\right] = 0.08332621$$

$$k_2 = 0.1\left[3\left(x_1 + \frac{h}{2}\right) + \frac{y_1 + \dfrac{k_1}{2}}{2}\right] = 0.10040925$$

$$k_3 = 0.1\left[3\left(x_1 + \frac{0.1}{2}\right) + \frac{y_1 + \dfrac{k_1}{2}}{2}\right] = 0.100836441$$

$$k_4 = 0.119497636$$

$$k = \frac{1}{6}[k_1 + 2k_2 + 2k_3 + k_4] = 0.10841655$$

$$y_2 = y_1 + k = 1.0665242 + 0.10841655 = 1.1749$$

Example 8.25: For the equation $\dfrac{dy}{dx} = x - y, y(0) = 1$, find the value of y when $x = 0.1, 0.2$ taking $h = 0.1$

Solution: Given $x_0 = 0,\ y_1 = 1, f(x_0, y_0) = -1$.

$$k_1 = hf(x_0, y_0) = -0.1$$

$$k_2 = 0.1\left(x_0 + \frac{h}{2}\right) - \left(y_0 + \frac{k_1}{2}\right) = 0.1\left[\frac{0.1}{2} - 1 + \frac{0.1}{2}\right] = -0.09$$

$$k_3 = 0.1\left(x_0 + \frac{h}{2}\right) - \left(y_0 + \frac{k_2}{2}\right) = 0.1\left[\frac{0.1}{2} - 1 + \frac{0.09}{2}\right] = -0.0905$$

$$k_4 = 0.1[0 + 0.1 - 1 + 0.0905] = -0.08085$$

$$k = \frac{1}{6}[k_1 + 2k_2 + 2k_3 + k_4] = -0.090325$$

$$y_1(0.1) = y_0 + k = 1 + (-0.090325) = 0.909675$$

When $x_1 = x_0 + h = 0 + 0.1$, $y_1 = 0.909675$, $h = 0.1$

$$k_1 = 0.1[0.1 - 0.90968] = -0.80968$$

$$k_2 = 0.1\left[0.1 + \frac{0.1}{2} - 0.90968 + \frac{0.80968}{2}\right] = -0.0719195$$

$$k_3 = 0.1\left[0.1 + \frac{0.1}{2} - 0.90968 + \frac{0.07192}{2}\right] = -0.07237$$

$$k_4 = 0.1[0.1 + 0.1 - 0.90968 + 0.07237] = -0.06373$$

$$k = -0.072213$$

$$y_2(0.2) = y_1 + k = 0.90968 - 0.072213 = 0.981893$$

8.7 PREDICTOR–CORRECTOR METHODS

Consider the differential equation

$$\frac{dy}{dx} = f(x, y); \, y(x_0) = y_0$$

By Euler's method, we have

$$y_i = y_{i-1} + hf[x_0 + (i-1)h, y_{i-1}], \, i = 1, 2, 3$$

or
$$= y_{i-1} + hf(x_{i-1}, y_{i-1}) \tag{8.14}$$

By modified Euler's method

$$y_i = y_{i-1} + \frac{h}{2}[f(x_{i-1}, y_{i-1}) + f(x_i, y_i)] \tag{8.15}$$

Substitute the value of y_i from Eq. (8.14) in Eq. (8.15), we get better approximate value of y_i and again substitute y_i in Eq. (8.2), we get still better approximation of y_i. This process is repeated till two values of y_i are same. This technique of refining an initially crude y_i by means of a more accurate formula is known as **Predictor–Corrector method.**

Equation (8.14) is called **Predictor** and Eq. (8.15) is called **Corrector.**

In predictor–corrector methods, four prior values are needed for finding the value of y at x_i. Though slightly complex, these methods have the advantage of giving an estimate of error from successive approximation of y_i.

In this section, we will study two such methods, namely

i. Milne's method

ii. Adams-Bashforth method

8.8 MILNE'S METHOD

Given $$\frac{dy}{dx} = f(x, y) \tag{8.16}$$

and $$y = y_0, x = x_0$$

To find an approximate value of y for $x = x_0 + nh$ by Milne's method, we proceed as follows.

The value $y_0 = y(x_0)$ being given, we compute
$$y_1 = y(x_0 + h), y_2 = y(x_0 + 2h), y_3 = y(x_0 + 3h)$$

by Picard's or Taylor's series method.

Next we calculate $f_0 = f(x_0, y_0), f_1 = f(x_0 + h, y_1)$
$$f_2 = f(x_0 + 2h, y_2), f_3 = f(x_0 + 3h, y_3)$$

Then to find $y_4 = y(x_0 + 4h)$, we substitute Newton's forward interpolation formula

$$f(x, y) = f_0 + n\Delta f_0 + \frac{n(n-1)}{2}\Delta^2 f_0 + \frac{n(n-1)(n-2)}{6}\Delta^3 f_0 + \dots$$

Integrating Eq. (8.16) with its limits (x_0, x_4)

$$\int_{x_0}^{x_4} \frac{dy}{dx} \cdot dx = \int_{x_0}^{x_4} f(x, y)dx$$

$$y_4 - y_0 = \int_{x_0}^{x_4} f(x, y)dx$$

$$y_4 = y_0 + \int_{x_0}^{x_0 + 4h} f(x, y)dx$$

$$= y_0 + \int_{x_0}^{x_0 + 4h}\left(f_0 + n\Delta f_0 + \frac{n(n-1)}{2}\Delta^2 f_0 + \dots \right)dx$$

Put $x = x_0 + nh, dx = hdn$

$$y_4 = y_0 + h\int_0^4 \left(f_0 + n\Delta f_0 + \frac{n(n-1)}{2}\Delta^2 f_0 + \dots \right)dn$$

$$= y_0 + h\left(4f_0 + 8\Delta f_0 + \frac{20}{3}\Delta^2 f_0 + \frac{8}{3}\Delta^3 f_0 + \dots \right)$$

Neglecting fourth and higher order differences and expressing * in terms of functiion value, we get

$$y_4 = y_0 + \frac{4h}{3}(2f_1 - f_2 + 2f_3)$$

[∵ we know that $\Delta f_0 = f_1 - f_0, \Delta^2 f = f_2 - 2f_1 + f_0, \Delta^3 f_0 = f_3 - 3f_2 + 3f_1 + f_0$]

$$= 4f_0 + 8\Delta f_0 + \frac{20}{3}\Delta^2 f_0 + \frac{8}{3}\Delta^3 f_0$$

$$= 4f_0 + 8(f_1 - f_0) + \frac{20}{3}(f_2 - 2f_1 + f_0) + \frac{8}{3}(f_3 - 3f_2 + 3f_1 + f_0)$$

$$= \frac{4}{3}(2f_1 - f_2 + 2f_3)$$

Hence $\qquad y_4 = y_0 + \dfrac{4h}{3}(2f_1 - f_2 + 2f_3)$ which is called predictor.

In general predictor

$$y_{n+1} = y_{n-3} + \frac{4h}{3}[2f_{n-2} - f_{n-1} + 2f_n]$$

Then an improved value of f_4 is computed and again the corrector method is applied to find a better value of y_4, we repeat this step until y_4 remains unchanged.

Hence, Milne's corrector formula is

$$y_{n+1} = y_{n-1} + \frac{h}{3}[f_{n-1} + 4f_n + f_{n+1}]$$

Then y_5 can be obtained from the predictor as

$$y_5 = y_1 + \frac{4h}{3}[2f_2 - f_3 + 2f_4]$$

Then f_5 is calculated and a better value of

$$y_5 = y_3 + \frac{h}{3}[f_3 + 4f_4 + f_5]$$

This step is to be repeated till two values of y_5 are same, then calculate y_6 and so on. This is Milne's predictor–corrector method.

Example 8.26: Find $y(2.0)$ if $y(x)$ is the solution of $\dfrac{dy}{dx} = \dfrac{1}{2}(x+y) = f(x, y)$, assuming $y(0) = 2$, $y(0.5) = 2.636$, $y(1.0) = 3.595$, $y(1.5) = 4.968$ using Milne's predictor–corrector method.

Solution: By Milne's predictor–corrector

$$y_{n+1} = y_{n-3} + \frac{4h}{3}[2f_{n-2} - f_{n-1} + 2f_n]$$

$$y_4 = y_0 + \frac{4h}{3}[2f_1 - f_2 + 2f_3]$$

x	y	$f(x, y) = (x+y)/2$
$x_0 = 0$	2	$1 = f_0$
$x_1 = 0.5$	2.636	$1.568 = f_1$
$x_2 = 1.0$	3.595	$2.2975 = f_2$
$x_3 = 1.5$	4.968	$3.234 = f_3$

$$y_4 = y_0 + \frac{4h}{3}[2f_1 - f_2 + 2f_3]$$

$$= 2 + \frac{4(0.5)}{3}[2(1.568) - 2.2975 + 2(3.234)] = 6.871$$

$$f_4 = \frac{(x+y)}{2} = \frac{2 + 6.871}{2} = 4.4355$$

By corrector $\qquad y_{n+1} = y_{n-1} + \dfrac{h}{3}[f_{n-1} + 4f_n + f_{n+1}]$

$$y_4 = y_2 + \frac{h}{3}[f_2 + 4f_3 + f_4]$$

$$= 3.595 + \frac{0.5}{3}[2.2975 + 4(3.234) + 4.4355] = 6.873346667$$

or $\qquad f_4 = 4.43658333$

Again by corrector method $y_4 = y_2 + \frac{h}{3}[f_2 + 4f_3 + f_4]$

$$= 3.595 + \frac{0.5}{3}[2.2975 + 4(3.234) + 4.43658]$$

$$= 6.873346667 \text{ a value which is same as before.}$$

Hence $\qquad y_4 = 6.8733$ correct to four decimal places.

Example 8.27: Using RK method of order four, find y for $x = 0.1, 0.2, 0.3$ given $\frac{dy}{dx} = x - y^2$, $y(0) = 1$, then continue the solution at $x = 0.4$ using Milne's method.

Solution: Given $f(x, y) = x - y^2 = \frac{dy}{dx}$, $x_0 = 0$, $y_0 = 1$, $h = 0.1$.

By *RK method of fourth order*

To find y_1: $k_1 = hf(x_0, y_0) = (0.1)(x_0 - y_0^2) = (0.1)(0 - 1^2) = -0.1$

$$k_2 = h\left[f\left(x_0 + \frac{h}{2}, y_0 + \frac{k_1}{2}\right)\right] = 0.1\left[\frac{h}{2} - \left(y_0 + \frac{k_1}{2}\right)^2\right]$$

$$= 0.1\left[\frac{0.1}{2} - \left(1 - \frac{0.1}{2}\right)^2\right] = -0.08525$$

$$k_3 = hf\left(x_0 + \frac{h}{2}, y_0 + \frac{k_2}{2}\right) = (0.1)\left[\frac{h}{2} - \left(y_0 + \frac{k_2}{2}\right)^2\right]$$

$$= (0.1)\left[\frac{0.1}{2}\left(1 - \frac{0.08525}{2}\right)^2\right] = -0.0866567$$

$$k_4 = hf(x_0 + h, y_0 + k_3) = (0.1)[1.0 - (1 - 0.0866567)^2] = -0.0734196$$

$$k = \frac{1}{6}(k_1 + 2k_2 + 2k_3 + k_4) = -0.0862054$$

$$y_1 = y_0 + k = 1 - 0.0862054 = 0.9137956 = 0.91375$$

To find y_2 $y_2 = y(0.2)$

$$x_1 = x_0 + h = 0 + 0.1 = 0.1$$

$$k_1 = (0.1)[x_1 - y_1^2] = (0.1)[0.1 - (0.9137946)^2] = -0.073502057$$

$$k_2 = (0.1)\left[x_1 + \frac{h}{2} - \left(y_1 + \frac{k_1}{2}\right)^2\right]$$

$$= (0.1)\left[\left(0.1 + \frac{0.1}{2}\right) - \left(0.913795 - \frac{0.073502}{2}\right)^2\right]$$

$$= -0.0619206$$

$$= y_0 + \frac{h}{24}[19(0.15342) + 9(0.208170368) - 5(0.10101) + 0.050125]$$

$$k_3 = (0.1)\left[x_1 + \frac{h}{2} - \left(y_1 + \frac{k_2}{2}\right)^2\right]$$

$$= (0.1)\left[\left(0.1 + \frac{0.1}{2}\right) - \left(0.913795 - \frac{0.0619206}{2}\right)^2\right]$$

$$= -0.0629398$$

$$k_4 = (0.1)\,[x_1 + h - (y_1 + k_3)^2]$$
$$= (0.1)\,[0.1 + 0.1 - (0.913795 - 0.0629398)^2] = -0.0522395$$

$$k = \frac{1}{6}(k_1 + 2k_2 + 2k_3 + k_4) = 0.0626025$$

$$y_2 = 0.913795 - 0.0626025 = 0.8511925$$

To find y_3: $\quad y_3 = y(0.3) = ?, \; x_2 = x_1 + h = 0.1 + 0.1 = 0.2$

$$k_1 = hf(x_2, y_2) = (0.1)\,[0.2 - (0.8511925)^2] = -0.052453$$

$$k_2 = h\left[x_2 + \frac{h}{2} - \left(y_2 + \frac{k}{2}\right)^2\right]$$

$$= (0.1)\left[0.2 + \frac{0.1}{2} - \left(0.8511925 - \frac{0.052453}{2}\right)^2\right] = -0.043057$$

$$k_3 = (0.1)\left[0.25 - \left(0.8511925 - \frac{0.043057}{2}\right)\right] = 0.0438342$$

$$k_4 = (0.1)\,[(0.2 + 0.1) - (0.8511925 - 0.0438342)^2] = -0.0351827$$

$$k = -0.0435697$$

$$y_3 = y_2 - 0.0435697 = 0.8511925 - 0.0435697 = 0.8076228$$

x	y	$f = x - y^2$
$x_0 = 0$	1	$-1 = f_0$
$x_1 = 0.1$	0.913795	$-0.73502 = f_1$
$x_2 = 0.2$	0.8511925	$-0.52453 = f_2$
$x_3 = 0.3$	0.8076228	$-0.35225 = f_3$

To find y_4 $\quad y_4 = y_0 + \frac{4h}{3}[2f_1 - f_2 + 2f_3]$

$$= 1 + \frac{4(0.1)}{3}[2(-0.73502) + 0.52453 + 2(-0.35225)] = 0.779999$$

$$f_4 = 0.4 - (0.779)^2 = -0.208398$$

By corrector

$$y_4 = y_2 + \frac{h}{3}[f_2 + 4f_3 + f_4]$$

$$= 0.8511925 + \frac{0.1}{3}[-0.52453 + 4(-0.35225) - 0.208398] = 0.779794$$

Hence $y(0.4) = 0.779$.

Example 8.28: Use Milne's method to find $y(0.3)$ from $y' = x^2 + y^2$, $y(0) = 1$, find initial value $y(0.1)$, $y(0.2)$ from Taylor's series method.

Solution: By Taylor's series

$$y = y_0 + (x - x_0)\, y_0' + \frac{(x - x_0)^2}{2!}\, y_0'' + \frac{(x - x_0)^3}{3!}\, y_0''' + \ldots$$

$$y' = x^2 + y^2,\ y'(0) = 1$$

$$y'' = 2x + 2yy',\ y''(0) = 0 + 2y_0 y_0' = 2 \times 1 \times 2 = 2$$

$$y''' = 2 + 2y'^2 + 2yy'',\ y'''(0) = 2 + 2 + 2 \times 1 \times 2 = 8$$

$$y = 1 + x(1) + \frac{x^2}{2!}(2) + \frac{x^3}{3!}(8) + \ldots = 1 + x + x^2 + \frac{4}{3}x^3 + \ldots$$

$$y(0.1) = 1 + (0.1) + (0.1)^2 + \frac{4(0.1)^3}{3} + \ldots = 1.11133$$

$$y(0.2) = 1 + (0.2) + (0.2)^2 + \frac{4(0.2)^3}{3} + \ldots = 1.25066$$

$$y(-0.1) = 1 + (-0.1) + (-0.1)^2 + \frac{4(-0.1)^3}{3} + \ldots = 0.908667$$

x	y	$f = x - y^2$
-1	$0.908667\ (y_1)$	$f_1 = 0.835675$
0	$1\ (y_0)$	$f_0 = 1$
0.1	$1.11133\ (y_2)$	$f_2 = 1.245054$
0.2	$1.25066\ (y_3)$	$f_3 = 1.6041504$

Using predictor formula

$$y_4 = y_0 + \frac{4h}{3}(2f_1 - f_2 + 2f_3)$$

$$= 1 + \frac{4(0.1)}{3}[2(0.835675) - (1.245054) + 2(1.6041504)] = 1.484612907$$

∴ $$f_4 = x_4^2 + y_4^2 = (0.3)^2 + (1.484612907)^2 = 2.298579613$$

Using corrector formula

$$y_4 = y_2 + \frac{h}{3}(f_2 + 4f_3 + f_4)$$

$$= 1.11133 + \frac{0.1}{3}(1.245054 + 4(1.6041504) + 2.298579613) = 1.44333784$$

$$f_4 = x_4^2 + y_4^2 = (0.3)^2 + (1.44333784)^2 = 2.173224122$$

Again using corrector formula

$$y_4 = 1.11133 + \frac{0.1}{3}[1.245054 + 4(1.6041504) + 2.173224122] = 1.439159324$$

$$f_4 = 2.16117956$$

Again applying corrector formula

$$y_4 = 1.438757839$$
$$f_4 = 2.160024118$$

Hence $y(0.3) = 1.43$.

Example 8.29: Using Milne's predictor–corrector method to obtain the solution of $\frac{dy}{dx} = x - y^2$ at $x = 0.8$ given

x	0	0.2	0.4	0.6
y	0.0000	0.0200	0.0795	0.1762

Solution: Given $f(x, y) = x - y^2$, $h = 0.2$, $x_0 = 0$, $y_0 = 0$, $f_0 = 0$, $f_1 = 0.1996$, $f_2 = 0.39368$, $f_3 = 0.56895$.

By using predictor formula

$$y_4 = y_0 + \frac{4h}{3}[2f_1 - f_2 + 2f_3]$$

$$= 0 + \frac{4(0.2)}{3}[2(0.1996) - 0.39368 + 2(0.56895)] = 0.304912$$

$$f_4 = x_4 - y_4^2 = 0.8 - (0.304912)^2 = 0.707029$$

By using corrector formula

$$y_4 = y_2 + \frac{h}{3}[f_2 + 4f_3 + f_4]$$

$$= 0.02 + \frac{0.2}{3}[0.39368 + 4(0.56895) + 0.707029] = 0.3576509$$

$$f_4 = (0.8) - (0.3576509)^2 = 0.672086$$

$$y_4 = 0.02 + \frac{0.2}{3}[0.39368 + 4(0.56895) + 0.672086] = 0.3541566$$

$$f_4 = (0.8) - (0.3541566)^2 = 0.674573$$

$$y_4 = 0.02 + \frac{0.2}{3}[0.39368 + 4(0.56895) + 0.674573] = 0.3544053$$

Hence $y(0.8) = 0.354$.

Example 8.30: Solve the initial value problem $\frac{dy}{dx} = 1 + xy^2$, $y(0) = 1$ for $x = 0.4$ using Milne's method, given

x	0.1	0.2	0.3
y	1.105	1.223	1.355

Solution: Using the predictor

$$y_4 = y_0 + \frac{4h}{3}[2f_1 - f_2 + 2f_3]$$

$$f_1 = 1 + x_1 \cdot y_1^2 = 1.1221$$

$$f_2 = 1 + (0.2)(1.223)^2 = 1.2991$$

$$f_3 = 1 + (0.3)(1.355)^2 = 1.5508$$

$$f_4 = ?$$

$$y_4 = 1 + \frac{4(0.1)}{3}[2(1.1221) - 1.2991 + 2(1.5508)] = 1.56593$$

$$f_4 = 1 + (0.4)(1.56593)^2 = 1.98085$$

By applying corrector formula

$$y_4 = y_2 + \frac{h}{3}[f_2 + 4f_3 + f_4]$$

$$= 1.223 + \frac{0.1}{3}[1.2991 + 4(1.5508) + 1.98085] = 1.539105$$

$$f_4 = 1 + (0.4)(1.539105)^2 = 1.94754$$

Again using the corrector formula

$$y_4 = 1.224 + \frac{0.1}{3}[1.2991 + 4(1.5508) + 1.94754] = 1.5385$$

$$f_4 = 1 + (0.4)(1.5385)^2 = 1.946879$$

Again using the corrector formula

$$y_4 = 1.537973$$
$$f_4 = 1.946144$$

Again using the corrector formula

$$y_4 = 1.5379 \approx 1.5$$

Hence $y(0.4) = 1.5$.

8.9 ADAMS-BASHFORTH METHOD

Given $\qquad \dfrac{dy}{dx} = f(x, y)$ and $y_0 = y(x_0)$

We calculate $\qquad y_{-1} = y(x_0 - h), y_{-2} = y(x_0 - 2h), y_{-3} = f(x_0 - 3h)$
by Taylor's series method or Euler's method.

Next we calculate $\qquad f_{-1} = f(x_0 - h, y_{-1}), f_{-2} = f(x_0 - 2h, y_{-2}), f_{-3} = f(x_0 - 3h, y_{-3})$

Then to find y_1, use Newton's backward interpolation formula, i.e.

$$f(x, y) = f_0 + n\nabla f_0 + \frac{n(n+1)}{2!}\nabla^2 f_0 + \frac{n(n+1)(n+2)}{3!}\nabla^3 f_0 + \dots$$

$$y_1 = y_0 + \int_{x_0}^{x_0+h} f(x, y)dx$$

$$= y_0 + \int_{x_0}^{x_0+h} \left(f_0 + n\nabla f_0 + \frac{n(n+1)}{2!} \nabla^2 f_0 + \dots \right) dx$$

[Put $x = x_0 + nh$ so that $dx = hdn$]

$$= y_0 + h\int_0^1 (f_0 + n\nabla f_0 + \dots) \, dn$$

$$= y_0 + h\left[f_0 + \frac{1}{2}\nabla f_0 + \frac{5}{12}\nabla^2 f_0 + \frac{3}{8}\nabla^3 f_0 + \dots \right]$$

Neglecting fourth and higher order differences expressing ∇f_0, $\nabla^2 f_0$, $\nabla^3 f_0$ in terms of function value, we get

$$y_1 = y_0 + \frac{h}{24}[55f_0 - 59f_{-1} + 37f_{-2} - 9f_{-3}] \tag{8.17}$$

This is called Adam–Bashforth predictor formula.

Having found y_1, we find

$$f_1 = f(x_0 + h, y_1)$$

Then to find a better value of y_1, we derive a corrector formula by substituting Newton's backward formula as f_1, i.e.

$$f(x, y) = f_1 + n\nabla f_1 + \frac{n(n+1)}{2!}\nabla^2 f_1 + \dots$$

$$y_1 = y_0 + \int_{x_0}^{x_1} f(x, y)dx$$

Put
$$x = x_1 + nh, \, dx = hdn$$

$$= y_0 + h\int_{-1}^0 \left(f_1 + n\nabla f_1 + \frac{n(n+1)}{2!}\nabla^2 f_1 + \dots \right) dn$$

$$= y_0 + h\left[f_1 - \frac{1}{2}\nabla f_1 - \frac{1}{12}\nabla^2 f_1 - \frac{1}{24}\nabla^3 f_1 + \dots \right]$$

Neglecting fourth and higher order differences and expressing ∇f_1, $\nabla^2 f_1$, $\nabla^3 f_1$ in terms of their function value, we have

$$y_1 = y_0 + \frac{h}{24}[9f_1 + 19f_0 - 5f_{-1} + f_{-2}] \tag{8.18}$$

which is called Adams–Moulton corrector formula.

Example 8.31: Using Adams–Bashforth formula, determine $y\,(0.4)$ given $\dfrac{dy}{dx} = \dfrac{1}{2}xy$ and the following data

x	0	0.1	0.2	0.3
y	1	1.0025	1.0101	1.0228

Solution: Here $f(x, y) = xy$, $h = 0.1$

$x = 0$	$y_{-3} = 1$	$f_{-3} = \frac{1}{2} \times 0 \times 1 = 0$
$x = 0.1$	$y_{-2} = 1.0025$	$f_{-2} = \frac{1}{2}\,(0.1)\,(1.0025) = 0.050125$
$x = 0.2$	$y_{-1} = 1.0101$	$f_{-1} = \frac{1}{2}\,(0.2)\,(1.0101) = 0.10101$
$x = 0.3$	$y_0 = 1.0228$	$f_0 = \frac{1}{2}\,(0.3)\,(1.0228) = 0.15342$

Using the predictor

$$y_1 = y_0 + \frac{h}{24}(55f_0 - 59f_{-1} + 37f_{-2} - 9f_{-3})$$

$$= 1.0228 + \frac{0.1}{24}[55(0.15342) - 59(0.10101) + 37(0.050125) - 9(0)] = 1.04085184$$

$x = 0.4, y_1 = 1.04085184$

$$f_1 = \frac{1}{2}(0.4)(1.04085184) = 0.208170368$$

Using the corrector

$$y_1 = y_0 + \frac{h}{24}[19(0.15342) + 9(0.208170368) - 5(0.10101) + 0.050125]$$

$$= 1.040856328$$

Hence $y(0.4) = 1.04085$.

Example 8.32: Given $y' = x^2 - y$, $y(0) = 1$ and the starting values $y(0.1) = 0.90516$, $y(0.2) = 0.82127$, $y(0.3) = 0.74918$, find $y(0.4)$ using Adams–Bashforth method.

Solution: Here $f(x, y) = x^2 - y$, $h = 0.1$

$x = 0.0$	$y_{-3} = 1.0$	$f_{-3} = (0)^2 - 1 = -1$
$x = 0.1$	$y_{-2} = 0.90516$	$f_{-2} = (0.1)^2 - 0.90516 = -0.89516$
$x = 0.2$	$y_{-1} = 0.82127$	$f_{-1} = (0.2)^2 - 0.82127 = 0.78127$
$x = 0.3$	$y_0 = 0.74918$	$f_0 = \frac{1}{2}(0.3)^2 - 0.74918 = -0.65918$

Using the predictor

$$y_1 = y_0 + \frac{h}{24}[55f_0 - 59f_{-1} + 37f_{-2} - 9f_{-3}]$$

$$= 0.74918 + \frac{(0.1)}{24}[55(-0.65918) - 59(0.78127) + 37(-0.89516) - 9(-1)]$$

$$= 0.6758013$$

$x = 0.4, y_1 = 0.6758013, f_1 = (0.4)^2 - 0.6758013 = -0.05158013$

Using the corrector

$$y_1 = y_0 + \frac{h}{24}[9f_1 + 19f_0 - 5f_{-1} + f_{-2}]$$

$$= 0.74918 + \frac{(0.1)}{24}[9(-0.5158013) + 19(-0.65918) - 5(0.78127) + (-0.89516)]$$

$$= 0.690199$$

$x = 0.4, y_1 = 0.690199, f_1 = (0.4)^2 - 0.690199 = -0.530199$

Using the corrector

$$y_1 = 0.74918 + \frac{(0.1)}{24}[9(-0.530199) + 19(-0.65918) - 5(0.78127) + (-0.89516)]$$

$$= 0.689659$$

$$= 0.4, y_1 = 0.689659, f_1 = (0.4)^2 - 0.689659 = -0.529659$

Using the corrector

$$y_1 = 0.74918 + \frac{(0.1)}{24}[9(-0.529659) + 19(-0.65918) - 5(0.78127) + (-0.89516)]$$

$$= 0.68968$$

Hence $y(0.4) = 0.6896$

Example 8.33: Using Adams–Bashforth method, obtain the solution $\frac{dy}{dx} = x - y^2$ at $x = 0.8$ using the given values:

x	0	0.2	0.4	0.6
y	0	0.0200	0.07915	0.1762

Solution: Here $f(x, y) = x - y^2$, $h = 0.2$

$x = 0.0$	$y_{-3} = 0.0$	$f_{-3} = 0 - (0)^2 = 0$
$x = 0.2$	$y_{-2} = 0.0200$	$f_{-2} = 0.2 - (0.0200)^2 = 0.1996$
$x = 0.4$	$y_{-1} = 0.07915$	$f_{-1} = 0.4 - (0.0795)^2 = 0.3937$
$x = 0.6$	$y_0 = 0.1762$	$f_0 = 0.6 - (0.1762)^2 = 0.5689$

Using the predictor

$$y_1 = y_0 + \frac{h}{24}[55f_0 - 59f_{-1} + 37f_{-2} - 9f_{-3}]$$

$$= 0.1762 + \frac{(0.2)}{24}[55(0.5689) - 59(0.3937) + 37(0.1996) - 9(0)] = 0.30492$$

$x = 0.8$, $y_1 = 0.30492$, $f_1 = 0.8 - (0.30492)^2 = 0.707023$

Using the corrector

$$y_1 = y_0 + \frac{h}{24}[9f_1 + 19f_0 - 5f_{-1} + f_{-2}]$$

$$= 0.1762 + \frac{(0.2)}{24}[9(0.7070237) + 19(0.5689) - 5(0.3937) + (0.1996)]$$

$$= 0.3056177$$

$x = 0.8$, $y_1 = 0.30456177$, $f_1 = 0.8 - (0.30456177)^2 = 0.701051$

Again using the corrector

$$y_1 = 0.1762 + \frac{(0.2)}{24}[9(0.701051) + 19(0.5689) - 5(0.3937) + (0.1996)]$$

$$= 0.304114$$

$$f_1 = 0.8 - (0.30414)^2 = 0.70751$$

Again using the corrector

$$y_1 = 0.1762 + \frac{(0.2)}{24}[9(0.70751) + 19(0.5689) - 5(0.3937) + (0.1996)]$$

$$= 0.304598$$

Hence $y(0.8) = 0.304$.

PROBLEMS 8.3

Euler's Method

1. Evaluate the integral of $y' - y^2 = 0$ by Euler's method at $x = 0.1, 0.2, 0.3$ given $y(0) = 1$.
[Ans. $y(0.1) = 1.1, y(0.2) = 1.221, y(0.3) = 1.37$]

2. Evaluate by Euler's method $y' = y - x$, $y(0) = 2$, at $x = 0.2, 0.4, 0.6$.
[Ans. $y(0.2) = 2.4, y(0.4) = 2.84, y(0.6) = 3.38$]

3. Apply Euler's method to solve $dy/dx = x + y$, $y(0) = 0$, find out $y(0.2)$.
[Ans. $y(0.2) = 1.18318081$]

4. Consider the initial value problem $dy/dx = y - x^2 - 1$, $y(0) = 0.5$. Find $y(0.2)$ by Euler's method.
[Ans. $y(0.2) = 0.8$]

Modified Euler's Method

1. Solve $\dfrac{dy}{dx} = 1 - y$, $y(0) = 0$ by modified Euler's method at $x = 0.1, 0.2, 0.3$. Compare your result with exact solution.
[Ans. $y(0.1) = 0.095, y(0.2) = 0.18098, y(0.3) = 0.258787$]

2. Use modified Euler's method to find y at $x = 0.1, 0.2$, given $\dfrac{dy}{dx} = x^2 + y^2$, $y(0) = 1$.
[Ans. $y(0.1) = 1.1105, y(0.2) = 1.25026$]

3. Solve $\dfrac{dy}{dx} = \log(x + y)$, $y(0) = 2$ by modified Euler's method and find y at $x = 1.2, 1.4$.
[Ans. $y(1.2) = 2.5351, y(1.4) = 2.6531$]

4. Solve the modified Euler's method $y' = x + y$, $y(0) = 1$, $h = 0.1$ at $x = 0.1, 0.2, 0.3$.
[Ans. $y(0.1) = 1.1105, y(0.2) = 1.2432, y(0.3) = 1.4004$]

5. Given $\dfrac{dy}{dx} = \dfrac{y - x}{y + x}$ with $y(0)$, find y for $x = 0.1$ by (a) Euler's method (b) Modified Euler's method.
[Ans. (a) $y(0.1) = 1.0928$, (b) $y(0.1) = 1.0928$]

6. Determine the value of y correct to 4 decimal places using Euler's method and modified Euler's method when $x = 0.4$, given that $y' = y + x^2$, $y(0) = 1$ with $h = 0.2$ and compare this result with the exact result.
[Ans. Euler's = 1.4480, Modified Euler's = 1.5141, exact = 1.5155]

7. Given $\dfrac{dy}{dx} = \sqrt{xy + 1}$, $y(0) = 1$, $h = 0.025$. Determine the value of y correct to three decimal places, using Euler's modified method when $x = 0.075$.   [Ans. $y = 1.076$]

Runge–Kutta Method

1. Given $\dfrac{dy}{dx} + y = 0$, $y(0) = 1$, find the value of y at $x = 0.1, 0.2$, using Runge-Kutta method of (i) second order (ii) third order (iii) fourth order.
[Ans. (i) 0.905, 0.819 (ii) 0.91, 0.8234 (iii) 0.9048, 0.818731]

2. Tabulate by RK method, the numerical solution of $\dfrac{dy}{dx} = \dfrac{y^2 - 2x}{y^2 + x}$ and $y(0) = 1$, find y

for $x = 0.1, 0.2, 0.3, 0.4, 0.5$.

    [Ans. $y(0.1) = 1.0911, y(0.2) = 1.1677, y(0.3) = 1.235, y(0.4) = 1.2902, y(0.5) = 1.338$]

3. Use RK method of order four to find y when $x = 0.4$ in steps of 0.2 given $\dfrac{dy}{dx} = 1 + y^2$,

$y(0) = 0$.   [Ans. $y(0.4) = 0.4$]

4. Solve $y' = 1 + \sqrt{xy}, y(0) = 1$ by RK method of second order chosing $h = 0.25$, find $y(1)$.   [Ans. $y(1) = 2.9118$]

5. For the initial value problem $y' = x + y^2$, $y(0) = 1$, $h = 0.2$. Compare the solution obtained for $y(0.4)$ by modified Euler's method, RK method of order four and exact solution.   [Ans. MEM = 1.51408, RKM = 1.51547, exact = 1.51547]

6. Solve $y' = \dfrac{1 + xy}{x + y}$, $y(1) = 1.2$ by RK method of order two choosing $h = 0.1$, find $y(1.2)$.

                                        [Ans. $y(1.2) = 1.4028$]

7. Using RK method of order four, solve $y'' - 2y' + 2y = e^{2x} \sin x$ with $y(0) = 0$, $y'(0) = -0.6$ to find $y(0.2)$.   [Ans. $y(0.2) = -0.5207$]

Milne's Predictor–Corrector Method

1. Solve $\dfrac{dy}{dx} = 2e^x - y$ at $x = 0.4$ and 0.5 by Milne's predictor-corrector method given values at the four points $x = 0, 0.1, 0.2$ and 0.3 as $y_0 = 2.0$, $y_1 = 2.010$, $y_2 = 2.040$, $y_3 = 2.090$.   [Ans. $y(0.4) = 2.1621, y(0.5) = 2.2447$]

2. Given $y'' + xy + y = 0$, $y(0) = 1$, obtain y for $x = 0, 0.1, 0.2$ and 0.3 by Taylor's series method and find the solution for $y(0.4)$ by Milne's predictor-corrector method.   [Ans. $y(0.4) = -0.3692$]

3. Consider the problem $y' = y + x^2$, $y(0) = 1$, using Milne's predictor-corrector method, find $y(0.8)$, $h = 0.2$.   [Ans. $y(0.8) = 2.4366$]

4. Using Milne's method, find $y(0.4)$, given $y' = xy + y^2$, $y(0) = 1$, use Taylor's series method or Euler's method or RK method to get $y(0.1)$, $y(0.2)$ and $y(0.3)$.   [Ans. $y(0.4) = 2.1119$]

5. Using Milne's method, find $y(0.8)$ and $y(1.0)$ given $y' = \dfrac{1}{x + y}$, $y(0) = 2$,

$y(0.2) = 2.0933, y(0.4) = 2.175, y(0.6) = 2.2493$. [Ans. $y(0.8) = 1.83299, y(1.0) = 1.83698$]

Adam–Bashforth Method

1. Given $y' = 1 + y^2$, $y(0) = 0$, find $y(0.2)$ by Taylor's method, $y(0.4)$ by modified Euler's method, $y(0.6)$ by RK method hence find $y(0.8)$ by both Adam's and Milne's method.   [Ans. $0.0035, 0.0069, 0.0104, 0.0139$]

2. Solve $2y' - x - y = 0$, given $y(0) = 2, y(0.5) = 2.636, y(1) = 3.595, y(1.5) = 4.968$ to get $y(2)$ by Adam's method.   [Ans. $y(2) = 6.8731$]

3. Using Adam's method to find $y(0.4)$, given $\dfrac{dy}{dx} = \dfrac{xy}{2}$, $y(0) = 1, y(0.1) = 1.01$,

$y(0.2) = 1.022, y(0.3) = 1.023$.   [Ans. $1.04081, 1.04101$]

4. Solve by ABM for $y(0.4)$, $y(0.5)$ given $y' = -2x - y$, $y(0) = -1$, $y(0.1) = -0.9145122$, $y(0.2) = -0.8561923$, $y(0.3) = -0.8224547$.

[Ans. at 0.4, predicted value $= -0.810968$, Correction value $= -0.810965$]

5. Solve $y' = x^3 + y^2$, $h = 0.1$, find y at $y(0.8)$, $y(0.9)$ initial values are $y(0.2) = 0.0004$, $y(0.4) = 0.0064$, $y(0.6) = 0.0325$.

[Ans $y(0.8) = 0.1035$, $y(0.9) = 0.1669$]

Numerical Solution of Partial Differential Equations

9.1 INTRODUCTION

Partial differential equations occur very frequently in science, engineering and applied mathematics. Many partial differential equations cannot be solved by analytical methods in closed form solution. In most of the research work in fields like, applied elasticity, theory of plates and shells, hydrodynamics, quantum mechanics, etc. the research problem is reduced to partial differential equations since analytical solutions are not available, we go in for numerical solutions of the partial differential equations by various methods. Certain types of boundary value problems can be solved by replacing the differential equation by the corresponding difference equation and then solving the latter by a process of iteration. This method was devised and first used by LF Richardson and it was later improved by H Liebmann.

9.2 FINITE DIFFERENCE APPROXIMATION TO VARIOUS DERIVATIVES

If $y(x)$ and its derivatives are single valued continuous function of x. Then by Taylor's expansion, we have

$$y(x + h) = y(x) + hy'(x) + \frac{h^2}{2!}y''(x) + \frac{h^3}{3!}y'''(x) + ... \tag{9.1}$$

and

$$y(x - h) = y(x) - hy'(x) + \frac{h^2}{2!}y''(x) - \frac{h^3}{3!}y'''(x) + ... \tag{9.2}$$

Equation (9.1) gives

$$y'(x) = \frac{1}{h}[y(x + h) - y(x)] - \frac{h^2}{2!}y''(x) + ...$$

i.e.

$$= \frac{1}{h}[y(x + h) - y(x)] + O(h)$$

which is forward difference approximation of $y'(x)$ with an error of order h.

Similarly Eq. (9.2) gives

$$y'(x) = \frac{1}{h}[y(x) - y(x - h)] + O(h)$$

which is backward difference approximation.

Subtracting Eq. (9.2) from Eq. (9.1), we have

$$y(x + h) - y(x - h) = 2h[y'(x)] + O(h^2)$$

or
$$y'(x) = y(x + h) - y(x - h)/2h + O(h^2)$$

which is central difference approximation of $y'(x)$ with an error of order h^2.

Adding Eqs (9.1) and (9.2)

$$y(x + h) + y(x - h) = 2\left[y(x) + \frac{h^2}{2!}y''(x)\right] + \frac{2h^4}{4!}y''''(x)$$

$$\frac{y(x + h) - 2y(x) + y(x - h)}{h^2} = y''(x) + O(h^2)$$

Central differences are better approximations than forward and backward approximations.

9.3 FINITE DIFFERENCE APPROXIMATION TO PARTIAL DERIVATIVES

Let xy plane be divided into a series of rectangles whose sides are parallel to x- and y-axes such that $\Delta x = h$ and $\Delta y = k$. The points of intersection of these families of lines are called mesh points, lattice points or grid points.

If (x_i, y_j) is any grid point, $x_i = x_0 + ih$, $y_j = y_0 + jk$, and we take one corner as origin, then
$$x_i = ih, \ y_j = jk, \ i, j = 0, 1, 2, \ldots$$

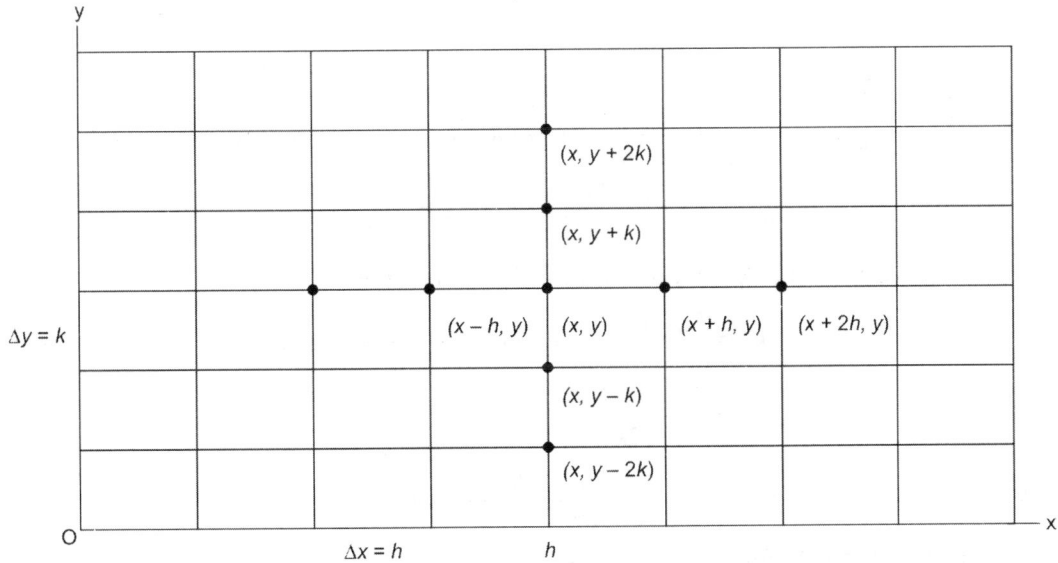

Thus finite difference approximation for the partial derivative in x direction

$$\frac{\partial u}{\partial x} = \frac{u(x + h, y) - u(x, y)}{h} + O(h) \qquad \text{[forward difference]}$$

$$= \frac{u(x, y) - u(x - h, y)}{h} + O(h) \qquad \text{[backward difference]}$$

$$= \frac{u(x + h, y) - u(x - h, y)}{h} + O(h^2) \qquad \text{[central difference]}$$

Coordinates of grid points

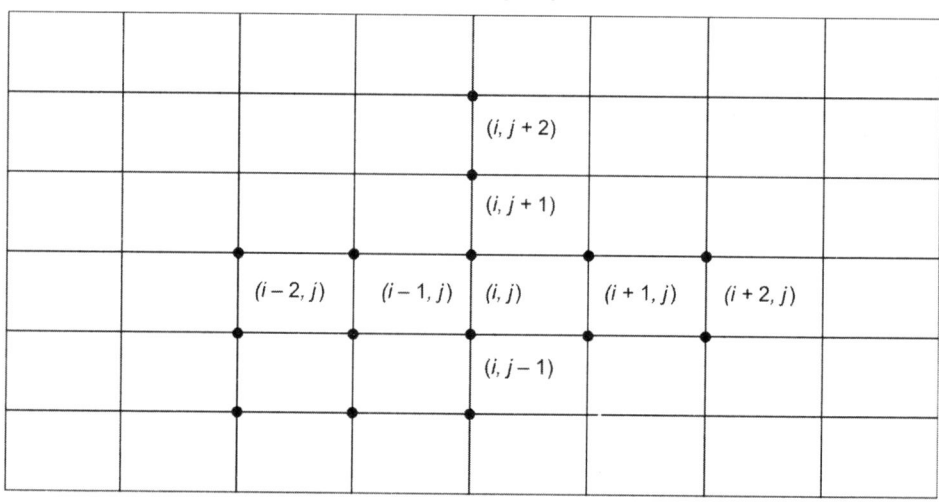

$$\frac{\partial^2 u}{\partial x^2} = \frac{u(x-h, y) - 2u(x, y) + u(x+h, y)}{h^2} + O(h^2)$$

Writing $u(x, y) = u(ih, jk)$ as simply $u_{i,j}$

$$\therefore \qquad u_x = \frac{u_{i+1,j} - u_{i,j}}{h} + O(h) \qquad \qquad \text{[forward difference]}$$

$$= \frac{u_{i,j} - u_{i-1,j}}{h} + O(h) \qquad \qquad \text{[backward difference]}$$

$$= \frac{u_{i+1,j} - u_{i-1,j}}{h} + O(h^2) \qquad \qquad \text{[central difference]}$$

$$u_{xx} = \frac{u_{i-1,j} - 2u_{i,j} + u_{i+1,j}}{h^2} + O(h^2)$$

Similarly, we have the approximation for the derivative w.r.t. y

$$u_y = \frac{u_{i,j+1} - u_{i,j}}{k} + O(k) = \frac{u_{i,j} - u_{i,j-1}}{k} + O(k)$$

$$= \frac{u_{i,j+1} - u_{i,j-1}}{2k} + O(k^2)$$

and
$$u_{yy} = \frac{u_{i,j-1} - 2u_{i,j} + u_{i,j+1}}{k^2} + O(k^2)$$

The third and fourth derivatives are as

$$y_1''' = \frac{1}{2h^3}(y_{i+2} - 2y_{i+1} + 2y_{i-1} - y_{i-2})$$

$$y_1'''' = \frac{1}{h^4}(y_{i+2} - 4y_{i+1} + 6y_i - 4y_{i-1} + y_{i-2})$$

Note: The accuracy of this method depends on the size of subinterval h and also the order of approximation. As we reduce h, the accuracy improves but the number of equations to be solved also increases.

9.4 CLASSIFICATION OF PARTIAL DIFFERENTIAL EQUATIONS

In this section, we will classify partial differential equation of the second order. The general linear partial differential equation of second order in two independent variables is of the form

$$A\frac{\partial^2 u}{\partial x^2} + B\frac{\partial^2 u}{\partial x \partial y} + C\frac{\partial^2 u}{\partial y^2} + D\frac{\partial u}{\partial x} + E\frac{\partial u}{\partial y} + Fu = 0$$

or $\qquad Au_{xx} + Bu_{xy} + Cu_{yy} + Du_x + Eu_y + Fu = 0 \qquad\qquad (9.3)$

where A, B, C, D, E, F are in general function of x and y.

The above equation of second order (linear) Eq. (9.3) is said to be

i. Elliptic at a point (x, y) in the plane if $B^2 - 4AC < 0$
ii. Parabolic if $B^2 - 4AC = 0$
iii. Hyperbolic if $B^2 - 4AC > 0$

Note: The same differential equation may be elliptic in one region, parabolic in another and hyperbolic in some other region. For example, $u_{xx} + u_{yy} = 0$ is elliptic if $x > 0$, hyperbolic if $x < 0$ and parabolic if $x = 0$.

Examples:

Elliptic type	Parabolic type	Hyperbolic type
1. $\dfrac{\partial^2 u}{\partial x^2} + \dfrac{\partial^2 u}{\partial y^2} = 0$	$\dfrac{\partial^2 u}{\partial x^2} = \dfrac{1}{\alpha^2}\dfrac{\partial u}{\partial t}$	$\dfrac{\partial^2 u}{\partial x^2} = \dfrac{1}{\alpha^2}\dfrac{\partial^2 u}{\partial t^2}$
(Laplace equation in two dimensions)	(One dimensional heat equation)	(One dimensional wave equation)

2. $\dfrac{\partial^2 u}{\partial x^2} + \dfrac{\partial^2 u}{\partial y^2} = f(x, y)$

(Poisson's equation)

9.5 ELLIPTIC EQUATION

An important equation of the ellipse is

$$\frac{\partial^2 u}{\partial x^2} + \frac{\partial^2 u}{\partial y^2} = 0, \text{ i.e. } u_{xx} + u_{yy} = 0 \qquad\qquad (9.4)$$

This equation is called Laplace's equation.

Replacing the derivatives by the corresponding difference expressions in Eq. (9.4), we get

$$\frac{u_{i-1,j} - 2u_{i,j} + u_{i+1,j}}{h^2} + \frac{u_{i,j-1} - 2u_{i,j} + u_{i,j+1}}{k^2} = 0$$

Taking a square mesh and putting $h = k$, we get

$$u_{i,j} = \frac{1}{4}[u_{i+1,j} + u_{i-1,j} + u_{i,j+1} + u_{i,j-1}] \qquad\qquad (9.5)$$

i.e. the value of u at any interior mesh point is the arithmetic mean of its values at the four neighbouring mesh points to the left, right, below and above. This is called standard five point formula (SFPF).

Instead of Eq. (9.5), we may also use the formula

$$u_{i,j} = \frac{1}{4}[u_{i-1,j-1} + u_{i+1,j-1} + u_{i-1,j+1} + u_{i+1,j+1}] \tag{9.6}$$

which shows that the value of $u_{i,j}$ is the arithmetic mean of its values at four neighbouring diagonal mesh points. This is called the diagonal five point formula (DFPF).

The SFPF and DFPF are represented in Figs 9.1 and 9.2.

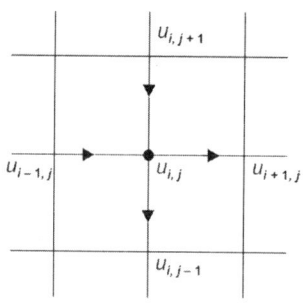

Fig. 9.1

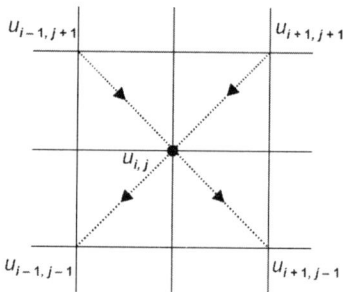

Fig. 9.2

Although Eq. (9.6) is less accurate than Eq. (9.5), yet is serves as a reasonably good approximation.

Note: The DFPF is valid since we know that Laplace equation remains invariant when coordinate axes are rotated through 45°, but the error in DFPF is four times the error in SFPF, therefore we prefer SFPF.

9.5.1 Jacobi's Method

Let (n) denote the nth iteration value of $u_{i,j}$, the iterative formula to solve Eq. (9.5) is

$$u_{i,j}^{(n+1)} = \frac{1}{4}\left[u_{i-1,j}^{(n)} + u_{i+1,j}^{(n)} + u_{i,j+1}^{(n)} + u_{i,j-1}^{(n)}\right] \tag{9.7}$$

It gives improved value of $u_{i,j}$ at interior mesh points is called five point Jacobi's formula.

9.5.2 Gauss–Seidal Method

The iterative formula is

$$u_{i,j}^{(n+1)} = \frac{1}{4}\left[u_{i-1,j}^{(n+1)} + u_{i+1,j}^{(n)} + u_{i,j+1}^{(n+1)} + u_{i,j-1}^{(n)}\right] \tag{9.8}$$

The method uses the latest iterative values available and scans the mesh points systematically from left to right along successive rows.

Gauss–Seidal method converges twice as fast as Jacobi's method.

9.5.3 Successive Over Relaxation Method (SOR Method)

Equation (9.8) can be written as

$$u_{i,j}^{(n+1)} = u_{i,j}^{(n)} + \frac{1}{4}\left[u_{i-1,j}^{(n+1)} + u_{i+1,j}^{(n)} + u_{i,j-1}^{(n+1)} + 4u_{i,j}^{(n)}\right]$$

$$= u_{i,j}^{(n)} + \frac{1}{4}R_{i,j}$$

which shows that $\dfrac{1}{4}R_{i,j}$ is the change in the value $u_{i,j}$ for one Gauss-Seidal iteration. In the SOR method, if there is a large change then this is given to $u_{i,j}^{(n)}$, and the iteration formula is written as

$$u_{i,j+1}^{(n+1)} = u_{i,j}^{(n)} + \frac{1}{4}wR_{i,j}$$

$$= \frac{1}{4}w\left[u_{i-1,j}^{(n+1)} + u_{i+1,j}^{(n)} + u_{i,j-1}^{(n+1)} + u_{i,j+1}^{(n)}\right] + (1-w)u_{i,j}^{(n)} \qquad (9.9)$$

The rate of convergence of Eq. (9.9) depends on the choice of w, which is called the accelerating factor and lie between 1 and 2.

9.6 SOLUTION OF LAPLACE EQUATION BY LIEBMANN'S ITERATION PROCESS

Consider the Laplace equation

$$\frac{\partial^2 u}{\partial x^2} + \frac{\partial^2 u}{\partial y^2} = 0$$

with the given boundary condition. For simplicity, we assume that function $u(x, y)$ is required over a rectangular R with boundary C. Let R be divided into a network of small squares of side h. Let the values of $u(x, y)$ on boundary C be given by C_i and the interior mesh points and boundary points be as shown in figure below.

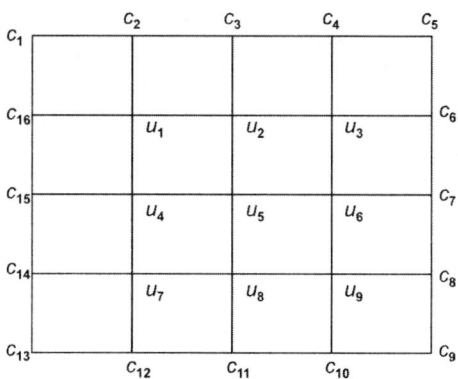

The value of $u(x, y)$ satisfying the Laplacian equation can be replaced by either SFPF or DFPF. To start the iteration process, initially we find rough values at interior points and then improve them by iterative process mostly using SFPF.

We first find u_5 at the centre of the square by taking average of four boundary values

$$u_5 = \frac{1}{4}[c_{15} + c_7 + c_3 + c_{11}]$$

Next, we find the initial values at the centre of the four large inner squares using DFPF

$$u_1 = \frac{1}{4}[c_1 + u_5 + c_3 + c_{15}]$$

$$u_3 = \frac{1}{4}[u_5 + c_5 + c_3 + c_7]$$

$$u_7 = \frac{1}{4}[u_5 + c_{15} + c_{13} + c_{11}]$$

$$u_9 = \frac{1}{4}[u_5 + c_9 + c_7 + c_{11}]$$

The values at the remaining interior points are obtained by SFPF

$$u_2 = \frac{1}{4}[u_1 + u_3 + c_3 + u_5]$$

$$u_4 = \frac{1}{4}[c_{15} + u_5 + u_1 + u_7]$$

$$u_6 = \frac{1}{4}[u_5 + c_7 + u_3 + u_9]$$

$$u_8 = \frac{1}{4}[u_7 + u_9 + u_5 + c_{11}]$$

Now we have got all the boundary values of u and rough values at every mesh grid points in the interior of region R, we proceed with an iteration process to improve their accuracy. We start with u_1 and iterate it using the latest available values of the four adjacent points. We iterate all the mesh points systematically from left to right along successive rows. The iterative formula is

$$u_{i,j}^{(n+1)} = \frac{1}{4}\left[u_{i-1,j}^{n+1} + u_{i+1,j}^{(n)} + u_{i,j-1}^{(n)} + u_{i,j+1}^{n+1}\right]$$

Example 9.1: Find the value of $u(x, y)$ satisfying Laplace equation $\Delta^2 u = 0$ at the pivotal point of system mesh with boundary values as shown below.

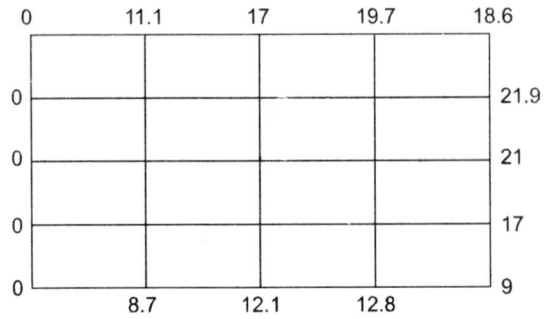

Solution:

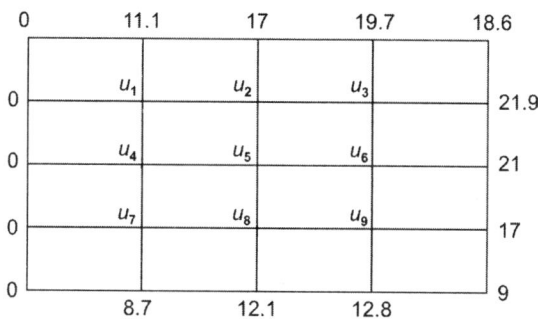

$$u_5^{(0)} = \frac{1}{4}[0 + 21 + 17 + 12.1] = \frac{49.1}{4} = 12.275 \text{ (SFPF)}$$

$$u_1^{(0)} = \frac{1}{4}[0 + 12.275 + 17 + 0] = \frac{29.275}{4} = 7.0687 \text{ (DFPF)}$$

$$u_3^{(0)} = \frac{1}{4}[18.6 + u_5 + 17 + 21] = \frac{1}{4}[18.6 + 12.275 + 38] = \frac{68.875}{4} = 17.219 \text{ (DFPF)}$$

$$u_7^{(0)} = \frac{1}{4}[0 + 12.1 + u_5 + 0] = \frac{24.375}{4} = 6.0937 \text{ (DFPF)}$$

$$u_9^{(0)} = \frac{1}{4}[9 + u_5 + 33.1] = \frac{54.375}{4} = 13.5937 \text{ (DFPF)}$$

$$u_2^{(0)} = \frac{1}{4}[u_1 + u_3 + 17 + u_5] = \frac{53.5627}{4} = 13.3907 \text{ (SFPF)}$$

$$u_4^{(0)} = \frac{1}{4}[0 + u_5 + u_1 + u_7] = \frac{1}{4}(25.4374) = 6.3593 \text{ (SFPF)}$$

$$u_6^{(0)} = \frac{1}{4}[u_5 + 21 + 19.7 + u_9] = \frac{1}{4}(68.5687) = 17.1422 \text{ (SFPF)}$$

$$u_8^{(0)} = \frac{1}{4}[u_5 + 12.1 + u_7 + u_9] = \frac{45.0624}{4} = 11.2656$$

Now by Gauss-Seidal method

$$u_{i,j}^{(n+1)} = \frac{1}{4}\left[u_{i-1,j}^{(n+1)} + u_{i+1,j}^{(n)} + u_{i,j+1}^{(n+1)} + u_{i,j-1}^{(n)}\right]$$

$$u_1^{(n+1)} = \frac{1}{4}\left[0 + u_2^{(n)} + 11.1 + u_4^{(n)}\right]$$

$$u_2^{(n+1)} = \frac{1}{4}\left[u_1^{(n+1)} + u_3^{(n)} + u_5^{(n)} + 17\right]$$

$$u_3^{(n+1)} = \frac{1}{4}\left[u_2^{(n+1)} + 21.9 + 19.7 + u_6^{(n)}\right]$$

$$u_4^{(n+1)} = \frac{1}{4}\left[0 + u_5^{(n)} + u_1^{(n+1)} + u_7^{(n)}\right]$$

$$u_5^{(n+1)} = \frac{1}{4}\left[u_4^{(n+1)} + u_6^{(n)} + u_2^{(n+1)} + u_8^{(n)}\right]$$

$$u_6^{(n+1)} = \frac{1}{4}\left[u_5^{(n)} + 21 + u_3^{(n+1)} + u_9^{(n)}\right]$$

$$u_7^{(n+1)} = \frac{1}{4}\left[0 + u_8^{(n)} + u_4^{(n+1)} + 8.7\right]$$

$$u_8^{(n+1)} = \frac{1}{4}\left[0 + u_7^{(n+1)} + u_5^{(n+1)} + 12.1 + u_9^{(n)}\right]$$

$$u_9^{(n+1)} = \frac{1}{4}\left[u_8^{(n+1)} + 17 + u_6^{(n+1)} + 12.8\right]$$

Put $n = 0$

$$u_1^{(1)} = \frac{1}{4}[0 + 13.3907 + 6.3593 + 11.1] = \frac{30.8500}{4} = 7.7125$$

$$u_2^{(1)} = \frac{1}{4}[7.7125 + 17.2190 + 12.275 + 17] = \frac{54.2065}{4} = 13.5515$$

$$u_3^{(1)} = \frac{1}{4}[13.5516 + 41.6 + 17.1422] = \frac{72.2938}{4} = 18.0734$$

$$u_4^{(1)} = \frac{1}{4}[0 + 12.275 + 7.7125 + 6.0937] = \frac{26.5203}{4} = 6.5203$$

$$u_5^{(1)} = \frac{1}{4}[6.5203 + 17.1422 + 13.5516 + 11.2656] = \frac{48.4979}{4} = 12.1199$$

$$u_6^{(1)} = \frac{1}{4}[12.1199 + 21 + 18.0734 + 13.5937] = \frac{64.7870}{4} = 16.1967$$

$$u_7^{(1)} = \frac{1}{4}[8.7 + 11.2656 + 6.5203] = \frac{26.4859}{4} = 6.6215$$

$$u_8^{(1)} = \frac{1}{4}[6.6215 + 12.1199 + 12.1 + 13.5937] = \frac{44.4351}{4} = 11.1088$$

$$u_9^{(1)} = \frac{1}{4}[11.1088 + 29.8 + 16.1967] = \frac{57.1055}{4} = 14.2764$$

Put $n = 1$

$$u_1^{(2)} = \frac{1}{4}[0 + 13.5516 + 6.5203 + 11.1] = \frac{31.1719}{4} = 7.793$$

$$u_2^{(2)} = \frac{1}{4}[7.793 + 18.0734 + 12.1199 + 17] = \frac{54.9863}{4} = 13.7466$$

$$u_3^{(2)} = \frac{1}{4}[41.6 + 13.7466 + 16.1967] = \frac{71.5433}{4} = 17.8858$$

$$u_4^{(2)} = \frac{1}{4}[12.1199 + 7.7130 + 6.6215] = \frac{26.5344}{4} = 6.6336$$

$$u_5^{(2)} = \frac{1}{4}[6.6336 + 16.1967 + 13.7466 + 11.1088] = \frac{47.5857}{4} = 11.8964$$

$$u_6^{(2)} = \frac{1}{4}[11.8964 + 21 + 17.8858 + 14.2764] = \frac{65.0586}{4} = 16.2646$$

$$u_7^{(2)} = \frac{1}{4}[8.7 + 11.1088 + 6.6336] = \frac{26.4424}{4} = 6.6106$$

$$u_8^{(2)} = \frac{1}{4}[6.6106 + 11.8964 + 12.1 + 14.2764] = \frac{44.8834}{4} = 11.2208$$

$$u_9^{(2)} = \frac{1}{4}[11.2208 + 29.8 + 16.2696] = \frac{57.2854}{4} = 14.3213$$

Put $n = 2$

$$u_1^{(3)} = \frac{1}{4}[0 + 13.7466 + 6.6336 + 11.1] = 7.8700$$

$$u_2^{(3)} = \frac{1}{4}[7.8700 + 17.8858 + 11.8964 + 17] = 13.6630$$

$$u_3^{(3)} = \frac{1}{4}[41.6 + 13.663 + 16.2646] = 17.8819$$

$$u_4^{(3)} = \frac{1}{4}[71.8964 + 7.8700 + 6.6106] = 6.5942$$

$$u_5^{(3)} = \frac{1}{4}[6.5942 + 16.2646 + 13.6630 + 11.2208] = 11.9356$$

$$u_6^{(3)} = \frac{1}{4}[11.9356 + 21 + 17.8819 + 14.3213] = 16.2597$$

$$u_7^{(3)} = \frac{1}{4}[8.7 + 11.2208 + 6.5942] = 6.6288$$

$$u_8^{(3)} = \frac{1}{4}[6.6288 + 11.9356 + 12.1 + 14.3213] = 11.2464$$

$$u_9^{(3)} = \frac{1}{4}[29.8 + 11.2464 + 16.2597] = 14.3265$$

Put $n = 3$

$u_1^{(4)} = 7.8393$	$u_1^{(5)} = 7.413$
$u_2^{(4)} = 13.6642$	$u_2^{(5)} = 13.6663$
$u_3^{(4)} = 17.8810$	$u_3^{(5)} = 17.8885$
$u_4^{(4)} = 6.6009$	$u_4^{(5)} = 6.6052$
$u_5^{(4)} = 11.9428$	$u_5^{(5)} = 11.9214$
$u_6^{(4)} = 16.2876$	$u_6^{(5)} = 16.2846$
$u_7^{(4)} = 6.6368$	$u_7^{(5)} = 6.6329$
$u_8^{(4)} = 11.2265$	$u_8^{(5)} = 11.2457$
$u_9^{(4)} = 14.3285$	$u_9^{(5)} = 14.3326$

The value of $u_1^{(4)}, u_2^{(4)}, u_3^{(4)}, u_5^{(4)}, u_6^{(4)}, u_7^{(4)}, u_8^{(4)}, u_9^{(4)}$, and $u_1^{(5)}, u_2^{(5)}, u_3^{(5)}, u_5^{(5)}, u_6^{(5)}, u_7^{(5)},$ $u_8^{(5)}, u_9^{(5)}$, are approximately the same.

Hence $u_1 = 7.84$, $u_2 = 13.66$, $u_3 = 17.88$, $u_4 = 6.60$, $u_5 = 11.92$, $u_6 = 16.28$, $u_7 = 6.63$, $u_8 = 11.29$, $u_9 = 14.33$.

Example 9.2: Solve the elliptic equation $u_{xx} + u_{yy} = 0$, for the following square mesh with boundary values as shown.

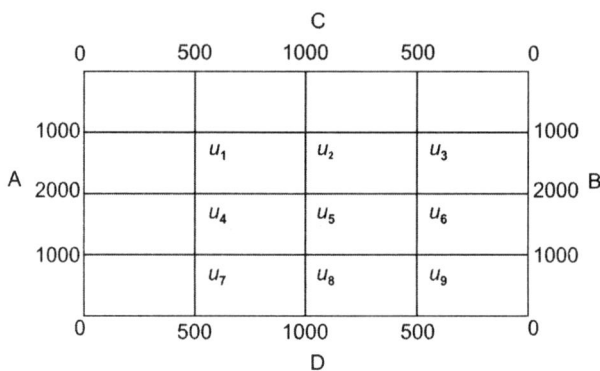

Solution: Let $u_1, u_2, ..., u_9$ be the values of the interior mesh points from the given figure, we can see that the boundary values of u are symmetrical about AB.

$\therefore$

$$u_7 = u_1, u_8 = u_2, u_9 = u_3$$

Also the values of u being symmetrical about CD are

$$u_3 = u_1, u_6 = u_4, u_9 = u_7$$

$\therefore$ It is sufficient to find the values u_1, u_2, u_4 and u_5, the initial values of these are given below.

$$u_5^{(0)} = \frac{1}{4}[2000 + 2000 + 1000 + 1000] = 1500 \text{ (SFPF)}$$

$$u_1^{(0)} = \frac{1}{4}[0 + 1500 + 1000 + 2000] = 1125 \text{ (DFPF)}$$

$$u_2^{(0)} = \frac{1}{4}[1125 + 1125 + 1000 + 1500] = 1187.5 \text{ (SFPF)}$$

$$u_4^{(0)} = \frac{1}{4}[2000 + 1500 + 1125 + 1125] = 1437.5 \text{ (SFPF)}$$

Now we carry out the iteration using the following formula by SFPF

$$u_1^{(n+1)} = \frac{1}{4}[1000 + u_2^{(n)} + 500 + u_4^{(n)}]$$

$$u_2^{(n+1)} = \frac{1}{4}[u_1^{(n+1)} + u_3^{(n+1)} + 1000 + u_5^{(n)}]$$

$$u_4^{(n+1)} = \frac{1}{4}[2000 + u_5^{(n)} + u_1^{(n+1)} + u_1^{(n+1)}]$$

$$u_5^{(n+1)} = \frac{1}{4}[u_4^{(n+1)} + u_4^{(n+1)} + u_2^{(n+1)} + u_2^{(n+1)}]$$

First iteration ($n = 0$)

$$u_1^{(1)} = \frac{1}{4}[1000 + 1187.5 + 500 + 1437.5] = 1031.25$$

$$u_2^{(1)} = \frac{1}{4}[1031.25 + 1031.25 + 1000 + 1500] = 1140.625$$

$$u_4^{(1)} = \frac{1}{4}[2000 + 1500 + 1031.25 + 1031.25] = 1390.625$$

$$u_5^{(1)} = \frac{1}{4}[1390.625 + 1390.625 + 1140.625 + 1140.625] = 1265.625$$

Second iteration ($n = 1$)

$$u_1^{(2)} = \frac{1}{4}[1000 + 1140.625 + 500 + 1390.625] = 1007.8125$$

$$u_2^{(2)} = \frac{1}{4}[1007.8125 + 1007.8125 + 1000 + 1265.625] = 1070.3125$$

$$u_4^{(2)} = \frac{1}{4}[2000 + 1265.625 + 1007.8125 + 1007.8125] = 1320.3125$$

$$u_5^{(2)} = \frac{1}{4}[1320.3125 + 1320.3125 + 1070.3125 + 1070.3125] = 1195.3125$$

Third iteration ($n = 2$)

$$u_1^{(3)} = \frac{1}{4}[1000 + 1070.3125 + 500 + 1320.3125] = 972.65625$$

$$u_2^{(3)} = \frac{1}{4}[972.65625 + 972.65625 + 1000 + 1195.3125] = 1035.1563$$

$$u_4^{(3)} = \frac{1}{4}[2000 + 1195.3125 + 972.65625 + 972.65625] = 1285.1563$$

$$u_5^{(3)} = \frac{1}{4}[2(1285.1563) + 2(1035.1563)] = 1160.1563$$

Fourth iteration ($n = 3$)

$$u_1^{(4)} = \frac{1}{4}[1000 + 1035.1563 + 500 + 1285.1563] = 955.07815$$

$$u_2^{(4)} = \frac{1}{4}[955.07815 + 955.07815 + 1000 + 1160.1563] = 1017.5782$$

$$u_4^{(4)} = \frac{1}{4}[2000 + 1160.1563 + 955.07815 + 955.07815] = 1267.5782$$

$$u_5^{(5)} = \frac{1}{4}[2 \times 1267.5782 + 2 \times 1017.5782] = 1142.5782$$

Fifth iteration ($n = 4$)

$$u_1^{(5)} = \frac{1}{4}[1000 + 1017.5782 + 500 + 1267.5782] = 946.2891$$

$$u_2^{(5)} = \frac{1}{4}[946.2891 + 946.2891 + 1000 + 1142.5782] = 1008.7891$$

$$u_4^{(5)} = \frac{1}{4}[2000 + 1142.5782 + 2(946.2891)] = 1258.7891$$

$$u_5^{(5)} = \frac{1}{4}[2(1258.7891) + 2(1008.7891)] = 1133.7891]$$

Sixth iteration ($n = 5$)

$$u_1^{(6)} = \frac{1}{4}[1000 + 1008.7891 + 500 + 1258.7891] = 941.89455$$

$$u_2^{(6)} = \frac{1}{4}[2(941.89455) + 1000 + 1133.7891] = 1004.3946$$

$$u_4^{(6)} = \frac{1}{4}[2000 + 1133.7891 + 2(941.89455)] = 1254.3496$$

$$u_5^{(6)} = \frac{1}{4}[1254.3946 + 1254.3946 + 2(1004.3946)] = 1129.3946$$

Seventh iteration ($n = 6$)

$$u_1^{(7)} = \frac{1}{4}[1000 + 1004.3946 + 500 + 1254.3946] = 939.6973$$

$$u_2^{(7)} = \frac{1}{4}[2(939.6973) + 1000 + 1129.3946] = 1002.1973$$

$$u_4^{(7)} = \frac{1}{4}[2000 + 1129.3946 + 2(939.6973)] = 1252.1973$$

$$u_5^{(7)} = \frac{1}{4}[2(1252.1973) + 2(1002.1973)] = 1127.1973$$

Eighth iteration ($n = 7$)

$$u_1^{(8)} = \frac{1}{4}[1000 + 1002.1972 + 500 + 1252.1973] = 938.59863$$

$$u_2^{(8)} = \frac{1}{4}[2(938.59863) + 1000 + 1127.1973] = 1001.0987$$

$$u_4^{(8)} = \frac{1}{4}[2000 + 1127.1973 + 2(938.59685)] = 1251.0987$$

$$u_5^{(8)} = \frac{1}{4}[2(1251.0987) + 2(1001.0987)] = 1126.0987$$

Ninth iteration ($n = 8$)

$$u_1^{(9)} = \frac{1}{4}[1000 + 1001.0987 + 500 + 12521.0987] = 938.04935$$

$$u_2^{(9)} = \frac{1}{4}[2(938.04935) + 1000 + 1126.09871] = 1000.5494$$

$$u_4^{(9)} = \frac{1}{4}[2000 + 1126.0987 + 2(938.04935)] = 1250.5494$$

$$u_5^{(9)} = \frac{1}{4}[2(1250.5494) + 2(1000.5494)] = 1125.5494$$

From the eighth and ninth iterations, we see there is a negligible difference between the values

∴ $u_1 = 939, u_2 = 1001, u_4 = 1251,$ and $u_5 = 1126$

⇒ $u_3 = 939, u_6 = 1251, u_7 = 939, u_8 = 1001, u_9 = 939$

Example 9.3: Given the values of $u(x, y)$ on the boundary of the square in the following figure. Evaluate the function $u(x, y)$ satisfying Laplace equation $\nabla^2 u = 0$ as the pivot point on this figure by (a) Jacobi method (b) Gauss-Seidel method.

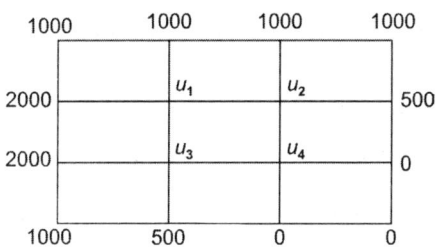

Solution: Let $u_4 = 0$ then

$$u_1 = \frac{1}{4}[1000 + 0 + 1000 + 2000] = 1000 \text{ (DFPF)}$$

$$u_2 = \frac{1}{4}[1000 + 500 + 1000 + 0] = 625 \text{ (SFPF)}$$

$$u_3 = \frac{1}{4}[2000 + 0 + 1000 + 500] = 875 \text{ (SFPF)}$$

$$u_4 = \frac{1}{4}[874 + 0 + 625 + 0] = 375 \text{ (SFPF)}$$

a. Using Jacobi formula

$$u_1^{(n+1)} = \frac{1}{4}[2000 + u_2^{(n)} + 1000 + u_3^{(n)}]$$

$$u_2^{(n+1)} = \frac{1}{4}[u_1^{(n)} + 5000 + 1000 + u_4^{(n)}]$$

$$u_3^{(n+1)} = \frac{1}{4}[2000 + u_4^{(n)} + u_1^{(n)} + 500]$$

$$u_4^{(n+1)} = \frac{1}{4}[u_3^{(n)} + 0 + u_2^{(n)} + 0]$$

First iteration $(n = 0)$

$$u_1^{(1)} = \frac{1}{4}[2000 + 625 + 1000 + 875] = 1125$$

$$u_2^{(1)} = \frac{1}{4}[1000 + 500 + 1000 + 375] = 719$$

$$u_3^{(1)} = \frac{1}{4}[2000 + 375 + 1000 + 500] = 969$$

$$u_4^{(1)} = \frac{1}{4}[875 + 0 + 625 + 0] = 375$$

Second iteration ($n = 1$)

$$u_1^{(2)} = \frac{1}{4}[2000 + 719 + 1000 + 969] = 1172$$

$$u_2^{(2)} = \frac{1}{4}[1125 + 500 + 1000 + 375] = 750$$

$$u_3^{(2)} = \frac{1}{4}[2000 + 375 + 1125 + 500] = 1000$$

$$u_4^{(2)} = \frac{1}{4}[969 + 0 + 719 + 0] = 422$$

Similarly

$u_1^{(3)} = 1188$	$u_2^{(3)} = 774$	$u_3^{(3)} = 1024$	$u_4^{(3)} = 438$
$u_1^{(4)} = 1200$	$u_2^{(4)} = 782$	$u_3^{(4)} = 1032$	$u_4^{(4)} = 450$
$u_1^{(5)} = 1204$	$u_2^{(5)} = 788$	$u_3^{(5)} = 1038$	$u_4^{(5)} = 454$
$u_1^{(6)} = 1206.5$	$u_2^{(6)} = 790$	$u_3^{(6)} = 1040$	$u_4^{(6)} = 456.6$
$u_1^{(7)} = 1208$	$u_2^{(7)} = 791$	$u_3^{(7)} = 1041$	$u_4^{(7)} = 458$
$u_1^{(8)} = 1208$	$u_2^{(8)} = 791.5$	$u_3^{(8)} = 1041.5$	$u_4^{(8)} = 458$

Hence $u_1 = 1208$, $u_2 = 791.5$, $u_3 = 1041.5$, $u_4 = 258$.

b. Using Gauss–Seidal method

$$u_1^{(n+1)} = \frac{1}{4}[2000 + u_2^{(n)} + 1000 + u_3^{(n)}]$$

$$u_2^{(n+1)} = \frac{1}{4}[u_1^{(n+1)} + 500 + 1000 + u_4^{(n)}]$$

$$u_3^{(n+1)} = \frac{1}{4}[2000 + u_4^{(n)} + u_1^{(n+1)} + 500]$$

$$u_4^{(n+1)} = \frac{1}{4}[u_3^{(n+1)} + 0 + u_2^{(n+1)} + 0]$$

First iteration ($n = 0$)

$$u_1^{(1)} = \frac{1}{4}[2000 + 625 + 1000 + 875] = 1125$$

$$u_2^{(1)} = \frac{1}{4}[1125 + 500 + 1000 + 375] = 750$$

$$u_3^{(1)} = \frac{1}{4}[2000 + 375 + 1125 + 500] = 1000$$

$$u_4^{(1)} = \frac{1}{4}[1000 + 0 + 750 + 0] = 438$$

Second iteration ($n = 1$)

$$u_1^{(2)} = \frac{1}{4}[2000 + 750 + 1000 + 1000] = 1188$$

$$u_2^{(2)} = \frac{1}{4}[1188 + 500 + 1000 + 438] = 782$$

$$u_3^{(2)} = \frac{1}{4}[2000 + 438 + 1188 + 500] = 1032$$

$$u_4^{(2)} = \frac{1}{4}[1032 + 0 + 782 + 0] = 454$$

Similarly

$u_1^{(3)} = 1204$	$u_2^{(3)} = 789$	$u_3^{(3)} = 1040$	$u_4^{(3)} = 458$
$u_1^{(4)} = 1207$	$u_2^{(4)} = 791$	$u_3^{(4)} = 1041$	$u_4^{(4)} = 458$
$u_1^{(5)} = 1208$	$u_2^{(5)} = 791.5$	$u_3^{(5)} = 1041.5$	$u_4^{(5)} = 458.2$

Hence $u_1 = 1208$, $u_2 = 791.5$, $u_3 = 1041.5$, $u_4 = 458.25$.

Example 9.4: Solve the elliptic equation $u_{xx} + u_{yy} = 0$ for the following square mesh with boundary values as shown in the figure. Iterate until the maximum difference between two successive values at any point is less than 0.001.

Solution:

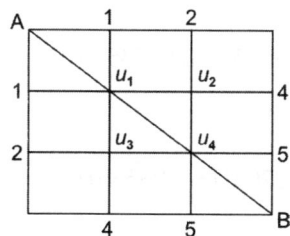

The figure is symmetrical about AB.

$$u_2 = u_3 = 0 \text{ (say)}$$

$$u_1^{(0)} = \frac{1}{4}[1 + u_2 + 1 + u_3] = 0.5 \text{ (SFPF)}$$

$$u_2^{(0)} = \frac{1}{4}[0.5 + 2 + 4 + 2.5] = 2.25 \text{ (SFPF)}$$

$$u_4^{(0)} = \frac{1}{4}[u_3 + 5 + u_2 + 5] = 2.5 \text{ (SFPF)}$$

Now using Gauss–Seidal formula

$$u_1^{(n+1)} = \frac{1}{4}[1 + u_2^n + 1 + u_3^n]$$

$$u_2^{(n+1)} = \frac{1}{4}[u_1^{(n+1)} + 4 + 2 + u_4^n]$$

$$u_4^{(n+1)} = \frac{1}{4}[u_3^{(n+1)} + 5 + u_2^{(n+1)} + 5]$$

First iteration $(n = 0)$

$$u_1^{(1)} = \frac{1}{4}[2 + 2 \times 2.25] = \frac{1}{2}[1 + 2.25] = 1.625$$

$$u_2^{(1)} = \frac{1}{4}[u_1^{(1)} + 4 + 2 + u_4^{(0)}] = \frac{1}{4}[1.625 + 6 + 2.5] = 2.53125$$

$$u_4^{(1)} = \frac{1}{4}[u_3^{(1)} + 5 + u_2^{(1)} + 5] = \frac{1}{4}[2.53125 + 5] = 3.765625$$

Second iteration $(n = 1)$

$$u_1^{(2)} = \frac{1}{4}[1 + u_2^1 + 1 + u_3^1] = \frac{1}{4}[2 + 2 \times 2.53125] = \frac{1}{2}[1 + 2.53125] = 1.765625$$

$$u_2^{(2)} = \frac{1}{4}[6 + 1.765625 + 3.765625] = 2.8828125$$

$$u_4^{(2)} = \frac{1}{2}[2.8828125 + 5] = 3.9414063$$

Third iteration $(n = 2)$

$$u_1^{(3)} = \frac{1}{2}[1 + 2.8828125] = 1.9414063$$

$$u_2^{(3)} = \frac{1}{6}[6 + 1.9414063 + 3.9414063] = 2.9707031$$

$$u_4^{(3)} = \frac{1}{2}[2.970703 + 5] = 3.9853516$$

Fourth iteration $(n = 3)$

$$u_1^{(4)} = \frac{1}{2}[1 + 2.9707031] = 1.9853516$$

$$u_2^{(4)} = \frac{1}{6}[6 + 1.9853516 + 3.9853516] = 2.9926758$$

$$u_4^{(4)} = \frac{1}{2}[2.9926758 + 5] = 3.9963379$$

Fifth iteration $(n = 4)$

$$u_1^{(5)} = \frac{1}{2}[1 + 2.9926578] = 1.9963289$$

$$u_2^{(5)} = \frac{1}{4}[6 + 1.99263289 + 3.9963379] = 2.9981667$$

$$u_4^{(5)} = \frac{1}{2}[2.9918667 + 5] = 3.9990834$$

Sixth iteration $(n = 5)$

$$u_1^{(6)} = \frac{1}{2}[1 + 2.9981667] = 1.9990834$$

$$u_2^{(6)} = \frac{1}{4}[6 + 1.9990839 + 3.9990834] = 2.9995417$$

$$u_4^{(6)} = \frac{1}{2}[2.9995417 + 5] = 3.9997709$$

Hence $u_1 = 1.999$, $u_2 = 2.999$, $u_4 = 3.999$.

Example 9.5: $u_{xx} + y_{yy} = 0$ in $0 \le x \le 4$, $0 \le y \le 4$, given that $u(0, y) = 0$; $u(4, y) = 8 + 2y$, $u(x, 0) = \dfrac{x^2}{2}$ and $u(x, 4) = x^2$, take $h = k = 1$, and obtain the result correct to one decimal.

Solution: Let us divide the region R, i.e. $0 \le x \le 4$, $0 \le y \le 4$ into 16 square meshes. The numerical values of the boundary using the given analytical expressions are calculated and exhibited in the figure below.

Let $u_1, u_2, u_3, ..., u_9$ be the values of u at the interior mesh points. Now the initial values of u's are calculated either by SFPF or DFPF as given below.

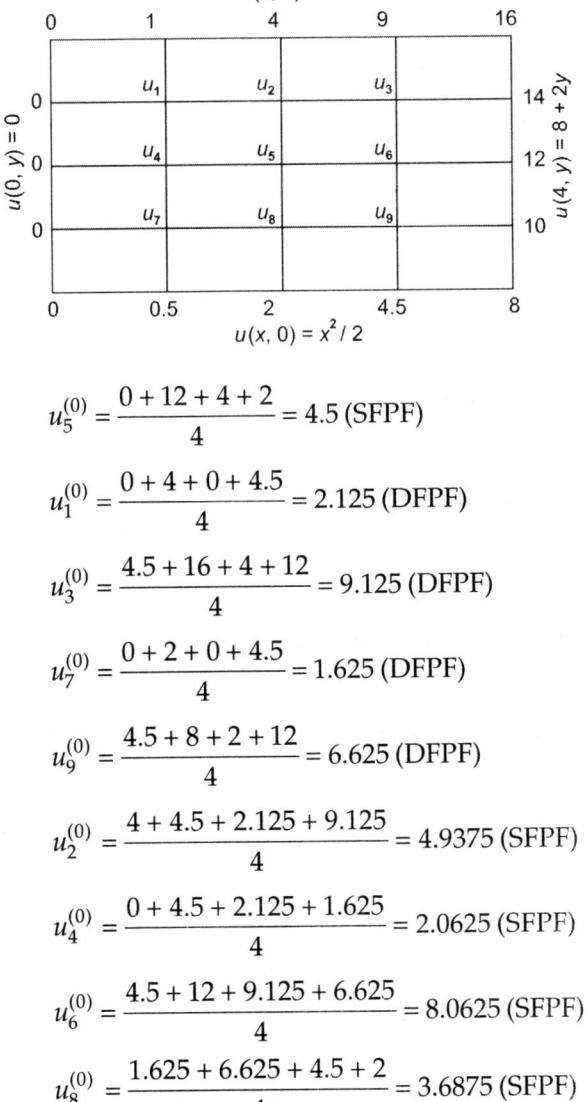

$$u_5^{(0)} = \frac{0 + 12 + 4 + 2}{4} = 4.5 \ (\text{SFPF})$$

$$u_1^{(0)} = \frac{0 + 4 + 0 + 4.5}{4} = 2.125 \ (\text{DFPF})$$

$$u_3^{(0)} = \frac{4.5 + 16 + 4 + 12}{4} = 9.125 \ (\text{DFPF})$$

$$u_7^{(0)} = \frac{0 + 2 + 0 + 4.5}{4} = 1.625 \ (\text{DFPF})$$

$$u_9^{(0)} = \frac{4.5 + 8 + 2 + 12}{4} = 6.625 \ (\text{DFPF})$$

$$u_2^{(0)} = \frac{4 + 4.5 + 2.125 + 9.125}{4} = 4.9375 \ (\text{SFPF})$$

$$u_4^{(0)} = \frac{0 + 4.5 + 2.125 + 1.625}{4} = 2.0625 \ (\text{SFPF})$$

$$u_6^{(0)} = \frac{4.5 + 12 + 9.125 + 6.625}{4} = 8.0625 \ (\text{SFPF})$$

$$u_8^{(0)} = \frac{1.625 + 6.625 + 4.5 + 2}{4} = 3.6875 \ (\text{SFPF})$$

Now we use Gauss–Seidal formula

$$u_{i,j}^{(n+1)} = \frac{1}{4}\left[u_{i-1,j}^{(n+1)} + u_{i+1,j}^{(n)} + u_{i,j-1}^{(n)} + u_{i,j+1}^{(n+1)}\right]$$

To improve the above result, we proceed as follows.

$$u_1^{(n+1)} = \frac{1}{4}[0 + u_2^{(n)} + u_4^{(n)} + 1]$$

$$u_2^{(n+1)} = \frac{1}{4}[u_1^{(n+1)} + u_3^{(n)} + u_5^{(n)} + 4]$$

$$u_3^{(n+1)} = \frac{1}{4}[u_2^{(n+1)} + 14 + u_6^{(n)} + 9]$$

$$u_4^{(n+1)} = \frac{1}{4}[0 + u_5^{(n)} + u_7^{(n)} + u_1^{(n+1)}]$$

$$u_5^{(n+1)} = \frac{1}{4}[u_4^{(n+1)} + u_6^{(n)} + u_8^{(n)} + u_2^{(n+1)}]$$

$$u_6^{(n+1)} = \frac{1}{4}[u_5^{(n+1)} + 12 + u_9^{(n)} + u_3^{(n+1)}]$$

$$u_7^{(n+1)} = \frac{1}{4}[0 + u_8^{(n)} + 0.5 + u_4^{(n+1)}]$$

$$u_8^{(n+1)} = \frac{1}{4}[u_7^{(n+1)} + u_9^{(n)} + 2 + u_5^{(n+1)}]$$

$$u_9^{(n+1)} = \frac{1}{4}[u_8^{(n+1)} + 10 + 4.5 + u_6^{(n+1)}]$$

First iteration ($n = 0$)

$$u_1^{(1)} = \frac{1}{4}[0 + u_2^{(0)} + u_4^{(0)} + 1] = \frac{1}{4}[0 + 4.9375 + 2.0625 + 1] = 2$$

$$u_2^{(1)} = \frac{1}{4}[u_1^{(1)} + u_3^{(0)} + u_5^{(0)} + 4] = \frac{1}{4}[2 + 9.125 + 4.5 + 4] = 4.90625$$

$$u_3^{(1)} = \frac{1}{4}[u_2^{(1)} + 14 + u_6^{(0)} + 9] = \frac{1}{4}[4.90625 + 14 + 8.0625 + 9] = 8.9921875$$

$$u_4^{(1)} = \frac{1}{4}[0 + u_5^{(0)} + u_1^{(0)} + u_7^{(0)}] = \frac{1}{4}[0 + 4.5 + 1.625 + 2] = 2.03125$$

$$u_5^{(1)} = \frac{1}{4}[u_4^{(1)} + u_6^{(0)} + u_5^{(0)} + u_2^{(1)}]$$

$$= \frac{1}{4}[2.03125 + 8.0625 + 3.6875 + 4.90625] = 4.671875$$

$$u_6^{(1)} = \frac{1}{4}[u_5^{(1)} + 12 + u_9^{(0)} + u_3^{(1)}]$$

$$= \frac{1}{4}[4.671875 + 12 + 6.625 + 8.9921875] = 8.0722656$$

$$u_7^{(1)} = \frac{1}{4}[0 + u_8^{(0)} + 0.5 + u_4^{(1)}] = \frac{1}{4}[0 + 3.6875 + 0.5 + 2.03125] = 1.5546875$$

$$u_8^{(1)} = \frac{1}{4}[u_7^{(1)} + u_9^{(0)} + 2 + u_5^{(1)}] = \frac{1}{4}[1.5546875 + 6.625 + 2 + 4.671875] = 3.7128906$$

$$u_9^{(1)} = \frac{1}{4}[u_8^{(1)} + 10 + 4.5 + u_6^{(1)}] = \frac{1}{4}[3.7128906 + 10 + 4.5 + 8.0722656] = 6.5712891$$

Second iteration ($n = 1$)

$$u_1^{(2)} = \frac{1}{4}[0 + u_2^{(1)} + u_4^{(1)} + 1] = \frac{1}{4}[0 + 4.90625 + 2.03135 + 1] = 1.984375$$

$$u_2^{(2)} = \frac{1}{4}[u_1^{(2)} + u_3^{(1)} + u_5^{(1)} + 4] = \frac{1}{4}[1.984375 + 8.9921875 + 4.645875 + 4] = 4.9121094$$

$$u_3^{(2)} = \frac{1}{4}[u_2^{(2)} + 14 + u_6^{(1)} + 9] = \frac{1}{4}[23 + 4.9121094 + 8.0722656] = 8.9960937$$

$$u_4^{(2)} = \frac{1}{4}[0 + u_5^{(1)} + u_7^{(1)} + u_1^{(2)}] = \frac{1}{4}[0 + 4.671875 + 1.5546875 + 1.984375] = 2.0527344$$

$$u_5^{(2)} = \frac{1}{4}[u_4^{(2)} + u_6^{(1)} + u_8^{(1)} + u_2^{(2)}]$$

$$= \frac{1}{4}[2.0527344 + 8.0722656 + 3.7128906 + 4.9121094] = 4.6875$$

$$u_6^{(2)} = \frac{1}{4}[u_5^{(2)} + 12 + u_9^{(1)} + u_3^{(2)}] = \frac{1}{4}[4.6875 + 6.7512891 + 8.9960937] = 8.063720$$

$$u_7^{(2)} = \frac{1}{4}[0 + u_8^{(1)} + 0.5 + u_4^{(2)}] = \frac{1}{4}[0.5 + 3.7128906 + 2.0527344] = 1.5664063$$

$$u_8^{(2)} = \frac{1}{4}[u_7^{(2)} + u_9^{(1)} + 2 + u_5^{(2)}] = \frac{1}{4}[1.5664063 + 6.571289 + 4.6875] = 3.7062988$$

$$u_9^{(2)} = \frac{1}{4}[u_8^{(2)} + 10 + 4.5 + u_6^{(2)}] = \frac{1}{4}[3.7062988 + 14.5 + 8.063207] = 6.5675049$$

Third iteration ($n = 2$)

$$u_1^{(3)} = 1.991211 \qquad u_2^{(3)} = 4.91181012 \qquad u_3^{(3)} = 8.9956055$$

$$u_3^{(3)} = 2.0612793 \qquad u_5^{(3)} = 4.6875 \qquad u_6^{(3)} = 8.0626526$$

$$u_7^{(3)} = 1.5668945 \qquad u_8^{(3)} = 3.7054749 \qquad u_9^{(3)} = 6.56707319$$

Hence $u_1 = 1.99$, $u_2 = 4.91$, $u_3 = 8.99$, $u_4 = 2.06$, $u_5 = 4.69$, $u_6 = 8.06$, $u_7 = 1.57$, $u_8 = 3.71$, $u_9 = 6.57$.

PROBLEMS 9.1

1. Solve the equation $u_{xx} + u_{yy} = 0$ for the square mesh with the boundary values as shown.

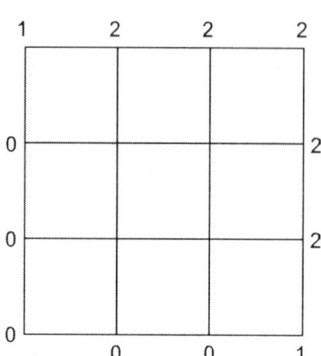 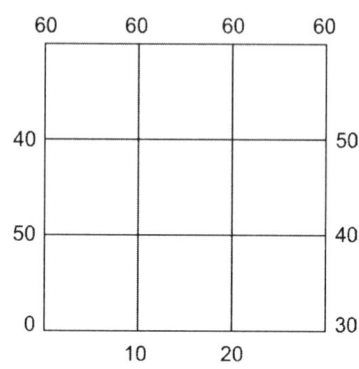

[Ans. (a) $u_1 = 1.999$, $u_2 = 2.999$, $u_3 = 3.999$, $u_4 = 2.999$, (b) $u_1 = 26.66$, $u_2 = 33.33$, $u_3 = 43.33$, $u_4 = 46.66$]

2. Solve the equation $\nabla^2 u = 0$ for the square mesh given below by (a) Jacobi method (b) Gauss–Seidel method.

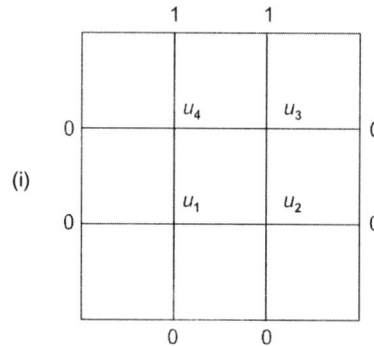 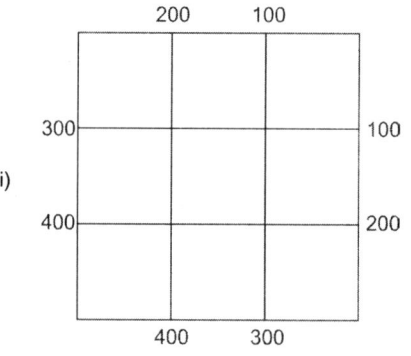

[Ans. (a) $u_1 = 0.126$, $u_2 = 0.126$, $u_3 = 0.376$, $u_4 = 0.376$, (b) $u_1 = u_4 = 250$, $u_2 = 175$, $u_3 = 3251$]

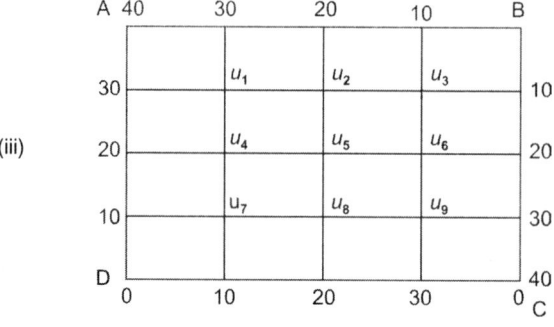

[Ans. (a) $u_1 = u_9 = 25$, $u_3 = u_7 = 15$, $u_2 = 20$, $u_5 = 20$]

9.7 WAVE EQUATION

Consider
$$\frac{\partial^2 u}{\partial t^2} = c^2 \frac{\partial^2 u}{\partial x^2} \tag{9.10}$$

Given the conditions $u(0, t) = 0$, $u(1, t) = 0$, $u(x, 0) = f(x)$ $u_t(x, 0) = 0$.

Solution: Consider a rectangular mesh in x-t plane spacing h in x-direction and k in t-direction. Denoting the mesh point

$$(x, t) = (ih, jk)$$
$$u(x, t) = u(ih, jk) = u_{i,j}$$

We have
$$\frac{\partial^2 u}{\partial x^2} = \frac{u_{i-1,j} - 2u_{i,j} + u_{i+1,j}}{h^2} \text{ (using difference approximation)}$$

$$\frac{\partial^2 u}{\partial t^2} = \frac{u_{i,j-1} - 2u_{i,j} + u_{i,j+1}}{k^2}$$

Substituting in Eq. (9.10), we have

$$u_{i,j-1} - 2u_{i,j} + u_{i,j+1} = c^2 \frac{k^2}{h^2} [u_{i-1,j} - 2u_{i,j} + u_{i+1,j}]$$

Put $\alpha = k/h$

$$u_{i,j-1} - 2u_{i,j} + u_{i,j+1} = c^2 \alpha^2 [4_{i-1,j} - 2u_{i,j} + u_{i+1,j}]$$
$$u_{i,j+1} = 2(1 - c^2 \alpha^2) u_{i,j} + \alpha^2 c^2 (u_{i-1,j} + u_{i+1,j}) - u_{i,j-1}$$

which is a difference equation where $\alpha = k/h$.

Now
$$u(0, t) = 0 \qquad\qquad u_{0,j} = 0 \qquad\qquad j = 0, 1, 2$$
$$u(1, t) = 0 \qquad\qquad u_{1,j} = 0$$
$$u(x, 0) = f(x) \qquad\qquad u_{i,0} = f(x) = f(ih) \quad i = 0, 1, 2$$

$$u_t(x, 0) = 0 \qquad\qquad \frac{\partial}{\partial t} u(x, 0) = 0$$

The boundary conditions are
$$u_{0,j} = 0, \quad u_{1,j} = 0, \quad j = 0, 1, 2, \ldots$$

The initial conditions are
$$u_{i,0} = f(ih), \quad i = 0, 1, 2, \ldots$$

Also
$$\frac{\partial u}{\partial t}(x, 0) = \frac{u_{i,j+1} - u_{i,j}}{k} = 0$$

or
$$u_{i,j+1} = u_{i,j}$$
$$u_{i,1} = u_{i,0} = f(ih)$$

Now the above equation gives the values of u on the first two rows $j = 0, j = 1$ which are the same.

Putting $j = 1$ in the difference equation

$$u_{i,j+1} = 2(1 - c^2 \alpha^2) u_{i,j} + \alpha^2 c^2 (u_{i+1,j} + u_{i-1,j}) - u_{i,j-1}$$

we have
$$u_{i,2} = 2(1 - \alpha^2 c^2) u_{i,1} + \alpha^2 c^2 (u_{i+1,1} + u_{i-1,1}) - u_{i,0}, \, i = 1, 2, 3, \ldots$$

we get the values in the third row. Proceeding in the same way

$$u_{i,j}$$

j \ i	0	1	2	3	......	1
0	0	0	0	0	......	0
1	0	0	0	0	......	0
2	0	0	0	0	......	0
...	...	...	...	...	......	...

Putting $j = 2$, we get

$$u_{i,2} = 2(1 - \alpha^2 c^2) u_{i,2} + \alpha^2 c^2 (u_{i+1,2} + u_{i-1,2}) - u_{i,1}$$

Putting $i = 1, 2, 3, ...$, we get the values of u in the fourth row. Hence proceeding in this way, we get all the rows filled.

This is the scheme for the solution of the wave equation.

Example 9.6: Evaluate the pivotal values of the equation $u_{tt} = 16u_{xx}$ taking $\Delta x = 1$ up to $t = 1.25$, the boundary conditions are $u(0, t) = u(5, t) = 0$, $u_t(x, 0) = 0 = x^2(5 - x)$.

Solution: Here $c^2 = 16$, $h = 1$, $t = 1.25$, $jk = 1.25$.

The difference equation is

$$u_{i,j+1} = 2(1 - 16\alpha^2) u_{i,j} + 16\alpha^2 (u_{i-1,j}) + u_{i+1,j} - u_{i,j-1}$$

We chose α in such a way that coefficient of $u_{i,j} = 0$

i.e. $\qquad 1 - 16\alpha^2 = 0$, $\alpha^2 = 1/16$

$$\alpha = \frac{1}{4} = \frac{k}{h} = \frac{k}{1} \quad \therefore k = \frac{1}{4}$$

Now $\qquad jk = \dfrac{5}{4}, k = \dfrac{1}{4}$

$$j \times \frac{1}{4} = \frac{5}{4}, j = 5$$

i goes from 0 to 5
j goes from 0 to 5

Now $\qquad \left. \begin{array}{l} u(0, t) = 0, \ u_{0,j} = 0 \\ u(5, t) = 0, \ u_{5,j} = 0 \end{array} \right\}$ for all j

The entries in the first and last columns are

$$u_{i,j}$$

j \ i	0	1	2	3	4	5
0	0	4	12	18	16	0
1	0	4	12	18	16	0
2	0	8	10	10	2	0
3	0	6	6	-6	-6	0
4	0	-2	-10	-10	-8	0
5	0	-16	-18	-12	-4	0

Also $\qquad u(x, 0) = x^2(5 - x)$

$$u_{i,0} = i^2(5 - i)$$

$$u_{1,0} = 4$$

$$u_{2,0} = 12$$

$$u_3 = 18$$

$$u_{4,0} = 16$$

Also $\qquad u_t(x, 0) = 0$ or $\dfrac{u_{i,j+1} - u_{i,j}}{k} = 0$

$$u_{i,1} = u_{i,0}$$

The entries in first two rows are the same.

Again we have the difference equation putting $\alpha^2 = \dfrac{1}{16}$

$$u_{i,j+1} = u_{i-1,j} + u_{i+1,j} - u_{i,j-1}$$

Putting $j = 1$, we get

$$u_{i,2} = u_{i-1,j} + u_{i+1,j} - u_{i,0}$$

Taking $i = 1, 2, 3, 4, \dots$ respectively, we obtain

$$i = 1 \; u_{1,2} = u_{0,1} + u_{2,1} - u_{1,0} = 0 + 12 - 4 = 8$$

$$= 2 \; u_{2,2} = u_{1,1} + u_{3,1} - u_{2,0} = 4 + 18 - 12 = 10$$

$$= 3 \; u_{3,2} = u_{2,1} + u_{4,1} - u_{3,0} = 12 + 16 - 18 = 10$$

$$= 4 \; u_{4,2} = u_{3,1} + u_{5,1} - u_{4,0} = 18 + 0 - 16 = 2$$

These are the entries in the third row.

Again putting $j = 2$, we get

$$u_{i,3} = u_{i-1,2} + u_{i+1,2} - u_{i,1}$$

Putting $i = 1, 2, 3, 4$ $\quad i = 1 \; u_{1,3} = u_{0,2} + u_{2,2} - u_{1,1} = 0 + 10 - 4 = 6$

$$= 2 \; u_{2,3} = u_{1,2} + u_{3,2} - u_{2,1} = 8 + 10 - 12 = 6$$

$$= 3 \; u_{3,3} = u_{2,2} + u_{4,2} - u_{3,1} = 10 + 2 - 18 = -6$$

$$= 4 \; u_{4,3} = u_{3,2} + u_{5,2} - u_{4,1} = 10 + 0 - 16 = -6$$

Similarly putting $j = 3, 4, 5$; we get the remaining rows filled as in table.

Example 9.7: Find the nodal values of the wave equation $\dfrac{\partial^2 u}{\partial t^2} = 16\dfrac{\partial^2 u}{\partial x^2}$ given that $u(0, t) = u(5, t) = 0$, $u(x, 0) = x^2(5 - x)$, $u_t(x, 0) = 0$, taking $h = 1$ and up to one half of the period of vibration.

Solution: Here $h = 1$, $c^2 = 16$, $c = 4$.

Period of vibration is $\dfrac{2l}{c} = \dfrac{2 \times 5}{4} = \dfrac{5}{2}$

Half period of vibration $= \dfrac{5}{4}$

up to $\qquad t = \dfrac{5}{4} = 1.25$

Follow the same procedure as in Example 9.6.

Example 9.8: Solve $u_{tt} = 4u_{xx}$ with the boundary conditions $u(0, t) = 0$, $u(4, t)$, $u_t(x, 0) = 0$, $u(x, 0) = x(4 - x)$, taking $h = 1$ and $k = 0.5$.

Solution: Given equation is $u_{tt} = 4u_{xx}$.

Here $c^2 = 4$, $c = 2$.

Now the initial condition are $u(0, t) = 0$, i.e. $u_{0, j} = 0$

$$u(4, t) = 0 \quad u_{4, j} = 0 \text{ for all } j$$

Hence entries in the first and last column are zero.

Also

$$u(x, 0) = x(4 - x)$$
$$u_{i, 0} = i(4 - i)$$
$$u_{1, 0} = 3$$
$$u_{2, 0} = 4$$
$$u_{3, 0} = 3$$
$$u_{4, 0} = 0$$

Hence the entries in the first row are 0, 3, 4, 3, 0.

Also $\quad u_t(x, 0) = 0 \Rightarrow u_{i, 1} = u_{i, 0}$

Entries in the first row and second row are the same.

$$u_{i, j}$$

j \ i	0	1	2	3	4
0	0	3	4	3	0
1	0	3	4	3	0
2	0	1	2	1	0
3	0	-1	-2	-1	0
4	0	-3	-4	-3	0

Also we have difference equation

$$u_{i, j+1} = u_{i+1, j} + u_{i-1, j} - u_{i, j-1}$$

Putting $j = 1$

$$u_{i, 2} = u_{i+1, 1} + u_{i-1, 1} - u_{i, 0}$$

Putting $i = 1, 2, 3$, we get

$$u_{1, 2} = u_{2,1} + u_{0,1} - u_{1,0} = 4 + 0 - 3 = 1$$
$$u_{2, 2} = u_{3,1} + u_{1,1} - u_{2,0} = 3 + 3 - 4 = 2$$
$$u_{3, 2} = u_{4,1} + u_{2,1} - u_{3,0} = 0 + 4 - 3 = 1$$

Putting $j = 2$ in difference equation, we get

Putting $i = 1, 2, 3;$ $\left\{ \begin{array}{l} u_{i, 3} = u_{i+1, 1} + u_{i-1, 2} - u_{i, 1} \\ u_{1, 3} = u_{2,2} + u_{0,2} - u_{1,1} = 2 + 0 - 3 = -1 \\ u_{2, 3} = u_{3,2} + u_{1,2} - u_{2,1} = 1 + 1 - 4 = -2 \\ u_{3, 3} = u_{4,2} + u_{2,2} - u_{3,1} = 0 + 2 - 3 = -1 \end{array} \right.$

Putting $j = 3$ in the difference equation, we get fourth row filled.

Example 9.9: Solve $u_u = u_{xx}$ up to $t = 0.5$ with spacing of 0.1, given $u(0, t) = 0 = u(1, t)$, $u_t(x, 0) = 0$, $u(x, 0) = 10 + x(1 - x)$.

Solution: Here $u_{tt} = u_{xx}$, $c^2 = 1$, $t = 0.5$, $k = 0.1$, j goes from 0 to 5.

We have the difference equation

$$u_{i,j+1} = 2(1 - \alpha^2 c^2) u_{i,j} + \alpha^2 c^2 (u_{i,-1,j} + u_{i+1,j}) - u_{i,j-1}$$

We choose t in such a way that the coefficient of $u_{i,j} = 0$

$$\therefore \qquad 1 - \alpha^2 c^2 = 0,\ 1 - \alpha^2 = 0,\ \alpha^2 = 1,\ \alpha = 1$$

But $\qquad\qquad \alpha = k/h$

$$1 = 0.1/h,\ h = 0.1,\ i \text{ goes from 0 to 10}$$

$u_{i,j}$

x	0	0.1	0.2	0.3	0.4	0.5	0.6	0.7	0.8	0.9	1
t j \ i	0	1	2	3	4	5	6	7	8	9	10
0 0	0	10.09	10.16	10.21	10.24	10.25	10.24	10.21	10.16	10.09	0
1 1	0	10.09	10.16	10.21	10.24	10.25	10.25	10.21	10.16	16.09	0
2 2	0	0.07	10.14	10.19	10.22	10.25	10.22	10.19	10.14	0.07	0
3 3	0	0.05	0.1	10.15	10.18	10.19	10.18	10.15	0.1	0.05	0
4 4	0	0.03	0.06	0.09	10.12	10.13	10.12	0.09	0.06	0.03	0
5 5	0	0.01	0.02	0.03	0.04	10.05	0.04	0.03	0.02	0.01	0

Initial conditions are

$$u(0, t) = 0 \quad u_{0,j} = 0$$
$$u(1, t) = 0 \quad u_{1,j} = 0 \text{ for all } j$$

$\therefore$ entries in the first and last columns are zero

$$[u(1, t) = u(10 \times 0.1, jk) = u(10h, jk) = u_{10,j}]$$

Boundary conditions are

$$u(x, 0) = 10 + x(1 - x)$$
$$u_{i,0} = 10 + i/10\,(1 - i/10)$$

$$u_{1,0} = 10 + \frac{1}{10}\left(1 - \frac{1}{10}\right) = 10.09$$

$$u_{2,0} = 10 + \frac{2}{10}\left(1 - \frac{2}{10}\right) = 10.16$$

$$u_{3,0} = 10 + \frac{3}{10}\left(1 - \frac{3}{10}\right) = 10.21$$

$$u_{4,0} = 10 + \frac{4}{10}\left(1 - \frac{4}{10}\right) = 10.24$$

$$u_{5,0} = 10 + \frac{5}{10}\left(1 - \frac{5}{10}\right) = 10.25$$

Similarly $\qquad u_{6,0} = 10.24$

$$u_{7,0} = 10.21$$
$$u_{8,0} = 10.16$$
$$u_{9,0} = 10.09$$

Also $\qquad u_t(x, 0) = 0 \Rightarrow u_{i,0} = u_{i,1}$

The entries in the first two rows are the same.

Now, the difference equation is

$$u_{i,j+1} = u_{i-1,j} + u_{i+1,j} - u_{i,j-1}$$

Putting $j = 1$ $\qquad u_{i,2} = u_{i-1,1} + u_{i+1,1} - u_{i,0}$

Taking $i = 1, 2, 3, 4, 5, 6, 7, 8, 9$

$$u_{1,2} = u_{0,1} + u_{2,1} - u_{1,0} = 0 + 10.16 - 10.07 = 0.07$$

$$u_{2,2} = u_{1,1} + u_{3,1} - u_{2,0} = 10.09 + 10.21 - 10.16 = 10.14$$

$$u_{3,2} = u_{2,1} + u_{4,1} - u_{3,0} = 10.16 + 10.24 - 10.21 = 10.19$$

$$u_{4,2} = u_{3,1} + u_{5,1} - u_{4,0} = 10.21 + 10.25 - 10.24 = 10.22$$

$$u_{5,2} = u_{4,1} + u_{6,1} - u_{5,0} = 10.24 + 10.24 - 10.25 = 10.23$$

Similarly we get $u_{6,2} = 10.22$, $u_{7,2} = 10.19$, $u_{8,2} = 10.14$, $u_{9,2} = 0.07$.

Put $j = 2, 3, 4$, we get all the rows filled.

PROBLEMS 9.2

1. Solve the equation $\dfrac{\partial^2 u}{\partial t^2} = \dfrac{\partial^2 u}{\partial x^2}$ given the boundary conditions $u(x, 0) =$ for $0 \le x \le 1$,

$$u(0, t) = u(1, t) = 0 \text{ for } t \ge 0$$

Use the explicit scheme to calculate u for

$$x = 0(0.1), 1 \text{ and } t = 0(0.1)0.5.$$

Hint:

$$u_{i,j}$$

j \ i	0	0.1	0.2	0.3	0.4	0.5
0.1	0	0.37	0.07	0.096	0.113	0.119
0.2	0	0.31	0.059	0.082	0.096	0.101
0.3	0	0.23	0.043	0.059	0.07	0.073
0.4	0	0.12	0.023	0.031	0.037	0.039
0.5	0	0	0	0	0	0

2. Solve the hyperbolic partial differential equation (vibration of strings) for one half period of oscillation taking $h = 1$, $u_{tt} = 25u_{xx}$, $u(0, t) = u(5, t) = 0$

$$u_i(x, 0) = 0, \; u(x, 0) = \begin{cases} 2x \text{ for } 0 \le x \le 2.5 \\ 10 - 2x \text{ for } 2.5 \le x \le 5 \end{cases}$$

Hint:

j \ i	0	1	2	3	4	5
0	0	2	4	4	2	0
1	0	2	3	3	2	0
2	0	1	1	1	1	0
3	0	-1	-1	-1	-1	0
4	0	-2	-3	-3	-2	0
5	0	-2	-4	-4	-2	0

3. Approximate the solution to the equation $u_{tt} = 16u_{xx}$ taking $\Delta x = 1$ up to $t = 1.25$. The boundary condition are $u(0, t) = u(5, t) = 0$, $u_t(x, 0) = 0$ and $u(x, 0) = x^2(5 - x)$.

Hint:

i / j	0	1	2	3	4	5
0	0	4	12	18	16	0
1	0	4	12	18	16	0
2	0	8	10	10	2	0
3	0	6	6	-6	-6	0
4	0	-2	-10	-10	-8	0
5	0	-16	-18	-12	-4	0

9.8 SOLUTION OF ONE-DIMENSIONAL HEAT EQUATION

Consider
$$\frac{\partial u}{\partial t} = c^2 \frac{\partial^2 u}{\partial x^2} \qquad (9.11)$$

where $c^2 = k/s\rho$ is the diffusivity of the substance (cm^2/sec).

9.8.1 Bender-Schmidt Method

Consider a mesh (rectangular) in x-t plane with spacing h along x-direction and k along t-direction, denoting mesh point

$$(x, t) = (ih, jk)$$

$\therefore \qquad u(x, t) = u(ih, jk) = u_{i, j}$

We have
$$\frac{\partial u}{\partial t} = \frac{u_{i, j+1} - u_{i, j}}{h}$$

$$\frac{\partial^2 u}{\partial x^2} = \frac{u_{i+1, j} - 2u_{i, j} + u_{i-1, j}}{h^2}$$

$i-1, j+1$	$i, j+1$	$i+1, j+1$ $(j+1)$th level
$i-1, j$	i, j	$i+1, j$ (j)th level
0	$i, j-1$	

Substituting in Eq. (9.11), we get

$$u_{i, j+1} = \frac{c^2 k}{h^2}[u_{i-1, j} - 2u_{i, j} + u_{i+1, j}] + u_{i, j}$$

$$= \alpha u_{i-1, j} + (1 - 2\alpha)u_{i, j} + \alpha u_{i+1, j} \qquad (9.12)$$

where
$$\alpha = \frac{c^2 k}{h^2}$$

This formula enables us to determine the value of u at the $(i, j + 1)$th mesh point in terms of the known values at the point x_{i-1}, x_i, x_{i+1} at the instant t_j. It is a relation between the values at the two time levels $j + 1, j$. Therefore, it is called a two-level function.

Equation (9.12) is called Schmidt explicit formula which is valid only for $0 < \alpha < \dfrac{1}{2}$.

If
$$\alpha = \frac{1}{2}$$

$$u_{i,j+1} = \frac{1}{2}[u_{i-1,j} + u_{i+1,j}] \tag{9.13}$$

This relation is called Bender–Schmidt recurrence relation.

9.8.2 Crank–Nicholson Method

According to this method $\dfrac{\partial^2 u}{\partial x^2}$ is replaced by the average of its finite difference approximation on the jth and $(j + 1)$th level of rows. Thus

$$\frac{\partial^2 u}{\partial x^2} = \frac{1}{2}\left[\frac{u_{i-1,j} - 2u_{i,j} + u_{i+1,j}}{h^2} + \frac{u_{i-1,j+1} - 2u_{i,j+1} + u_{i-1,j+1}}{h^2}\right]$$

$$\frac{u_{i,j+1} - u_{i,j}}{h} = c^2 \frac{1}{2}\left[\frac{u_{i-1,j} - 2u_{i,j} + u_{i+1,j}}{h^2} + \frac{u_{i-1,j+1} - 2u_{i,j+1} + u_{i-1,j+1}}{h^2}\right]$$

or
$$-\alpha u_{i-1,j+1} + (2 + 2\alpha)u_{i,j+1} - \alpha u_{i+1,j+1} = \alpha u_{i-1,j} + (2 - 2\alpha)u_{i,j} + \alpha u_{i+1,j} \tag{9.14}$$

where
$$\alpha = \frac{c^2 k}{h^2}$$

$i-1, j+1$	$i, j+1$		$i+1, j+1$	$(j+1)$th level
	i, j			(j)th level
$i-1, j$			$i-1, j$	

Clearly the left side of Eq. (9.14) contains three unknown values of u at the $(j + 1)$th level while all the three values on the right are known at the jth level.

Thus Eq. (9.14) is a two-level implicit relation and is known as Crank-Nicholson formula.

If there are n internal mesh point on each row then the relation in Eq. (9.14) gives n simultaneous equations for the n unknown values in terms of known boundary values. These equations can be solved to obtain the values at these mesh points.

9.8.3 Du Fort and Frankel Method

If we replace derivatives in Eq. (9.11) by the central difference approximations

$$\frac{\partial u}{\partial t} = \frac{u_{i,j+1} - u_{i,j-1}}{2k}$$

and
$$\frac{\partial^2 u}{\partial x^2} = \frac{u_{i-1,j} - 2u_{i,j} + u_{i+1,j}}{h^2}$$

we obtain
$$u_{i,j+1} - u_{i,j-1} = \frac{2kc^2}{h^2}[u_{i-1,j} - 2u_{i,j} + u_{i+1,j}]$$

$$u_{i,j+1} = u_{i,j-1} + 2\alpha[u_{i-1,j} + 2u_{i,j} + u_{i+1,j}] \tag{9.15}$$

where $\alpha = \dfrac{c^2 k}{h^2}$. This difference equation is called the *Richardson scheme* which is a three-level method.

If we replace $u_{i,j}$ by means of the values $u_{i,j-1}$ and $u_{i,j+1}$, i.e.

$$u_{i,j} = \frac{1}{2}[u_{i,j-1} + u_{i,j+1}) \text{ in Eq. (9.15)}$$

we get
$$u_{i,j+1} = u_{i,j-1} + 2\alpha[u_{i,j-1} - (u_{i,j-1} + u_{i,j+1}) + u_{i+1,j}]$$

On simplification, we have

$$u_{i,j+1} = \frac{1-2\alpha}{1+2\alpha}u_{i,j-1} + \frac{2\alpha}{1+2\alpha}(u_{i-1,j} + u_{i+1,j})$$

This difference scheme is called Du Fort–Frankel scheme which is a three level explicit method, its computational model is given below.

Example 9.10: Find the value of $u(x, t)$ satisfying the parabolic equation $\dfrac{\partial u}{\partial t} = 4\dfrac{\partial^2 u}{\partial x^2}$ and the boundary condition $u(0, t) = 0 = u(8, t)$ and $u(x, 0) = 4x - \dfrac{1}{2}x^2$ at the points $x = i, i = 0,$ 1, 2, ..., 7 and $t = \dfrac{1}{8}j; j = 0, 1, 2, ..., 5$.

Solution: The equation is $\dfrac{\partial u}{\partial t} = 4\dfrac{\partial^2 u}{\partial x^2}$ with boundary conditions $u(0, t) = 0, u(8, t) = 0$

$$u(x, 0) = 4x - \frac{1}{2}x^2 \text{ at the points } x = i, i = 0, 1, 2, 3, ..., 7$$

and
$$t = \frac{1}{8}j; j = 0, 1, 2, 3, ..., 5$$

Here
$$c^2 = 4, t = kj = \frac{1}{8}j; k = \frac{1}{8}$$

$$x = ih = i, h = 1 \text{ also } \alpha = \frac{kc^2}{\alpha^2} = \frac{1}{2}$$

We have Bendre–Schmidt recurrence relation

$$u_{i,j+1} = \frac{1}{2}[u_{i-1,j} + u_{i+1,j}]$$

Now since $u(0, t) = 0$, $u_{0,j} = 0$, $u(8, t) = 0$, $u_{8,j} = 0$ for all values of j. The entries in the first and last column are zero.

Also $\qquad u(x, 0) = 4x - \frac{1}{2}x^2$

$$u_{i,0} = 4i - \frac{1}{2}i^2$$

Putting $i = 1, 2, 3, 4, 5, 6, 7$, $u_{1,0} = 3.5$, $u_{2,0} = 6$, $u_{3,0} = 7.5$, $u_{4,0} = 8$, $u_{5,0} = 7.5$, $u_{6,0} = 6$, $u_{7,0} = 3.5$.

$$u_{i,j}$$

j \ i	0	1	2	3	4	5	6	7	8
0	0	3.5	6	7.5	8	7.5	6	3.5	0
1	0	3	5.5	7	7.5	7	5.5	3	0
2	0	2.75	5	6.5	7	6.5	5	2.75	0
3	0	2.5	4.625	6	6.5	6	4.625	2.5	0
4	0	2.3125	4.25	5.562	6	5.5625	4.25	2.3125	0
5	0	2.125	3.9375	5.125	5.562	5.125	3.9375	2.125	0

This means $\qquad c = \dfrac{a+b}{2}$

Now by Bendre–Schmidt recurrence relation, we have

$$u_{i,j+1} = \frac{1}{2}[u_{i-1,j} + u_{i+1,j}]$$

Put $j = 0$ $\qquad u_{i,1} = \frac{1}{2}[u_{i-1,0} + u_{i+1,0}]$

Putting $i = 1, 2, 3, 4, 5, 6, 7$

$$u_{1,1} = \frac{1}{2}[u_{0,0} + u_{2,0}] = \frac{1}{2}[0 + 6] = 2$$

$$u_{2,1} = \frac{1}{2}[u_{1,0} + u_{3,0}] = \frac{1}{2}[3.5 + 7.5] = 5.5$$

$$u_{3,1} = \frac{1}{2}[u_{2,0} + u_{4,0}] = \frac{1}{2}[6 + 8] = 7$$

$$u_{4,1} = \frac{1}{2}[u_{3,0} + u_{5,0}] = \frac{1}{2}[7.5 + 7.5] = 7.5$$

$$u_{5,1} = 7, \; u_{6,1} = 5.5, \; u_{7,1} = 3$$

There are entries in the second row.

Put $j = 1$, we have $\quad u_{i,2} = \dfrac{1}{2}[u_{i-1,1} + u_{i+1,1}]$

Putting $i = 1, 2, 3, 4, 5, 6, 7$, we get $u_{1,2} = 2.275$, $u_{2,2} = 5$, $u_{3,2} = 6.5$, $u_{4,2} = 7$, $u_{5,2} = 5$, $u_{6,2} = 5$, $u_{7,2} = 2.75$.

Put $j = 2$ $\qquad\qquad u_{i,3} = \dfrac{1}{2}[u_{i-1,2} + u_{i+1,2}]$

$$u_{1,3} = \frac{1}{2}[0+5] = 2.5$$

$$u_{2,3} = \frac{1}{2}[2.75+6.5] = 4.625$$

$$u_{3,3} = \frac{1}{2}[5+7] = 6$$

$$u_{4,3} = \frac{1}{2}[6.5+6.5] = 6.5$$

$$u_{5,3} = \frac{1}{2}[7+5] = 6$$

$$u_{6,3} = \frac{1}{2}[6.5+2.75] = 4.625$$

$$u_{7,3} = \frac{1}{2}[5+0] = 2.5$$

Similarly we can obtain entries in fourth and fifth rows.

Example 9.11: Solve the equation $\dfrac{\partial u}{\partial t} = \dfrac{\partial^2 u}{\partial x^2}$ subject to the condition $u(x, 0) = \sin x\pi$, $0 \le x \le 1$: $u(0, t) = u(1, t) = 0$ using (1) Schmidt method (2) Crank–Nicholson method (3) Dufort-Frankel method. Carry out computation for two levels taking $h = 1/3, k = 1/36$.

Solution: Here $c^2 = 1$, $h = \dfrac{1}{3}$, $k = \dfrac{1}{36}$, $\alpha = c^2 \dfrac{k}{h^2} = \dfrac{1 \times \dfrac{1}{36}}{\dfrac{1}{9}} = \dfrac{1}{4}$.

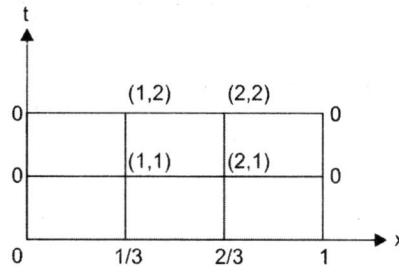

$$u(x, 0) = \sin \pi x, \quad u(ih, 0) = \sin \pi i/3$$

$$u_{1,0} = \sin \frac{\pi}{3} = \sqrt{\frac{3}{2}}, u_{i,0} = \sin \pi \frac{i}{3}$$

$$u_{2,0} = \sin \frac{2\pi}{3} = \sqrt{\frac{3}{2}}, u_{3,0} = 0$$

and all the other boundary values are zero.

1. Schmidt formula

$$u_{i,j+1} = \alpha u_{i-1} + (1 - 2\alpha) u_{i,j} + 2u_{i+1,j}$$

$$= \frac{1}{4}(u_{i-1,j} + 2u_{i,j} + u_{i+1,j}) \tag{i}$$

Put $j = 0$
$$u_{i,1} = \frac{1}{4}(u_{i-1,0} + 2u_{i,0} + u_{i+1,0})$$

Putting $i = 1, 2$
$$u_{1,1} = \frac{1}{4}(u_{0,0} + 2u_{1,0} + u_{2,0}) = 0.65$$

$$u_{2,1} = \frac{1}{4}(u_{1,0} + 2u_{2,0} + u_{3,0}) = 0.65$$

Put $j = 1$ in Eq. (i)

$$u_{i,2} = \frac{1}{4}(u_{i-1,1} + 2u_{i,0} + u_{i+1,1})$$

Putting $i = 1, 2$
$$u_{1,2} = \frac{1}{4}(u_{0,1} + 2u_{1,1} + u_{2,1}) = 0.49$$

$$u_{2,2} = \frac{1}{4}(u_{1,1} + 2u_{2,1} + u_{3,1}) = 0.49$$

2. Crank–Nicholson formula

$$-\frac{1}{4}u_{i-1,j+1} + \frac{5}{2}u_{i,j+1} - \frac{1}{4}u_{i+1,j+1} = \frac{1}{4}u_{i-1,j} + \frac{3}{2}u_{i,j} + \frac{1}{4}u_{i+1,j}$$

Put $j = 0$
$$-\frac{1}{4}u_{i-1,1} + \frac{5}{2}u_{i,1} - \frac{1}{4}u_{i+1,1} = \frac{1}{4}u_{i-1,0} + \frac{3}{2}u_{i,0} + \frac{1}{4}u_{i+1,0}$$

Putting $i = 1, 2$
$$-\frac{1}{4}u_{0,1} + \frac{5}{2}u_{1,1} - \frac{1}{4}u_{2,1} = \frac{1}{4}u_{0,0} + \frac{3}{2}u_{1,0} + \frac{1}{4}u_{2,0}$$

$$-u_{0,1} + 10u_{1,1} - u_{2,1} = u_{0,0} + 6u_{1,0} + u_{2,0}$$

$$10u_{1,1} - u_{2,1} = u_{0,0} + 6u_{1,0} + u_{2,0}$$

$$10u_{1,1} - u_{2,1} = 6\frac{\sqrt{3}}{2} + \frac{\sqrt{3}}{2} = 7\frac{\sqrt{3}}{2} \tag{ii}$$

$$-\frac{1}{4}u_{1,1} + \frac{5}{2}u_{2,1} - \frac{1}{4}u_{3,1} = \frac{1}{4}u_{1,0} + \frac{3}{2}u_{2,0} + \frac{1}{4}u_{1,0}$$

$$-u_{1,1} + 10u_{2,1} = 7\frac{\sqrt{3}}{2} \tag{iii}$$

Solving Eqs (ii) and (iii), we get $\qquad u_{1,1} = u_{2,1} = 0.67$

Put $j = 1$
$$-\frac{1}{4}u_{i-1,2} + \frac{5}{2}u_{i,2} - \frac{1}{4}u_{i+1,2} = \frac{1}{4}u_{i-1,2} + \frac{3}{2}u_{i,1} + \frac{1}{4}u_{i-1,1}$$

Putting $i = 1, 2$, we get $\quad -u_{0,2} + 10u_{1,2} - u_{2,2} = u_{0,1} + 6u_{1,1} + u_{2,1}$

$$10u_{1,2} - u_{2,2} = 4.69 \tag{iv}$$

$$-u_{1,2} + 10u_{2,2} - u_{3,2} = u_{1,1} + 6u_{2,1} + u_{3,1}$$

$$-u_{1,2} + 10u_{2,2} = 4.69$$

Solving Eqs (iii) and (iv), we get $\qquad u_{1,2} = u_{2,2} = 0.52$

3. Du Fort–Frankel formula

$$u_{i,j+1} = \frac{1-2\alpha}{1+2\alpha} u_{i,j-1} + \frac{2\alpha}{1+2\alpha}(u_{i-1,j} + u_{i+1,j})$$

Now

$$\alpha = \frac{1}{4}$$

∴

$$u_{i,j+1} = \frac{1}{3}[u_{i,j-1} + u_{i-1,j} + u_{i+1,j}], \, j \ne 0$$

Now put $j = 1$

$$u_{i,2} = \frac{1}{3}[u_{i,0} + u_{i-1,1} + u_{i+1,1}]$$

Putting $i = 1, 2$ $i = 1, u_{1,2} = \frac{1}{3}[u_{1,0} + u_{0,1} + u_{2,1}] = \frac{1}{3}\left[\frac{\sqrt{3}}{2} + 0 + u_{2,1}\right]$

$$= 2, u_{2,2} = \frac{1}{3}[u_{2,0} + u_{1,1} + u_{3,1}]$$

$$= \frac{1}{3}\left[\frac{\sqrt{3}}{2} + u_{1,1} + 0\right] = \frac{1}{3}\left[\frac{\sqrt{3}}{2} + u_{1,1}\right]$$

Now $u_{2,1}$ and $u_{1,1}$ will be obtained by Schmidt method as $u_{2,1} = u_{1,1} = 0.65$

$$u_{1,2} = \frac{1}{3}\left[\frac{\sqrt{3}}{2} + 0.65\right] = 0.5$$

$$u_{2,2} = \frac{1}{3}\left[\frac{\sqrt{3}}{2} + 0.65\right] = 0.5$$

Example 9.12: Find the solution of the parabolic equation $u_{xx} = 2u_t$

where $u(0, t) = u(4, t) = 0$

$u(x, 0) = x(4 - x), h = 1.$

Find the values up to $t = 5$.

Solution: Here $c^2 = \frac{1}{2}, t = 5, jk = 5.$

Let us choose k such that $\alpha = 1/2$

$$\alpha = \frac{c^2 k}{h^2} = \frac{1}{2}k, \frac{1}{2} = \frac{1}{2}k, k = 1, h = 1$$

j goes from 0 to 5
i goes from 0 to 4
Also $u(0, t) = 0, u_{0,j} = 0$
$\quad u(4, t) = 0, u_{4,j} = 0$ for all j
$\quad\quad u(x, 0) = x(4 - x) \quad u_{i,0} = i(4 - i)$
Since $u_{0,j} = 0, u_{4,j} = 0$ for all j

∴ all the entries in the first and last columns are zero as shown.

i \ j	0	1	2	3	4
0	0	3	4	3	0
1	0	2	3	2	0
2	0	1.5	2	1.5	0
3	0	1	1.5	1	0
4	0	0.75	1	0.75	0
5	0	0.5	0.75	0.5	0

Also $u_{i,0} = i(4 - i)$

$u_{1,0} = 3$

$u_{2,0} = 4$

$u_{3,0} = 3$

$u_{4,0} = 0$

Therefore, the entries in the first row are 3, 4, 3, 0.

Since $\alpha = \dfrac{1}{2}$, Bendre–Schmidt recurrence relation can be applied, i.e.

$$u_{i,j+1} = \frac{1}{2}[u_{i-1,j} + u_{i+1,j}]$$

Put $j = 0$ $$u_{i,1} = \frac{1}{2}[u_{i-1,0} + u_{i+1,0}]$$

Putting $i = 1, 2, 3$ $$u_{1,1} = \frac{1}{2}[u_{0,0} + u_{2,0}] = \frac{1}{2}[0 + 4]$$

$$u_{2,1} = \frac{1}{2}[u_{1,0} + u_{3,0}] = \frac{1}{2}[3 + 3] = 3$$

$$u_{3,1} = \frac{1}{2}[u_{2,0} + u_{4,0}] = \frac{1}{2}[4 + 0] = 2$$

Put $j = 1$ $$u_{i,2} = \frac{1}{2}[u_{i-1,1} + u_{i+1,1}]$$

Putting $i = 1, 2, 3$ $$u_{1,2} = \frac{1}{2}[u_{0,1} + u_{2,1}] = \frac{1}{2}[0 + 3] = 1.5$$

$$u_{2,2} = \frac{1}{2}[u_{1,1} + u_{3,1}] = \frac{1}{2}[2 + 2] = 2$$

$$u_{3,2} = \frac{1}{2}[u_{2,1} + u_{4,1}] = \frac{1}{2}[3 + 0] = 1.5$$

Similarly, we can find other values and put in the table as shown.

Put $j = 2$ $$u_{1,3} = \frac{1}{2}[0 + 2] = 1$$

$$u_{2,3} = \frac{1}{2}[1.5 + 1.5] = 1.5$$

$$u_{3,3} = \frac{1}{2}[2 + 0] = 1$$

Put $j = 3$ $\qquad u_{1,4} = \dfrac{1}{2}[0 + 1.5] = 0.75$

$$u_{2,4} = \dfrac{1}{2}[1 + 1] = 1$$

$$u_{3,4} = \dfrac{1}{2}[0 + 1.5] = 0.75 \text{ etc.}$$

Example 9.13: Solve the equation $\dfrac{\partial^2 u}{\partial x^2} = \dfrac{\partial u}{\partial t}$ with the conditions $u(0, t) = u(1, t) = 0$, $u(x, 0) = x(1 - x)$. Assume $h = 0.1$. Tabulate u for $t = k, 2k, 3k$ choosing an appropriate value of k.

Solution:

$u_{i,j}$

x	0	0.1	0.2	0.3	0.4	0.5	0.6	0.7	0.8	0.9	1
i j	0	1	2	3	4	5	6	7	8	9	10
0	0	0.09	0.16	0.21	0.24	0.25	0.24	0.21	0.16	0.09	0
1	0	0.08	0.15	0.20	0.23	0.24	0.23	0.20	0.15	0.08	0
2	0	0.075	0.14	0.19	0.22	0.23	0.22	0.19	0.14	0.075	0
3	0	0.07	0.1325	0.18	0.21	0.22	0.21	0.18	0.1325	0.07	0

Here $c^2 = 1, h = 0.1$

$\qquad t = k, 2k, 3k$

$\qquad jk = k, 2k, 3k$

$\qquad j = 1, 2, 3$

$\qquad x = ih = 0.1 \times i$

i.e. $\qquad x = 0, 0.1, 0.2, 0.3, 0.4, \ldots, 1$

and j goes from 0 to 3.

Also $u(0, t) = 0$, i.e. $u_{0,j} = 0$

$\qquad u(1, t) = 0, u_{10,j} = 0$ for all j

$\therefore$ the entries in the first and last column are zero.

Since $u(x, 0) = x(1 - x)$, $u(ih, 0) = ih(i - ih)$

$$u_{i,0} = \frac{i}{10}\left(1 - \frac{i}{10}\right)$$

$$u_{1,0} = \frac{1}{10}\left(1 - \frac{1}{10}\right) = \frac{1}{10} \times \frac{9}{10} = 0.09$$

$$u_{2,0} = \frac{2}{10}\left(1 - \frac{2}{10}\right) = \frac{2}{10} \times \frac{8}{10} = 0.16$$

$$u_{3,0} = \frac{3}{10}\left(1 - \frac{3}{10}\right) = \frac{1}{10} \times \frac{7}{10} = 0.21$$

$$u_{4,0} = \frac{4}{10}\left(1 - \frac{4}{10}\right) = \frac{1}{10} \times \frac{6}{10} = 0.24$$

$$u_{5,0} = \frac{5}{10}\left(1 - \frac{5}{10}\right) = \frac{1}{10} \times \frac{5}{10} = 0.25$$

$$u_{6,0} = \frac{6}{10}\left(1 - \frac{6}{10}\right) = 0.24$$

$$u_{7,0} = 0.21, u_{8,0} = 0.16, u_{9,0} = 0.09$$

Let us choose $\qquad \alpha = \frac{1}{2}$

$$\frac{1}{2} = 100k, k = \frac{1}{200} = 0.005$$

We choose $\qquad k = 0.05$

Since $\alpha = \frac{1}{2}$, we apply Bendre-Schmidt recurrence relation,

i.e. $\qquad u_{i,j+1} = \frac{1}{2}[u_{i-1,j} + u_{i+1,j}]$

Put $j = 0$ $\qquad u_{i.1} = \frac{1}{2}[u_{i-1,0} + u_{i+1,0}]$

Put $i = 1, 2, 3, 4, 5 \quad u_{1,1} = \frac{1}{2}[u_{0,0} + u_{2,0}] = \frac{1}{2}[0 + 0.16] = 0.08$

$$u_{2,1} = \frac{1}{2}[u_{1,0} + u_{3,0}] = \frac{1}{2}[0.09 + 0.21] = 0.15$$

$$u_{3,1} = \frac{1}{2}[u_{2,0} + u_{4,0}] = \frac{1}{2}[0.16 + 0.24] = 0.20$$

$$u_{4,1} = \frac{1}{2}[u_{3,0} + u_{5,0}] = \frac{1}{2}[0.21 + 0.25] = 0.23$$

$$u_{5,1} = \frac{1}{2}[u_{4,0} + u_{6,0}] = \frac{1}{2}[0.24 + 0.24] = 0.24$$

Going in the reverse order, we get
$\qquad i = 6 \qquad u_{6,1} = 0.23$
$\qquad\qquad\qquad i = 7 \qquad u_{7,1} = 0.20$
$\qquad\qquad\qquad i = 8 \qquad u_{8,1} = 0.15$
$\qquad\qquad\qquad i = 9 \qquad u_{9,1} = 0.08$

Putting $j = 1$ $\qquad u_{i,2} = \frac{1}{2}[u_{i-1,1} + u_{i+1,1}]$

Put $\quad i = 1 \qquad u_{1,2} = \frac{1}{2}[u_{0,1} + u_{2,1}] = \frac{1}{2}[0 + 0.15] = 0.075$

$\qquad i = 2 \qquad u_{2,2} = \frac{1}{2}[u_{1,1} + u_{3,1}] = \frac{1}{2}[0.08 + 0.20] = 0.14$

$\qquad i = 3 \qquad u_{3,2} = \frac{1}{2}[u_{2,1} + u_{4,1}] = \frac{1}{2}[0.15 + 0.23] = 0.19$

$$i = 4 \qquad u_{4,2} = \frac{1}{2}[u_{3,1} + u_{5,1}] = \frac{1}{2}[0.20 + 0.24] = 0.22$$

$$i = 5 \qquad u_{5,2} = \frac{1}{2}[u_{4,1} + u_{6,1}] = \frac{1}{2}[0.23 + 0.23] = 0.23$$

Going in the reverse order, we get

$$
\begin{aligned}
i &= 6 \qquad u_{6,2} = 0.22 \\
i &= 7 \qquad u_{7,2} = 0.19 \\
i &= 8 \qquad u_{8,2} = 0.14 \\
i &= 9 \qquad u_{9,2} = 0.075
\end{aligned}
$$

Putting $j = 2$ and $i = 1, 2, 3, 4, 5, 6, 7, 8, 9, 10$, we get the last row filled.

PROBLEMS 9.3

1. Given $\dfrac{\partial^2 f}{\partial x^2} = \dfrac{\partial f}{\partial t}$, $f(0, t) = f(5, t) = 0$, $f(x, 0) = x^2(25 - x^2)$, find the values of f for

$x = ih$ $(i = 0, 1, ..., 5)$, $t = jk$ $(j = 0, 1, 2, ..., 6)$ with $h = 1$ and $k = \dfrac{1}{2}$ using the explicit method.

Ans. The values of $f(i = 0, 1, 2, 3, 4, 5$ and $j = 1, 2, 3, 4, 5, h = 1, k = \frac{1}{2})$ are

$u_{i,j}$

j \ i	0	1	2	3	4	5
0	0	24	84	144	144	0
1	0	42	78	57	57	0
2	0	39	67.5	39	39	0
3	0	30	49.5	33.75	33.75	0
4	0	26.6	43.5	24.75	24.75	0
5	0	19.9	32.25	21.75	21.75	0

2. Solve boundary value problem $u_t = u_{xx}$ under the conditions $u(0, t) = u(1, t) = 0$ and $u(x, 0) = \sin \pi x$ $0 \le x \le 1$ by taking $h = 0.2$ by Schmidt method $\alpha = \dfrac{1}{2}$.

Ans.

$u_{i,j}$

j \ i	0	1	2	3	4	5
0	0	0.59	0.95	0.95	0.59	0
1	0	0.475	0.77	0.77	0.475	0
2	0	0.38	0.62	0.62	0.38	0
3	0	0.31	0.50	0.50	0.31	0
4	0	0.25	0.41	0.41	0.25	0
5	0	0.20	0.33	0.33	0.20	0

3. Using Crank–Nicholson method, solve $u_{xx} = 16u_t$, $0 < x < 1$, $t > 0$ given $u(x, 0) = 0$, $u(1, t) = 50$ and $u(0, t) = 0$. Compute u for two steps in t-direction taking $h = \dfrac{1}{4}$.

Ans.

j \ i	0	0.25	0.50	0.75	1
0	0	0	0	0	0
1	0	41	42	43	50
2	0	43	44	45	100

4. Solve $u_t = u_{xx}$ subjected to the conditions $u(0, t) = u(1, t) = 0$ and $u(x, 0) = 2x$ for $0 \le x \le \dfrac{1}{2} = 2(1 - x)$ for $\dfrac{1}{2} \le x \le 1$, take $h = \dfrac{1}{4}$ and k according to Bender–Schmidt equation.

Ans.

j \ i	0	1	2	3	4
0	0	5	1	5	0
1	0	5	5	5	0
2	0	25	5	25	0
3	0	25	25	25	0

10

Computational Techniques (Programming Using C)

10.1 GAUSS ELIMINATION METHOD

Note: The variables used in this programme have their specific meaning as given below.

n: no. of unknowns
a: an array which holds the augmented matrix
x: an array which contain values of unknowns
i, j, k: loop control variables
multi: the multiplier

Programme Demonstrating Gauss Elimination Method

/*Solution of a system of linear equations using Gauss elimination method. The programme is for $m \times n$ system*/

Let system of equations be:

$3.15x_1 + 1.96x_2 + 3.85x_3 = 12.95$
$2.13x_1 + 5.12x_2 - 2.89x_3 = -8.61$
$5.92x_1 + 3.05x_2 + 2.15x_3 = 6.88$

/*Gauss Elimination Method*/

```
#include<stdio.h>
#include<conio.h>
#include<math.h>
#define n 3
void main ( )
{
clrscr ( )
float a[n] [n + 1], x[n], multi, s;
int i, j, k;
printf ("enter the element of aug. matrix row wise\n");
for (i=0;i<n;i++)
for (j=0;j<n+1;j++)
scanf ("%f",&a[i] [j]);//reading the elements of augmented
matrix for (j=0;j<n-1;j++)
for (i=j+1;i<n;i++)//loops for obtaining upper triangular
matrix
{
```

Flow Chart 10.1

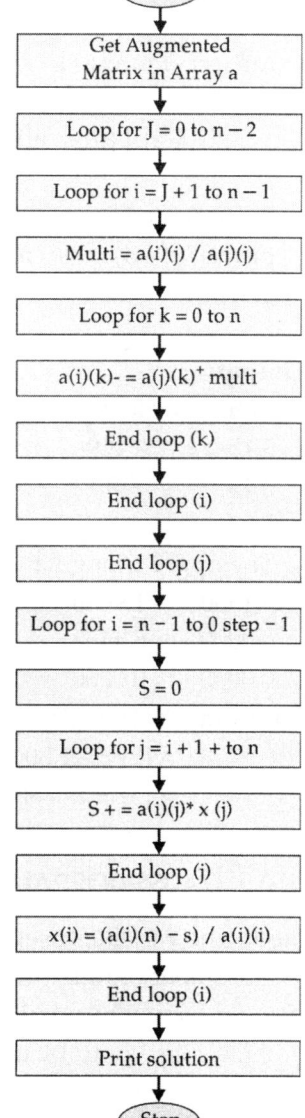

Start
↓
Get Augmented Matrix in Array a
↓
Loop for J = 0 to n − 2
↓
Loop for i = J + 1 to n − 1
↓
Multi = a(i)(j) / a(j)(j)
↓
Loop for k = 0 to n
↓
a(i)(k)- = a(j)(k)⁺ multi
↓
End loop (k)
↓
End loop (i)
↓
End loop (j)
↓
Loop for i = n − 1 to 0 step − 1
↓
S = 0
↓
Loop for j = i + 1 + to n
↓
S + = a(i)(j)* x (j)
↓
End loop (j)
↓
x(i) = (a(i)(n) − s) / a(i)(i)
↓
End loop (i)
↓
Print solution
↓
Stop

```
    multi=a[i][j]/a[j][j];
    for (k=0;k<n+1;k++)
    a[i] [k]==a[j] [k]*multi;//row operation is applied
}
printf ("the upper triangular matrix|n");
for (i=0;i<n;i++)
{
    for(j=0;j<n+1;j++)
    printf ("%f", a[i] [j]);//printing the upper triangular matrix printf ("\n");
}
for (i=n-1;i>=0;i--)//loop for back substitution
{
s=0;
for (j=i+1;j<n;j++]
    s+=a[i] [j]*x[j];
    x[i]=(a[i] [n]-s/a[i] [i];
}
for (i=0;i<n;i++)
printf("x(%d) = %f\n", i+1,x[i]);//printing the values of unknowns
getch( );
}
```

Output:

Enter the elements of augmented matrix rowwise

3.15	1.96	3.85	12.95
2.13	5.12	-2.89	-8.61
5.92	3.05	2.15	6.88

The upper triangular matrix is

3.1500	-1.9600	3.8500	12.9500
0.0000	6.4453	-5.433	-17.3666
0.0000	0.0000	0.6534	0.6853

Solution: $X[1] = 1.7089$
$X[2] = -1.8005$
$X[3] = 1.0488$

10.2 GAUSS–JORDAN METHOD

Note: The variables used in this programme have their specific meaning as given below.

n: number of unknown variables in a system of equations

x: an array which will contain values of unknowns

i, j, k: loop control variables

a: an array which holds the augmented matrix

multi: the multiplier.

Flow Chart 10.2

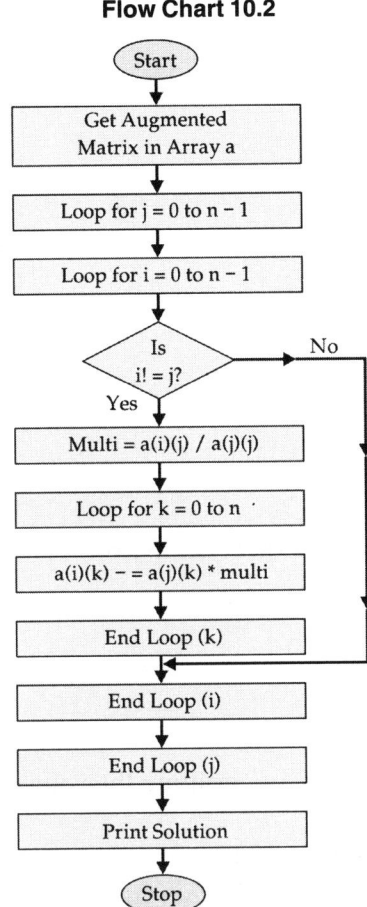

Start

Get Augmented
Matrix in Array a

Loop for j = 0 to n − 1

Loop for i = 0 to n − 1

Is
i! = j?   No

Yes

Multi = a(i)(j) / a(j)(j)

Loop for k = 0 to n

a(i)(k) − = a(j)(k) * multi

End Loop (k)

End Loop (i)

End Loop (j)

Print Solution

Stop

Programme Demonstrating Gauss–Jordan Method

Solution of $m \times n$ system of linear equations using Gauss–Jordan method.

Let system of equations be:

$$5x_1 + x_2 + x_3 + x_4 = 4$$
$$x_1 + 7x_2 + x_3 + x_4 = 12$$
$$x_1 + x_2 + 6x_3 + x_4 = -5$$
$$x_1 + x_2 + x_3 + x_4 = -6$$

//Gauss–Jordan Method

```
#include<stdio.h>
#include<conio.h>
#include<math.h>
#define  n 4
void main ( )
{
    float a[n] [n+1], multi,x[n];
```

```
int i, j, k;
clrscr ( );
printf ("enter the element of aug. matrix row wise\n");
for (i=0;i<n;i++)
    for (j=0;j<n+1;j++)
    scanf ("%f",&a[i] [j]);
    for (j=0;j<n;j++)
    for (i=j;i<n;i++)
    if (i!=j)
{
    multi=a[i][j]/a[j][j];
    for (k=0;k<n+1;k++)
    a[i] [k]-=a[j] [k]*multi;
}
    for (i=0;i<n;i++)//printing the diagonal matrix
{
    for(j=0;j<n+1;j++)
    printf ("%f", a[i] [j]);
    printf ("\n");
}
printf("solution of the equations");//printing the values of unknowns
for (i=0;i<n;i++)
printf("x(%d)=%f\n",i+1,a(i) [n]/a[i] [i];
getch( );
}
```

Output:

Enter the elements of augmented matrix rowwise

5	1	1	1	4
1	7	1	1	12
1	1	6	1	-5
1	1	1	4	-6

Diagonal matrix:

1.00	0.00	0.00	0.00	1.00
0.00	1.00	0.00	0.00	2.00
0.00	0.00	1.00	0.00	-1.00
0.00	0.00	0.00	-1.00	2.00

Solution: $X[1] = 1.00$
$X[2] = 2.00$
$X[3] = -1.00$
$X[4] = -2.00$

10.3 GAUSS–SEIDAL METHOD

Note: The variable used in this program have their specific meaning as given below.

N: the number of unknowns a is the array which holds the augmented matrix is an array which will hold the values of unknowns

aerr: allowed error

maxitr: the maximum number of iterations to be performed

itr: the counter which keeps track of number of iterations performed

err: error in value of x_1

maxerr: maximum error in any value of xi after an iteration

multi: the multiplier.

Program Demonstrating Gauss–Seidal Method

Program to solve a system of equations using Gauss–Seidal iteration method. Order of matrix is N maximum number of iterations, maxitr, error tolerance.

Flow Chart 10.3

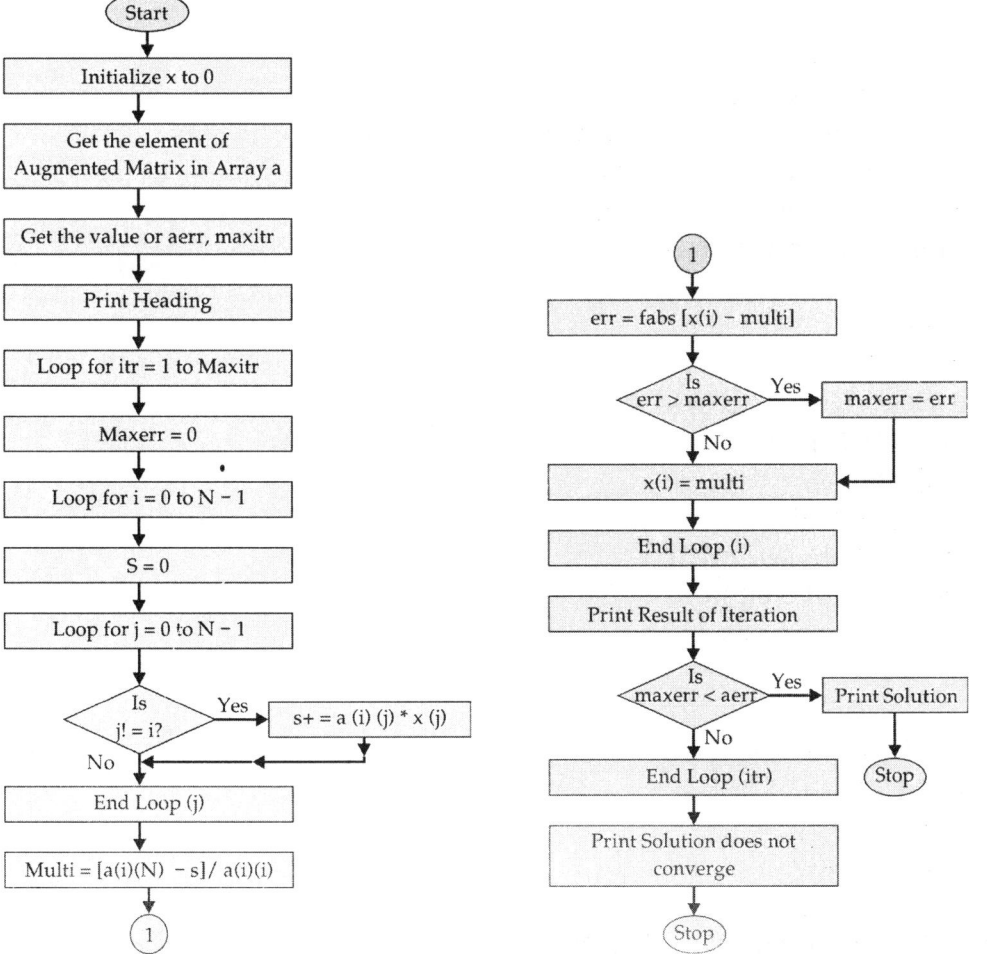

Let system of equations be:

$$10x_1 - 2x_2 - x_3 - x_4 = 3$$
$$-2x_1 + 10x_2 - x_3 - x_4 = 15$$
$$-x_1 - x_2 + 10x_3 + 2x_4 = 27$$
$$-x_1 - x_2 + 2x_3 + 10x_4 = -9$$

//Gauss–Seidal Method

```c
#include<stdio.h>
#include<math.h>
#define N 4

main ( )
{
float a[N] [N+1], x[N], aerr,maxerr,multi,s,err;
int i,j,itr,maxitr;
/* first initializing the array x */
for (i = 0; i<N;i++) x[i] = 0;
printf("\n Enter the elements of the augmented matrix row wise \n");
    for (i=0; i<N; i++)
    for (j=0; j<N+1;j++);
    scanf ("%f", &a[i] [j];
printf("\n Enter the allowed error, maximum iterations \n");
scanf ("%f%d", &aerr,&maxitr);
printf("\n iteration x[1] x[2] x[3] \n");
for (itr=1; itr<=maxitr;itr++)
    {
    maxerr=0;
    for (i=0; i<N; i++)
    s=0;
    for(j=0;j<N;j++)
    if (j!=i)
    s+=a[i] [j]*x[j];
    multi=a[i] [N]-s)/a[i] [i];
    err = fab(x[i]-multi);
    if (err>maxerr) maxerr = err;
x[i] = multi;
    }
printf ("%d",itr);
for (i=0;i<N;i++)
printf("%f", x[i]);
if (maxerr < aerr)
{
printf("\n Converges is %d" iterations \n",itr);
```

```
for (i=0;i<N;i++)
printf("\n x[%d]=%f\n", i+1, x[i]);
return 0;
}
}
printf ("\n solution does not converge, iterations not sufficient
\n");
return 1;
}
```

Output:

Enter the elements of augmented matrix rowwise

10	-2	-1	-1	3
-2	10	-1	-1	15
-1	-1	10	-2	27
-1	-1	-2	10	-9

Enter the allowed error, maximum iteration 0.000115

Iteration	$x[1]$	$x[2]$	$x[3]$	$x[4]$
1	0.3000	1.5600	2.8860	-0.1368
2	0.8869	1.9523	2.9566	-0.0248
3	0.9836	1.9899	2.9924	-0.0042
4	0.9968	1.9982	2.9927	-0.0008
5	0.9994	1.9997	2.9998	-0.0001
6	0.9999	1.9999	3.0000	-0.0000
7	1.0000	2.0000	3.0000	-0.0000

Converge in 7 iterations

$$x[1] = 1.0000$$
$$x[2] = 2.0000$$
$$x[3] = 3.0000$$
$$x[4] = -0.0000$$

10.4 BISECTION METHOD

Note: The variables used in this program have their specific meaning as given below.

x_1, x_2: initial approximation enclosing the root

err: allowed error

x_3: new approximation to the root in each iteration

i: counter for iteration

Program Demonstrating Bisection Method

Program for finding out a real root of equation $f(x) = 0$ by bisection method.

Let $f(x) = \exp(x) - x*3$

Flow Chart 10.4

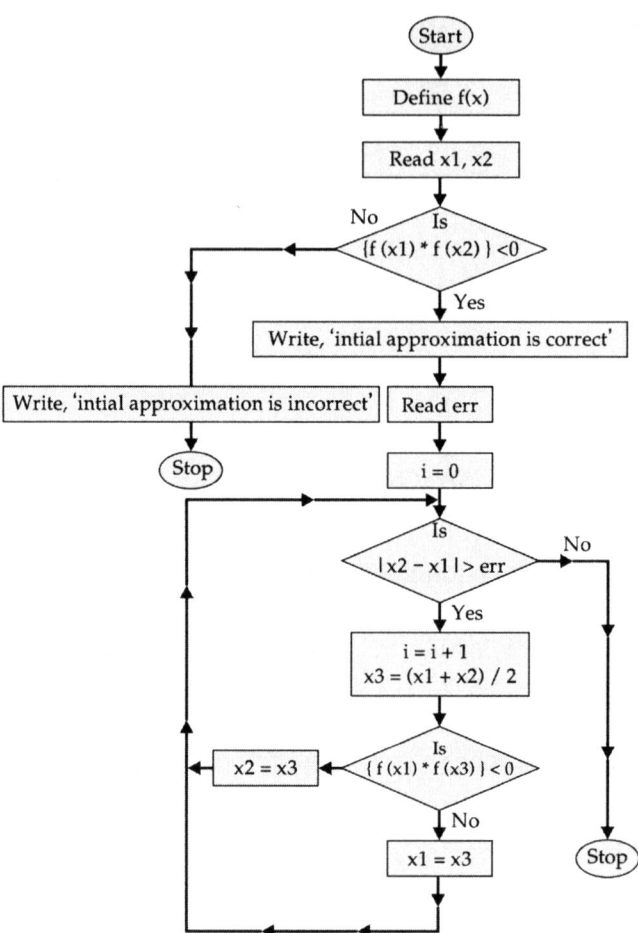

//**Bisection Method**
```c
#include<stdio.h>
#include<conio.h>
#include<math.h>
#include<stdlib.h>
float f (float x)
{
    return (exp(x)-x*3);
}
void main ( )
{
    int i=0;
    float x1,x2,x3,err;
    clrscr ( );
    printf("enter the first approximation x1\t");
    scanf ("%f",&x1);
    printf ("enter the second approximation x2\t");
```

```
      scanf ("%f",&x2);
      if (f(x1) *f (x2)<0)
      printf("the approximations are correct");
   else
      {
      printf("the approximations are wrong")
      getch ( );
      exit(0);
      }
   printf ("enter the allowed error for the solution\n");
   scanf ("%f",&err);
   while(fabs(x2-x1)>err)
   {
      i++;
      x3=(x1+x2)/2;
      if (f(x1)*f(x3)<0)
   {
      x2=x3;
   printf("in iteration %d x1=%f & x2=%f\n",i,x1,x2);
   printf("\n");
   }
   else
      x1=x3;
   printf("in iteration %d x1=%f & x2=%f\n",i,x1,x2);
   printf ("/n");
   }
   printf ("solution converges in iteration %d \n",i);
   printf ("the solution is %f",x3);
   getch ( );
   }
```

Output:

```
   enter the first approximation x1 1
   enter the second approximation x2 2
   the approximations are correct
   enter the allowed error for the solution 0.001
   Solution converges
       in iteration  1, x1 = 1.50000 and x2 = 2.000000
                     2, x1 = 1.500000 and x2 = 1.750000
                     3, x1 = 1.500000 and x2 = 1.625000
                     4, x1 = 1.500000 and x2 = 1.562500
                     5, x1 = 1.500000 and x2 = 1.531250
                     6, x1 = 1.500000 and x2 = 1.515625
                     7, x1 = 1.507812 and x2 = 1.515625
```

8, $x1 = 1.511719$ and $x2 = 1.515625$

9, $x1 = 1.511719$ and $x2 = 1.51367$

10, $x1 = 1.511719$ and $x2 = 1.512695$

solution converges in iteration 10, the solution is 1.512695.

10.5 NEWTON–RAPHSON METHOD

Note: The variables used in this program have their specific meaning as given below.

$x_0 x_1$: initial approximation enclosing the root

err: allowed error

df: derivative of $f(x)$

Flow Chart 10.5

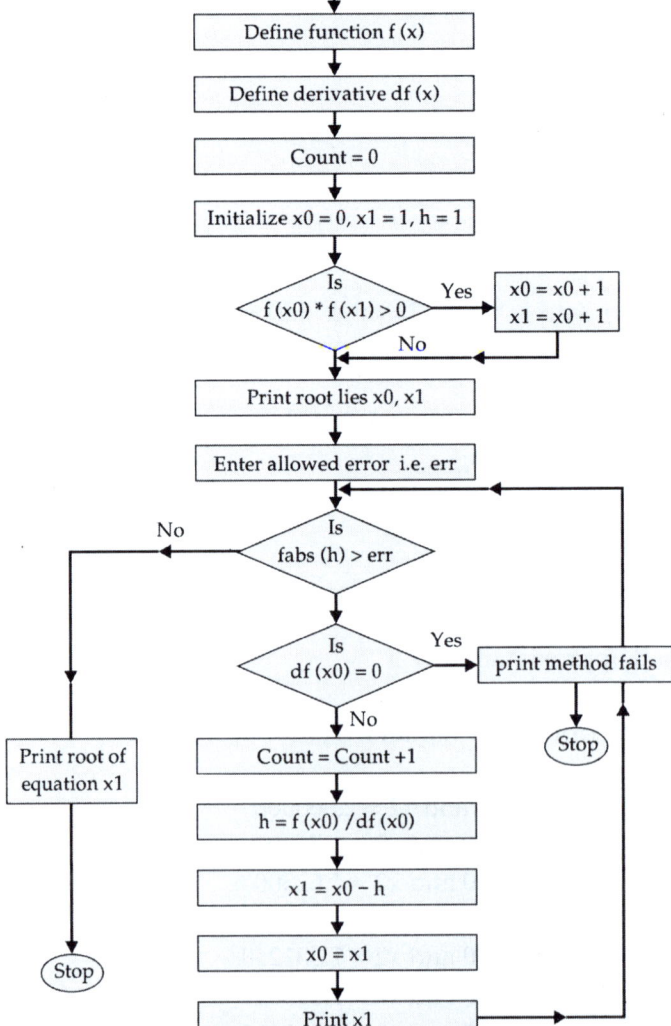

Program Demonstrating Newton–Raphson Method

Program for finding out a real root of equation $f(x) = 0$ by Newton–Raphson method.

Let $f(x) = x*x*x - 2*x - 5$

```
//Newton–Raphson Method
#include<stdio.h>
#include<conio.h>
#include<math.h>
float f (float x)
{
    return (x*x*x –2*x –1);
}
float df(float x)
{
    return (3*x*x-2);
}
void main ( )
{
    int count=0;
    float x0=0, x1=1, h=1, aerr;
    clrscr ( );
    while (f(x0)*f(x1)>0,0)
{
    x0++;
    x1=x0+1;
}
    printf ("the interval in which root lies is %f, %f", x0, x1);
    printf ("enter the aerr", err);
    scanf ("%f",&aerr);
    while (fabs(h) > err)
    {
    if (df(x0) = = 0.0)
    {
    printf ("method fails \n");
    break;
    }
    else
    {
    count++;
    h=f (x0)/df(x0);
    x1=x0-h;
    x0=x1;
    }
```

```
printf ("iteration %d the value of x1=%f", count,x1);
}
printf ("the root of equation is %f",x1);
getch ( );
}
```

Output is

The interval in which root lies is (1.000000, 2.000000)

enter the aerr 0.001

iteration 1 the value of $x1 = 3.000000$

iteration 2 the value of $x1 = 2.200000$

iteration 3 the value of $x1 = 1.780831$

iteration 4 the value of $x1 = 1.636303$

iteration 5 the value of $x1 = 1.618305$

iteration 6 the value of $x1 = 1.618034$

the root of equation is 1.618034.

10.6 REGULA–FALSI METHOD

Note: The variable used in this program have their specific meaning as given below.

x_0, x_1: initial approximation enclosing the root

n: number of iteration

err: allowed error

x_2: new approximation to the root in each iteration

i: counter for iteration

Program Demonstrating Regula–Falsi Method

Program for finding out a real root of nonlinear equation $f(x) = 0$ by Regula–Falsi method.
Let $f(x) = \cos x - x * \exp(x)$

```
//Regula–Falsi Method
#include<stdio.h>
#include<conio.h>
#include<math.h>
#include<process.h>
float f (float x)
{
return (cos(x)-x*exp(x));
}
void main ( )
{
int i,n;
float x0, x1, x2, err;
printf ("\n Enter the initial approximation x0, x1");
```

Flow Chart 10.6

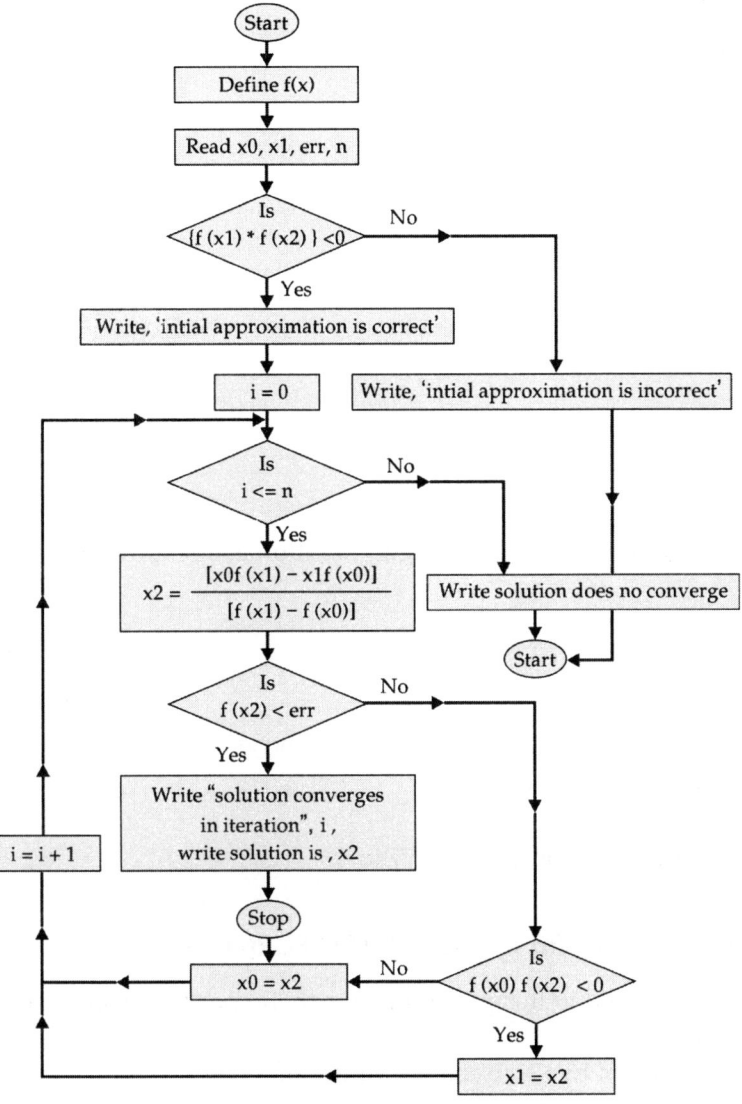

```
scanf ("%f %f", &x0, &x1);
printf ("\n Enter the allowed error, number of iteration");
scanf ("%f%d", &err, &n);
if (f(x0)*f(x1)<0)
    printf ("\n initial approximations are correct");
else
{
printf ("\n the initial approximations are wrong");
getch ( );
exit (0);
}
```

```
for (i=1; i<=n;i++)
{
x2=(x0*f(x1)-x1*f(x0))/f(x1)-f(x0));
if (fabs (f(x2))<=err)
{
printf ("\n Solution converges");
printf ("\n The solution after iteration %d is %f",i,x2);
getch ( );
exit (0);
}
if ((f(x0)*f(x2))<0)
x1=x2
else
x0=x2;
}
printf ("\n Solution does not converges, iteration is not sufficient");
getch ( );
}
```

Output:
Enter the initial approximation x0, x1 0 1
Enter the allowed error, number of iteration 0.00005 20
initial approximations are correct
Solution converges
The solution after iteration 10 is 0.517748

10.7 NEWTON'S FORWARD INTERPOLATION FORMULA

Note: The variables used in this program have their specific meaning as given below.

x_0: initial value of xi

h: difference between two consecutive values of x

n: the number of subinterval

x: value at which we have to find the value of y

Program Demonstrating Newton's Forward Interpolation Formula

Let the given set of data is

x	0	1	2	3	4	5
y	3	18	79	222	483	898, find y (0.4)

//Newton Forward Interpolation Formula

```
#include<stdio.h>
#include<conio.h>
#include<process.h>
#include<math.h>
void main ( )
```

Flow Chart 10.7

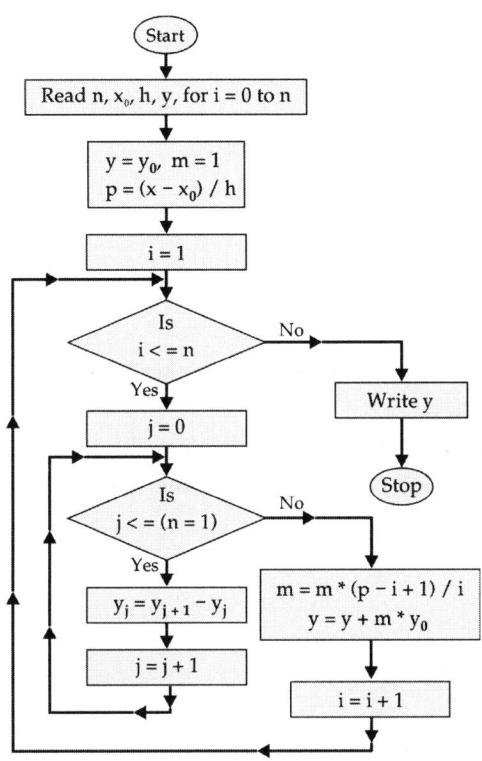

```
{
    int i,j,n;
    float x0,ay[20],p,m=1,x,y,h;
    clrscr ( );
    printf ("\n Enter the number of Subinterval in the given data, n");
    scanf("%d", &n);
    printf ("\n Enter the initial value x0");
    scanf ("%f", &x0);
    printf ("\n Enter the length of interval h");
    scanf ("%f", &h);
    printf ("\n Enter the value of x at which we have to find the value of y");
    scanf ("%f", &x);
    for (i=0;i<=n;i++)
{
printf ("\n Enter the value of y %d=",i);
scanf ("%f", &ay[i]);
}
y=ay[0];
p=(x-x0)/h;
for (i=      =n;i++)
```

```
{
    for (j=0;j<=n-i;j++)
    ay[j]=ay [j+1]-ay[j];
    m=m* (p-i+1)/i;
    y=y+m*ay [0];
}
printf ("\n The value of y at x=%f is %f",x,y);
getch ( );
}
```

Output:

Enter the number of Subinterval in the given data i.e. n 5

Enter the initial value $x0$ 0

Enter the length of interval h 1

Enter the value of x at which we have to find the value of y 0.5

Enter the value of $y0 = 3$

Enter the value of $y1 = 18$

Enter the value of $y2 = 79$

Enter the value of $y3 = 222$

Enter the value of $y4 = 483$

Enter the value of $y5 = 898$

The value of y at $x = 0.500000$ is 7.000000

10.8 NEWTON'S BACKWARD INTERPOLATION FORMULA

Note: The variables used in this program have their specific meaning as given below.

x_n: last value of xi's

x: the value of x at which we have to find the value of y

h: the interval of differencing

n: number of subinterval

y: the value of y at x

Program Demonstrating Newton's Backward Interpolation Formula

Let the given set of data is

x	0	1	2	3	4	5
y	3	18	79	222	483	898, find $y (4.5)$

//Newton Backward Interpolation Formula

```
#include<stdio.h>
#include<conio.h>
#include<process.h>
#include<math.h>
void main ( )
{
int i,j,n;
```

Flow Chart 10.8

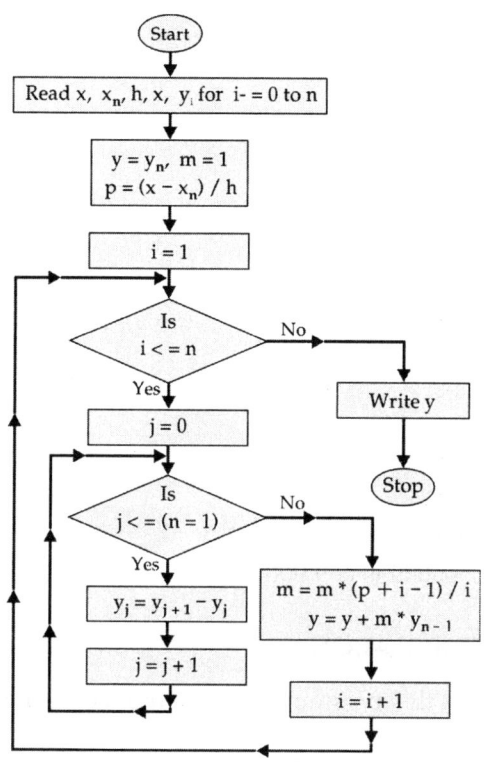

```
float xn,ay[20]p,m=1,x,y,h;
clrscr ( );
printf ("\n Enter the number of subinterval in the given data i.e. n");
scanf ("%d", &n);
printf ("\n Enter the last value xn");
scanf ("%f", &xn);
printf ("\n Enter the length of interval h");
scanf ("%f", &h);
printf ("\n Enter the value of x at which we have to find the value of y");
scanf ("%f", &x);
for (i=0; i<=n; i++)
{
    printf ("\n Enter the value of y %d=",i);
    scanf ("%f", &ay[i]);
}
y=ay[n];
p=(x-xn)/h;
for (i=1; i<=n; i++)
{
    for (j=0; j<=n-i; j++)
    ay[j]=ay[j+1]-ay[j];
    m=m* (p+i-1)/i;
```

```
    y=y+m*ay[n-i];
  }
  printf ("\n The value of y at x=%f is %f",x,y);
  getch ( );
  }
```

Output:
  Enter the number of subinterval in the given datai.e.n 5
  Enter the last value xn 5
  Enter the length of interval h 1
  Enter the value of x at which we have to find the value of y 4.5
  Enter the value of $y0 = 3$
  Enter the value of $y1 = 18$
  Enter the value of $y2 = 79$
  Enter the value of $y3 = 222$
  Enter the value of $y4 = 483$
  Enter the value of $y5 = 898$
  The value of y at x = 4.500000 is 669.000000

10.9 LAGRANGE'S INTERPOLATION

Note: The variables used in this program have their specific meaning as given below.
  MAX: the maximum value of n
  ax: an array containing values of x ($x0, x1, ..., xn$)
  ay: an array containing values of y ($y0, y1, ..., yn$)
  x: the value of x at which value of y is required
  y: the calculated value of y
  nr: numerator of the terms in expansion of y
  dr: denominator of the terms in expansion of y

Program Demonstrating Lagrange's Interpolation Formula

Let the given set of data is

x	5	6	9	11
y	12	13	14	16, find y when $x = 10$

```
/*Lagrange's Interpolation*/
  #include<stdio.h>
  #define MAX 100
  main ( )
  {
  float ax [Max + 1], ay[Max + 1], nr, dr, x, y = 0;
  int i, j, n;
  printf ("Enter the value of n\n");
  scanf ("%d", &n);
  printf ("Enter the set of values\n");
  for (i=0; i<=n; i++)
     scanf ("%f%f", &ax[i], &ay[i]);
```

Flow Chart 10.9

```
                    Start
                      │
                      ▼
          ┌───────────────────────┐
          │    Get value of n     │
          └───────────────────────┘
                      │
                      ▼
          ┌───────────────────────────┐
          │ Get value of elements of ax, ay │
          └───────────────────────────┘
                      │
                      ▼
          ┌───────────────────────┐
          │    Get value of x     │
          └───────────────────────┘
                      │
                      ▼
          ┌───────────────────────┐
          │        y = 0          │
          └───────────────────────┘
                      │
                      ▼
          ┌───────────────────────┐
          │  Loop for i = 0 to n  │
          └───────────────────────┘
                      │
                      ▼
          ┌───────────────────────┐
          │     nr = dr = 1       │
          └───────────────────────┘
                      │
                      ▼
          ┌───────────────────────┐
          │  Loop for j = 0 to n  │
          └───────────────────────┘
                      │
                      ▼
                   ╱  Is  ╲        No
                  ◁  j! = i  ▷────────┐
                   ╲      ╱           │
                      │ Yes           │
                      ▼               │
          ┌───────────────────────┐   │
          │   nr* = x − ax (i)    │   │
          └───────────────────────┘   │
                      │               │
                      ▼               │
          ┌───────────────────────┐   │
          │ dr* = ax(i) − ax(j)   │   │
          └───────────────────────┘   │
                      │◄──────────────┘
                      ▼
          ┌───────────────────────┐
          │     End loop (j)      │
          └───────────────────────┘
                      │
                      ▼
          ┌───────────────────────┐
          │ y + = (nr / dr) * ay (i) │
          └───────────────────────┘
                      │
                      ▼
          ┌───────────────────────┐
          │     End loop (i)      │
          └───────────────────────┘
                      │
                      ▼
          ┌───────────────────────┐
          │  Print x, y as solution │
          └───────────────────────┘
                      │
                      ▼
                    Stop
```

```c
printf ("Enter the value of x for which" "values of y is required");
scanf ("%f", &x);
for (i=0; i<=n; i++)
{
nr=dr=1;
for (j=0; j<=n; j++)
if (j! = i)
{
    nr * = x = ax [j];
    dr * = ax [i] - ax [j];
}
y+=(nr/dr)*ay[i];
}
printf ("when x = %f y=%f\n",x,y);
}
```

Output:

Enter the value of n

4

Enter the set of values

5	12
6	13
9	14
11	16

Enter the value of x for which value of y is wanted

10 when $x = 10y = 14.67$

10.10 METHOD OF LEAST SQUARE (to Fit a Straight Line)

Note: The variables used in this program have their specific meaning as given below.

aug: augmented matrix

n: number of data points

multi: multiplier

Flow Chart 10.10

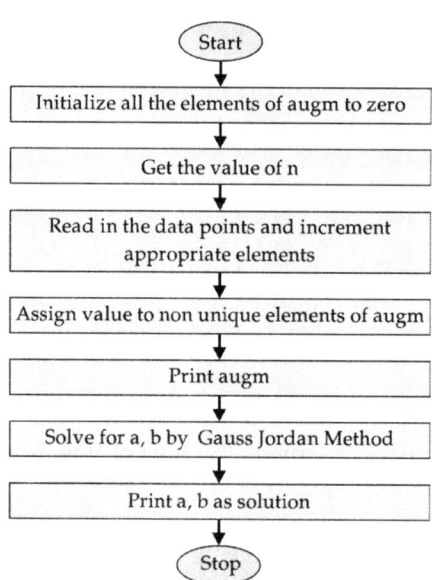

Program Demonstrating Method of Least Square

To fit a straight line to given data by least square approximation.

Let the data be:

x	1	2	3	4	5
y	14	13	9	5	2

//**Method of Least Square**

```
#include<stdio.h>
#include<conio.h>
```

```
#include<math.h>
void main ( )
{
float aug[2] [3] = {{0,0,0}, {0,0,0}};
float multi,a,b,x,y,xsq;
int i,j,k,n;
clrscr ( );
printf ("Enter the pair of observed values");
scanf ("%d", &n);
aug [0] [0]=n;
for (i=0;i<n;i++)
{
printf ("pair no.%d", i+1);
scanf ("%f %f"6x,6y);
xsq=x*x;
//sumation of x,y,xsq,x*y
aug[0] [1]+=x;
aug[0] [2]+=y;
aug[1] [1]+=xsq;
aug[1] [2]+=x*y;
}
ang[1] [0]=aug[0] [1];
printf ("enter the value of aug.matrix");
for (i=0;i<2;i++)
{
for (j=0;j<3;j++)
printf ("%f", aug[i] [j]);
printf ("\n");
}
//applying gauss jordan method to reduce the coefficient matrix to diagonal matrix
for (j=0;j<2;j++)
for (i=0;i<2;i++)
if (i!=j)
   {
   multi=aug [i] [j]/aug [j] [j];
   for (k=0;k<3;k++)
   aug [i] [k]-=aug[j] [k]*multi;
}
a=aug[0] [2]/aug [0] [0];
b=aug[1] [2]/aug [1] [1];
printf("a=%f,b=8f", a,b);
getch ( );
}
```

Output:

Enter the pair of observed values

5

pair no. 1

1 14

pair no. 2

2 13

pair no. 3

3 9

pair no. 4

4 5

pair no. 5

5 2

Augmented matrix is

$$\begin{bmatrix} 1 & 15 & : & 43 \\ 15 & 55 & : & 97 \end{bmatrix}$$

$a = 18.52, b = -3.2$

10.11 METHOD OF LEAST SQUARE (to Fit a Parabola)

Note: The variables used in this program have their specific meaning as given below.

n: number of pair of observed values

multi: the multiplier factor

xsq: the square of x

Flow Chart 10.11

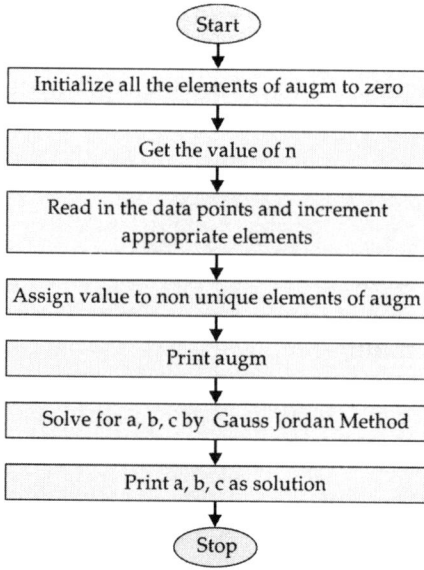

Program Demonstrating Method of Least Square

To fit a parabola by least square approximation.

Let the data be:

x	1	2	3	4	5
y	5	12	26	60	97

//Method of Least Square

```
#include<stdio.h>
#include<conio.h>
#include<math.h>
void main ( )
{
float augm[3] [4] ={(0,0,0,0), (0,0,0,0), (0,0,0,0)};
float multi, a,b,c,x,y,xsq;
int i,j,k,n;
printf ("Enter the number of pairs of observed values");
scanf ("%d", &n);
for (i=0;i<n;i++)
{
print ("pair number %d\n",i+1);
scanf ("%f%f", &x, &y);
xsq=x*x; //summation of x,y,.xsq,x*xsq, xsq*xsq,x*y,xsq*y
augm [0] [0]=n;
augm [0] [1]+=x;
augm [0] [2]+=xsq;
augm [1] [2]+=x*xsq;
augm [2] [2]+=xsq*xsq;
augm [0] [3]+=y;
augm [2] [3]+=xsq*y;
augm [1] [3]+=x*y;
}
augm [1] [1]=augm [0] [2];
augm [2] [1]=augm [1] [2];
augm [1] [0]=augm [0] [1];
augm [2] [0]=augm [1] [1];
printf ("the augmented matrix is :");
for (i=0;i<3;i++)
{
   for (j=0;j<4;j++)
   printf ("%f",augm [i] [j]);
   printf ("\n");
}
//applying the gauss jordan method to reduce the coefficient matrix to diagonal matrix
for [j=0;j<3;j++)
```

```
for [i=0;i<3;i++)
if (i!=j)
{
multi+augm [i] [j]/augm [j] [j];
for (k=0;k<4;k++)
augm [i] [k]-=augm [j] [k]*multi;
}
a=augm [0] [3]/augm [0] [0]
b=augm [1] [3]/augm [1] [1]
c=augm [2] [3]/augm [2] [2]
printf ("a=%f" "b=%f" "c=%f",a,b,c);
getch ( );
}
```

Output is

Enter the number of pairs of observed values

5

pair number 1

1.00   5.00

pair number 2

2.00   12.00

pair number 3

3.00   26.00

pair number 4

4.00   60.00

pair number 5

5.00   97.00

Augmented matrix:

55.00	15.00	5.00	200.00
225.00	55.00	15.00	832.00
979.00	225.00	55.00	3672.00

$a = 5.7143, b = -11.0858, c = 10.4001$

10.12 TRAPEZOIDAL RULE OF INTEGRATION

Note: The variables used in this program have their specific meaning as given below.

n: number of subintervals

s: sum

$x0$: the lower limit and xn is upper limit

Program Demonstrating Trapezoidal Rule of Integration

Program to evaluate the integral of $f(x)$ between the limits $x0$ to xn using trapezoidal rule of integration based on n subintervals.

Let $\int_0^6 \frac{1}{1+x*x} dx$ then $f(x) = \frac{1}{1} + x*x, x0 = 0, xn = 6$

Flow Chart 10.12

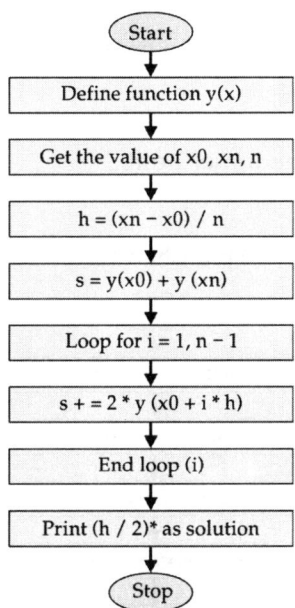

//**Trapezoidal rule of integration**

```c
#include<stdio.h>
#include<conio.h>
#include<math.h>
float y (float x)
{
return 1/(1+x*x);
}
void main ( )
{
clrscr ( );
int i,n;
float x0,xn,h,s;
printf ("enter the value of x0,xn,n");
scanf ("%f%f%d", &x0, &xn, &n);
h=(xn-x0)/n;
s=y(x0)+y(xn);
for (i=1;i<=n-1;i++)
s+=2*y(x0+i*h);
printf("the value of integral %f", (h/2)*s);
getch ( );
}
```

Output:

Enter $x0$, xn, number of subintervals

0   6   6

value of integral is 1.4108

10.13 SIMPSON'S 1/3 RULE

Note: The variables used in this program have their specific meaning as given below:

$y(x)$: function to be integrating

n: number of subintervals

$x0$: the lower limit and xn is the upper limit

s: the sum

Flow Chart 10.13

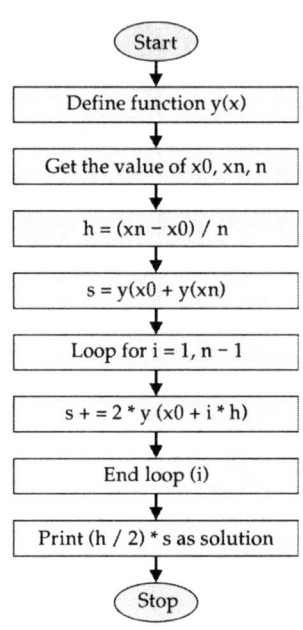

Program Demonstrating Simpson's 1/3 Rule

Program to evaluate the integral of $f(x)$ between the limits $x0$ to xn using Simpson's rule of integration based on intervals $\int_0^4 e^x dx$ then $f(x) = \exp(x)$, $x0 = 0$, $xn = 4$.

//Simpson's 1/3 rule

```
#include<stdio.h>
#include<conio.h>
#include<math.h>
float y (float x)
{
    return exp(x);
}
void main ( )
{
int i,n;
float x0,xn,h,s;
printf ("enter the value of x0, xn, n");
```

```
scanf(%f%f%d";&x0,&xn,&n);
h=(xn-x0)/n;
s=y(x0)+y(xn)+4*y(x0+h);
for(i=3;i<=n-1;i+=2)
s+=4*y(x0+1*h)+2*y (x0 + (i-1)*h);
printf ("the integral is equal to %f\n", (h/3)*s);
getch ( );
}
```

Output:

enter $x0$, xn, number of intervals

0   4   4

The value of integral is 53.87.

10.14 SIMPSON'S 3/8 RULE

Note: The variables used in this program have their specific meaning as given below

n: number of subdivision

$x0$: lower limit of integral

xn: upper limit of integral

h: length of each subinterval

Flow Chart 10.14

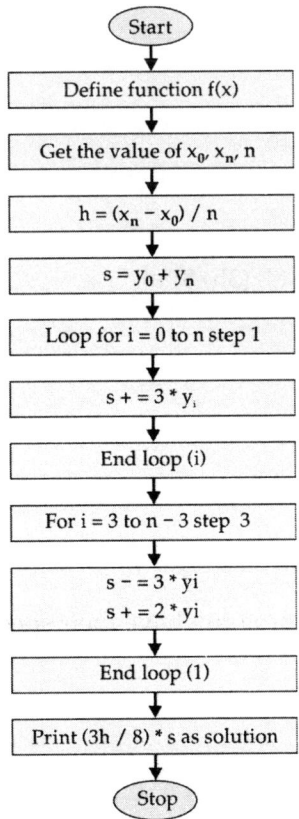

Start

Define function f(x)

Get the value of x_0, x_n, n

$h = (x_n - x_0) / n$

$s = y_0 + y_n$

Loop for i = 0 to n step 1

$s + = 3 * y_i$

End loop (i)

For i = 3 to n − 3 step 3

$s - = 3 * yi$
$s + = 2 * yi$

End loop (1)

Print (3h / 8) * s as solution

Stop

Program Demonstrating Simpson's 3/8 Rule

Program to evaluate the integral of $f(x)$ between the limits $x0$ to xn using Simpson's rule of integration based on subintervals.

Let $\int_0^4 1/(1 + x * x)$, then $f(x) = 1/(1 + x * x)$, $x0 = 0$, $xn = 1$.

```
//Simpson's 3/8 rule
#include<stdio.h>
#include<conio.h>
#include<math.h>
float y (float x)
{
return 1/(1+x*x);
}
void main ( )
{
clrscr ( );
int i,n;
float x0,xn,h,s;
printf ("enter the value of x0,xn,n");
scanf ("%f%f%d",&x0, &xn, &n);
h=(xn-x0)/n;
s=y(x0)+y(xn);
for (i=1;i<=n-1;i++)
s+=3*y(x0+i*h);
for (i=2;i<=n-3;i+3)
s+=2*y(x0+I*h);
s-=3*y(x0+I*h);
printf("the value of integral %f", (3h/8)*s);
getch ( );
}
```

Output is
Enter the value of $x0$, xn, n 0 1 6
The value of integral is 0.785396.

10.15 EULER'S METHOD

Note: The variables used in this program have their specific meaning as given below.

$df(x, y)$: dy/dx

$x0$: initial value of x

$y0$: initial value of y

$x1$: the next value of x

$y1$: the next value of y

Flow Chart 10.15

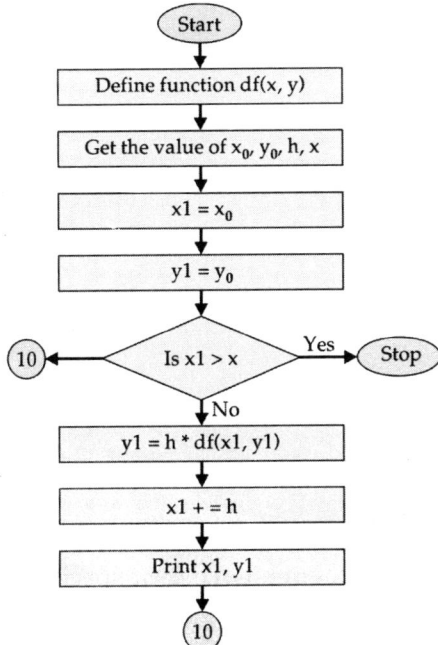

Program Demonstrating Euler's Rule

Program for Euler's method to find the value of y at x for the equation $dy/dx = f(x, y)$ with the initial condition $y(x0) = y0$ by dividing the interval into n equal parts.

Let $dy/dx = x*x + y*y;\ y(0) = 0,\ h = 0.1,\ 0 < x < 0.5$.

```
//Euler's method
#include<stdio.h>
#include<conio.h>
#include<math.h>
float df (float x, float y)
{
    return x*x+y*y;
}
main ( )
{
    float x0,y0,h,x,x1,y1;
    printf ("enter the values of x0,y0,h);
    scanf ("%f%f%f", &xo, &y0, &h);
    x1=x0;y1=y0;
    while (1)
{
if (x1>x)
return;
```

```
y1+=h*df (x1,y1);
x1+=h;
printf ("when x=%f", y=%f\n", x1,y1);
}
}
```

Output is

Enter the values of x0, y0, h, x

0 0 0.1 0.5

when $x = 0.1$, $y = 0.001$

when $x = 0.2$, $y = 0.005$

when $x = 0.3$, $y = 0.014$

when $x = 0.4$, $y = 0.03$

when $x = 0.5$, $y = 0.055$

10.16 RUNGE–KUTTA METHOD

Note: The variables used in this program have their specific meaning as given below.

$x0$: starting value of x, i.e. $x0$

xn: the value of x for which y is to be determined

Flow Chart 10.16

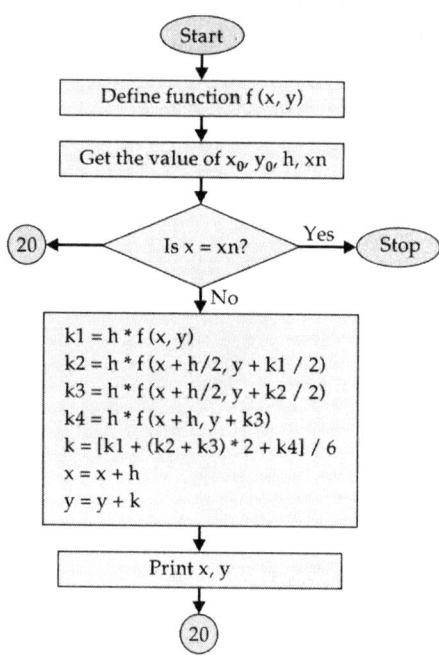

Program Demonstrating Runge-Kutta Method

Program to find an approximate value of y for the given value of x, given that

$$dy/dx = f(x, y), y(x0) = y0 \text{ by using Runge-Kutta method}$$

Let $\qquad dy/dx = x*x + y*y; y(0) = 1.$

//Runge-Kutta method

```
#include<stdio.h>
#include<conio.h>
#include<math.h>
float f (float x, float y)
{
    return (x*x+y*y);
}
void main ( )
{
float x0,y0,h,xn,x,y,k1,k2,k3,k4,k;
clrscr ( );
printf ("Enter the values of x0,y0,h,xn");
scanf ("%f%f%f",&*xo,&yo,&h,&xn);
x=x0;y=y0;
while (1)
{
if (x==xn) break;
k1=h*f(x,y);
k2=h*f(x+h/2,y+k1/2);
k3=h*f(x+h/2,y+k2/2);
k4=h*f(x+h,y+k3);
k=(k1+(k2+k3)*2+k4)/6;
x+=h;
y+=k;
printf ("whenx=%f,y=%f",x,y);
}
getch ( );
}
```

Output is

Enter the value of $x0, y0, h, xn$

0.0   1.0   0.2   0.2

when $x = 0.2, y = 0.19599$

10.17 PROGRAM DEMONSTRATING SOLUTION OF LAPLACE'S EQUATION

Note: The variables used in this program have their specific meaning as given below.

ssm: size of a square of mesh

counter: used to count the number of iteration to achieve the desired level of accuracy

flag: a control variable for the program that control the iteration $ti, s = f(xi, yi)$

temp: a variable used to store the calculated values of fi, j in each iteration for

temporarily: to find the absolute error from the previous iterate

Flow Chart 10.17

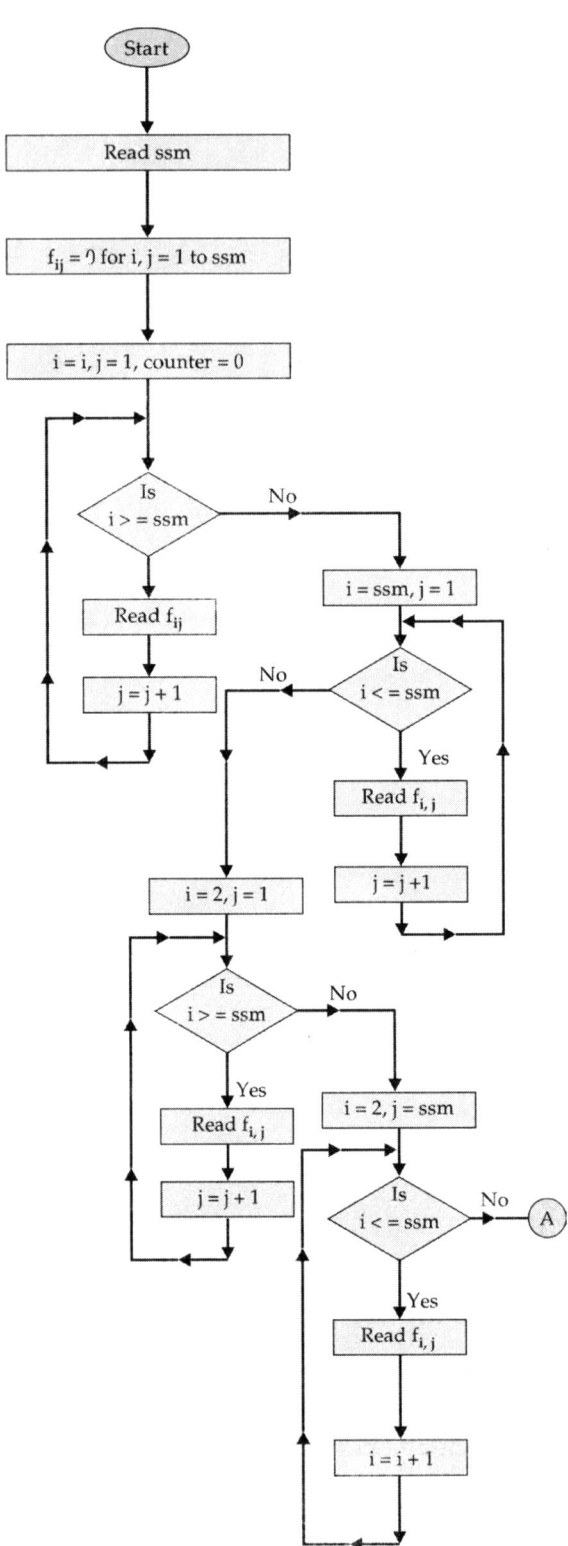

Flow Chart 10.17 (Contd.)

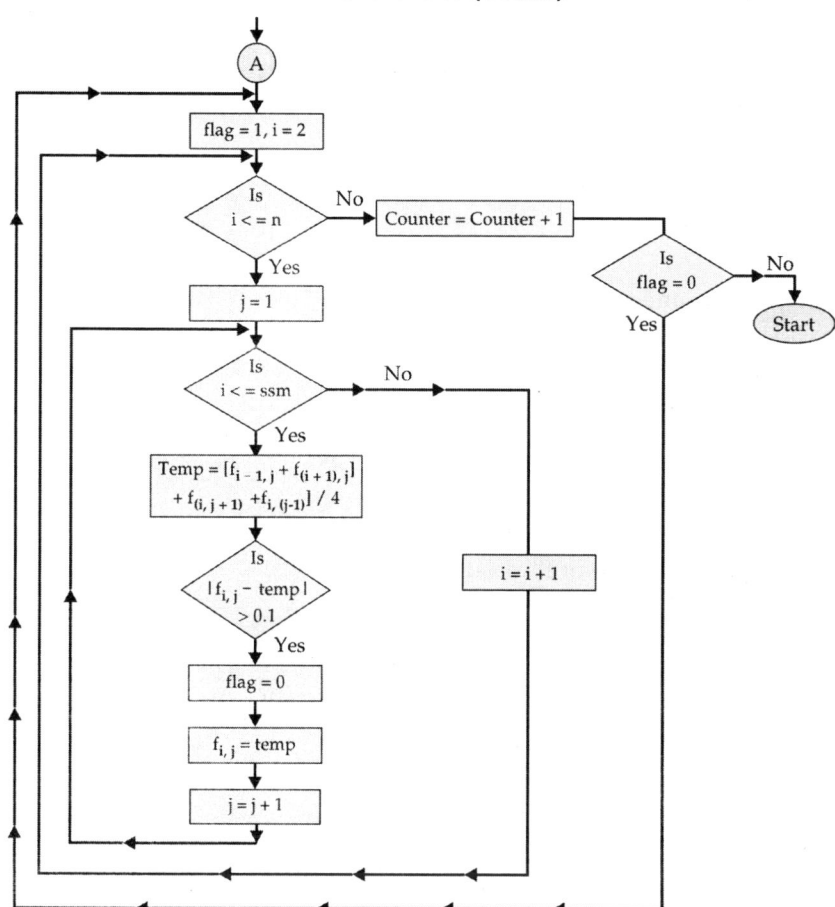

Program Demonstrating Solution of Laplace Equations

Program to find the solution of partial differential equation (Laplace equations) $fxx + fyy = 0$ under given boundary conditions by Jacobi's method using five point formula.

//Laplace's equations

```
#include<stdio.h>
#include<conio.h>
#include<math.h>
#include<process.h>
void main ( )
{
int ssm, computer=0,i,j,k,flag;
float f[10] [10], temp;
clrscr ( );
printf ("\n Enter the size of ssm");
scanf ("%d",&ssm);
for (i=1;i<=ssm;i++)
```

```
for (j=1;j<=ssm;j++)
    f[i] [j]=0;
print ("\n Enter the boundary conditions");
for (i=1,j=1;j<=ssm;j++)
    {
printf ("\n Enter the value of f%d%d=",i,j);
scanf ("%f",&f[i] [j]);
    }
for (i=ssm,j=1;j<=ssm;j++)
    {
printf ("\n Enter the value of f%d%d=",i,j);
scanf ("%f", &f [i] [j]);
    }
for (i=2, j=ssm;i<ssm;i++)
    {
printf ("\n Enter the value of f%d%d=",i,j);
scanf ("%f", &f [i] [j]);
    }
do
{
    flag=1;
    for (i=2;i<ssm;i++)
    for (j=2;j<ssm;j++)
        {
    temp=(f[i 1] [j]+f[i+1] [j]+f[i] [j]+f[i] [j+1]+f[i] [j-1]])/4;
    if (fabs (f[i] [j]-temp>0.1)
        flag=0;
    f[i] [j]=temp;
        }
    counter++;
        }
while (flag= =0);
printf ("\n the value of f at various mesh points");
printf ("\n after iteration %d are",counter);
for (i=1;i<=ssm;i++)
        {
printf("\n");
for (j=1;j<=ssm;j++)
printf ("%f",f[i] [j]);
        }
getch ( );
        }
```

Output:

Enter the size of sum 4

Enter the boundary conditions

Enter the value of f11 = 100

Enter the value of f12 = 100

Enter the value of f13 = 100

Enter the value of f14 = 100

Enter the value of f41 = 100

Enter the value of f41 = 50

Enter the value of f43 = 0

Enter the value of f44 = 0

Enter the value of f21 = 200

Enter the value of f31 = 200

Enter the value of f24 = 50

Enter the value of f34 = 0

The value of f at various mesh points after iteration 7 are

100.0000	100.0000	100.00000	100.000000
200.00000	120.82214	79.16107	50.000000
200.00000	104.161072	45.83053	0.000000
100.00000	50.0000	0.00000	0.000000

10.18 SOLUTION OF ONE-DIMENSIONAL HEAT FLOW EQUATION BY SCHMIDT METHOD

Note: The variables used in this program have their specific meaning as given below.

csquare is square of c

xe is end value of x i.e. 1

te is end value of t

$fxbt = f(0, t)$

$fxet = f(xe, t)$

$r = k*(csquare)/h^2$

h = stepsize along x direction in x-t plane

k = stepsize along t direction in x-t plane

$fx = f(x, o)$

Program Demonstrating Solution of One-Dimensional Heat Flow Equation by Schmidt Method

Program to find the solution of one-dimensional heat flow equation by Schmidt method.

//Solution to one-dimensional heat flow equation by Schmidt method

```
#include<stdio.h>
#include<conio.h>
#include<process.h>
```

Flow Chart 10.18

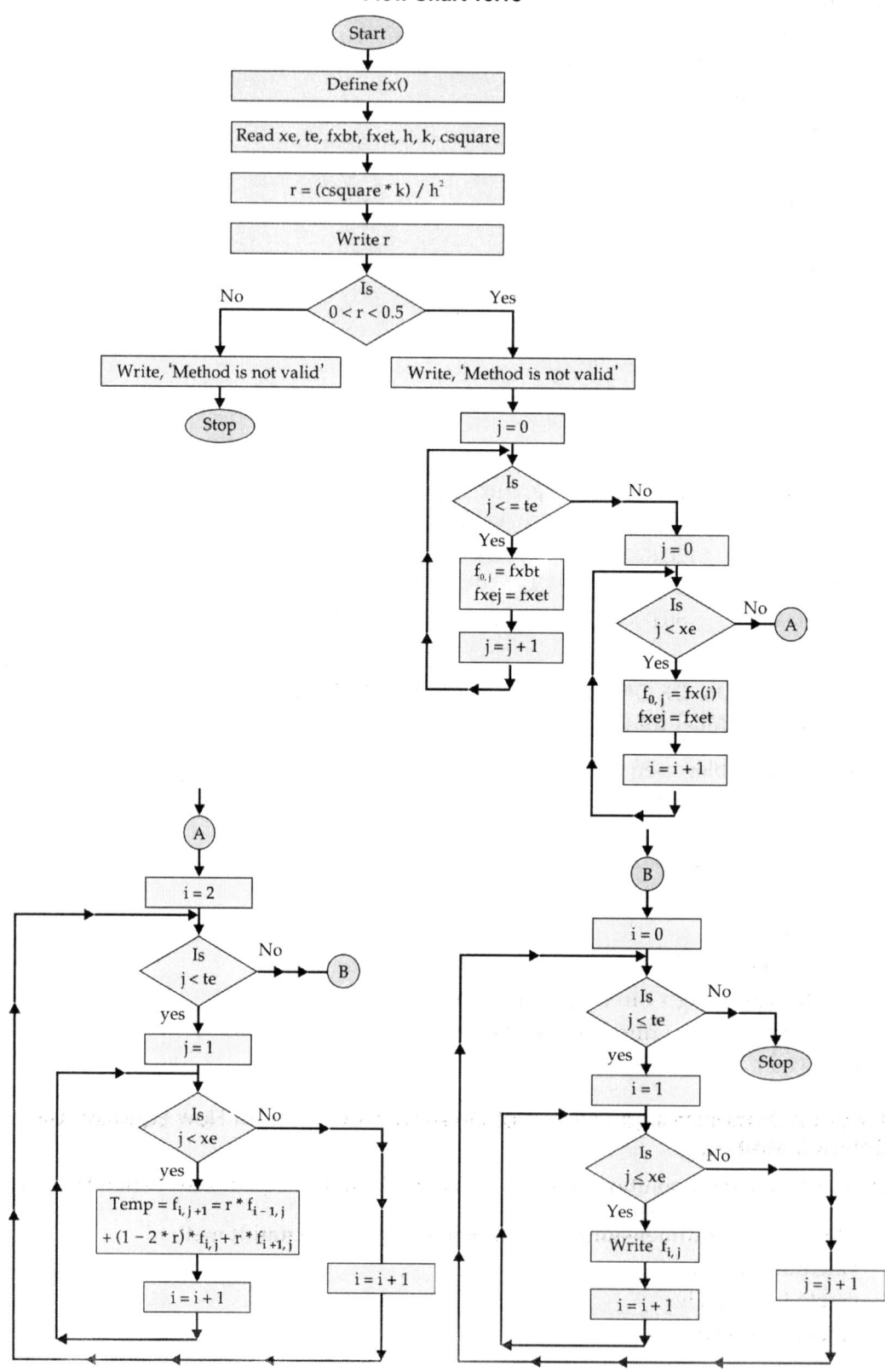

```c
#include<math.h>
float fx(float x)
{
    return 4*x-x*x;
}
void main ( )
{
    int i,j,xe,te;
    float h,k,f[10] [10,\],r,cs,fxbt,fxet;
    clrscr ( );
    printf ("\n Enter the end value of x and t i.e. xeand te");
    scanf ("%d%d", &xe,&te);
    printf ("\n Enter the value of fxbt and fxet");
    scanf ("%f%f", &fxbt, %fxet);
    printf ("\n Enter the value of h, k and cs");
    scanf ("%f%f%f", &h,&k,&cs);
        r=cs*k/h*h;
    printf ("\n the value of r is %f",r);
    if (0<r&&r<=0.5)
        printf("\n the method is applicable");
else
    {
    printf ("\n this method is not applicable");
    getch ( );
    exit (0);
}
for (j=0;j<=te;j++)
{
f[0] [j]=fxbt;
f[xe] [j]=fxet;
}
for (i=1;i<xe;i++)
f[i] [0]=fx(i);
for (j=0;j<te;j++)
for (i=1;i<xe;i++)
f[i] [j+1]=r*f[i-1] [j]+(1-2*r)*f[i] [j]+r*f[i+1] [j];
printf ("\n the value of fij are");
for (j=0;j<=te;j++)
{
printf ("\n\n");
for (i=0;i<=xe;i++)
printf("\t %f",f[i] [j]);
}
getch ( );
}
```

Output:

Enter the end value of x and t i.e. xe and te 4 5

Enter the value of fxbt and fxet 0 0

Enter the value of h,k and cs 1 1 0.5

the value of r is 0.500000

the method is applicable

the value of fij are

0.000000	3.000000	4.000000	3.000000	0.00000
0.000000	2.000000	3.000000	2.000000	0.00000
0.000000	1.500000	2.000000	1.500000	0.00000
0.000000	1.000000	1.500000	1.000000	0.00000
0.000000	0.750000	1.000000	0.750000	0.00000
0.000000	0.500000	0.750000	0.500000	0.00000

11

Computational Techniques (Programming Using MATLAB)

11.1 GAUSS–JACOBI METHOD

Note: The variables used in this program have their specific meaning as given below.

f1, f2, f3: simultaneous algebraic equations

x, y, z: unknowns which is to be calculated.

Program:

Let us take the equations

$$20x + y - 2z = 17$$
$$3x - 20y - z = -18$$
$$2x - 3y + 20z = 25$$

```
clc;
f1=@(x,y,z) (1/20)*(17-y+2*z);
f2=@(x,y,z) (1/20)*(-18-3*x+z);
f3=@(x,y,z) (1/20)*(25-2*x+3*y);
x=0; y=0; z=0;

for i=1:6
    f1(x,y,z);
    f2(x,y,z);
    f3(x,y,z);
    a=f1(x,y,z);
    b=f2(x,y,z);
    c=f3(x,y,z);
    x=a;y=b;z=c;
end
x
y
z
```

Command Window

X =

    1.0000

|

y=

    -1.0000

z=

    1.0000

fx>>

11.2 LAGRANGE'S INTERPOLATION METHOD

Note: The variables used in this program have their specific meaning as given below.

x: array corresponding to values of x

y: array corresponding the values of function

x0: value of x at which value of function to be determined

n: number of values of x

Program:

We have to find the value of f(x) at x=3 from the following data:

x	0	1	2	5
y=f(x)	2	3	12	147

```
clc;
x=input('Enter the values of x:');
y=input('Enter the values of y:');
x0=input('Enter the value of x at which y is to be determined:');
n=length(x);
if   length(y)~=n
      error("x and y must be of same length');
end
s=0;
for i=1:n
      p=y(i);
for j=1:n
      if i~=j
          p=p*((x0-x(j))/(x(i)-x(j)));
      end
end
      s=s+p;
end
fprintf('The value of y at x=%d is %f\n', x0,s);
```

Command Window

Enter the values of x: [0 1 2 5]

Enter the values of y: [2 3 12 147]

Enter the value of x at which y is to be determined:3

The value of y at x=3 is 35.000000

fx>>

11.3 NEWTON–RAPHSON METHOD

Note: The variables used in this program have their specific meaning as given below.

f: function (may be algebraic/transcendental)

df: derivative of function

x: initial solution

e: maximum allowed error

Program:

Let us take the equation $x^3 - 21x - 3500 = 0$

```
clc;
f=input ('enter the function:');
df=input ('enter the derivative of the function:');
x=input ('Enter the initial solution:');
e=input ('Enter the allowed error:');
disp ('iteration root');
for i=1:10
    x=x-feval(f,x)/feval(df,x);
    error=abs(x-xold);
    if error<=e
        break
    end
    fprintf('%d %f\n',i,x);
end
```

Command Window

enter the function:inline('x^3-21*x+3500')

enter the derivative of the function:inline("3*x^2-21')

Enter the initial solution:-15.8

Enter the allowed error:0.001

iteration   root

1   -15.645434

2   -15.643851

fx>>

11.4 SECANT METHOD

Note: The variables used in this program have their specific meaning as given below.

f: function (may be algebraic/transcendental)

a,b: initial value of interval

maxerr: maximum allowed error

Program:

Let us take the equation $x^3 - 18 = 0$

```
clc;
f=@(x)x^3-18
   a=2;
   b=3;
   maxerr=0.0001;
c=(a*f(b)-b*f(a))/f(b)-f(a));
disp("Xn-1 f(Xn-1) Xn f(Xn) Xn+1 f(Xn+1');
disp([a f(a) b f(b) c f(c)]);
while abs(f(c)) > maxerr
   a=b;
   b=c;
   c=(a*f(b) - b*f(a))/(f(b) - f(a));
   disp([1 f(a) b f(b) c f(c)]);
   flag=flag+1;
   if(flag==100)
      break;
   end
end
display (['Root is x = ' num2str(c)]);
y=c;
```

Command Window

```
f =
```

@(x)x^3-18|

Xn-1	f(Xn-1)	Xn	f(Xn)	Xn+1	f(Xn+1)
2.0000	−10.0000	3.0000	9.0000	2.5263	−1.8764
3.0000	9.0000	2.5263	−1.8764	2.6080	−0.2606
2.5263	−1.8764	2.6080	−0.2606	2.6212	0.0097
2.6080	−0.2606	2.6212	0.0097	2.6207	−0.0000

Root is x = 2.6207

fx>>

11.5 EULER METHOD

Note: The variables used in this program have their specific meaning as given below:

f:f(x,y): the first order differential equation $\dfrac{dy}{dx} = f(x, y)$

x0,y0: initial value of x and y

h: step size

x: value of x at which solution to be found

Program:

Let us take the differential equation $\dfrac{dy}{dx} = x + y + xy$, $y(0) = 1$. Find y (0.025).

```
clc;
f=input ('Enter the function:');
x0=input ('Enter the initial value of x0:');
y0=input ('Enter the value of y0:');
x=input ('Enter the value of x:');
h=x-x0;
y1=y0+h*f (x0, y0);
fprintf ('Value of y at %d is %d', x,y1);
```

Command Window

Enter the function:inline ('x+y+x*y')

Enter the initial value of x0:0 |

Enter the value of y0:1

Enter the value of x:0.025

fx Value of y at 2.500000e-02 is 1.025000e+00>> |

11.6 RUNGE–KUTTA METHOD

Note: The variables used in this program have their specific meaning as given below.

f:f(x,y) is the first order differential equation $\dfrac{dy}{dx} = f(x, y)$

x0,y0: initial value of x and y

h: step size

x: value of x at which solution to be found

Program:

Let us take the differential equation $\dfrac{dy}{dx} = 3x + \dfrac{y}{2}$, $y(0) = 1$. Find y (0.2).

```
clc;
f=input ('Enter the function:');
x0=input ('Enter the initial value of x0:');
y0=input ('Enter the value of y0:');
```

```
x=input ('Enter the value of x:');
h=x-x0;
k1=h*f (x0, y0);
k2=h*f(x0+h/2,y0+k1/2);
k3=h*f(x0+h/2,y0+k2/2);
k4=h*f(x0+h,y0+k3);
y=y0+(1/6)*(k1+2*(k2+k3) + k4);
fprintf (%d/t',y);
```

Command Window

Enter the function:inline ('3*x+y/2')
Enter the initial value of x0:0 |
Enter the value of y0:1
Enter the value of x:0.1
fx 1.066524e+00 >> |

11.7 REGULA–FALSI METHOD (Method of False Position)

Note: The variables used in this program have their specific meaning as given below.
    f: function (may be algebraic/transcendental)
    a,b: initial value of interval
    maxerr: maximum allowed error

Program:
Let us take the equation $x^3 - 2x - 5 = 0$

```
clc;
f=input ('Enter the function:');
a=input ('Enter the value of a:');
b=input ('Enter the value of b:');
e=input ('Enter the allowed error:');
n=round ((log(b-a)-log(e))/log(2));
disp('iteration a b x');
for i=1:n
    fa=feval(f,a);
    fb=feval(f,b);
    c=(a*fb-b*fa)/(fb-fa);
    x=feval(f,c);
    if fa*x<0
        b=c;
    else
        a=c;
    end
    fprintf('%d %f %f\n',i,a,b,c);
end
```

Command Window

Enter the function:inline ('x^3-2*x-5')

Enter the value a:2

Enter the value of b:3

Enter the allowed error:0.001

Iteration	a	b	x
1	2.058824	3.000000	2.058824
2	2.081264	3.000000	2.081264
3	2.089639	3.000000	2.089639
4	2.092740	3.000000	2.092740
5	2.093884	3.000000	2.093884
6	2.094305	3.000000	2.094305
7	2.094461	3.000000	2.094461
8	2.094518	3.000000	2.094518
9	2.094539	3.000000	2.094539
10	2.094547	3.000000	2.094547

fx>>

11.8 TRAPEZOIDAL RULE

Note: The variables used in this program have their specific meaning as given below.

f: integrable function

a: lower limit of integration

b: upper limit of integration

n: number of sub intervals

Program:

Let us take the equation $\int_0^6 \dfrac{dx}{(1+x^2)}$

```
clc;
f=input ('Enter the function:');
a=input ('Enter the lower limit of integration:');
b=input ('Enter the upper limit of integration:');
n=input ('Enter the number of subintervals:');
h=(b-a)/n;
s=0
for i=a+h:h:b-h
    s=s+feval (f, i);
    y=h/2*(feval (f,a) + feval (f,b) + 2*s);
end
fprintf('the integration of the given function by trapezoidal rule is %f\n',y);
```

Command Window

Enter the function:inline ('1/(1+x^2)')
Enter the lower limit of integration:0
Enter the upper limit of integration:6
Enter the number of subintervals:6
The integration of the given function by trapezoidal rule is 1.410799|
fx>>

11.9 SIMPSON'S 1/3 RULE

Note: The variables used in this program have their specific meaning as given below.
f: integrable function
a: lower limit of integration
b: upper limit of integration
n: number of sub intervals

Program:

Let us take the equation $\int_0^6 \frac{dx}{(1+x^2)}$

```
clc;
f=input ('Enter the function:');
a=input ('Enter the lower limit of integration:');
b=input ('Enter the upper limit of integration:');
n=input ('Enter the number of subintervals:');
if  gcd(n,2)~=2
    error('n must be an even number');
end
h=(b-a)/n;
s=0;
for  i=a+h:2*h:b-h
    s=s+4*feval (f, i);
end
for  j=a+2*h:2*h:b-2*h
    s=s+2*feval(f,j);
    y=(h/3)*(feval(f,a)+feval(f,b)+s;
end
fprintf('value of integral by Simpson's 1/3 rule is %f',y);
```

Command Window

Enter the function:inline ('x')
Enter the lower limit of integration:0
Enter the upper limit of integration:6
Enter the number of subintervals:5

Error using Simpsons1 (line 7)

n must be an even number

fx>>

Command Window

Enter the function:inline ('1/(1+x^2)')

Enter the lower limit of integration:0

Enter the upper limit of integration:6

Enter the number of subintervals:6

fx Value of integral by Simpson's 1/3 rule is 1.366173>>|

11.10 SIMPSON'S 3/8 RULE

Note: The variables used in this program have their specific meaning as given below.

f: integrable function

a: lower limit of integration

b: upper limit of integration

n: number of sub intervals

Program:

Let us take the equation $\int_0^6 \frac{dx}{(1+x^2)}$

```
clc;
f=input ('Enter the function:');
a=input ('Enter the lower limit of integration:');
b=input ('Enter the upper limit of integration:');
n=input ('Enter the number of subintervals:');
if  gcd(n,3)~=3
    error('n must be a multiple of 3');
end
h=(b-a)/n;
s=0;
for  i=a+h:h:b-h
if   gcd(i,3)==3
    s=s+2*feval (f, i);
else
    s=s+3*feval(f,i);
end
y=3*(h/8)*(feval(f,a)+feval(f,b))+s;
end
fprintf('value of integral by Simpson's 3/8 rule is %f',y);
```

Command Window

Enter the function:inline ('1/(1+x^2)')
Enter the lower limit of integration:0
Enter the upper limit of integration:6 |
Enter the number of subintervals:6
fx Value of integral by Simpson's 3/8 rule is 1.357081>> |

Command Window

Enter the function:inline ('x')
Enter the lower limit of integration:0
Enter the upper limit of integration:6
Enter the number of subintervals:5
Error using Simpsons3 (line 7)
n must be a multiple of 3
fx>>

11.11 BOOLE'S 3/8TH RULE

Note: The variables used in this program have their specific meaning as given below.
f: integrable function
a: lower limit of integration
b: upper limit of integration
n: number of sub intervals
Program:

Let us take the equation $\int_0^4 e^x dx$

```
clc;
f=input ('Enter the function:');
a=input ('Enter the lower limit of integration:');
b=input ('Enter the upper limit of integration:');
n=input ('Enter the number of subintervals:');
if gcd(n,4)~=4
    error('n must be a multiple of 4');
end
h=(b-a)/n;
s=0;
for i=a+h:h:b
if   gcd(i,4)==1
    s=s+32*feval (f, i);
end
if   gcd(i,4)==4
    s=s+14*feval (f, i);
```

```
end
y=2*(h/45)*(7*feval(f,a)+s);
end
fprintf('value of integral by Boole's rule is %f',y);
```

Command Window

```
    Enter the function:inline ('exp(x)')
    Enter the lower limit of integration:0
    Enter the upper limit of integration:4
    Enter the number of subintervals:4
fx Value of integral by Boole's rule is 70.656221>>|
```

11.12 WEDDLE'S RULE

Note: The variables used in this program have their specific meaning as given below.
  f: integrable function
  a: lower limit of integration
  b: upper limit of integration
  n: number of sub intervals

Program:

Let us take the equation $\int_0^6 e^x dx$

```
clc;
f=input ('Enter the function:');
a=input ('Enter the lower limit of integration:');
b=input ('Enter the upper limit of integration:');
n=input ('Enter the number of subintervals:');
if  gcd(n,6)~=6
    error('n must be a multiple of 6');
end
h=(b-a)/n;
s=0;
for i=a+h:h:b-h
if   gcd(i,6)==1
      s=s+5*feval (f, i);
end
if gcd(i,6)==2
s=s+feval (f, i);
end
if   gcd(i,6)==3
      s=s+6*feval (f, i);
end
```

```
if   gcd(i,6)==6
       s=s+2*feval (f, i);
end
y=3*(h/10)*(feval(f,a)+feval(f,b)+s);
end
fprintf('value of integral by Weddle's rule is %f',y);
```

Command Window

Enter the function:inline ('exp(x)')
Enter the lower limit of integration:0
Enter the upper limit of integration:6
Enter the number of subintervals:6
fx Value of integral by Weddle's rule is 402.775928>> |

Model Question Papers and Previous Years Question Papers*

Model Question Paper I

1. Write down the normal equations to fit a quadratic curve by least square method.
2. Write the distributive, commutative and index laws of operator Δ.
3. Show that $(1 + \Delta)(1 - \nabla) = 1$ with usual notation.
4. Evaluate $\dfrac{\Delta^2}{E} x^3$.
5. State Newton's formula on interpolation.
6. State Simpson's rule.
7. Derive the difference equation by eliminating a and b from the relation
$$y_n = a2^x + b(-2)^x$$
8. Solve $U_x + 2 - 6U_{x+1} + 9U_x = 0$.
9. Explain the terms *round off error; truncation error.*
10. Explain briefly Gauss-Jordon iteration to solve simultaneous equation.
11. Show that Newton-Raphson formula to find $\sqrt{2}$ can be expressed in the form
$$x_{n+1} = \frac{1}{2}\left(x_n + \frac{a}{x_n}\right) n = 0, 1, 2.$$
12. Explain convergency of the relaxation method.
13. State the algorithm of Runge-Kutta method of order two to solve
$$y' = f(x, y), y(x_0) = y_0 \text{ at } x = x_0 + h$$
14. State Milne's predictor formula.
15. Write the merits and demerits of the Taylor method of solution.
16. State the Adams-Bashforth predictor–corrector formula.
17. State the finite difference scheme of $u_{xx} + u_{yy} = 0$.
18. Identify the equation $f_{xx} + 2f_{xy} + f_{yy} = 0$.
19. Write the Crank-Nicholson difference scheme to solve $u_{xx} = au_t$ with $u(0, t) = T_0$, $u(l, t) = T_l$ and the initial condition as $u(x, 0) = f(x)$.
20. Write down the finite difference form of the equation $\nabla^2 u = f(x, y)$.

---

* Based on papers of different universities

21. (a) Fit a second degree parabola to the following:

$x =$	0	1	2	3	4
$y =$	1	1.8	1.3	2.5	6.3

(b) (i) Evaluate $\Delta^2(\cos 2x)$.

(ii) Explain the difference between $\left(\dfrac{\Delta^2}{E}\right)u_x$ and $\dfrac{\Delta^2 u_x}{E u_x}$, find the value of these when $u_x = x^3$.

22. (a) The following are data from the steam table

Temp (°C)	140	150	160	170	180
Pressure (kg/cm²)	3.685	4.854	6.302	8.076	10.225

Using Newton's formula, find the pressure of the stream for a temperature of 142°.

(b) Dividing the range into 10 equal parts, find the approximate value of $\int_0^\pi \sin x \, dx$ by (i) Trapezoidal rule (ii) Simpson's rule.

23. (a) (i) Use Newton-Raphson method to find the roots of the equation $e^x = 2x + 21$.
   (ii) By Gauss-Seidal method, solve $4x_1 + x_2 + 2x_3 = 4$, $3x_1 + 5x_2 + x_3 = 7$ and $x_1 + x_2 + 3x_3 = 3$.

(b) (i) Solve $9x - y + 2z = 9$, $x + 10y - 2z = 15$, $2x - 2y - 13z = 17$ to three decimal places by relaxation method.
   (ii) Solve $x^3 - 8x^2 + 17x - 10 = 0$ by Graffe's method.

24. (a) Using Euler's modified method, solve $\dfrac{dy}{dx} = x + y$ and $y(0) = 1$ at $x = 0.05$ to 0.20

(b) Solve $y' = x^2 + y - 2$ using Milne's predictor-corrector method for $x = 0.3$ given the initial value $x = 0$, $y = 1$, the values of y for $x = -0.1$, 0.1 and 0.2 should be computed by Taylor series expansion.

25. (a) Given the values of $u(x, y)$ on the boundary of the square given in the figure, evaluate the function $u(x, y)$ satisfying Laplace's equation at the pivotal points of the figure.

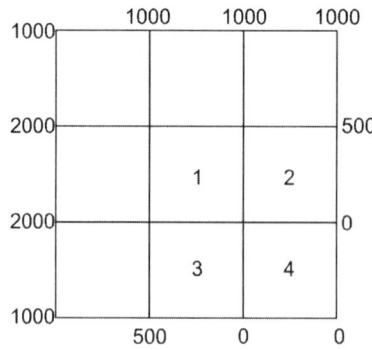

(b) Solve by Crank-Nicolson's method

$$\frac{\partial u}{\partial t} = \frac{1}{16}\frac{\partial^2 u}{\partial x^2} \quad 0 < x < 1, t > 0$$

$$u(x, 0) = 0, \ (0, t) = 0, \ u(1, t) = 100t$$

Compute u for one step with $h = \dfrac{1}{4}$.

Model Question Paper II

State True of False

1. The number of real roots of an odd degree algebraic equation with real coefficient is odd.
2. If a curve is fit to the points (x_1, y_1), $i = 1, 2, ..., n$ by least square method, then the curve passes through the points $(\bar{x}, \bar{y})$, i.e. $\left(\dfrac{\sum x_1}{n}, \dfrac{\sum y_1}{n} \right)$.
3. Gauss-Seidal iteration converges only if the coefficient matrix is diagonally dominant.
4. For any root, the order of convergence of Newton-Raphson method is two.
5. Crank-Nicholson's difference formula is used to solve wave equation.

Fill in the Blanks

6. The sum of the roots of the equation $x^2 - |x| - 12 = 0$ is ......
7. $\Delta^m f_n = \nabla^m f_p$ where p is ......
8. The solution of the difference equation $Y_{n+1} - 2y_n = 1, y_0 = 1$ is $y_n = $ ......
9. The number of real roots of $x^4 + x^3 + x^2 + x + 1 = 0$ is ......
10. In terms of factorial powers, $\dfrac{1}{x^2 - 1}$ is ......
11. If the roots of $x^3 - px + q = 0$ are in AP, then ......
    (a) $p = 0$ (b) $q = 0$ (c) $p = q$ (d) $p + q = 0$.
12. $\dfrac{1}{\Delta} - \dfrac{1}{\nabla}$ is equal to ......
    (a) 1   (b) -1 (c) Δ (d) ∇
13. $\dfrac{\Delta}{\nabla} - \dfrac{\nabla}{\Delta}$ is equal to ......
    (a) 0   (b) $\Delta - \nabla$ (c) $\nabla - \Delta$ (d) $\nabla + \Delta$
14. If α, β, γ are the roots of $x^3 - 2x - 3 = 0$, then $\alpha^2 + \beta^2 + \gamma^2$ is ...
    (a) 1   (b) 2   (c) 3   (d) 4
15. Simpson's rule of integration is exact for all polynomials of degree not exceeding ...
    (a) 1   (b) 3   (c) 4   (d) 5

Short Answer Questions

16. If α, β, γ are the roots of $x^3 + 2x - 1 = 0$, what is the value of $\alpha^3 + \beta^3 + \gamma^3$?
17. What is the solution of difference equation $y_{n+2} - 2y_{n+1} - 24y_n = 0$?
18. What is the equation whose roots are $\pm 1, \pm i, 1 \pm i$?
19. State Schmidt's explicit formula for solving heat equation.
20. State a difference formula for solving wave equation.
21. (a) (i) Fit the least square straight line $y = a + bx$ to the data: $f(0) = 6, f(1) = 1, f(1) = 2, f(2) = 3, f(4) = 0, f(5) = -3$.
    (ii) Solve the reciprocal equation $4x^4 - 20x^3 + 33x^2 - 20x + 4 = 0$.
    (b) (i) Fit the least square parabola $y = a + bx + cx^2$ to the data $f(-1) = -2, f(0) = 1, f(1) = 2, f(2) = 4$.
    (ii) Solve $x^3 - 19x^2 + 114x - 216 = 0$ given that the roots are in GP.

22. (a) (i) Compute the positive root of $x^4 - x - 10 = 0$ correct to two decimal places.

(ii) By Gauss elimination method invert the matrix $\begin{pmatrix} 10 & 1 & 0 \\ 1 & 0 & 0 \\ 0 & 0 & 1 \end{pmatrix}$.

(b) (i) Compute the root $4x = e^x$ near 2 correct to two decimal places.
(ii) Solve by Crout's method: $2x + 5y - z = 10$, $8x - y + 3z = 12$, $x + 3y + 6z = -1$.

23. (a) (i) Compute the missing value:

$x:1$	2	3	4	5
$f:0$	7	26	–	124

(ii) Find the sum $\displaystyle\sum_{x=1}^{\infty} \frac{1}{x(x+2)}$.

(b) (i) Express x^3 in factorial powers.
(ii) Compute the missing value:

$x:-1$	0	1	2	3
$f:5$	2	–	0	1

24. (a) (i) If $f(-1) = -2, f(0) = -1, f(2) = 7, f(5) = 124$, compute $f(1)$.

(ii) Compute $\int_0^1 e^x dx$ by Simpson's 1/3 rule with 10 subdivisions.

(b) (i) Compute $f'(1)$ using the following data:

x	1.0	1.5	2.0	2.5	3.0
f	27.00	106.75	324.00	783.75	1621.00

(ii) Compute $\int_0^1 \frac{dx}{1+x^2}$ by Simpson's 1/3 rule with six subdivisions.

25. (a) (i) Given that $y' = y, f(0) = 1$, express $y(h)$ as a polynomial in h by Runge-Kutta method.
(ii) Obtain the five point formula for solving Laplace equation.
(b) (i) Compute $y(0, 2)$ correct to four decimal places from the Taylor series solution of the equation $yy' = y^2 - 2x$, $y(0) = 1$.

(ii) Solve the Poisson's equation $\dfrac{\partial^2 u}{\partial x^2} + \dfrac{\partial^2 u}{\partial y^2} = -100|x| \le 1, |y| = 1$ given that $u = 0$ on the boundary of the square. Take $h = \dfrac{1}{2}$.

Model Question Paper III

State True or False

1. By the method of least square, a curve of the form $y = ax^2 + bx + c$ is fitting to the data $(xi, yi), i = 1, 2, ..., n$. This curve will pass through all the points.
2. Iteration method is a self-correcting method.
3. The nth differences of a polynomial of degree n are zeros.
4. Liebmann's iteration process is used to solve one-dimensional heat equation.
5. If α, β, γ are the roots of $x^3 - x + 1 = 0$ then $\Sigma\alpha^3$ is ......

6. The order of convergence of Newton-Raphson method is ....

7. "Whenever trapezoidal rule is not applicable, Simpson's rule can be applied". This is ......

8. **The particular integral of** $u_{x+2} + 2u_{x+1} + u_x = 2^x$ is ......

9. Write down the normal equations in the case of fitting a parabola $y = ax^2 + bx + c$.

10. Find the values of $\Sigma\alpha^3$ and $\Sigma\alpha^6$ if α, β, γ are the roots of $x^3 - 4 = 0$.

11. If $\Delta f(x) = x^3 + 2x^2 + x - 1$, find $f(x)$.

12. Define $\Delta, \nabla, \delta, \mu$.

13. State Simpson's 1/3 rule and 3/8 rule.

14. Solve $y_{x+2} + y_{x+1} - 2y_x = 2x + 7$.

15. What are the methods you use to solve one-dimensional wave equation?

16. Write down the algorithm of Runge-Kutta method of fourth order.

17. In solving $u_t = \alpha^2 u_{xx}$ by Crank-Nicholson method, to simplify the equation, we take $\dfrac{(\Delta x)^2}{\alpha^2 k}$ as

(i) 1/2   (ii) 2   (iii) 1   (iv) -1

18. Express $a^2 u_{xx} = u_{tt}$ in terms of difference quotients.

19. Using Euler's method find $y(0.1)$ given $y' = -y, y(0) = 1$.

20. State Bessel's and Everett's formulae.

21. (a) Fit a straight line to the data by the method of least squares.

x	0	5	10	15	20
y	7	10	15	21	25

(b) Solve $6x^5 + x^4 - 43x^3 - 43x^2 + x + 6 = 0$.

(c) Fit a straight line to the data by the method of moments.

x	1	2	3	4
y	16	19	23	26

(d) If α, β, γ are the roots of $x^3 + 2x^2 + x - 1 = 0$, form the equation whose roots are $\alpha + \beta - \gamma, \beta + \gamma - \alpha, \gamma + \alpha - \beta$.

22. (a) Find the positive root of $x^3 + x - 1 = 0$ correct to two decimal places by Horner's method.

(b) Using Gauss-Seidal method, solve:

$$28x + 4y - z = 32$$
$$x + 3y - 10z = 24$$
$$2x + 17y - 4z = 35$$

(c) Solve by Crout's method $2x - 6y + 8z = 24, 3x + y + 2z = 6, 5x + 4y - 3z = 2$.

(d) Find the inverse of $\begin{bmatrix} 2 & 1 & -1 \\ 0 & 2 & 1 \\ 5 & 2 & -3 \end{bmatrix}$ by Gauss's method.

23. (a) Sum the series $1.2.3 + 2.3.4 + 3.4.5 + ...$ to n terms by finite integration.

(b) Find the missing value from the table

x	2	4	6	8	10
y	5.6	8.6	13.9	–	35.6

(c) If $u_x = x^3 + 3x^2 - 5x + 1$, find Δu_x, $\Delta^2 u_x$, $\Delta^3 u_x$ and $\Delta^{-1} u_x$.

(d) Apply Gauss's forward formula to find $y(3.75)$ from the table.

x	2.5	3.0	3.5	4.0	4.5	5.0
y	24.14	22.04	20.22	18.64	17.26	16.04

24. (a) Given the table

x	14	17	31	35
y	68.7	64.1	44.2	39.6

Find $f(27)$.

(b) Using the table below, find $f'(0)$ and $\int_0^9 f(x)dx$

x	0	2	3	4	7	9
y	4	26	58	110	460	920

(c) Solve $u_{x+2} + 6u_{x+1} + 9u_x = 3^x + x \cdot 2^x + 7$.

(d) By means of Newton's divided difference formula, find $f(8)$ from the following data.

x	4	5	7	10	11	13
$f(y)$	46	100	290	900	1200	2020

25. (a) If $y' = \dfrac{y - x}{y + x}$, $y(0) = 1$, find $y(0, 1)$ by Picard's method.

(b) Solve the ellipse equation $u_{xx} + u_{yy} = 0$ for the following square mesh.

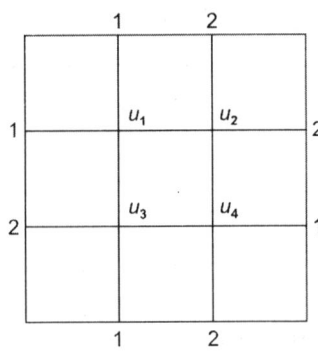

(c) Apply Runge-Kutta method of fourth order to calculate $y(0.2)$ given $y' = x + y$, $y(0) = 1$ taking $h = 0.1$.

(d) Evaluate the pivotal values of the equation $25u_{xx} = u_{tt}$ for one half period of oscillation given.

$$u(0, t) = u(5, t) = 0$$
$$u(x, 0) = 2x \text{ for } 0 \le x \le 2.5$$
$$= 10 - 2x \text{ for } 2.5 \le x \le 5$$

and $\qquad \dfrac{\partial u}{\partial t}(x, 0) = 0$

Model Question Paper IV

State True or False

1. A nonlinear relation between x and y cannot be transformed into linear relation always.
2. To solve the equation $f(x) = 0$ by the simple method of iteration, the equation can be rewritten in the form $x = f(x)$ in only one way.
3. $\sum_{r=1}^{n} y_r = (\Delta^{-1} y_r)_1^n$.
4. To find the interpolating polynomial for the given data (x_r, y_r), $r = 1, 2, ..., n$. Lagrange's method can be applied only when the $x_r's$ are not equally spaced.
5. Euler's formula for the solution of $\dfrac{dy}{dx} = f(x, y)$, $y(x_0) = y_0$ provides a pair of predictor-corrector formula.

Fill in the Blanks

6. If $\alpha, \beta, \gamma, \delta, \in$ are the roots of $ax_5 + bx_4 + cx + d = 0$, the equation whose roots are $-\alpha, -\beta, -\gamma, -\delta, -\in$ is ......
7. If we start with zero values for x, y, z while solving the equations, $10x + y + z = 12$, $x + y + 10z = 12$ by Gauss–Seidel iteration, the values for x, y, z after one iteration will be ......
8. The particular integral of the solution of the difference equation $y_{n+2} - 6y_{n+1} + 9y_n = 3_n$ is ......
9. To compute $\nabla^3 u_5$, we require, apart from u_5, the value of ......
10. To get the simplest explicit difference formula for the parabolic equaion $u_t = \alpha^2 u_{xx}$, we should take $\dfrac{\Delta x^2}{\alpha^2 \Delta t} = $ ......

Multiple Choice Questions

11. If α, β, γ are the roots of the equation $x^3 + px^2 + qx + 1 = 0$, the equation whose roots are $-\dfrac{1}{\alpha}, \dfrac{1}{\beta}, -\dfrac{1}{\gamma}$ is
    (a) $x^3 + qx^2 - px - 1 = 0$
    (b) $x^3 + qx^2 - px + 1 = 0$
    (c) $x^3 - qx^2 + px - 1 = 0$
    (d) $x^3 - qx^2 + px + 1 = 0$
12. If an approximate value of the root of the equation $x^x = 1000$ is 4.5, a better approximation of the root got by Newton-Raphson method is
    (a) 4.44
    (b) 4.56
    (c) 5.17
    (d) none of the above
13. If $u_1 = 1, u_3 = 17, u_4 = 43$ and $u_5 = 89$, the value of u_2 is
    (a) 5
    (b) 10
    (c) 12
    (d) 15
14. Given $f(0) = -1, f(1) = 1$ and $f(2) = 4$, the root of the polynomial equation $f(x) = 0$ is
    (a) 0.35
    (b) 0.44
    (c) 0.56
    (d) 0.62

15. Which of the following formulae is a particular case of Runge–Kutta formula of the second order?

    (a) Taylor series formula   (b) Picard's formula

    (c) Euler's modified formula   (d) Milne's predictor formula

Short Answer Questions

16. Write down the normal equations to be used for finding a and b, when fitting a straight line $y = ax + b$ by the method of moments.

17. Under what conditions are the group and block relaxations not necessary for solving a set of simultaneous algebraic equations?

18. State Laplace–Everett interpolation formula.

19. State Simpson's 1/3rd and 3/8th rules of numerical integration.

20. Write down the general and the simplest forms of the difference equation corresponding to the hyperbolic equation $u_{tt} = c^2 u_{xx}$.

21. (a) (i) Fit the least square straight lines $y = a + bx$ to the following data:

x	-5	-3	-1	0	1	2	4
y	0.4	-0.1	-0.2	-0.3	-0.3	0.1	0.4

    (ii) Solve $x^4 - 2x^3 + 4x^2 + 6x - 21 = 0$, given that the sum of two of its roots is zero.

    (b) (i) Find a and b so that $y = ab^x$ best fits the following data:

x	0.2	0.3	0.4	0.5	0.6	0.7
y	3.16	2.38	1.75	1.34	1.00	0.74

    (ii) Solve the equation $x^5 - 5x^4 + 9x^3 - 9x^2 + 5x - 1 = 0$.

22. (a) (i) Find the root of the equations $x^3 - 2x - 5 = 0$ lying between 2 and 3, correct to three places of decimal, using Regula-Falsi method.

    (ii) Using Gauss elimination method, find the inverse of the matrix $\begin{pmatrix} 2 & 1 & 1 \\ 1 & 0 & -1 \\ 2 & -1 & 2 \end{pmatrix}$.

    (b) (i) Compute the positive root of the equation $x - \cos x = 0$, correct to two decimal places using the bisection method.

    (ii) Solve the following set of equations, correct to three decimal places using relaxation method $28x + 4y - z = 32$, $x + 3y + 10z = 24$, $2x + 17y + 4z = 35$.

23. (a) (i) Express $x^3 - 2x + 1$ in terms of factorial polynomials.

    (ii) Obtain the missing term in the following table:

x	1	2	3	4	5
$f(x)$	0	7	–	63	124

    (b) (i) Show that $\displaystyle\sum_{n=1}^{\infty} \frac{1}{n(n+3)} = \frac{1}{18}$.

    (ii) Find $f(2.36)$ from the following table:

x	1.6	1.8	2.0	2.2	2.4	2.6
y	4.95	6.05	7.39	9.03	11.02	13.46

24. (a) (i) Find $f(2)$, if $f(-1) = 2, f(0) = 1, f(1) = 0$ and $f(3) = -1$.

    (ii) Solve $y_{n+2} + 5y_{n+1} + 6y_n = 2^n + 1$.

(b) (i) Evaluate $\int_0^{0.8} e^{-x^2} dx$ by Romberg's method with $h = 0.1$ and 0.2.

(ii) Solve $y_{n+2} - 2y_{n+1} + 2y_n = 2^n + n^2$.

25. (a) (i) Solve $y' = y - x^2$, $y(0) = 1$, by Picard's method, up to the third approximation.

(ii) Using Liebmann's method, solve the equation $\nabla^2 u = 0$ for the following square mesh with boundary values as shown in the figure.

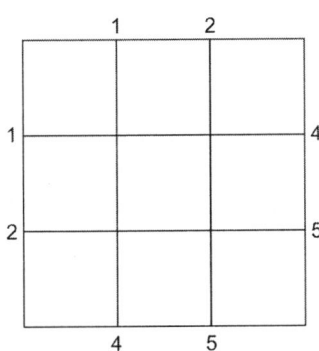

(b) (i) Compute y at $x = 0.25$ by by modified Euler's method given that $y' = 2xy$, $y(0) = 1$.

(ii) Compute u for 4 time steps with $h = 1$, given that $u_n = 4u_{xx}$, $u(0, t) = 0$, $u(4, t) = 0$, $u_t(x, 0) = 0$ and $u(x, 0) = x(4 - x)$.

Model Question Paper V

State True or False

1. By the method of least square, a curve of the form $y = ax^2 + bx + c$ is fitted to the data (x_i, y_i), $i = 1, 2, ..., n$. This curve will pass through all the points.

2. Iteration method is a self-correction method.

3. The third differences of a polynomial of degree 4 are zeros.

4. Liebmann's iteration process is used to solve two-dimensional heat equation.

5. Milne's corrector formula is given by $y_{n+1} = y_{n-1} + \dfrac{h}{3}(y'_{n-1} + 2y'_n + y'_n + y'_{n+1})$.

Fill in the Blanks

6. If α, β, γ are the roots of the equation $x^3 + x^2 + x + 1 = 0$ then the equation whose roots are $\dfrac{1}{\alpha}, \dfrac{1}{\beta}, \dfrac{1}{\gamma}$ is ......

7. Gauss-elimination and Gauss–Jordan are direct methods while ...... and ...... are iterative methods.

8. The difference equation which satisfies $y_x = A \cdot 2^x + B \cdot 3^x$ is ......

9. $E - \Delta = $ ......

10. The finite difference formula equivalent to Laplace's equation $u_{xx} + u_{yy} = 0$ with $\Delta x = \Delta y = h$ is given by $u_{ij} = $ ......

Multiple Choice Questions

11. If the roots of the equation $x^3 - 19x^2 + 114x - 216 = 0$ are in GP, then the product of the two extreme roots equals
    (a) 24   (b) 54
    (c) 36   (d) 35

12. Which of the following is true?
    (a) $\Delta r^x = r x^{r-1}$ (b) $\Delta x^{(r)} = r x^{(r-1)}$
    (c) $\Delta^n e^x = e^x$ (d) $\Delta \sin x = \cos x$

13. If $u_1 = 1$, $u_3 = 17$, $u_4 = 43$ and $u_5 = 89$, the value of u_2 is
    (a) 12   (b) 15
    (c) 5   (d) 10

14. By evaluating $\int_0^1 \dfrac{dx}{1+x^2}$ by numerical integration method, we can obtain an approximate value of
    (a) $\log_e 2$ (b) π
    (c) e (d) $\log_{10} 2$

15. In solving equation $u_t = \alpha^2 u_{xx}$ by Crank-Nicholson method, and to simplify the method, we take $\dfrac{(\Delta x)^2}{\alpha^2 K}$ as
    (a) $1/2$ (b) 2
    (c) 1   (d) 0

Short Answer Questions

16. Write down the normal equations to be used for finding a, b and c when fitting a parabola $y = ax^2 + bx + c$ by the method of least squares.
17. State the condition for convergence of Gauss–Seidel method.
18. Define μ and δ.
19. State Lagrange's interpolation formula.
20. Define a difference quotient.
21. (a) (i) Fit a straight line to the data by the method of least squares:

x	0	5	10	15	20
y	7	10	15	21	25

    (ii) Solve $6x^5 + x^4 - 43x^3 - 43x^2 + x + 6 = 0$.
    (b) (i) Fit a curve $y = ae^{bx}$ to the data.

x	0	2	4
y	5.1	10	31.1

    (ii) Diminish the roots of the equation $x^4 - 4x^3 - 7x^2 + 22x + 24 = 0$ by 1 and hence solve it.

22. (a) (i) Find the positive root of $x^3 + x - 1 = 0$ correct to two decimal places by Horner's method.
    (ii) Using Gauss–Seidel method, solve
$$28x + 4y - z = 35$$
$$x + 3y + 10z = 24$$
$$2x + 17y - 4z = 35$$

(b) (i) Solve the Crout's method

$$2x - 6y + 8z = 24$$
$$3x + y + 2z = 6$$
$$5x + 4y - 3z = 2$$

(ii) Find the inverse of $\begin{bmatrix} 2 & 1 & -1 \\ 0 & 2 & 1 \\ 5 & 2 & -3 \end{bmatrix}$ by Gauss's method.

23. (a) (i) Sum the series $1.2.3 + 2.3.4 + 3.4.5 + ... + n$ terms by finite integration.

(ii) Find the missing value from the table:

x	0	1	2	3	4
y	1	3	9	–	81

(b) (i) If $u_x = x^3 + 3x^2 - 5x + 1$, find Δu_x, $\Delta^2 u_x$, $\Delta^3 u_x$ and $\Delta^{-1} u_x$.

(ii) Apply Gauss's forward formula to find $y(3.75)$ from the table.

x	2.5	3.0	3.5	4.0	4.5	5.0
y	24.14	22.04	20.22	18.64	17.26	16.04

24. (a) (i) Given the table:

x	14	17	31	35
y	68.7	64.1	44.2	39.6

Find $f(27)$.

(ii) Find $f'(5)$ if

x	0	2	3	4	5	9
y	5	25	55	100	560	900

(b) (i) Solve $u_{x+2} + 6u_{x+1} + 9u_x = 3^x + x \cdot 2^2 + 7$.

(ii) Fit a polynomial of least degree to fit the data by Lagrange's formula:

x	0	1	3	4
y	-4	1	29	52

25. (a) (i) Apply Runge–Kutta method of fourth order to calculate $y(0, 2)$ given $\dfrac{dy}{dx} = x + y, y(0)$ taking $h = 0.1$.

(ii) Solve $\nabla^2 u = 0$ in the region $0 \le x \le 4, 0 \le y \le 4$ under the conditions ($h = 1, k = 1$) $u(0, y) = 0$, $u(4, y) = 12 + y$, $0 \le y \le 4$, $u(x, 0) = 3x$, $u(x, 4) = x^2$, $0 \le x \le 4$.

(b) (i) Derive Bender-Schmidt formula for one-dimensional heat equation.

(ii) Evaluate the pivotal values of the equation $25u_{xx} = u_{tt}$ for one half period of oscillation given

$$u(0, t) = u(5, t) = 0$$
$$u(x, 0) = 2x, 0 \le x \le 2.5$$
$$= 10 - 2x, 2.5 \le x \le 5$$

Model Question Paper VI

1. If $g(x)$ is continuous in $[a, b]$, then under what condition the iterative method $x = g(x)$ has a unique solution in $[a, b]$?

2. Compare Gauss–Jacobi and Gauss–Seidel methods for solving linear systems of the form $AX = B$.

3. Construct a linear interpolating polynomial given the points (x_0, y_0) and (x_1, y_1).

4. Write down the range for p for which Stirling's formula gives the most accurate result.

5. Find the error in the derivative of $f(x) = \cos x$ by computing directly and using the approximation $f'(x) = \dfrac{f(x+h) - f(x-h)}{2h}$ at $x = 0.8$ choosing $h = 0.01$.

6. What are the errors in trapezoidal and Simpson's rules of numerical integration?

7. What is predictor-corrector method?

8. What do we mean by saying that a method is self-starting? Not self-starting?

9. What is the truncation error of the central difference approximation of $y'(x)$?

10. For what value of λ, the explicit method of solving the hyperbolic equation $\dfrac{\partial^2 u}{\partial x^2} = \dfrac{1}{c^2}\dfrac{\partial^2 u}{\partial y^2}$ is stable, where $\lambda = \dfrac{C\Delta t}{\Delta x}$?

11. (i) Consider the nonlinear system $x^2 - 2x - y + 0.5 = 0$ and $x^2 - 4y^2 - 4 = 0$. Use Newton-Raphson method with starting value $(x_0, y_0) = (2.00, 0.25)$ and compute (x_1, y_1), (x_2, y_2) and (x_3, y_3).

(ii) Find all eigen values of the matrix $\begin{bmatrix} 2 & -1 & 0 \\ -1 & 2 & -1 \\ 0 & -1 & 2 \end{bmatrix}$ by Jacobi method (apply only three iterations).

12. (a) Find an approximate polynomial for $f(x)$ which agrees with the data:

k	xk	$f(x_k)$	$f'(xk)$
0	1.3	0.62009	-0.52202
1	1.6	0.45540	-0.56990
2	1.9	0.28182	-0.58116

using Hermite's interpolation. Hence find the approximate value of $f(1.5)$.

(b) (i) Use Newton's backward difference formula to construct an interpolating polynomial of degree 3 for the data $f(-0.75) = -0.07181250$, $f(-0.5) = -0.024750$, $f(-0.25) = 0.33493750$ and $f(0) = 1.10100$. Hence find $f(-1/3)$.

(ii) Given:

x	0°	5°	10°	15°	20°	25°	30°
f(x)	0	0.0875	0.1763	0.2679	0.3640	0.4663	0.5774

Using Stirling's formula, find $f(16°)$.

13. (a) (i) Consider the following table of data:

x	0.2	0.4	0.6	0.8	1.0
$f(x)$	0.9798652	0.9177710	0.8080348	0.6386093	0.3843735

Find $f'(0.25)$ using Newton's forward difference approximation, $f(0.6)$ using Stirling's approximation and $f'(0.95)$ using Newton's backward difference approximation.

(ii) For the given data:

x	0.0	0.9	1.1	1.3
$f(x)$	0.64835	0.91360	1.16092	1.36178
x	1.5	1.7	1.9	2.1
$f(x)$	1.49500	1.55007	1.52882	1.44573

Use Simpson's 1/3 rule for the first six intervals and trapezoidal rule for the last interval to evaluate $\int_{0.7}^{2.1} f(x)\,dx$. Also use trapezoidal rule for the first interval and Simpson's 1/3 rule for the rest of the intervals to evaluate $\int_{0.7}^{2.1} f(x)\,dx$. Comment on the obtained values by comparing with the exact value of the integral which is equal to 1.81759.

(b) (i) Evaluate $\int_{0.2}^{1.5} e^{-x^2}\,dx$ using the three point Gaussian quadrature.

(ii) Using trapezoidal rule evaluate $\int_{1.4}^{2.0}\int_{0.2}^{1.5} \ln(x + 2y)\,dy\,dx$ choosing $\Delta x = 0.15$ and $\Delta y = 0.25$.

14. (a) Consider the initial value problem $\dfrac{dy}{dx} = y - x^2 + 1,\ y(0) = 0.5$.

(i) Using the modified Euler method, find $y(0.2)$

(ii) Using 4th order Runge-Kutta method, find $y(0.6)$ and $y(0.4)$.

(iii) Using Adam-Bashforth predictor–corrector method, find $y(0.8)$.

(b) Consider the second order initial value problem $y'' - 2y' = e^{2t}\sin t$, with $y(0) = -0.4$ and $y'(0) = -0.6$.

(i) Using Taylor series approximation, find $y(0.1)$.

(ii) Using fourth order Runge-Kutta method, find $y(0.2)$.

15. (a)   Solve $\dfrac{\partial^2 u}{\partial x^2} = \dfrac{\partial u}{\partial t},\ 0 < x < 2,\ t > 0,\ u(0, t) = u(2, t) = 0,\ t > 0$ and $u(x, 0) = \sin\dfrac{\pi x}{2}$,
$0 \le x \le 2$ using $\Delta x - 0.5$ and $\Delta t = 0.25$ for two times steps by Crank–Nicholson implicit finite difference method.

(b) Approximate the solution to the wave equation $\dfrac{\partial^2 u}{\partial x^2} = \dfrac{\partial^2 u}{\partial t^2},\ 0 < x < 1,\ i > 0,$

$u(0, t) = u(1, t) = 0,\ t > 0,\ u(x, 0) = \sin 2\pi x,\ 0 < x < 1$ and $\dfrac{\partial u}{\partial t}(x, 0) = 0,\ 0 < x < 1$ with $\Delta x = 0.25$ and $\Delta t = 0.25$ for 3 steps.

Model Question Paper VII

1. State order of convergence and convergence condition for Newton-Raphson method.

2. Find the dominant eigen value of $A = \begin{pmatrix} 1 & 2 \\ 3 & 4 \end{pmatrix}$ by power method.

3. If $u = \dfrac{x - x_0}{h}$, then specify the range for u to obtain better results using Stirling's and Bessel's formulae.

4. Show that $\overset{3}{\underset{bcd}{\Delta}}\left(\dfrac{1}{a}\right) = \dfrac{1}{abcd}$.

5. State three point Gaussian quadrature formula.

6. Using trapezoidal rule, evaluate $\int_0^\pi \sin x \, dx$ by dividing the range into six equal parts.

7. Using modified Euler's method, find $y(0.1)$ if $\dfrac{dy}{dx} = x^2 + y^2$, $y(0) = 1$.

8. Write down the formula to solve second order differential equation using Runge–Kutta method of fourth order.

9. Write down the diagonal five point formula to solve the equation $u_{xx} + u_{yy} = 0$.

10. Write down the implicit formula to solve one-dimensional heat flow equation
$$u_{xx} = \dfrac{1}{c^2} u_t.$$

11. (i) Derive explicit scheme to solve parabolic equation $u_{xx} = \dfrac{1}{c^2} u_t$.

    (ii) Solve $y_{tt} = t_{xx}$ up to $t = 0.5$ with a spacing of 0.1 subject to $y(0, t) = 0$, $y(1, t) = 0$, $y_t(x, 0) = 0$ and $y(x, 0) = 10 + x(1 - x)$.

12. (a) (i) Find a root of $x \log_{10} x - 1.2 = 0$ by Newton's method correct to three decimal places.

    (ii) Solve $10x + y + z = 12$, $2x + 10y + z = 13$, $x + y + 5z = 7$ by Gauss–Jordan method.

    (b) (i) Find all the eigen values and eigen vectors of $A = \begin{pmatrix} 2 & 3 & 1 \\ 3 & 2 & 2 \\ 1 & 2 & 1 \end{pmatrix}$ by Jacobi's method.

    (ii) Using Gauss-Jordan method, find the inverse of $A = \begin{pmatrix} 1 & 1 & 3 \\ 1 & 3 & -3 \\ -2 & -4 & -4 \end{pmatrix}$.

13. (a) (i) Obtain the root of $f(x) = 0$ by Lagrange inverse interpolation, given that $f(30) = -30, f(34) = -13, f(38) = 3, f(42) = 18$.

    (ii) Find $e^{0.644}$ from the following data, using Bessel's formula

x	0.61	0.62	0.63	0.64	0.65	0.66	0.67
e^x	1.8404	1.8589	1.8776	1.8965	1.9155	1.9348	1.9542

(b) (i) Find $f(x)$ as a polynomial in x for the following data by Newton's divided difference formula.

x	-4	-1	0	2	5
$f(x)$	1245	33	5	9	1335

(ii) Find $y\,(35)$ by Stirling's formula from the following data:

x	20	30	40	50
y	512	439	346	243

14. (a) (i) Evaluate $\int_0^1\int_1^2 \dfrac{2xy}{(1+x^2)}\dfrac{dxdy}{(1+y^2)}$ by trapezoidal rule with $h = k = 0.25$.

(ii) Find the value of sec $31°$ from the following data:

$\theta°$	$31°$	$32°$	$33°$	$34°$
$\tan \theta$	0.6008	0.6249	0.6494	0.6745

(b) (i) Using Simpson's 3/8th rule evaluate $\int_0^6 \dfrac{dx}{1+x^2}$, by dividing the range into six equal parts.

(ii) Find the maximum and minimum value of y tabulated below:

x	-2	-1	0	1	2	3	4
y	2	-0.25	0	-0.25	2	15.75	56

15. (a) (i) Using Taylor series method find y at $x = 0.1$, if $\dfrac{dy}{dx} = x^2y - 1$.

(ii) Given $\dfrac{dy}{dx} = x^2(y + 1)$, $y(1) = 1$, $y(1.1) = 1.233$, $y(1.2) = 1.548$, $y(1.3) = 1.979$, evaluate $y\,(1.4)$ by Adams-Bashforth method.

(b) (i) Using Runge-Kutta method of fourth order, solve $\dfrac{dy}{dx} = \dfrac{y^2 - x^2}{y^2 + x^2}$ with $y\,(0) = 1$ at $x = 0.2$.

Model Question Paper VIII

State True or False

1. By the method of least square, a curve of the form $y = ax^2 + bx + c$ is fitted to the data (x_k, y_k), $k = 1, 2, ..., n$. This curve will pass through all the points.
2. The root of the transcendental equation can be obtained by Horner's method.
3. $\Delta + \nabla = \Delta\nabla$.
4. Lagrange's interpolation formula can be used whether arguments are equally spaced or not.
5. Liebmann's iteration process is used to solve one-dimensional heat flow equation.

Fill in the Blanks

6. When the roots of the equation $x^3 - 15x^2 + 71x - 105 = 0$ are in AP, one of the roots equal to ......
7. When Gauss-Jordan elimination method is used to solve $AX = B$, A is transferred into a ...... matrix.

8. The particular integral of $u_{n+2} + 2u_{n+1} + u_n = 2^n$ is ......

9. If $h = 1$, $\Delta^3(1-x)(1-2x)(1-3x) = $ ......

10. The simplest form of the explicit formula to solve $a^2 u_{xx} = u_{tt}$ can be got if we select λ as ......

Multiple Choice Questions

11. If $1, 1, 3$ are the roots of $x^3 + ax^2 + bx + c = 0$, then ab equals to
    (a) 32
    (b) -32
    (c) -18
    (d) 18

12. Under the conditions $f(a)$ and $f(b)$ having opposite signs and $a < b$, the first approximation of one of the roots $f(x) = 0$, by Regula-Falsi method is given by

    (a) $\dfrac{af(b) + bf(a)}{f(b) + f(a)}$

    (b) $\dfrac{af(a) - bf(b)}{f(a) - f(b)}$

    (c) $\dfrac{af(b) - bf(a)}{f(b) - f(a)}$

    (d) $\dfrac{af(a) + bf(b)}{f(a) + f(b)}$

13. The sixth term of the sequence $8, 12, 19, 29, 42, \ldots$ is
    (a) 48
    (b) 58
    (c) 49
    (d) 57

14. By trapezoidal rule, the value of $\int_0^6 \dfrac{1}{1+x}\,dx$, by dividing the range into six equal parts is
    (a) 1.94591015
    (b) 1.95873016
    (c) 1.96607143
    (d) 2.02142857

15. Bender–Schmidt recurrence equation is given by

    (a) $U_{i,j} = \dfrac{1}{4}(U_{i-1,j-1} + U_{i-1,j+1} + U_{i+1,j-1} + U_{i+1,j+1})$

    (b) $U_{i,j} = \dfrac{1}{4}(U_{i-1,j} + U_{i+1,j} + U_{i,j-1} + U_{i,j+1})$

    (c) $U_{i,j+1} = (U_{i-1,j} + U_{i+1,j} - U_{i,j+1})$

    (d) $U_{i,j+1} = \dfrac{1}{2}(U_{i-1,j} + U_{i+1,j})$

Short Answer Questions

16. Define a reciprocal equation.

17. By Gauss elimination method, solve $x + y = 2$, $2x + 3y = 5$.

18. State Everette's interpolation formula.

19. Evaluate $\dfrac{1}{(E-a)^3} a^x$.

20. In solving $\dfrac{dy}{dx} = f(x, y)$, $y(x_0) = y_0$, write down Taylor's series for $y(x_1)$.

Part B

21. (a) (i) Use the method of moments to fit a straight line to the given data.

x	1	3	5	7	9
y	1.5	2.5	4.0	4.7	6.0

(ii) Solve the equation $x^5 - 5x^4 + 9x^3 - 9x^2 + 5x - 1 = 0$.

(b) (i) From the table given below, find the best value of a and b in the law $y = ae^{bx}$ by the method of least squares.

x	0	3	4
y	5.012	10	31.62

(ii) Remove the second term in $x^4 - 12x^3 + 48x^2 - 72x + 35 = 0$ and hence solve it.

22. (a) (i) Find the positive root of $x - \cos x = 0$ by bisection method.

(ii) Solve $x + y + z = 3$, $2x - y + 3z = 16$, $3x + y - z = -3$ by Crout's method.

(b) (i) Find the cube root of 24, by Newton's method.

(ii) Solve the following system of equation $30x - 2y + 3z = 75$, $2x + 2y + 18z = 30$, $x + 17y - 2z = 48$ by Gauss-Seidel method.

23. (a) (i) Prove that $\Delta = \mu\delta + \frac{1}{2}\delta^2$.

(ii) Sum the series: $1.2 + 2.3x + 3.4x^2 + \dots \infty$.

(iii) From the following table, find the value of $\tan 45°15'$.

$x°$	45	46	47	48	49	50
$\tan x°$	1.00000	1.03553	1.07237	1.11061	1.15037	1.19175

(b) (i) $\frac{1}{2}\delta = E^{1/2}$.

(ii) Find the missing value from the table:

x	2	4	6	8	10
y	5.6	8.6	13.9	35.6	–

24. (a) (i) Using Lagrange's formula, fit a polynomial to the given data:

x	0	1	3	4
y	-12	0	6	12

Also find y at $x = 2$.

(ii) Find the first and second derivative of the function tabulated below at $x = 0.6$.

x	0.4	0.5	0.6	0.7	0.8
$f(x)$	1.5836	1.7974	2.0442	2.3275	2.6511

(b) (i) Find the value of $\log 2^{1/2}$ from $\int_0^1 \frac{x^2}{1+x^3}$ using Simpson's one-third rule with $h = 0.25$.

(ii) Solve $y_x + 2 - y_{x+1} + yx = 0$ given $y_0 = 1$, $y_1 = \frac{\sqrt{3}+1}{2}$.

25. (a) (i) If $\frac{dy}{dx} = \frac{y-x}{y+x}$ $y(0) = 1$, find $y(0.1)$ by Picard's method.

(ii) Solve $\Delta^2 u = 0$ in the square region bounded by $x = 0$, $x = 4$, $y = 0$, $y = 4$ and with boundary condition, $u(0, y) = 0$, $y(4, y) = 8 + 2y$, $u(x, 0) = \dfrac{1}{2}x^2$, taking $h = k = 1$.

(b) Obtain the values of y at $x = 0.1, 0.2$ using RK method of fourth order for the differential equation $y' = -y$, given $y(0) = 1$.

Model Question Paper IX

State True or False

1. The cubic equation with real coefficients always has one real root.
2. Gauss-Jacobi's method converges faster than Gauss-Seidel method.
3. $\left[\dfrac{\Delta^2}{E}\right] U_x = \dfrac{\Delta^2 U_x}{EU_x}$.
4. The solution of the difference equation $Y_{x+2} - 5y_{x+1} + 6y_x = 0$ is $Y_x = Ae^{2x} + Be^{3x}$.
5. Adam–Bashforth predictor formula is given by
$$Y_{n+1, p} = Y_n + \dfrac{h}{24}[55y_n + 59_{n-1} - 37y_{n-2}, ..., 9y_{n-3}].$$

Fill in the Blanks

6. To fit a straight line by least square method, the normal equations are ......
7. In Gauss elimination method, the coefficient matrix is transformed to ...... form.
8. $\Delta + E = $ ......
9. Newton divided difference formula is ......
10. The type of the partial differential equation $\dfrac{\partial^2 f}{\partial x^2} + 2\dfrac{\partial^2 f}{\partial x \partial y} + 4\dfrac{\partial^2 f}{\partial y^2} = 0$ is ......

Multiple Choice Questions

11. If the roots of the equation $x^3 + px^2 + qx + r = 0$ are in AP, then the condition is
    (a) $2q^3 - 9pqr + 27r^2 = 0$ (b) $p^3 r = q^3$
    (c) $2p^3 - 9pq + 27r = 0$ (d) None
12. Newton's formula converges if
    (a) $|f'(x), f''(x)| < \{f'(x)\}^2$ (b) $|f(x), f'(x)| < \{f,(x)\}2$
    (c) $|f(x), f'(x)| < f'(x)^2$ (d) None

13. $\Delta\left[\dfrac{1}{f(x)}\right]$ is equal to
    (a) $\dfrac{\Delta f(x)}{f(x)\, f(x+1)}$ (b) $\dfrac{\Delta f(x+1)}{f(x+1)\, f(x)}$
    (c) $\dfrac{\Delta f(x)}{f(x)\, f(x+1)}$ (d) None

14. The error in the trapezoidal rule is of the order
    (a) h^3 (b) h
    (c) h^2 (d) None

15. The partial differential equation $\dfrac{\partial^2 u}{\partial x^2} + \dfrac{\partial^2 u}{\partial y^2} = f(x, y)$ is called
    (a) Heat equation   (b) Wave equation
    (c) Laplace equation   (d) Poisson's equation

Short Answer Questions

16. If α, β, γ are the roots of $x^3 + qx + r = 0$ from the equation whose roots are $\dfrac{1}{\alpha}, \dfrac{1}{\beta}, \dfrac{1}{\gamma}$.

17. Compare Gauss elimination and Gauss–Seidal methods.

18. Write the formula for $\dfrac{dy}{dx}$ at $x = x_n$ using backward difference operator.

19. State Simpson's 3/8th rule.

20. State Milne's predictor formula.

Part B

21. (a) (i) Fit a straight line by the method of moments for the following data.

x	1	3	5	7	9
y	8	14	20	26	32

(ii) Solve $6x^4 + 5x^3 - 38x^2 + 5x + 6 = 0$.

(b) (i) Fit a curve of the form for the following data by the method of least squares.

x	10	20	30	40	50
y	1.581	2.236	2.739	3.162	3.536

22. (a) (i) Using Gauss–Seidel method, solve
$$4x + 2y + z = 14$$
$$x + 5y - z = 10$$
$$x + y + 8z = 20,$$
correct to three decimals.

(b) (i) Find the positive root of $x^3 + x - 1 = 0$ correct to two decimals by Horner's method.

(ii) Solve by Crout's method
$$x + y + 2x = 7$$
$$3x + 2y + 4x = 13$$
$$4x + 3y + 2z = 8$$

23. (a) (i) By finite integration, sum to n terms of the series $\dfrac{1}{1.2.3} + \dfrac{1}{2.3.4} + \dfrac{1}{3.4.5} + \cdots$

(ii) The following are the data from the steam table.

Temp °C	140	150	160	170	180
Pressure kg/cm²	3.685	4.854	6.302	8.076	10.225

Using Newton's formula, find the pressure of the steam for a temperature of 142°.

(b) (i) Express $f(x) = 2x^3 - 3x^2 + 3x - 10$ is factorial notation and hence find $\Delta^3 f(x)$.

(ii) Given the values

x	20	23	26	29
y	0.3420	0.3907	0.4384	0.4848

Find y at $x = 28$.

24. (a) (i) Using Lagrange's interpolation formula, find $y(2)$ from the following data.

x	0	1	3	4
y	0	1	81	256

(ii) Find $y(1.05)$ if

x	1	1.05	1.1	1.15	1.2	1.25	1.3
y	1	1.025	1.049	1.072	1.095	1.118	1.14

(b) (i) Find the value of $\log_2^{1/2}$ from $\int_0^1 \frac{x^2}{1+x^2} dx$ using Simpson's one-third rule with $h = 0.25$.

(ii) Solve $y_{x+2} - 5y_{x+1} - 8y_x = x(x-1)^{2x}$.

25. (a) (i) Solve $\frac{dy}{dx} = \frac{3x+y}{x+2y}$, $y(1) = 1$ at $x = 1.1$ using Runge-Kutta fourth order method.

(ii) Solve $u_{xx} + u_{yy} = 0; 0 \le x, y \le 1$ with $u(0, y) = 10 = u(1, y)$ and $u(x, 0) = 20 = u(x, 1)$. Take $h = 0.25$ and apply Liebmann method to three decimal places accuracy.

(b) (i) Compute at $y = 0.25$ by modified Euler method, given $\frac{dy}{dx} = 2xy$, $y(0) = 1$.

(ii) Using Crank-Nicholson's scheme, solve $U_{xx} = 16U_{tt}$, $0 < x < 1$, $t > 0$ given $u(x, 0) = 0, u(0, t) = 0$, of $(1, t) = 100t$.

Compute u for one step in t-direction taking $h = \frac{1}{4}$.

Model Question Paper X

1. Write the observation equations when the equation $y = ax + b$ is fit by the method of moments.

2. State true or false: If a curve is fit to the points (x, y), $I = 1, 2, ..., n$ by least square method, then the curve passes through the point $\left(\dfrac{\sum x_i}{n}, \dfrac{\sum y_i}{n} \right)$.

3. Fill in the blanks: If α, β, γ are the roots of $x^3 + x^2 + x + 1$, then the equation whose roots are $\dfrac{1}{\alpha}, \dfrac{1}{\beta}, \dfrac{1}{\gamma}$ is ......

4. Choose the correct answer: If the roots of equation $x^3 - 19x^2 + 114x - 216 = 0$ are in GP, then product of the two extreme roots equals to
   (a) 24   (b) 54
   (c) 36   (d) 35

5. What is the condition for the convergence of Gauss–Seidel iterative method?

6. State the condition for convergence of Jacobi's iteration method for solving a system of simultaneous algebraic equations.

7. When does relaxation method succeed in solving a set of simultaneous equations?

8. How to reduce the number of iterations while finding the root of an equation by Regula-Falsi method?

9. Solve the difference equation $u_{x+2} - 6u_{x+1} + 9u_x = 0$.

10. Write the difference table for the data given.

x	3	5	7	9
y	6	24	58	108

11. What is the Lagrange's formula to find y if three sets of values (x_0, y_0), (x_1, y_1) and (x_2, y_2) are given?

12. Prove that $\dfrac{D}{\nabla} - \dfrac{\nabla}{D} = E - E^{-1}$.

13. State Newton's forward interpolation formula.

14. Choose the correct answer: Simpson's rule of integration is exact for all polynomials of degree not exceeding

(a) 1   (b) 3

(c) 4   (d) 5

15. Enumerate Newton's divided difference interpolation formula.

16. What is the disadvantage in Taylor series method?

17. What is a predictor-corrector method of solving a differential equation?

18. Is Euler's modified formula, a particular case of second order Runge-Kutta method?

19. Give the Crank-Nicolson difference scheme formula to solve $u_1 = \alpha^2 u_{xx}$.

Part B

20. Fit a straight line to the data by the method of least squares.

x	0	5	10	15	20
y	7	10	15	21	25

21. Diminish the roots of the equation $x^4 - 4x^3 - 7x^2 + 22x + 24 = 0$ by 1, hence solve it.

22. Find the positive root of $x^3 + x - 1$ correct to two decimal places by Horner's method.

23. Solve by Crout's method:
$$2x - 6y + 8z = 24$$
$$3x + y + 2x = 6$$
$$5x + 4y - 3z = 2$$

24. Explain the difference between $\left(\dfrac{\Delta^2}{E}\right) u_x$ and $\dfrac{\Delta^2 u_x}{E u_x}$, and find the values of these when $u_x = x^3$.

25. Find the cube polynomial which takes the following set of values $(0, 1)$, $(1, 2)$, $(2, 1)$ and $(3, 10)$.

26. Compute $\int_0^4 e^x dx$ by Simpson's 1/3rd rule with 10 subdivisions.

27. Solve $u_{x+2} + 5u_{x+1} + 9u_x = 3x + x^{2x} + 7$.

28. Apply Runge-Kutta method of fourth order to calculate $y(0.2)$ given $\dfrac{dy}{dx} = x + y$, $y(0) = 1$ taking $h = 0.1$.

29. Solve $\dfrac{\partial^2 u}{\partial x^2} + \dfrac{\partial^2 u}{\partial y^2} = 0$, subject to

(a) $u(0, y) = 0$, for $0 \leq y \leq 4$ (b) $u(4, y) = 12 + y$, for $0 \leq x \leq 4$
(c) $u(x, 0) = 3x$, for $0 \leq x \leq 4$ (d) $u(x, 4) = x^2$, for $0 \leq x \leq 4$
by dividing the square into 16 square meshes of side 1.

Model Question Paper XI

State True or False

1. Bisection method is an iterative method.
2. Eighth order difference of a polynomial of degree five is always zero.
3. The order of the truncation error of Simpson's 1/3rd rule is higher than that of trapezoidal rule.
4. Crank-Nicolson formula is an explicit formula.

Fill in the Blanks

5. Crout's method is a ...... (direct/iterative) method.
6. The number of normal equations required to fit a curve of the form $y = ax^b$ is ......
7. The nth order difference of the polynomial $p = a_0 x^n + a_1 x^{n-1} + ... + a_n$ is ......
8. Stirling's formula is based on ...... and ...... formulae.
9. Laplace equation is ...... partial differential equation.

Multiple Choice Questions

10. The rational root of $x^3 - 11x^2 + 37x - 35 = 0$, given that $(3 + \sqrt{2})$ is a root, is
   (a) -17 (b) 7
   (c) 5   (d) -5
11. Horner's method is based on (i) diminishing of roots (ii) multiplication of roots.
   (a) (i)   (b) (ii)
   (c) (i) or (ii)   (d) (i) and (ii)
12. The function whose first order difference is $3x^{(2)} - 2x^{(1)}$ is

   (a) $x^{(3)} - x^{(2)}$ (b) $\dfrac{x^{(3)}}{3} - x^2$

   (c) $3x^{(3)} - 2x^{(2)}$ (d) $3x^{(3)} - x^{(1)}$
13. The solution of $u_{n+1} - 5u_{n+1} + 6u_n = 0$ is
   (a) $A(-3)^n + B(2)^n$ (b) $A(-3)^n + B(-2)^n$
   (c) $A(-3)^n + B(2)^n$ (d) $A(3)^n + B(-2)^n$
14. The equation $u_t = a^2 u_{xx}$ is
   (a) Elliptic   (b) Hyperbolic
   (c) Parabolic   (d) Poisson
15. Find the sum of the squares of the roots of $x^3 - x = 0$.

16. What is the other name for Regula-Falsi method?

17. Write Everett's formula.

18. Solve $u_{n+1} - 3u_n = 0$ given $u_n = 2$.

19. Whether Picard's method can be applied to any first order differential equation with an initial value? Explain.

Part B

20. (a) (i) Fit a curve of the form $I = aD^n$ for the data

D	1720	2300	3200	4100
I	655	789	1000	1164

(ii) Explain the method of moments.

(b) (i) Solve $3x^3 + 8x + 12 = 0$ by diminishing its roots by 4.

(ii) Show that the condition that $x^3 + 3px + q = 0$ should have repeated roots is $4p^3 + q^2 = 0$.

21. (a) Solve by Crout's method $2x + y + 3z = 13$, $3x + y + 4z = 17$, $x + 5y + z = 14$.

(b) (i) Solve, to find a real root of $x^3 - x - 2 = 0$ using Newton-Raphson method.

(ii) Use Gauss-Seidal method to solve $x + 5y - z = 10$, $4x + 2y + z = 14$, $x + y + 8z = 20$.

22. Prove that $u_0 + u_1 x + u_2 x^2 + \ldots \infty$ is equal to

$$\frac{u_0}{1-x} + \frac{x\Delta^2 u_0}{(1-x)^2} + \frac{x^2 \Delta u_0}{(1-x)^2} + \ldots$$

Model Question Paper XI

1. What is the principle of least squares?

2. If $1 + i$ and $2 + \sqrt{3}$ are two roots of a biquadratic equation, find the equation.

3. If $x^4 + ax^3 + bx^2 - x - 1 = 0$ is a reciprocal equation, then values of a and b are ...... and ......

4. If α, β, γ are the roots $x^3 + x^2 - 2x + 1 = 0$, find the value of $\dfrac{1}{\alpha} + \dfrac{1}{\beta} + \dfrac{1}{\gamma}$.

5. State Simpson's 1/3rd and 3/8th rules.

6. State Newton-Raphson iteration formula.

7. The order of convergence of Newton-Raphson method is ......

8. Write down the values of $\Delta, \nabla, \delta, \mu$ in terms of E.

9. Find the sum of n terms of $1.2 + 2.3 + 3.4 + \ldots$

10. Find $y(x = 1)$ given

X	1	2	3	4	5
Y	1	-1	1	-1	1

11. Also find $\int_1^5 y\,dx$ in question 10.

12. State Laplace-Everett's and Lagrange's interpolation formulae.

13. Solve $u_{x+2} + 5u_{x+1} + 6u_x = 2^x + 7$.

14. If $f(x) = \dfrac{1}{x}$, find the divided differences $f(a, b)$ and $f(a, b, c)$.

15. Given $y^2 = x + y$, $y(0) = 1$, find $y(0.1)$ by Taylor series.

16. Write down the Milne's predictor and corrector algorithm.

17. State Bender-Schmidt recurrence formula.

18. For the following mesh in solving $\nabla^2 u = 0$, find one set of rough values of u at interior and mesh points.

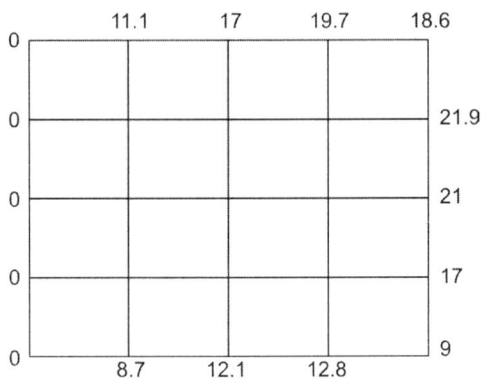

19. Write down Runge-Kutta algorithm (4th order).

20. Find $y(0.1)$ give $y' = \dfrac{1}{2}(x + y)$, $y(0) = 1$ by modified Euler's method.

Part B

21. (a) Find a curve to the data

x	0	2	4
y	5.1	10	31.1

(b) If α, β, γ are the roots of $x^3 + x + 5 = 0$, find the values of $\Sigma\alpha^2$, $\Sigma\alpha^3$, $\Sigma\alpha^5$.

(c) Find a parabola $y = ax^2 + bx + c$ given

x	1	2	3	4	5
y	10	12	8	10	14

(d) Solve $x^4 - 3x^3 + 8x + 35 = 0$ given $2 + \sqrt{3}i$ is a root.

22. (a) Find the positive roots of $x^3 + 2x - 5 = 0$ correct to two decimal places by Newton-Raphson method.

(b) By Gauss elimination, find A^{-1} if $A = \begin{bmatrix} 4 & 1 & 2 \\ 2 & 3 & -1 \\ 1 & -2 & 2 \end{bmatrix}$

(c) By relaxation method, solve $10x - 2y - 2z = 6$, $x - 10y + 2z = -7$, $x + y - 10z = -8$.

(d) Solve the equation in (c) by Gauss-Seidel method.

23. (a) Prove that $\left(\dfrac{\Delta^2}{E}\right)e^x \neq \dfrac{\Delta^2 e^x}{Ee^x}$.

(b) From the following data, find $y(42)$ and $y(85)$

x	40	50	60	70	80	90
y	184	204	226	250	276	304

(c) Sum to n terms of $2.5 + 5.8 + 8.11 + \dots$

(d) Using proper formula, find $y(65)$ in Question 23(b).

24. (a) Solve $u_{x+2} - 6u_{x+1} + 8u_x = 2^x + x^{2x} + 2.5$.

(b) For the table below, find $f'(1.76)$ and $f'(1.72)$.

x	1.72	1.73	1.74	1.75	1.76
$f(x)$	0.17907	0.17728	0.17552	0.17377	0.17204

(c) Find a polynomial of least degree to fit the data given by Lagrange's formula

x	0	1	3	4
y	−4	1	29	52

(d) Find $f'(5)$ if

x	0	2	3	4	5	9
$f(x)$	5	25	55	100	460	900

25. (a) Using modified Euler method, find $y(0.2)$ given $y' = \log(x + y)$, $y(0) = 1$ if $h = 0.2$.

(b) Solve $\nabla^2 u = 0$ in the region $0 \le x \le 4, 0 \le y \le 4$ under the conditions ($h = 1, k = 1$)
$$u(0, y) = 0, u(4, y) = 12 + y \text{ for } 0 \le y \le 4$$
$$u(x, 0) = 3x, u(x, 4) = x^2 \text{ for } 0 \le x \le 4$$

(c) Derive Bender-Schmidt formula for one-dimensional heat equation.

(d) Solve $4u_{xx} = u_{tt}$ given $(0, t) = 0 = u(4, t)$, $u_{tt}(x, 0) = 0$ and $u(x, 0) = x(4 - x)$, taking $h = 1, k = 1/2$.

MD University, Rohtak
BE 4th Semester (ECE) Examination
NUMERICAL METHODS

Time allowed: 3 hours   *Maximum marks*: 100

Note: Attempt any five questions taking at least two from each part.

Part A

1. (a) Find the cubic polynomial which takes the following values:

x	0	1	2	3
$f(x)$	1	2	1	10

   Hence or otherwise evaluate $f(4)$.   (10)

   (b) Fit a second degree parabola to the following data:

x	1.0	1.5	2.0	2.5	3.0	3.5	4.0
y	1.1	1.3	1.6	2.0	2.7	3.4	4.1

   (10)

2. (a) Find a real root of the equation $\cos x = 3x - 1$ correct to three decimal places using fixed point method.   (10)

   (b) Evaluate $\sqrt{12}$ to four decimal places by Newton's method.   (10)

3. (a) Solve the following equations by Gauss elimination method:

$$2x + y + z = 10$$
$$3x + 2y + 3z = 18$$
$$x + 4y + 9z = 16$$

   (10)

   (b) Solve the following equations by relaxation method:

$$10x - 2y - 2x = 6$$
$$-x + 10y - 2z = 7$$
$$-x - y + 10z = 8$$

   (10)

4. (a) Given that

x	1.0	1.1	1.2	1.3	1.4	1.5	1.6
y	7.989	8.403	8.781	9.129	9.451	9.750	10.031

   Find $\dfrac{dy}{dx}$ and $\dfrac{d^2y}{dx^2}$ at $x = 1.6$.   (10)

   (b) Evaluate $\int_0^4 e^x dx$ by Simpson's rule and compare it with the actual value.

Part B

5. Solve Euler's modified method $\dfrac{dy}{dx} = \log(x + y)$, $y(0) = 2$ at $x = 1.2$ and 1.4 with $h = 0.2$.   (10)

6. Using Runge-Kutta method of order 4, find y for $x = 0.1, 0.2, 0.3$ given that $\dfrac{dy}{dx} = xy + y^2$, $y(0) = 1$. Continue the solution at $x = 0.4$ using Milne's method.   (20)

7. Solve the equation $\dfrac{\partial u}{\partial t} = \dfrac{\partial^2 u}{\partial x^2}$ subject to the condition $u(x, 0) = \sin \pi x$, $0 \le x \le 1$; $u(0, t) = u(1, t) = 0$ using:

   (a) Schmidt method.

   (b) Crank-Nicolson method taking $h = \dfrac{1}{3}$, $k = \dfrac{1}{36}$. $\hspace{2cm}$ (20)

8. Solve Laplace equation $\nabla^2 u = 0$ for the following square mesh with boundary values as shown.

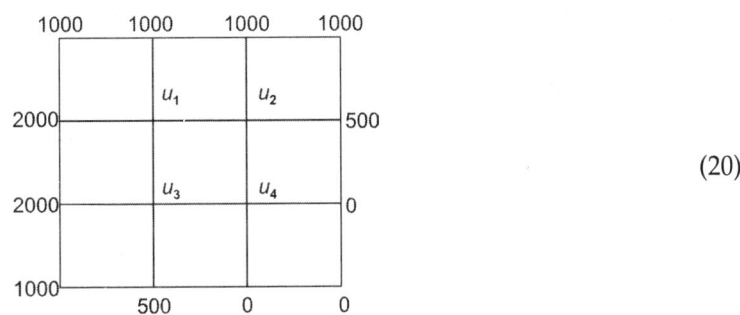

$\hspace{1cm}$ (20)

MD University, Rohtak
BE 5th Semester (ECE) Examination

APPLIED NUMERICAL TECHNIQUES

Time allowed: 3 hours $\hspace{4cm}$ *Maximum marks*: 100

Note: Attempt five questions, taking at least two questions from each part. Before answering the question paper, candidates should ensure that they have been supplied the correct and complete question paper. Complaints in this regard, if any, will not be entertained after the examination.

Part A

1. (a) By using the method of least squares, find a relation of the form $y = ax^b$ that fits the data:

x	2	3	4	5
$f(x)$	27.8	62.1	110	161

$\hspace{10cm}$ (10)

   (b) Find a cubic polynomial in x which takes on the following values $-3, 3, 11, 27, 57$ and 107, when $x = 0, 1, 2, 3, 4$ and 5 respectively. $\hspace{2cm}$ (10)

2. (a) Find by Newton–Raphson method, the real root of the equation $3x = \cos x + 1$ correct to four decimal places. $\hspace{3cm}$ (10)

   (b) Find by Regula–Falsi method, the real root of the equation $\log x - \cos x = 0$ correct to four decimal places. $\hspace{3cm}$ (10)

3. (a) Solve by relaxation method, the following equations:
$$3x + 9y - 2z = 11$$
$$4x + 2y + 13z = 24$$
$$4x - 4y + 3z = -8$$
$\hspace{10cm}$ (10)

(b) Solve the following equations by Gauss–Seidal method.
$$20x + y - 2z = 17$$
$$3x + 20y - z = -18$$
$$2x - 3y + 20z = 25 \qquad (10)$$

4. (a) Evaluate $\int_0^1 \dfrac{dx}{1+x^2}$ by using

(i) Simpson's 1/3rd rule taking $h = \dfrac{1}{4}$.

(ii) Simpson's 3/8th rule taking $h = \dfrac{1}{6}$. And compare the results with actual values
$$\qquad (10)$$

(b) Given that

x	1.0	1.1	1.2	1.3	1.4	1.5	1.6	
$f(x)$	7.989	8.403	8.781	9.129	9.451	9.750	10.031	(10)

Find $\dfrac{dy}{dx}$ and $\dfrac{d^2y}{dx^2}$ at $x = 1.6$. $\qquad (10)$

Part B

5. Given $\dfrac{dy}{dx} = \dfrac{y-x}{y+x}$ with boundary conditions $y = 1$ when $x = 0$, find approximately y
for $x = 0.1$ by (i) Euler's method (ii) Modified Euler's method. $\qquad (20)$

6. (a) Solve the boundary value problem defined by $y'' - x = 0$ and $y(0) = 0$, $y'(1) = -\dfrac{1}{2}$
by Rayleigh-Ritz method. $\qquad (10)$

(b) Determine the largest eigen value and the corresponding eigen vector of the
matrix using power method $\begin{bmatrix} 2 & -1 & 0 \\ -1 & 2 & -1 \\ 0 & -1 & 2 \end{bmatrix}$. $\qquad (10)$

7. (a) Solve the equation $\nabla^2 u = -10(x^2 + y^2 + 10)$ over the square with sides $x = 0 = y$,
$x = 3 = y$ with $u = 0$ on the boundary and mesh length $= 1$. $\qquad (14)$

(b) Derive a difference equation to represent a Poisson's equation. $\qquad (6)$

8. Solve the wave equation $\dfrac{\partial^2 u}{\partial t^2} = c^2 \dfrac{\partial^2 u}{\partial x^2}$ subject to initial condition $u = f(x)$, $\dfrac{\partial u}{\partial t} = g(x)$,
$0 < x < 1$ at $t = 0$ and the boundary conditions $u(0, t) = \phi(t)$, $u(1, t) = \psi(t)$. $\qquad (20)$

MD University, Rohtak
BE (Computer Science Engineering) Examination
(Fourth Semester)

COMPUTATIONAL TECHNIQUES

Time allowed: 3 hours   *Maximum marks*: 100

Note: Attempt any five questions taking at least two questions from each part. Before answering the question paper, candidates should ensure that they have been supplied the correct and complete question paper. Complaints in this regard, if any, will not be entertained after the examination.

Part A

1. (a) Estimate the missing term in the following table.

x	0	1	2	3	4	
$y = f(x)$	1	3	9	?	81	(10)

(b) Find $f(6)$, given $f(0) = -3$, $f(1) = 6$, $f(12) = 8$, $f(3) = 12$ third differences being constant. (6)

(c) With the usual notations, show that $\Delta = 1 - e^{-hD}$. (4)

2. (a) Estimate the values of $f(22)$ and $f(42)$ from the following data:

x	20	25	30	35	40	45	
$f(x)$	354	332	291	260	231	204	(10)

(b) Given the values

x	5	7	11	13	17	
$f(x)$	150	392	1452	2306	5202	(10)

Find $f(9)$ using Lagrange's and Newton's divided difference formula.

3. (a) Write a computer program to find the real root of the equation $y = f(x)$ using Newton-Raphson method. Also find the smallest root of the equation $e^{-x} = \sin x$. (12)

(b) Solve the equations by Gauss-Seidal method. (8)
$$83x + 11y - 4z = 95$$
$$7x + 52y + 13z = 104$$
$$3x + 8y + 29z = 71$$

4. (a) Find the first order derivative of $f(x)$ at $x = 1.5$ if

x	1.5	2.0	2.5	3.0	3.5	4.0	
$f(x)$	3.375	7.000	13.625	24.000	38.875	59.000	(10)

(b) Evaluate $\int_0^1 \dfrac{dx}{1 + x^2}$ using (i) Simpson's 1/3rd rule taking $h = \dfrac{1}{4}$, (ii) Simpson's 3/8th rule taking $h = \dfrac{1}{6}$. (10)

Part B

5. (a) Use Taylor's method to obtain approximate value of y at $x = 0.2$ for the differential equation $\dfrac{dy}{dx} = 2y + 3e^x$, $y(0) = 0$. Compare the numerical solution obtained with exact solution. (10)

(b) Using simple Euler's method, solve for y at $x = 0.1$ from $\dfrac{dy}{dx} = x + y + xy$, $y(0) = 1$ taking step size $h = 0.025$. (10)

6. (a) Using Runge-Kutta method of order 4, find $y(0.2)$ given that $\dfrac{dy}{dx} = 3x + \dfrac{1}{2}y$, $y(0) = 1$, taking $h = 0.1$. (10)

(b) Using Adams-Bashforth formula, find $y(0.4)$ for the differential equation $\dfrac{dy}{dx} = \dfrac{1}{2}xy$ and the data:

x	0	0.1	0.2	0.3	
y	1	1.0025	1.0101	1.0.228	(10)

7. (a) Obtain Picard's second approximation solution of the initial value problem $y' = x^2/(y^2 + 1), y(0) = 0.$ (8)

(b) Use Milne's method to find $y(0.3)$ from $y' = x^2 + y^2, y(0) = 1$. Find the initial values $y(-0.1), y(0.1),$ and $y(0.2)$ from Taylor's series method.

8. Find the values of $\mu(x, y)$ satisfying the Laplace's equation $\nabla^2\mu = 0$ at the pivotal points of a square region, with boundary values are shown in the figure below. (20)

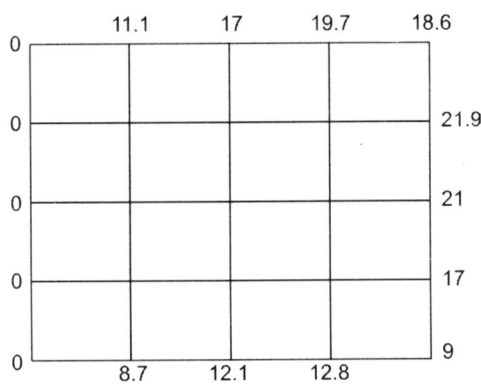

MD University, Rohtak
BE (Computer Science Engineering) Examination
(Fourth Semester)

COMPUTATIONAL TECHNIQUES

Time allowed: 3 hours   *Maximum marks*: 100

Note: Attempt five questions in all, taking at least two from each part.

Part A

1. (a) The area A of a circle of diameter d is given as:

d	80	85	90	95	100
A	5026	5674	6362	7088	7854

Find A, when $d = 82$. (10)

(b) Evaluate (i) $\Delta^n \log ax$, (ii) $\Delta\left(\dfrac{1}{1+x^2}\right)$. (10)

2. (a) Apply Newton's method to obtain the root of $x \log_{10}x = 4.7772393$ correct to four decimal places. (10)

(b) Using Regula–Falsi method, compute the real root of $x^3 - 4x - 9 = 0$ correct to three decimal places. (10)

3. Solve the following by (i) Gauss–Seidel (ii) Jacobi's iterative methods.
$$27x + 6y + z = 85$$
$$6x + 15y + 2z = 72$$
$$x + y + 54z = 110$$
(20)

4. (a) Find $f''(7.50)$ from the following table:

x	7.47	7.48	7.49	7.50	7.51	7.52	7.53	
$f(x)$	0.193	0.195	0.198	0.201	0.203	0.206	0.208	(10)

(b) Calculate the value of $\int_0^{\frac{\pi}{2}} \sin x \, dx$ by Simpson's 1/3rd rule, using 11 ordinates.

(10)

Part B

5. (a) Solve $\dfrac{dy}{dx} = -xy^2$, $y(0)$ by modified Euler's method to obtain $y(0.2)$ in two steps of 0.1 each. (10)

(b) Solve by Picard's method, the differential equation $\dfrac{dy}{dx} = x^2 + y^2$, $y(0) = 0$ to obtain $y(0.4)$ correct to four decimal places. (10)

6. Using Runge–Kutta method of order 4, compute $y(0.2)$ from

$$10\frac{dy}{dx} = x^2 + y^2, \ y(0) = 1 \ (\text{taking } h = 0.1)$$

7. Find the values of $u(x, y)$ satisfying the Laplace equation $\nabla^2 u = 0$ at the pivotal points of a square region, with boundary values as shown in figure below. (20)

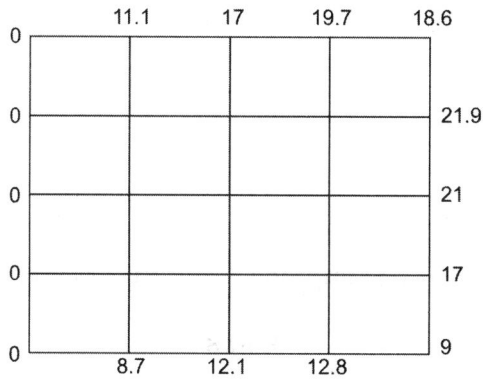

8. Solve the equation $\dfrac{\partial u}{\partial t} = \dfrac{\partial^2 u}{\partial x^2}$, given that $u(x, 0) = \sin \pi x$, $0 < x < 1$, $u(0, t) = (1, t) = 0$, using (i) Schmidt method (ii) Crank-Nicolson method.

MD University, Rohtak
BE
(Fourth Semester)

NUMERICAL METHODS

Time allowed: 3 hours   *Maximum marks*: 100

Note: Attempt five questions in all, taking at least two from each part.

Part A

1. (i) Estimate the values of $f(22)$ and $f(42)$ from the following data.

x	20	25	30	35	40	45	
$f(x)$	354	332	291	260	231	204	(10)

(ii) Use Newton's divided difference formula to find $f(x)$ from the following data.

x	0	1	2	4	5	6	
$f(x)$	1	14	15	5	6	19	(10)

2. (i) Write the condition for convergence of iteration method and hence solve
$$x^3 + x^2 - 100 = 0 \tag{10}$$

(ii) Use Newton-Raphson method to find the root of the equation correct to three decimal places which is near to 4.5.
$$e^x = x^3 + \cos 25x \tag{10}$$

3. (i) Solve the following equations by Gauss-Seidel iterative method.
$$0.2x + 2.1y + 4.2z = 9.9$$
$$4.3x + 6.1y + 4.7z = 21.6$$
$$9.2x + 8.3y + z = 15.2 \tag{10}$$

(ii) Solve the following system by relaxation method.
$$10x - 2y - 2z = 6$$
$$-x + 10y - 2z = 7$$
$$-x - y + 10z = 8 \tag{10}$$

4. (i) Given $\sin 0° = 0.0$, $\sin 10° = 0.1736$, $\sin 20° = 0.3420$, $\sin 40° = 0.6428$. Find the value of $\dfrac{dy}{dx}$ at $x = 10°$ for $y = \sin x$. (10)

(ii) Given that

x	4.0	4.2	4.4	4.6	4.8	5.0	5.2
$\log x$	1.3863	1.4351	1.481	1.5261	1.5686	1.094	1.6487

Evaluate $\int_4^{5.2} \log x \, dx$ by (a) Simpson's 1/3rd rule (b) Weddle's rule. (10)

Part B

5. Using Runge-Kutta method of order four, find $y(2)$ and $y(4)$ from
$$\frac{dy}{dx} = x^2 + y^2, y(0) = 1; \text{ Take } h = 0.1 \tag{20}$$

6. Using Milne's predictor–corrector method, find $f(0.3)$ from $\dfrac{dy}{dx} = x^2 + y^2$, $y(0) = 1$.
Find the initial values $y(-0.1)$, $y(0.1)$ and $y(0.2)$ from Taylor's series method. (20)

7. (i) Using power method, find the largest eigen value of the matrix. Also find corresponding eigen vector.
$$\begin{bmatrix} 25 & 1 & 2 \\ 1 & 3 & 0 \\ 2 & 0 & -4 \end{bmatrix} \tag{10}$$

(ii) Write various order finite difference approximation for $\dfrac{\partial u}{\partial x}, \dfrac{\partial u}{\partial y}, \dfrac{\partial^2 u}{\partial x^2}, \dfrac{\partial^2 u}{\partial y^2}$. Also derive the standard and diagonal five-point formula for the solution of Laplace equation. (10)

8. Solve the Laplace equation $u_{xx} + u_{yy} = 0$ in the domain given below, by Jacobi method.

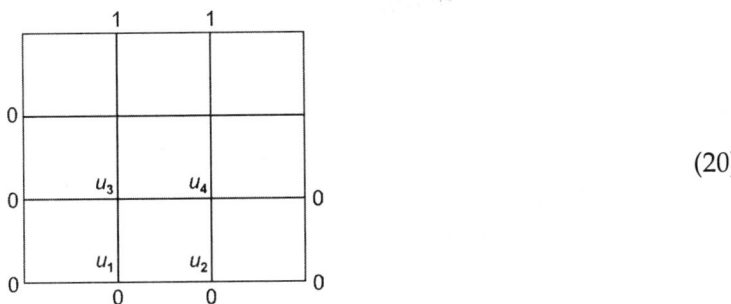

(20)

MD University, Rohtak
BE (Mechanical Engineering Examination)
(Fifth Semester)

APPLIED NUMERICAL METHODS

Time allowed: 3 hours   *Maximum marks*: 100

Note: Attempt five questions in all, taking at least two from each part.

Part A

1. (a) Using the following table, find $f(x)$ as a polynomial in x.

x	-1	0	3	6	7
$f(x)$	3	-6	39	822	1611

(10)

(b) Fit a second degree parabola to the following data:

x	1.0	1.5	2.0	2.5	3.0	3.5	4.0
$f(x)$	1.1	1.3	1.6	2.0	2.7	3.4	4.1

(10)

2. (a) Use the fixed-point iteration method to find a root of the following equation, correct to three decimal places.
$$x^3 + x^2 - 1 = 0$$
(10)

(b) Find by Newton's method, the real root of the equation $3x = \cos x + 1$.   (10)

3. (a) Solve, by relaxation method, the equations.
$$9x - 2y + z = 50$$
$$x + 5y - 3z = 18$$
$$-2x + 2y + 7z = 19$$
(10)

(b) Apply Gauss elimination method to solve the following equations.
$$x + 4y - z = -5$$
$$x + y - 6z = -12$$
$$3x - y - z = 4$$
(10)

4. (a) Given that

x	1.0	1.1	1.2	1.3	1.4	1.5	1.6
y	7.989	8.403	8.781	9.129	9.451	9.750	10.031

Find $\dfrac{dy}{dx}$ and $\dfrac{d^2y}{dx^2}$ at $x = 1.1$.   (10)

(b) Evaluate $\int_0^6 \frac{dx}{1+x^2}$ by using (i) Trapezoidal rule (ii) Simpson's 1/3rd rule, and
(iii) Simpson's 3/8th rule. (10)

Part B

5. Solve the equation $\frac{dy}{dx} = \log(x+y), y(0) = 2$ by Euler's modified method at $x = 1.2$
and 1.4 with $h = 0.2$. (20)

6. Using Runge-Kutta method of order 4, find y for $x = 0.1, 0.2, 0.3$ given that

$$\frac{dy}{dx} = xy + y^3, y(0) = 1$$

Continue the solution at $x = 0.4$, using Milne's method. (20)

7. (a) Solve the boundary value problem $y'' - 64y + 10 = 0, y(0) = y(1) = 0$ by shooting
method. (10)

(b) Use Rayleigh-Ritz method to solve the following boundary value problem

$$\frac{d^2y}{dx^2} + 2x = 0, y(0) = y(1) = 0$$ (10)

8. Solve the equation $u_{xx} + u_{yy} = 0$ for the square mesh with the boundary values as
shown below in the figure.

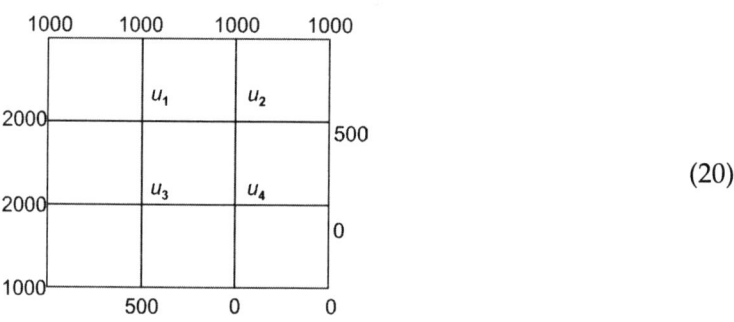

(20)

MD University, Rohtak
BE (Mechanical Engineering Examination)
(Fifth Semester)

APPLIED NUMERICAL METHODS

Time allowed: 3 hours   *Maximum marks*: 100

Note: Attempt five questions in all, taking at least two questions from each part. Assume
any missing data suitably, if any.

Part A

1. (a) Solve the following system of equations by Gauss–Seidal method:
$$83x + 11y - 4z = 95$$
$$7x + 52y + 13z = 104$$
$$3x + 8y + 29z = 171$$ (10)

(b) Using Gauss elimination method with partial pivoting, solve the following equations.

$$2.5x - 3y - 4.6z = -1.05$$
$$-3.5x + 2.6y + 1.5z = -14.46$$
$$-6.5x - 3.5y + 7.3z = -17.735 \tag{10}$$

2. (a) Using Simpson's 1/3rd rule, evaluate the integral $\int_0^{\frac{\pi}{2}} \dfrac{dx}{\sin^2 x + \dfrac{1}{4}\cos^2 x}$. (10)

(b) Find the value of cos 1.82, using the values given in table.

x	1.70	1.74	1.78	1.82	1.86
$\sin x$	0.9916	0.9857	0.9781	0.9691	0.9584

(10)

3. (a) If $y(1) = -3$, $y(3) = 9$, $y(4) = 30$, $y(6) = 132$, find the Lagrange's interpolation polynomial that takes the same values as y at the given points. (10)

(b) Using principle of least squares, find an equation of the form $y = ae^{bx}$, that fits the following data.

x	1	2	3	4	5
$f(x)$	0.6	1.9	4.3	7.6	12.6

(10)

4. (a) Using Regula-Falsi method, compute the real root of equation to four decimal places.

$$\cos x = 3x - 1 \tag{10}$$

(b) Find the root of the equation $x \log_{10} x = 4.77$ by Newton–Raphson method correct to four decimal places.

Part B

5. (a) Using Runge-Kutta method of order 4, compute $y(0.2)$ and $y(0.4)$ from $10\dfrac{dy}{dx} = x^2 + y^2$, $y(0) = 1$, taking $h = 1.0$. (8)

(b) Use Milne's method to find $y(0.3)$ from $y' = x^2 + y^2$, $y(0) = 1$, find the initial values of $y(-0.1)$. $y(0.1)$ and $y(0.2)$ from the Taylor's series. (12)

6. Solve the boundary value problem $\dfrac{d^4y}{dx^4} + 81y = 729x^2$, $y(0) = y'(0) = y''(1) = 0$, $y'''(1) = 0$. Use $n = 3$. (20)

7. (a) Solve the Poisson's equation $u_{xx} + u_{yy} = -10(x^2 + y^2 + 5)$ in the domain $0 \le x, y \le 1$; subject to conditions $u = 1$ at $y = 1$ for $0 < x < 1$, taking $h = 1/3$. (17)

(b) How are the partial differential equations classified. Give an example from real life for each type. (3)

8. The transverse displacement of a point at a distance x from one end and at any time t of a vibrating string satisfies the equation $\dfrac{\partial^2 u}{\partial t^2} = 25\dfrac{\partial^2 u}{\partial x^2}$, with the boundary conditions $u(0, t) = u(5, x) = 0$ and initial conditions

$$u(x, 0) = 20x \text{ for } 0 < x < 1$$
and
$$u_t(x, 0) = 5(5 - x) \text{ for } 1 < x < 5$$
$$u_t(x, 0) = 0$$

Solve this equation numerically for one half period of vibration, taking $h = 1$, $k = 0.1$. (20)

MD University, Rohtak
BE (Mechanical Engineering Examination)
(Fifth Semester)

APPLIED NUMERICAL METHODS

Time allowed: 3 hours   *Maximum marks*: 100

Note: Attempt five questions in all, taking at least two questions from each part. Assume any missing data suitably, if any.

Part-A

1. (a) Solve the following equations by relaxation method.
$$5x - 2y + z = 13$$
$$3x + 7y - 11z = 2$$
$$x + 20y - 2z = 8 \tag{12}$$

(b) Using Gauss–Jordon elimination method, solve the following equations:
$$x + y - z = 7$$
$$3x + 3y + 4z = 24$$
$$2x - y + 3z = 16 \tag{8}$$

2. (a) Evaluate the integral $\int_{1.0}^{1.8} \dfrac{c^2 + e^{-x}}{2} dx$ using Simpson's 1/3rd rule, taking $h = 0.2$.

$\hspace{12cm}$ (10)

(b) A rod is rotating in a plane. The following table gives the angle θ in radians through which the rod has turned for various values of the time t seconds:

t	0	0.2	0.4	0.6	0.8	1.0	1.2	
$f(x)$	0	0.12	0.49	1.12	2.02	3.2	4.67	(10)

Calculate the angular velocity of the rod when $t = 0.5$ second. (10)

3. (a) Find the root of the equation $2x = \cos x + 3$ correct to three decimal places using iteration method. (10)

(b) Find the root of the equation $x^2 - x - 1 = 0$ using Muller's method. (10)

4. (a) Using principle of least squares, find an equation of the form $y = a^{xb}$ from the following data.

x	1	2	3	4	5	
$f(x)$	0.5	2.0	4.5	8.0	12.5	(10)

(b) In the following table, the values of y are consecutive terms of a series of which 12.5 is the 5th term. Find the first and tenth terms of the series:

x	3	4	5	6	7	8	9	
$f(x)$	2.7	6.4	12.5	21.6	34.3	51.2	72.9	(10)

Part B

5. (a) Given that $\dfrac{dy}{dx} = x + y^2$ and $y + x = 0$. Find an approximate value of y at $x = 0.5$ by modified Euler's method. (14)

(b) Find an approximate value of y when $x = 0.1$ if $\dfrac{dy}{dx} = x - y^2$ at $x = 0$ using Picard's method. (6)

6. (a) Solve the equation $y^{11} + x = 0$ $(0 \le x \le 1)$ with the conditions $y(0) = y(1) = 0$ by Galerkin method. (10)

(b) Find the largest eigen value and corresponding eigen vectors of the matrix

$$\begin{bmatrix} 1 & 3 & -1 \\ 3 & 2 & 4 \\ -1 & 4 & 10 \end{bmatrix}.$$ (10)

7. Solve $u_{xx} + u_{yy} = 0$ for the square mesh with boundary values as shown in the figure. Calculate the mesh value correct to three decimal places. (20)

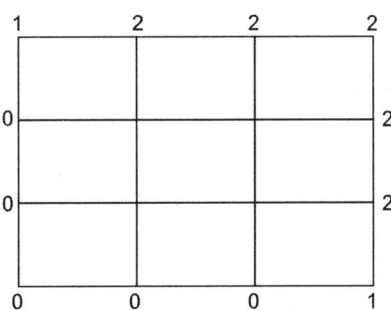

8. The transverse displacement u of a point at a distance x from one end and at any time t of a vibrating string satisfies the equation $\dfrac{\partial^2 u}{\partial t^2} = 4\dfrac{\partial^2 u}{\partial x^2}$, with boundary conditions $u = 0$ at $x = 0$, $t > 0$ and $u = 0$ at $x = 4$, $t > 0$ and initial conditions $u = x(4 - x)$ and $\dfrac{\partial u}{\partial t} = 0$, Solve this equation numerically for one half period of vibration taking $h = 1$ and $k = \dfrac{1}{2}$. (20)

MD University, Rohtak
BE (ECE Examination)
(Fifth Semester)

APPLIED NUMERICAL TECHNIQUES AND COMPUTING

Time allowed: 3 hours   *Maximum marks*: 100

Note: Attempt any five questions.

Part-A

1. (a) Define the term absolute error. Given

$$a = a\ 10.00 \pm 0.05$$
$$b = 0.0356 \pm 0.0002$$
$$c = 15300 \pm 100$$
$$d = 62000 \pm 500$$

Find the maximum value of the absolute error in (i) $a + b + c + d$ and (ii) c^3. (12)

(b) Find the number of terms of the exponential series such that their sum gives the value of e^x correct to five decimal places for all values of x in the range $0 \leq x \leq 1$. (10)

2. (a) By the method of least squares, find the straight line that best fits the following data:

x	1	2	3	4	5	
$f(x)$	14	27	40	35	68	(10)

(b) Following values of x and y are given

x	1	2	3	4
$f(x)$	1	2	5	11

Find the cubic splines and evaluate $y(1.5)$ and $y(3)$. (10)

3. (a) Find the first and second derivatives of $f(x)$ at $x = 1.5$, if

x	1.5	2.0	2.5	3.0	3.5	4.0	
$f(x)$	3.375	7.000	13.625	24.000	38.875	59.000	(10)

(b) Evaluate $\int_0^6 \dfrac{dx}{1+x^2}$ by using (i) Trapezoidal rule (ii) Simpson's 1/3rd rule (iii) Simpson's 3/8th rule and compare the results with its actual value. (10)

4. (a) Find by Newton's method, the real root of the equation $3x = \cos x + 1$. (10)

(b) Find the root of the equation $xe^x = \cos x$ using the Secant method correct to four decimal places. (10)

5. (a) Solve the following equations by Gauss elimination method.
$$2x + y + z = 10$$
$$3x + 2y + 3z = 18$$
$$x + 4y + 9z = 16$$

(b) Apply Gauss–Seidal iteration method to solve the following equations.
$$20x + y + 2z = 17$$
$$3x + 20y - z = -18$$
$$2x - 3y + 20z = 25 \tag{10}$$

6. (a) Using Jacobi's method, find all the eigen values and eigen vectors of the matrix.
$$\begin{bmatrix} 2 & 3 & 1 \\ 1 & 4 & 2 \\ 3 & 2 & 3 \end{bmatrix} \tag{10}$$

(b) Using Given's method, reduce the following matrix to the tri-diagonal form.
$$\begin{bmatrix} 2 & 1 & 2 \\ 3 & 2 & 2 \\ 1 & 2 & 1 \end{bmatrix} \tag{10}$$

7. Using Runge-Kutta method of order 4, find y for $x = 0.1, 0.2, 0.3$, given that $\dfrac{dy}{dx} = xy + y^3$, $y(0) = 1$. Continue the solution at $x = 0.4$, using Milne's method. (20)

8. Solve Laplace's equation $\nabla^2 u = 0$ for the following square mesh with boundary values as shown below.

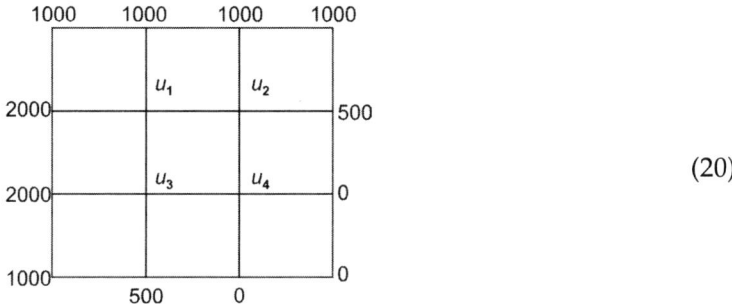

(20)

Subject Index

Programming Using C

Programming using MATLAB

READER'S NOTES

....
....
....
....
....
....
....
....
....
....
....
....
....
....
....
....
....
....
....
....
....
....
....
....
....
....
....
....
....

READER'S NOTES

READER'S NOTES

....
....
....
....
....
....
....
....
....
....
....
....
....
....
....
....
....
....
....
....
....
....
....
....
....
....
....
....

READER'S NOTES